Applied Linear Statistical Methods

Applied Linear Statistical Methods

Donald F. Morrison

The Wharton School
University of Pennsylvania

PRENTICE-HALL, INC., Englewood Cliffs, New Jersey 07632

Library of Congress Cataloging in Publication Data

Morrison, Donald F.
Applied linear statistical methods.

Includes bibliographies and index.
1. Multivariate analysis. 2. Analysis of variance.
3. Time-series analysis. I. Title.
QA278.M677 1983 519.5 82–18539
ISBN 0–13–041020–9

Editorial/production supervision
and interior design: *Paula Martinac and Kathleen M. Lafferty*
Cover design: *Edsal Enterprises*
Manufacturing buyer: *John Hall*

Printed in the United States of America

10 9 8 7 6 5 4 3 2 1

ISBN 0-13-041020-9

Prentice-Hall International, Inc., *London*
Prentice-Hall of Australia Pty. Limited, *Sydney*
Editora Prentice-Hall do Brasil, Ltda., *Rio de Janeiro*
Prentice-Hall Canada Inc., *Toronto*
Prentice-Hall of India Private Limited, *New Delhi*
Prentice-Hall of Japan, Inc., *Tokyo*
Prentice-Hall of Southeast Asia Pte. Ltd., *Singapore*
Whitehall Books Limited, *Wellington, New Zealand*

For Phyllis, Norman,
and Stephen

Contents

chapter 2
Regression and Correlation Models with Several Independent Variables *57*

chapter 3
Further Inference for Regression Models *120*

chapter 4
Polynomial Models for Time Series *181*

chapter 5
Choosing a Set of Independent Variables 207

chapter 6
Some Methods for Time Series 228

chapter 7
Some Further Topics in Regression Analysis 255

chapter 8
The Analysis of Variance for the One-Way Layout 281

Preface

Linear statistical inference is that large domain of modern statistical analysis that encompasses the fitting of lines and planes by least squares, the analysis of variance for experimental designs, correlation, traditional multivariate analysis, and, to some extent, the analysis of time series observations. In this book I have selected some techniques from those areas that have proved useful with data from many of the applied sciences. Emphasis has been placed upon the underlying assumptions, mathematical models, and applications of the methods. Except for some fundamental theorems on quadratic form distributions, the reader has been referred to more advanced sources for derivations and theoretical results. Wherever possible the methods have been illustrated with real experimental or observational data from psychology, the physical, life, and social sciences, economics, or the author's everyday experiences.

The first chapter begins with the linear regression and correlation models for relating two variables. In the second and third chapters those models are extended to the case of several independent variables, and tests, confidence intervals, and other inferential methods for multiple regression are described. The next two chapters address polynomial models for trends in time series and the choice of best sets of independent variables. Chapter 6 contains an extension of least squares to correlated observations on the dependent variable, and some limited methods for dealing with autocorrelation in time series. Chapter 7 treats such statistical techniques as linear discrimination and principal components in the multiple regression context. Chapters 8 and 9 are concerned with the analysis of variance for one- and two-way layouts; Chapter 10 gives the analysis of covariance in the one-way case. I have presented the analysis of variance and covariance methods directly from their underlying linear models without recourse to design matrices, singular estimation equations, or generalized inverse matrices. As in the regression chapters, my emphasis is on assumptions and applications of the basic analyses of variance.

I have tried to set a minimal amount of mathematical and statistical prerequisites for the use of this book. Students are assumed to have had a semester's introductory course through hypothesis testing and confidence intervals, so that they have encountered the normal, t, and other basic distributions. Those distributions are reviewed in Appendix B. Similarly, those who have not studied the essentials of matrix algebra should read Appendix C carefully. Because statistical computing packages and programming languages are constantly changing and often not portable, they are not treated in the text; for the implementation of the methods the reader is assumed to have access to one of the standard packages or elementary skill in APL or FORTRAN. Finally, I have attempted to keep the notation as informal as possible; separate symbols will not be used for random variables and their observed values. Wherever possible Greek letters are used for parameters and Roman for variables. Matrices and vectors are set in boldface; their dimensions are given as required in the text.

Many investigators have kindly permitted the use of their original data for examples and exercises; their contributions are gratefully acknowledged at each instance. I am also indebted to the journal editors, publishers, and other copyright holders for permitting the use of data, statistics, and mathematical tables and charts from their publications.

I am indebted to several persons for assistance and encouragement with the preparation of this book. The project was originally proposed by Harry Gaines, then executive editor, mathematics and science, at Prentice-Hall, and I am grateful for his motivation and support. Robert Sickles, my current executive editor, has continued in that same helpful manner. Some anonymous reviewers commented constructively on different versions of the manuscript. Mary Blue typed the manuscript expeditiously and carefully. Nita Innes typed an initial version of the first chapter, while Aleksandra Hall typed revisions and additions to the final copy. Shyamala Nagaraj computed solutions for a number of exercises. And a special acknowledgment is due my wife, Phyllis, for her support and encouragement while the work ground slowly on; the dedication needs no further elaboration.

Donald F. Morrison

chapter 1

Regression and Correlation Models for Two Variables

1-1 INTRODUCTION

In this book we describe statistical techniques for analyzing the relationships among variables. We begin by developing in this chapter two statistical models for the case of two variables. The first method will permit us to fit a straight line to pairs of observations on the variables. We assume that the first, or independent, variable can be measured exactly, whereas the other, dependent, variable is subject to some random variation. As a criterion for fitting the straight line through the data points we use the sum of squared vertical deviations from the line to the dependent variable values. For that reason the method is called *least squares estimation*. It is also known as *regression analysis*, a commonplace but rather curious name whose origin we explain in Section 1-6. The other approach consists of treating the data more symmetrically as pairs of observations from a population of two random variables. We assume that the relationship of the two variates can be described in terms of their correlation coefficient. In the correlation model we estimate the correlation coefficient from the data and draw inferences about it. For the special bivariate normal population correlation analysis will essentially lead to the same results as we obtained by the least squares model. We shall refer to the methods as "simple" regression and correlation analysis, for they have only one variable in the independent role.

In Section 1-2 we motivate least squares estimation through an example of fitting a trend line to a short series of annual observations. In Section 1-3 we develop the straight-line model and the estimation of its parameters more generally and formally. Sampling properties, significance tests, and confidence intervals for the parameters are given in Section 1-4, whereas statistical inference involving predictions from the linear model is discussed in Section 1-5. In Section 1-6 we shift to the correlation model. We establish the close connection with the least squares linear model when the joint population is bivariate normal. In that section we give hypothesis tests and confidence intervals for the correlation coefficient. In Section 1-7 we survey some history of simple least squares and correlation.

1-2 FITTING A STRAIGHT LINE BY LEAST SQUARES

Let us begin with a simple example. The placement office of a large university reported these mean starting salaries for graduates of its undergraduate business school for five years.

Year	Mean salary
1	$ 9,749
2	10,122
3	10,805
4	11,376
5	12,025

Such data are an example of a *time series*, for their values have been obtained sequentially in time. The salaries are shown in Figure 1-1. The nature of the data and their pictorial representation suggest these initial questions:

1. Can the mean-salary trend over the five years be represented by a straight-line, or linear, model?
2. If a straight line is a reasonable model for the trend, how should we fit it to the data in some "rational" or "optimal" way?
3. How can we measure the "goodness" of the fit of the straight-line trend to our data?
4. What other inferences about the salary time series can we make by statistical methods?

A glance at Figure 1-1 indicates that a straight line would summarize the trend in mean salaries very well—in fact, one could be easily fitted visually with a transparent straightedge. But we wish an analytical method that will work for large data plots with much dispersion in their values and ultimately for data that cannot be represented in only two- or three-dimensional space. For an answer to the second question we turn to a mathematical model for our data points and the method of least squares for the estimation of its parameters.

In our model denote the *dependent* variable, mean salary, at the tth year by y_t, where $t = 1, 2, \ldots, 5$ for the successive years. We propose this linear

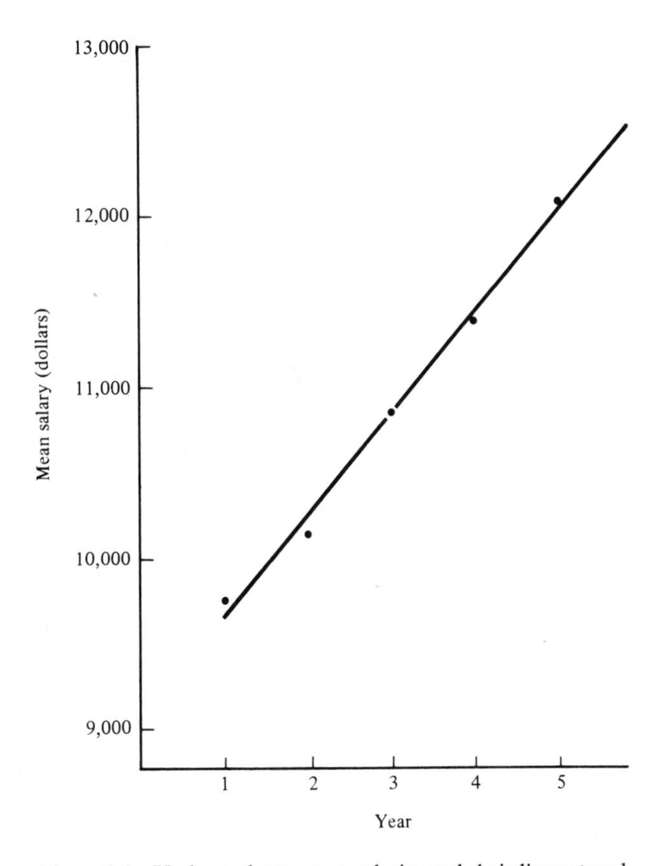

Figure 1-1 Undergraduate mean salaries and their linear trend.

model for mean salary:

$$y_t = \alpha + \beta(t - \bar{t}) + e_t \tag{1}$$

In this model and its generalizations we prefer to express the *independent* variable time in terms of its deviations from its mean

$$\bar{t} = \frac{1}{N} \sum_{t=1}^{N} t \tag{2}$$

In the present example the number N of observations is five. The "centering" of the time values about their mean will lead to algebraic simplifications and better computational and statistical properties of the estimates. The model consists of a straight-line component, $\alpha + \beta(t - \bar{t})$, plus an error or disturbance e_t that accounts for the failure of the salary means to increase with identical annual increments. The e_t are random variables or variates with some kind of probability distribution. For the estimation of the intercept parameter α and the slope β of the straight line we need not specify the distribution, although we should make some assumptions about its properties:

1. The expectation,* or population mean, of the e_t is zero, that is,

$$E(e_t) = 0 \tag{3}$$

   for all t.

2. The variance, or population second moment, of the e_t is the same for all t. We shall represent that variance as

$$\sigma^2 = \text{Var}(e_t) \tag{4}$$

   Such e_t are said to be *homoscedastic*, a word whose Greek roots mean "the same scatter."

3. The e_t are uncorrelated. In terms of the covariance operator

$$\text{Cov}(e_r, e_s) = 0 \qquad r \neq s \tag{5}$$

   In addition, the value of the independent variable must be fixed, as opposed to observations taken randomly on the time scale. This assumption is met in the present example: The mean starting salary of each year is considered to hold as of some uniform date, perhaps June 30.

Now we are ready to define the least squares criterion for the estimation of α and β. We begin with the sum of squared error terms†

$$\begin{aligned} S(\alpha, \beta) &= \sum_{t=1}^{N} e_t^2 \\ &= \sum [y_t - \alpha - \beta(t - \bar{t})]^2 \end{aligned} \tag{6}$$

Least squares estimation consists of finding estimators $\hat{\alpha}$ and $\hat{\beta}$ of the parameters that make $S(\alpha, \beta)$ as small as possible. Because $S(\alpha, \beta)$ is a second-degree function of α and β with a unique minimum, we can locate that point by the differential calculus in two variables. A necessary condition for the minimum is

$$\frac{\partial S(\alpha, \beta)}{\partial \alpha} = 0 \qquad \frac{\partial S(\alpha, \beta)}{\partial \beta} = 0 \tag{7}$$

The derivatives are

$$\begin{aligned} \frac{\partial S(\alpha, \beta)}{\partial \alpha} &= -2 \sum [y_t - \alpha - \beta(t - \bar{t})] \\ \frac{\partial S(\alpha, \beta)}{\partial \beta} &= -2 \sum [y_t - \alpha - \beta(t - \bar{t})](t - \bar{t}) \end{aligned} \tag{8}$$

Because $\sum (t - \bar{t}) = 0$ the least squares estimators follow directly from the individual equations in (8) as

*The expectation, variance, and covariance operators $E(e_t)$, $\text{Var}(e_t)$, and $\text{Cov}(e_r, e_s)$ are defined in Section C-1.

†The limits of the summation operator will usually be omitted when they are clear from the context of the problem or earlier summations.

$$\hat{\alpha} = \frac{1}{N}\sum_{t=1}^{N} y_t$$
$$= \bar{y} \tag{9}$$
$$\hat{\beta} = \frac{\sum_{t=1}^{N} (y_t - \bar{y})(t - \bar{t})}{\sum_{t=1}^{N} (t - \bar{t})^2}$$

We note that $\hat{\alpha}$ is the intercept of the fitted line

$$\hat{y}_t = \hat{\alpha} + \hat{\beta}(t - \bar{t}) \tag{10}$$

at the point $t = \bar{t}$. The use of the deviation $t - \bar{t}$ in the model (1) caused the simultaneous equations for the estimators to reduce to individual equations in the single unknown parameters.

Now we are ready to estimate the parameters of the linear trend for the salary means. The computations can be carried out easily on a hand calculator:

$$\sum y_t = 54{,}077 \qquad \sum t = 15$$
$$\bar{y} = 10{,}815.4 \qquad \bar{t} = 3$$
$$\sum (y_t - \bar{y})^2 = 3{,}395{,}525.2 \qquad \sum (t - \bar{t})^2 = 10$$
$$\sum (y_t - \bar{y})(t - \bar{t}) = 5806$$

The estimated trend line

$$\hat{y}_t = 10{,}815.4 + 580.6(t - 3)$$

is shown in Figure 1-1.

Finally, let us give one answer to the third question of the "goodness of fit" of our linear model to the data. We begin by computing the value of the minimized sum of squared error terms, which we shall call SSE:

$$\text{SSE} = \min_{\alpha,\beta} S(\alpha, \beta)$$
$$= \sum_{t=1}^{N} [y_t - \hat{\alpha} - \hat{\beta}(t - \bar{t})]^2 \tag{11}$$

or after some simplification,

$$\text{SSE} = \sum (y_t - \bar{y})^2 - \frac{[\sum (y_t - \bar{y})(t - \bar{t})]^2}{\sum (t - \bar{t})^2} \tag{12}$$

The notation SSE stands for *sum of squares due to error*. For the salary data SSE = 24,561.6.

The first term of (12) is merely the sum of squares of the salary means about their average. The fraction

$$1 - r^2 = \frac{\text{SSE}}{\sum (y_t - \bar{y})^2} \tag{13}$$

is the proportion of the total variation in the y_t which is not explained by the fitted linear trend, whereas its complement

$$r^2 = \frac{[\sum (y_t - \bar{y})(t - \bar{t})]^2}{[\sum (y_t - \bar{y})^2][\sum (t - \bar{t})^2]} \tag{14}$$

is the proportion of the y_t variation due to the linear function. In the present example

$$r^2 = 0.9928$$

or 99.28 % of $\sum (y_t - \bar{y})^2$ is due to the least squares trend line. The changes in mean salaries are represented very well by a linear function.

The quantity r defined by the positive or negative square root of the expression in (14) is the *correlation coefficient*, whereas r^2 is called the *coefficient of determination*. We shall return to a discussion of r in Section 1-6. The decomposition of the y_t sum of squares into two components has implications of great import: We shall study them in increasing depth in the later sections.

1-3 THE LINEAR MODEL AND ITS PROPERTIES

In the example we used to introduce curve fitting by least squares the independent variable values were equally spaced on the time scale. In this section we give the general straight-line model, estimates of its parameters, and their statistical properties under the assumption that the error component is a normally distributed variate.

Let us assume that N pairs of observations $(x_1, y_1), \ldots, (x_N, y_N)$ have been collected on a dependent variable Y and an independent variable X. The x_i may be distinct, or some values may occur more than once. However, as in the previous salary data collected annually, the x_i are fixed, nonrandom quantities under the control of the experimenter, economist, or other investigator. The linear model relating Y to X is

$$y_i = \alpha + \beta(x_i - \bar{x}) + e_i \qquad i = 1, \ldots, N \tag{1}$$

where

$$\bar{x} = \frac{1}{N}\sum_{i=1}^{N} x_i$$

As in Section 1-2, the e_i are random variables with zero expectation, a common variance σ^2, and zero correlation. Since the y_i are linear functions of the e_i, they have these moments:

$$\begin{aligned} E(y_i) &= \alpha + \beta(x_i - \bar{x}) \qquad i = 1, \ldots, N \\ \text{Var}(y_i) &= \sigma^2 \\ \text{Cov}(y_r, y_s) &= 0 \qquad r \neq s \end{aligned} \tag{2}$$

It is important to note that the y_i have the same variance for all i and that it is equal to the variance of the residual term; the linear trend does not affect the variances and covariances of the y_i.

The least squares estimators of α and β follow as before: We require that the estimators minimize

$$S(\alpha, \beta) = \sum_{i=1}^{N} [y_i - \alpha - \beta(x_i - \bar{x})]^2 \tag{3}$$

Since the e_i are uncorrelated and have the same variance, an unweighted sum of squared deviations seems a reasonable minimand. Partial differentiation leads to the estimators

$$\hat{\alpha} = \frac{1}{N} \sum_{i=1}^{N} y_i$$

$$= \bar{y} \tag{4}$$

$$\hat{\beta} = \frac{\sum (y_i - \bar{y})(x_i - \bar{x})}{\sum (x_i - \bar{x})^2} \tag{5}$$

If $\hat{\beta}$ must be computed with a hand calculator, the formula

$$\hat{\beta} = \frac{\sum y_i x_i - [(\sum y_i)(\sum x_i)]/N}{\sum x_i^2 - (\sum x_i)^2/N} \tag{6}$$

is more convenient, although it is sensitive to round-off error if the x_i are large and nearly equal.

We have preferred to introduce the least squares model with its independent variable expressed as deviations about its average. Such a centered variable permits direct solutions of the least squares equations for the estimators (4) and (5). The sums of squares and products for centered models are much smaller, and computational accuracy is easier to attain. For those reasons we shall use centered independent variables in most of the linear models in the sequel. However, the model

$$y_i = \beta_0 + \beta x_i + e_i \tag{7}$$

is often more appropriate, for it expresses the dependent variable directly as a linear function of the x_i rather than their deviations. The parameter β_0 is the intercept of the line at the origin of the x_i scale. The estimator of the slope β is still (5), but that of the intercept is

$$\hat{\beta}_0 = \bar{y} - \hat{\beta}\bar{x} \tag{8}$$

We should use the model (7) if we wish to compare least squares lines fitted to several sets of data, for its intercept parameter is located at the same origin rather than the means of the independent variables in the different sets.

The least squares predicted value of Y corresponding to $X = x_i$ is

$$\hat{y}_i = \bar{y} + \hat{\beta}(x_i - \bar{x}) \tag{9}$$

The deviations of the observed dependent variable values from their predictions will be denoted by

$$z_i = y_i - \hat{y}_i \tag{10}$$

The z_i are called the *residuals* of the fitted model. Their sum is zero:

$$\begin{aligned} \sum z_i &= \sum (y_i - \hat{y}_i) \\ &= \sum [y_i - \bar{y} - \hat{\beta}(x_i - \bar{x})] \\ &= \sum (y_i - \bar{y}) - \hat{\beta} \sum (x_i - \bar{x}) \\ &= 0 \end{aligned} \tag{11}$$

As we shall see presently, the sum of squares of the z_i provides a way of estimating the variance σ^2 of the disturbances.

The predicted values and residuals are illustrated in the plot of some hypothetical data in Figure 1-2. The role of Y as dependent variable determined

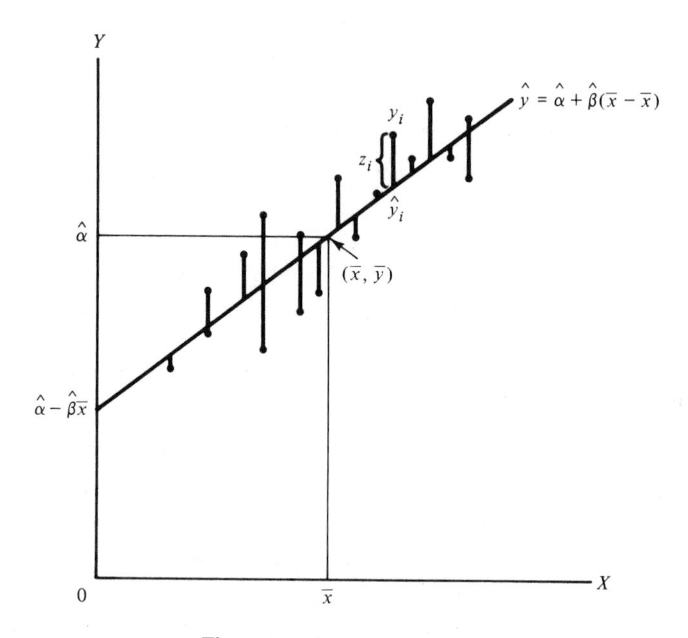

Figure 1-2 Least squares line.

that we minimize the sum of squared vertical distances: The line found in this case is called the least squares fit of Y on X. Minimization of the sum of squared horizontal distances would lead to the linear model for X on Y: Its line would be different from that of Y on X unless the (x_i, y_i) pairs lie on a straight line.

Now, what desirable qualities should our intercept and slope estimators possess, and to what extent do $\hat{\alpha}$ and $\hat{\beta}$ fulfill those expectations? First, we would prefer the estimators to be unbiased or to have expected values equal to the parameters they are estimating. Then the averages of the estimators over many repetitions or simulations of the same data set would converge to the respective population parameters. We may verify that $\hat{\alpha}$ and $\hat{\beta}$ are unbiased by applying the expectation operator and some of its properties for linear functions of random variables described in Section C-1:

$$\begin{aligned} E(\hat{\alpha}) &= E\left(\frac{1}{N}\sum y_i\right) \\ &= \frac{1}{N}\sum [\alpha + \beta(x_i - \bar{x})] \\ &= \alpha + \frac{\beta}{N}\sum (x_i - \bar{x}) \\ &= \alpha \end{aligned} \tag{12}$$

$$E(\hat{\beta}) = \frac{\sum (y_i - \bar{y})(x_i - \bar{x})}{\sum (x_i - \bar{x})^2}$$

$$= \frac{1}{\sum (x_i - \bar{x})^2} E\{\sum [\alpha + \beta(x_i - \bar{x}) + e_i - \bar{y}](x_i - \bar{x})\}$$

$$= \frac{1}{\sum (x_i - \bar{x})^2} \sum [\alpha + \beta(x_i - \bar{x}) - E(\bar{y})](x_i - \bar{x})$$

$$= \beta \tag{13}$$

Through the covariance operator we can verify that $\hat{\alpha}$ and $\hat{\beta}$ are uncorrelated. For convenience first write

$$\bar{y} = \alpha + \bar{e}$$

$$\bar{e} = \frac{1}{N} \sum e_i$$

Then

$$\text{Cov}\,(\hat{\alpha}, \hat{\beta}) = \text{Cov}\left\{\alpha + \bar{e}, \frac{\sum [\beta(x_i - \bar{x}) + e_i - \bar{e}](x_i - \bar{x})}{\sum (x_i - \bar{x})^2}\right\}$$

$$= \frac{1}{\sum (x_i - \bar{x})^2} \text{Cov}\,[\bar{e}, \sum e_i (x_i - \bar{x}) - \bar{e} \sum (x_i - \bar{x})]$$

$$= \frac{1}{N \sum (x_i - \bar{x})^2} [\sigma^2 \sum (x_i - \bar{x}) - \sigma^2 \sum (x_i - \bar{x})]$$

$$= 0$$

so that the two estimators are uncorrelated. As we shall soon learn, this property will permit independent statistical inferences about α and β.

The variances of $\hat{\alpha}$ and $\hat{\beta}$ can be calculated by the variance operator. Clearly,

$$\text{Var}\,(\hat{\alpha}) = \text{Var}\,(\bar{y}) = \frac{\sigma^2}{N} \tag{14}$$

or the variance of the average of N independent observations. The second variance is

$$\text{Var}\,(\hat{\beta}) = \frac{1}{[\sum (x_i - \bar{x})^2]^2} \text{Var}\,[\sum (y_i - \bar{y})(x_i - \bar{x})]$$

$$= \frac{1}{[\sum (x_i - \bar{x})^2]^2} \text{Var}\,[\sum y_i (x_i - \bar{x}) - \bar{y} \sum (x_i - \bar{x})]$$

$$= \frac{\sigma^2}{\sum (x_i - \bar{x})^2} \tag{15}$$

We note immediately that the variance of the estimator decreases with increasing dispersion $\sum (x_i - \bar{x})^2$ of the x_i about their mean; good least squares estimators require a wide range of values for the independent variable.

Apart from the unlikely case of errors with a known variance, the use of the variances of $\hat{\alpha}$ and $\hat{\beta}$ will require an estimator of σ^2. Let us derive one estimator from intuitive considerations of the variance of the e_i. Recall that we defined the minimized sum of squares, SSE, in (11), Section 1-2. In the present more general model this quantity is

$$\begin{aligned} \text{SSE} &= \sum_{i=1}^{N} [y_i - \hat{\alpha} - \hat{\beta}(x_i - \bar{x})]^2 \\ &= \sum (y_i - \bar{y})^2 - \frac{[\sum (y_i - \bar{y})(x_i - \bar{x})]^2}{\sum (x_i - \bar{x})^2} \\ &= \sum (y_i - \bar{y})^2 - \hat{\beta}^2 \sum (x_i - \bar{x})^2 \end{aligned} \tag{16}$$

Since SSE is the sum of squared deviations of the predicted values $\hat{y}_i$ from their actual observations, a reasonable estimator of σ^2 must be SSE divided by some function of the sample size. Let us determine if this is so by calculating the expected value of SSE. The first component of $E(\text{SSE})$ is

$$\begin{aligned} E[\sum (y_i - \bar{y})^2] &= E \sum [\beta(x_i - \bar{x}) + e_i - \bar{e}]^2 \\ &= \beta^2 \sum (x_i - \bar{x})^2 + E[\sum (e_i - \bar{e})^2] \\ &= \beta^2 \sum (x_i - \bar{x})^2 + (N - 1)\sigma^2 \end{aligned} \tag{17}$$

where the second expectation should be familiar from the calculation of the expected value of the sample variance of N independent observations. For the expectation of the second term recall that

$$\begin{aligned} E(\hat{\beta}^2) &= \text{Var}(\hat{\beta}) + [E(\hat{\beta})]^2 \\ &= \frac{\sigma^2}{\sum (x_i - \bar{x})^2} + \beta^2 \end{aligned} \tag{18}$$

from (11). Hence,

$$E(\text{SSE}) = \sigma^2(N - 2) \tag{19}$$

and an unbiased estimator of the variance of the disturbance terms would be

$$\hat{\sigma}^2 = \frac{\text{SSE}}{N - 2} \tag{20}$$

These expectations, variances, and covariance of the estimators are the practical extent of our inferences about the linear model if we do not specify a distribution for the disturbance terms. In Section 1-4 we shall make a commitment to normally distributed error terms and thereby obtain hypothesis tests and confidence statements for the model parameters.

Example 1-1

The graduate division of the same business school mentioned in Section 1-2 announced these average starting salaries for students receiving M.B.A. degrees in seven years:

Year	Average starting salary
1	$14,200
2	15,645
3	16,452
4	16,884
5	18,159
6	20,650
7	21,890

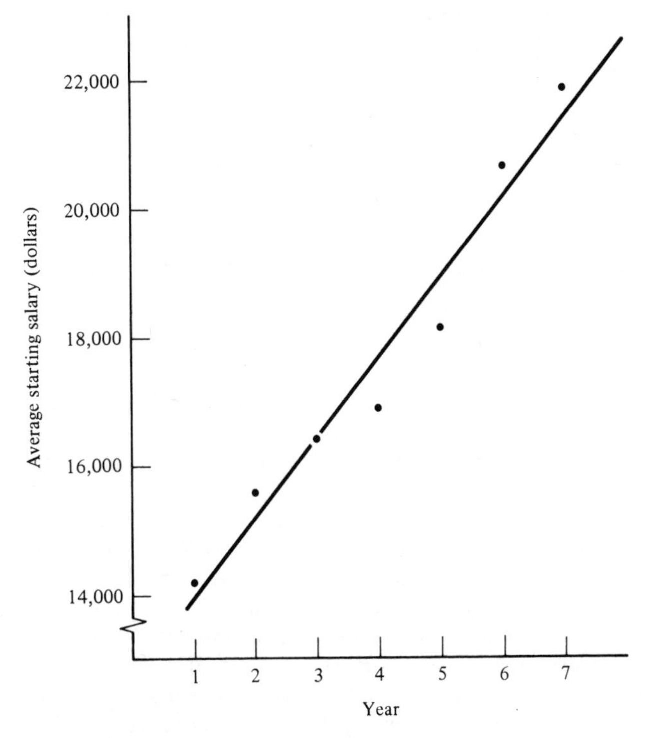

Figure 1-3 Graduate starting salaries and their linear trend.

The plot of the averages in Figure 1-3 exhibits a strong trend. Let us estimate the linear component by least squares. The necessary statistics are

$$\hat{\alpha} \doteq \bar{y} = 17{,}697.14 \qquad \sum (t - \bar{t})^2 = 28 \qquad \sum (y_t - \bar{y})(t - \bar{t}) = 34{,}787$$

$$\sum (y_t - \bar{y})^2 = 45{,}165{,}608.86$$

Then the trend line equation is

$$\hat{y}_t = 17{,}697.14 + 1242.39(t - \bar{t}) \qquad t = 1, \ldots, 7$$

The predicted mean salaries and their residuals from the actual salaries had these values:

Year	$\hat{y}_t$	$y_t - \hat{y}_t$
1	$13,970	230.04
2	15,212	432.65
3	16,455	−2.75
4	17,697	−813.14
5	18,940	−780.53
6	20,182	468.07
7	21,424	465.68

The sum of the squared residuals is

$$\text{SSE} = \sum (y_t - \hat{y}_t)^2 = 1{,}946{,}483$$

and the proportion of the sum of squares $\sum (y_t - \bar{y})^2$ attributable to the linear trend is $r^2 = 0.9569$. Average starting salaries have tended to increase linearly at a rate of \$1242 each year. The linear trend explains all but 4.31% of the variation of the seven averages about their mean.

Let us estimate the variance of the disturbance terms by expression (20):

$$\hat{\sigma}^2 = \frac{\text{SSE}}{7 - 2} = 389{,}297$$

and the estimated standard deviation is $\hat{\sigma} = 623.94$. The poorest predicted value is only $1.3\hat{\sigma}$ away from the actual salary for that year. The estimated variance of the slope is

$$\widehat{\text{Var}(\hat{\beta})} = \frac{\hat{\sigma}^2}{28} = 13{,}903.48$$

and the standard deviation is 117.91. We might gain a sense of the importance of the linear trend by noting that the slope is more than 10 standard deviations away from zero. Later we shall see how significance tests for the slope and intercept parameters can be made from such ratios.

Frequently, we think of the annual changes in prices or incomes as percentages of the previous year's value. Then the model for the y_t would not be a linear function of time but would follow a "compound interest," or exponential, trend. The exponential model with the usual additive disturbance terms leads to nonlinear simultaneous equations for the least squares estimators of its parameters. Those equations must be solved by successive approximations and are outside the scope of our treatment of least squares. As a rough approxmation to an exponential model for the trend in average starting salaries we might assume that the disturbances are *multiplicative* and adopt this model:

$$y_t = \gamma \exp \{\delta(t - \bar{t}) + e_t\}$$

The notation $\exp\{\cdot\}$ denotes the exponential function, or $e = 2.71828\ldots$ raised to the power $\delta(t - \bar{t}) + e_t$. If we take natural logarithms of both sides of the equation, we have

$$\begin{aligned} u_t &= \ln y_t \\ &= (\ln \gamma) + \delta(t - \bar{t}) + e_t \end{aligned}$$

or a linear trend in the logs of the annual average salaries. We can apply least squares to estimate $\alpha_1 = \ln \gamma$ and δ. The necessary statistics are

$$\sum u_t = 68.3974 \qquad \sum (u_t - \bar{u})(t - \bar{t}) = 1.95221 \qquad \sum (u_t - \bar{u})^2 = 0.140168$$

$$\hat{\alpha}_1 = \bar{u} = 9.77106 \qquad \hat{\delta} = 0.06972$$

The least squares trend line for the log salaries is

$$\hat{u}_t = \widehat{\ln y_t} = 9.77106 + 0.06972(t - \bar{t})$$

The line accounts for

$$100r^2\% = \frac{100[(1.95221)^2/28]}{0.140168} = 97.11\%$$

of the variation in the u_t-values. The equation can be expressed directly in dollars by taking antilogs of each side of the $\hat{u}_t$ equation, or by making each side a power of e:

$$\begin{aligned} \hat{y}_t = e^{\hat{u}_t} &= \exp\{9.77106 + 0.06972(t - \bar{t})\} \\ &= 17{,}519.29 \exp\{0.06972(t - \bar{t})\} \end{aligned}$$

These annual salaries were predicted by that equation:

Year	$\hat{y}_t$	$y_t - \hat{y}_t$
1	\$14,212	−12.72
2	15,239	405.98
3	16,339	112.58
4	17,519	−635.29
5	18,784	−625.36
6	20,141	509.23
7	21,595	294.86

The sum of the squared residuals is $\sum (y_t - \hat{y}_t)^2 = 1{,}318{,}581$, so that $100r^2 = 97.08\%$ of the variation is explained by the exponential model. That function appears to explain some of the curvature in the salary trend. Of course we might also fit second- and third-degree polynomials to the data; we see how that is done in Chapter 4.

1-4 STATISTICAL INFERENCE FROM THE LINEAR MODEL

We introduced the linear model under the assumption that its error terms could be represented by uncorrelated random variables with a mean of zero and a constant variance. Those requirements were adequate for obtaining least squares estimators of the intercept and slope of the true line and an intuitively appealing estimator of the common variance. However, we would also like to make inferences about the unknown line: We wish significance tests for hypothesized values of its parameters, confidence intervals for α, β, σ^2, and perhaps estimates of the uncertainty in predicted values of the dependent variable for a new value of the predictor. For those tests and intervals we must postulate some kind of distribution for the disturbance variates. Tradition, the properties of measurement errors, and the powerful result on sums of variates known as the *central limit theorem* have led us to assume that the disturbances $e_1, \ldots, e_N$ are distributed as independent normal random variables* with zero mean and variance σ^2. The earlier weaker assumption of zero correlation follows from the independence of normal variates. The properties of linear functions of normal variates given in Section C-2 imply these results:

1. The y_i are independently normally distributed with mean $\alpha + \beta(x_i - \bar{x})$ and common variance σ^2.
2. $\hat{\alpha}$ is normally distributed with mean α and variance σ^2/N.
3. $\hat{\beta}$ is normally distributed with mean β and variance

$$\frac{\sigma^2}{\sum (x_i - \bar{x})^2} \tag{1}$$

*The normal distribution and some of its properties are described in Section C-2.

In addition, these results follow from the properties of sums of squares of normal variates described in Appendix C and Section 3-2:

4. The quantity

$$\frac{\text{SSE}}{\sigma^2} = \frac{(N-2)\hat{\sigma}^2}{\sigma^2} \tag{2}$$

has the chi-squared distribution (see Section C-3) with $N - 2$ degrees of freedom.

5. $\hat{\alpha}$, $\hat{\beta}$, and SSE are independently distributed.

Results 2 through 5 provide hypothesis tests and confidence statements for α and β. For the intercept parameter we may show that

$$t = \frac{(\hat{\alpha} - \alpha)\sqrt{N}}{\hat{\sigma}} \tag{3}$$

has the t-distribution (see Section C-3 for further details) with $N - 2$ degrees of freedom. We may test the hypothesis

$$H_0: \quad \alpha = \alpha_0 \tag{4}$$

that the intercept has some specified value α_0 against the one-sided alternative

$$H_1: \quad \alpha > \alpha_0 \tag{5}$$

of a higher value by computing the statistic (3) with $\alpha = \alpha_0$, and referring it to the critical values $t_{\gamma;\nu}$ of the t-distribution given in Table 3, Appendix A. For a test at level γ we reject H_0 in favor of H_1 if

$$t > t_{\gamma;N-2} \tag{6}$$

and otherwise do not reject H_0. For a test against the two-sided alternative

$$H_1': \quad \alpha \neq \alpha_0 \tag{7}$$

t should be referred to the critical value $t_{\gamma/2;N-2}$: If $|t| \leq t_{\gamma/2;N-2}$ H_0 is tenable at the γ level. If $|t| > t_{\gamma/2;N-2}$ H_0 should be rejected in favor of a higher or lower intercept.

Alternatively, we can make inferences about α by confidence intervals. If we begin with the probability statement

$$P\left(\frac{|\hat{\alpha} - \alpha|\sqrt{N}}{\hat{\sigma}} \leq t_{\gamma/2;N-2}\right) = 1 - \gamma \tag{8}$$

and expand its left-hand term, we are led to the $100(1 - \gamma)\%$ confidence interval for α:

$$\bar{y} - t_{\gamma/2;N-2}\frac{\hat{\sigma}}{\sqrt{N}} \leq \alpha \leq \bar{y} + t_{\gamma/2;N-2}\frac{\hat{\sigma}}{\sqrt{N}} \tag{9}$$

Tests and the confidence interval for β follow in the same manner. For the test of

$$H_0: \quad \beta = \beta_0 \tag{10}$$

we compute the t-statistic

$$t = \frac{\hat{\beta} - \beta_0}{\sqrt{\widehat{\operatorname{Var}}(\hat{\beta})}} = \frac{(\hat{\beta} - \beta_0)\sqrt{\sum (x_i - \bar{x})^2}}{\hat{\sigma}} \tag{11}$$

and again refer it to Table 3 of Appendix A with $N - 2$ degrees of freedom. For the alternative

$$H_1: \quad \beta > \beta_0 \tag{12}$$

that the slope parameter is greater than the value hypothesized under H_0 we would reject H_0 if $t > t_{\gamma;N-2}$ and accept otherwise. For tests against the two-sided alternative

$$H_1: \quad \beta \neq \beta_0 \tag{13}$$

of a slope different from β_0, H_0 would be rejected if $|t| > t_{\gamma/2;N-2}$.

Of particular importance is the null hypothesis

$$H_0: \quad \beta = 0 \tag{14}$$

that Y is not linearly related to X. Although we may test that hypothesis by the statistic (11), an alternative method is also available. Write the sum of squares of the y_i about their mean as

$$\begin{aligned} \sum (y_i - \bar{y})^2 &= \sum [y_i - \hat{y}_i) + (\hat{y}_i - \bar{y})]^2 \\ &= \sum (y_i - \hat{y}_i)^2 + \hat{\beta}^2 \sum (x_i - \bar{x})^2 \\ &\quad + 2\hat{\beta} \sum [y_i - \bar{y} - \hat{\beta}(x_i - \bar{x})](x_i - \bar{x}) \end{aligned} \tag{15}$$

The third term vanishes upon simplification, and we can write

$$\sum (y_i - \bar{y})^2 = \text{SSE} + \text{SSR} \tag{16}$$

where SSE is the usual error sum of squares defined by (16), Section 1-3, and

$$\text{SSR} = \hat{\beta}^2 \sum (x_i - \bar{x})^2 = \frac{[\sum (y_i - \bar{y})(x_i - \bar{x})]^2}{\sum (x_i - \bar{x})^2} \tag{17}$$

SSE and SSR are independently distributed variates. When $H_0: \quad \beta = 0$ is true, SSR/σ^2 is distributed as a chi-squared variate with one degree of freedom. H_0 may be tested by computing

$$F = (N - 2)\frac{\text{SSR}}{\text{SSE}} \tag{18}$$

and referring it to the upper critical value $F_{\alpha;1,N-2}$ of the F-distribution (see Section C-3) with one and $N - 2$ degrees of freedom. Since such an F-variate is equal to the square of a t-variate with $N - 2$ degrees of freedom, the F-test is identical with a two-sided test made with the t-statistic. This F-test is an example of the simplest *analysis of variance* in linear models. In Chapter 3 we develop the theory of that analysis of variance more extensively.

The $100(1-\gamma)\%$ confidence interval for β is

$$\hat{\beta} - t_{\gamma/2;N-2} \frac{\hat{\sigma}}{\sqrt{\sum (x_i - \bar{x})^2}} \leq \beta \leq \hat{\beta} + t_{\gamma/2;N-2} \frac{\hat{\sigma}}{\sqrt{\sum (x_i - \bar{x})^2}} \tag{19}$$

We note that if the interval contains the value β_0, the hypothesis (10) is tenable at the γ level. If β_0 is not contained in the confidence interval, the two-sided alternative should be accepted.

Example 1-2

Let us apply these methods to the undergraduate starting salary means of Section 1-2. Recall that

$$\hat{\alpha} = \bar{y} = 10{,}815.4 \qquad \hat{\beta} = 580.6$$

$$\sum (x_i - \bar{x})^2 = 10 \qquad \text{SSE} = 24{,}561.6$$

Then

$$\hat{\sigma}^2 = 8187.2 \qquad \hat{\sigma} = 90.4831 \qquad \frac{\hat{\sigma}}{\sqrt{N}} = 40.465$$

$$\sqrt{\widehat{\text{Var}\,(\hat{\beta})}} = 28.6133 \qquad t_{0.025;3} = 3.182$$

The 95% confidence interval for the intercept parameter is

$$10{,}686 \leq \alpha \leq 10{,}944$$

The 95% confidence interval for the slope is

$$489.55 \leq \beta \leq 671.65$$

This means that in a frequency sense 95% of the time the interval (489.55, 671.65) will enclose the true slope β. Because the interval does not enclose zero the hypothesis $H_0\colon \beta = 0$ is not tenable at the 0.05 level. The slope in the line is clearly greater than any trend caused by random variation in the salary averages.

Example 1-3

Let us test the hypothesis (14) that the slope of the M.B.A. salary trend line found in Example 1-1 is zero. The statistic (11) has the value

$$t = \frac{1242.39}{117.91} = 10.54$$

The two-sided 0.01 critical value of the t-distribution is $t_{0.005;5} = 4.032$. We should conclude that the upward trend is real—an inference most persons would make even without the analysis. Alternatively, we might compute the F-ratio (18) of the error and explained mean squares:

$$F = (7-2)\left(\frac{43{,}219{,}122}{1{,}946{,}487}\right) = 111.02$$

Since F greatly exceeds the 0.01 critical value $F_{0.01;1,5} = 16.3$, we should reject the hypothesis that the slope is zero. We note that F is the square of the t-statistic 10.54, which was computed directly from the estimated slope.

Frequently, we wish to test the more general hypothesis that the slope has a particular value. For example, suppose that someone stated that the average annual increment in a certain population of M.B.A. starting salaries for the seven-year period was \$1500. Do the present data support such a slope, or is the hypothesis

$$H_0\colon \beta = 1500$$

tenable for the hypothetical universe giving rise to our data? The test statistic is

$$t = \frac{(1242.39 - 1500)\sqrt{28}}{623.94}$$
$$= -2.18$$

Because the estimated slope is less than the hypothesized one we shall use a one-sided 0.05-level test with critical value $-t_{0.05;5} = -2.015$. The hypothesis is just barely rejected: The observed series of average salaries appears to have come from a population with slope less than 1500.

1-5 CONFIDENCE AND PREDICTION INTERVALS

In our development of the least squares estimates for the linear model parameters we expressed our uncertainty about the parameter values in terms of confidence intervals. In this section we extend these intervals to the least squares predictions of future values and their population means.

Let us begin with an example of some data.* Plasma free fatty acid (FFA) has been shown to decrease with age in prepubescent boys (Heald et al., 1967). In a study of changes in FFA before and during adolescence blood samples were obtained from $N = 41$ normal boys judged prepubescent by the Tanner sexual maturation scale. The values y_i of the dependent variable FFA (milliequivalent/liter) have been plotted against the ages x_i (months) of the subjects in Figure 1-4. Actual values of the observations are deferred until Table 2-1. These statistics were computed from the data.

$$\sum (x_i - \bar{x})^2 = 6052.49 \qquad \sum (y_i - \bar{y})^2 = 2.59153$$
$$\bar{x} = 113.707 \qquad \bar{y} = 0.6028$$
$$\sum (y_i - \bar{y})(x_i - \bar{x}) = -47.3316$$

The least squares prediction of FFA for age x is given by

$$\hat{y} = 0.6028 - 0.00782(x - 113.707)$$
$$= 1.4920 - 0.00782x \qquad (1)$$

The sum of squared deviations about that line is

$$\text{SSE} = 2.59153 - \frac{(-47.3316)^2}{6052.49}$$
$$= 2.22139$$

and the estimate of the standard deviation of the error term in the model is

$$\hat{\sigma} = \sqrt{\frac{\text{SSE}}{41 - 2}}$$
$$= 0.23866$$

*I am indebted to Felix P. Heald, M.D., for the use of these data.

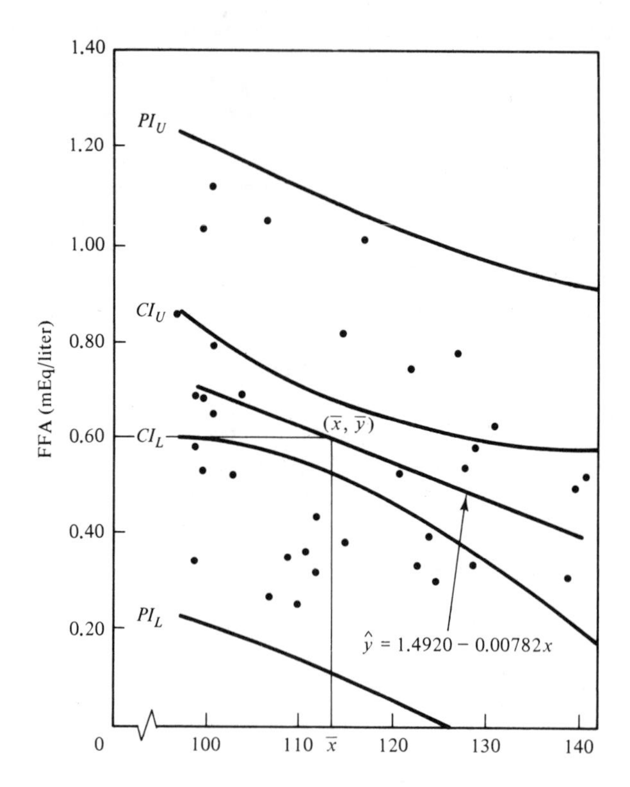

Figure 1-4 Plasma free fatty acid related to age.

Examination of the scatter of the points about the fitted line in Figure 1-4 shows that any relationship of FFA to age must be a tenuous one. Let us determine if a linear relationship is probably present by testing the hypothesis $H_0\colon \ \beta = 0$ that age has no effect on FFA. The test statistic is

$$t = -\frac{0.007820\sqrt{6052.49}}{0.23866}$$

$$= -2.55$$

Since $t_{0.01;39}$ is approximately 2.43, the hypothesis of a zero regression coefficient can be rejected at the 1% level in favor of the alternative that FFA decreases linearly with age in prepubescent boys eight years or older.

Nevertheless, given that a linear relationship probably exists in the population, what is the utility of equation (1) for predicting (a) FFA of a boy of specified age, and (b) the mean FFA for the population of all boys of a given age? Let us answer the first question by means of the *prediction interval* for a future observation on the dependent variable Y for the new value x of the independent variable. By the linear model,

$$Y = \alpha + \beta(x - \bar{x}) + e \tag{2}$$

where e (and thereby Y) is a random variable distributed independently of the error terms $e_1, \ldots, e_N$ in the original sample from which the least squares prediction equation

$$\hat{y} = \bar{y} + \hat{\beta}(x - \bar{x}) \tag{3}$$

of Y was estimated. The construction of a prediction interval for Y consists of using probability to measure the magnitude of the difference, called the prediction error,

$$Y - \hat{y} \tag{4}$$

of the realized and predicted future values. We begin by computing

$$E(Y - \hat{y}) = 0 \tag{5}$$

$$\begin{aligned} \operatorname{Var}(Y - \hat{y}) &= \operatorname{Var}(Y) + \operatorname{Var}(\hat{y}) \\ &= \sigma^2 + \operatorname{Var}[\bar{y} + \hat{\beta}(x - \bar{x})] \\ &= \sigma^2 + \operatorname{Var}(\bar{y}) + \operatorname{Var}[\hat{\beta}(x - \bar{x})] \\ &= \sigma^2\left[1 + \frac{1}{N} + \frac{(x - \bar{x})^2}{\sum(x_i - \bar{x})^2}\right] \end{aligned} \tag{6}$$

Furthermore, since Y and $\hat{y}$ are independently normally distributed variates, $Y - \hat{y}$ is normally distributed with mean zero and variance given by (6). Results 4 and 5 of Section 1-4 imply that the usual disturbance variance estimator

$$\hat{\sigma}^2 = \frac{\text{SSE}}{N - 2} \tag{7}$$

is distributed independently of $Y - \hat{y}$, and that

$$t = \frac{Y - \hat{y}}{\hat{\sigma}\sqrt{1 + \frac{1}{N} + \frac{(x - \bar{x})^2}{\sum(x_i - \bar{x})^2}}} \tag{8}$$

has the t-distribution with $N - 2$ degrees of freedom. From the symmetry of the t density function and the definition of its $100(1 - \alpha/2)$ percentage point $t_{\alpha/2; N-2}$,

$$P(-t_{\alpha/2; N-2} \leq t \leq t_{\alpha/2; N-2}) = 1 - \alpha \tag{9}$$

If we replace t by its definition (8) and rewrite the inequality within the probability statement (9) to encompass the unknown future value, we have the $100(1 - \alpha)\%$ prediction interval for Y:

$$\begin{aligned} \hat{y} - t_{\alpha/2; N-2}\hat{\sigma}\sqrt{1 + \frac{1}{N} + \frac{(x - \bar{x})^2}{\sum(x_i - \bar{x})^2}} &\leq Y \\ &\leq \hat{y} + t_{\alpha/2; N-2}\hat{\sigma}\sqrt{1 + \frac{1}{N} + \frac{(x - \bar{x})^2}{\sum(x_i - \bar{x})^2}} \end{aligned} \tag{10}$$

Let us find the 95% prediction interval for the FFA level of a boy known only to be of age 108 months. The critical value of the t-distribution is approximately

$$t_{0.025; 39} = 2.023 \tag{11}$$

The least squares estimate of FFA level is

$$\begin{aligned} \hat{y} &= 0.6028 - 0.007820(108 - 113.707) \\ &= 0.6474 \end{aligned} \tag{12}$$

The estimate of the variance (6) is

$$\widehat{\text{Var}\,(Y - \hat{y})} = \frac{2.23139}{39}\left[1.02439 + \frac{(108 - 113.707)^2}{6052.49}\right]$$

$$= 0.05892$$

and the prediction interval is

$$0.156 < Y < 1.138 \tag{13}$$

In the frequency sense 95% of the time this interval would cover the FFA level of a randomly selected boy aged nine years.

The upper and lower prediction limits can be plotted as a function of x to give a $100(1 - \alpha)\%$ prediction band for the least squares line. The prediction limits are the two branches of a hyperbola. When the term $(x - \bar{x})^2/\sum (x_i - \bar{x})^2$ is small, as in the free fatty acid example, the curvature in the hyperbola branches is only slight. Nevertheless, ignoring that term in expression (10) will lead to prediction intervals that are too narrow when x is distant from the average of the x_i. The hyperbolic shape of the prediction bands emphasizes the hazards of extrapolating the straight line beyond the range of the x_i and of predicting Y when its x-value is far from the mean. The bands also show that the most precise prediction can be made at $x = \bar{x}$. All the prediction intervals are made narrow by large values of $\sum (x_i - \bar{x})^2$: As we have stated before for the variance of $\hat{\beta}$, the values of the x_i used to fit the linear function should be widely dispersed about their mean.

The wide prediction intervals for FFA level indicate that the data and the estimated linear function are not especially useful for predicting the FFA of a single child from his age. The lower limit is in fact zero for ages beyond 126 months. However, instead of predicting individual values we might set confidence bounds on the population mean FFA level of all boys of age x. That mean is

$$E(Y \mid x) = \alpha + \beta(x - \bar{x}) \tag{14}$$

and its estimate is of course the least squares prediction

$$\hat{y} = \bar{y} + \hat{\beta}(x - \bar{x}) \tag{15}$$

Under the assumption of normally distributed error terms $\hat{y}$ is normally distributed with mean (14) and variance

$$\text{Var}\,(\hat{y}) = \sigma^2\left[\frac{1}{N} + \frac{(x - \bar{x})^2}{\sum (x_i - \bar{x})^2}\right] \tag{16}$$

As in the prediction interval development,

$$t = \frac{\hat{y} - [\alpha + \beta(x - \bar{x})]}{\hat{\sigma}\sqrt{\dfrac{1}{N} + \dfrac{(x - \bar{x})^2}{\sum (x_i - \bar{x})^2}}} \tag{17}$$

has the t-distribution with $N - 2$ degrees of freedom. If we write the probability statement (9) again for this t random variable, the $100(1 - \alpha)\%$ confidence interval for the conditional mean $E(Y|x) = \alpha + \beta(x - \bar{x})$ is

$$\hat{y} - t_{\alpha/2;N-2}\hat{\sigma}\sqrt{\frac{1}{N} + \frac{(x - \bar{x})^2}{\sum (x_i - \bar{x})^2}} \leq E(Y|x) \leq \hat{y} + t_{\alpha/2;N-2}\hat{\sigma}\sqrt{\frac{1}{N} + \frac{(x - \bar{x})^2}{\sum (x_i - \bar{x})^2}} \tag{18}$$

Because we are predicting a population mean rather than a single value from that distribution the variance (16) is much smaller than that required for the single value, and the confidence interval will always be narrower. The confidence intervals define the branches of a hyperbola when plotted as a function of x.

The 95% confidence bands for the conditional mean FFA level are shown in Figure 1-4. The curved nature of the bands and their narrowness relative to the prediction bands is apparent. Unlike the prediction intervals, the confidence bands approach one another with increasing sample size N.

Some properties of the two families of hyperbolas may help in sketching and interpreting the bands. Each branch of a particular hyperbola is equidistant vertically from the fitted line, but its major and minor axes must be calculated by other means [see, for example, any text on analytical geometry, or at a more advanced level, the treatment through matrix algebra by Hohn (1964, Chapter 9)]. The asymptotes of the left side of the lower branch and the right side of the upper branch of the prediction and confidence band hyperbolas are given by the line

$$Y = \bar{y} + \left[\hat{\beta} + t_{\alpha/2;N-2}\frac{\hat{\sigma}}{\sqrt{\sum (x_i - \bar{x})^2}}\right](x - \bar{x}) \tag{19}$$

Similarly, the asymptotes of the right lower and left upper branches are given by

$$Y = \bar{y} + \left[\hat{\beta} - t_{\alpha/2;N-2}\frac{\hat{\sigma}}{\sqrt{\sum (x_i - \bar{x})^2}}\right](x - \bar{x}) \tag{20}$$

The prediction interval hyperbolas approach these limits too slowly in most applications to be of any use, but the asymptotes may be a useful check on the confidence interval bands for large values of the independent variables.

In the use of prediction and confidence bands it is essential to bear in mind that the confidence coefficient $1 - \alpha$ is the probability that *one particular* interval covers the future value or conditional mean. If we wish to make inferences about the averages of several intervals for different values of x, our probability will be less than $1 - \alpha$. Such multiple inferences by tests and confidence intervals require protection at the *experiment* or *family* level. That protection can be achieved by various methods of *simultaneous inference*. We discuss those techniques at length in Chapter 3 and subsequent chapters as they are required in our development of linear models.

1-6 THE CORRELATION OF TWO VARIATES

In the preceding sections we have developed the method of fitting a straight line to a random sample of pairs of observations (x_i, y_i) on two variables under the restriction that the independent variable values $x_1, \ldots, x_N$ can be controlled by the investigator. In many cases, such as a laboratory experiment where temperatures, pressures, or amounts of reagents can be set precisely, or with time series data recorded at consistent and accurate times, this assumption is reasonable. However, in most sets of multivariate observations the investigator draws *sampling units* randomly from some population and records the values of the dependent and independent variables. Then the pairs (x_i, y_i) are observations on the two-dimensional random variable (X, Y) that is described by some bivariate distribution function. The study of the relationship of the y_i to the x_i in this sampling model is known as *correlation analysis*.

Example 1-4

Before turning to a model for two correlated random variables, let us examine a sample of observation pairs. A university consumer organization distributed questionnaires to students asking for their ratings of their off-campus apartments and their landlords.* From the 800 questionnaires returned by tenants two indices on a 1 (poor) to 10 (excellent) scale were computed for each of the $N = 29$ landlords:

X: landlord attitude

Y: apartment quality

The values of the indices are given in Table 1-1 and plotted in Figure 1-5. The sum of crossproducts of the centered observations is

$$\sum (y_i - \bar{y})(x_i - \bar{x}) = 46.11$$

The estimated linear function relating the two indices is

$$\hat{y}_i = 4.315 + 0.501x_i$$

Strictly speaking, the least squares estimates were computed as if the landlord attitude index values x_i were fixed numbers under the investigator's control. Although that is hardly the case, we shall presently see that the usual best-fitting line still has a valid interpretation in the correlation model.

The coefficient of determination, or squared correlation coefficient, can be computed from expression (14) of Section 1-2:

$$r^2 = \frac{(46.11)^2}{(28)^2(3.287)(1.171)} = 0.7046$$

*I am indebted to the Penn Consumers' Board for kindly permitting the use of these data.

TABLE 1-1 Landlord and Apartment Indices

Landlord	Attitude, X	Apartment quality, Y	Landlord	Attitude, X	Apartment quality, Y
1	9.60	9.32	16	6.90	8.15
2	9.53	9.20	17	6.89	8.15
3	8.83	8.52	18	6.86	7.09
4	8.83	7.82	19	6.79	7.60
5	8.77	9.62	20	6.73	6.95
6	8.64	8.90	21	6.66	8.26
7	8.43	8.71	22	6.09	6.77
8	8.32	8.33	23	5.87	7.49
9	8.16	8.80	24	5.69	6.92
10	7.96	7.81	25	5.49	6.72
11	7.86	8.62	26	4.14	4.76
12	7.86	8.37	27	4.08	7.04
13	7.64	7.97	28	3.09	7.21
14	7.48	8.10	29	2.55	5.51
15	6.91	7.95			
			Mean	6.99	7.82
			Variance	3.287	1.171
			S.D.	1.813	1.082

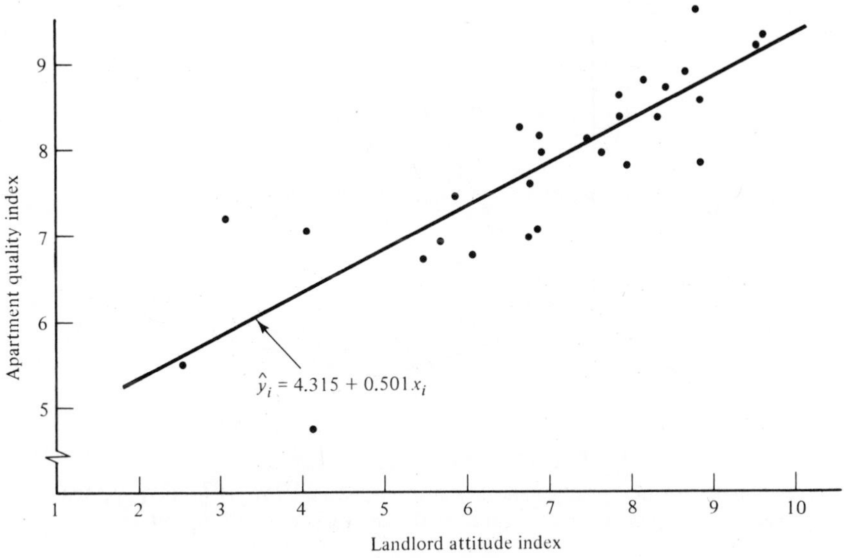

Figure 1-5 Attitude and quality indices.

and the coefficient of correlation for the 29 pairs of ratings is $r = 0.8394$. We chose the positive square root because $\hat{\beta} = 0.501$ is positive. The fitted straight line accounts for 70.46% of the variation in the apartment quality indices. It is apparent that landlord attitude is a good predictor of quality in most cases.

Our treatment of the correlation of two random variables is based on the population model of the bivariate normal distribution. This distribution is the extension of the normal distribution to two dimensions. It can be described by the *bivariate normal density function*

$$f(x, y) = \frac{1}{2\pi\sigma_1\sigma_2\sqrt{1-\rho^2}} \exp\left\{-\frac{1}{2}\frac{1}{1-\rho^2}\left[\left(\frac{x-\mu_1}{\sigma_1}\right)^2 - 2\rho\left(\frac{x-\mu_1}{\sigma_1}\right)\left(\frac{y-\mu_2}{\sigma_2}\right) + \left(\frac{y-\mu_2}{\sigma_2}\right)^2\right]\right\} \tag{1}$$

which defines a surface over the entire plane of the X- and Y-values. Probability statements about X and Y correspond to volumes under their density function. For example, the *joint distribution function*

$$\begin{aligned} F(x, y) &= \int_{-\infty}^{y}\int_{-\infty}^{x} f(t, u)\,dt\,du \\ &= P(X \leq x,\ Y \leq y) \end{aligned} \tag{2}$$

gives the probability that X and Y are simultaneously less than the respective values x and y.

One bivariate normal surface is shown in Figure 1-6. The density has a single mode, or maximum, at the means

$$E(X) = \mu_1 \qquad E(Y) = \mu_2$$

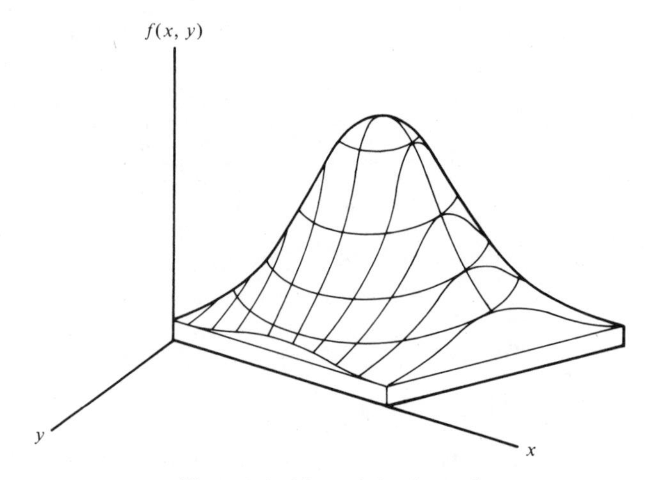

Figure 1-6 Normal density surface.

*John E. Freund, Mathematical Statistics, 2nd ed., Copyright 1971, p. 376. Reprinted by permission of Prentice-Hall, Inc., Englewood Cliffs, N.J.

As the slices through the base of the density surface suggest, the conditional densities of the variates are normal. The marginal densities of X and Y are also normal. We shall return to those densities shortly.

In addition to the means μ_1 and μ_2 the bivariate normal density depends upon the standard deviations σ_1 and σ_2 of X and Y, and the correlation coefficient

$$\rho = \frac{\text{Cov}(X, Y)}{\sqrt{\text{Var}(X)\,\text{Var}(Y)}} \tag{3}$$

ρ is a number between -1 and 1 that measures the degree of dependence of X and Y. If $\rho = 0$, the crossproduct term in the exponent of the density vanishes, and the function is the product of two univariate normal densities, thus X and Y are independent variates. Independence of X and Y also implies that $\rho = 0$, so that zero correlation is a necessary and sufficient condition for the independence of two normal random variables. If ρ is positive, large values of one variable tend to be associated with large values of the other. A negative ρ means that the relationship is reversed: Small values of X generally portend large values of Y, and vice versa. Subsequently, we shall express those notions of positive and negative correlation in terms of the conditional distribution of one variate with the other held constant. If X is proportional to Y, ρ is either 1 or -1. Then the density function (1) is undefined, for its constant term and exponent would contain divisors of zero. Such bivariate normal distributions are called *singular*. We shall deal only with nonsingular normal variates in this text.

Planes parallel to the X, Y plane cut the density surface in elliptical contours. If the cutting plane is at the height c, the ellipse has the equation

$$\begin{aligned} -2(1-\rho^2)\ln(2\pi\sigma_1\sigma_2 c\sqrt{1-\rho^2}) \\ = \left(\frac{x-\mu_1}{\sigma_1}\right)^2 - 2\rho\left(\frac{x-\mu_1}{\sigma_1}\right)\frac{y-\mu_2}{\sigma_2} + \left(\frac{y-\mu_2}{\sigma_2}\right)^2 \end{aligned} \tag{4}$$

The ellipse is centered at the means of X and Y. The orientation and relative lengths of the major and minor axes depend on σ_1^2, σ_2^2, and ρ. If ρ is positive, and if the variates have been standardized as

$$z_1 = \frac{x-\mu_1}{\sigma_1} \qquad z_2 = \frac{y-\mu_2}{\sigma_2} \tag{5}$$

the major axis is the line $z_2 = z_1$, and the minor axis is $z_2 = -z_1$. Four ellipses of the standardized bivariate normal density with $\rho = 0.8$ are shown in Figure 1-7. The ellipses are the contours of the surface at the proportions 0.2, 0.4, 0.6, and 0.8 of the maximum of the density.

The marginal densities of X and Y are

$$\begin{aligned} f_1(x) &= \int_{-\infty}^{\infty} f(x, y)\,dy \\ &= \frac{1}{\sqrt{2\pi}\sigma_1}\exp\left\{-\frac{1}{2}\left(\frac{x-\mu_1}{\sigma_1}\right)^2\right\} \\ f_2(y) &= \int_{-\infty}^{\infty} f(x, y)\,dx \\ &= \frac{1}{\sqrt{2\pi}\sigma_2}\exp\left\{-\frac{1}{2}\left(\frac{y-\mu_2}{\sigma_1}\right)^2\right\} \end{aligned} \tag{6}$$

X and Y are normally distributed with the same mean and variance parameters as in the bivariate density.

We shall develop the linear model for observations drawn from the bivariate normal population in terms of the conditional density function of Y for X

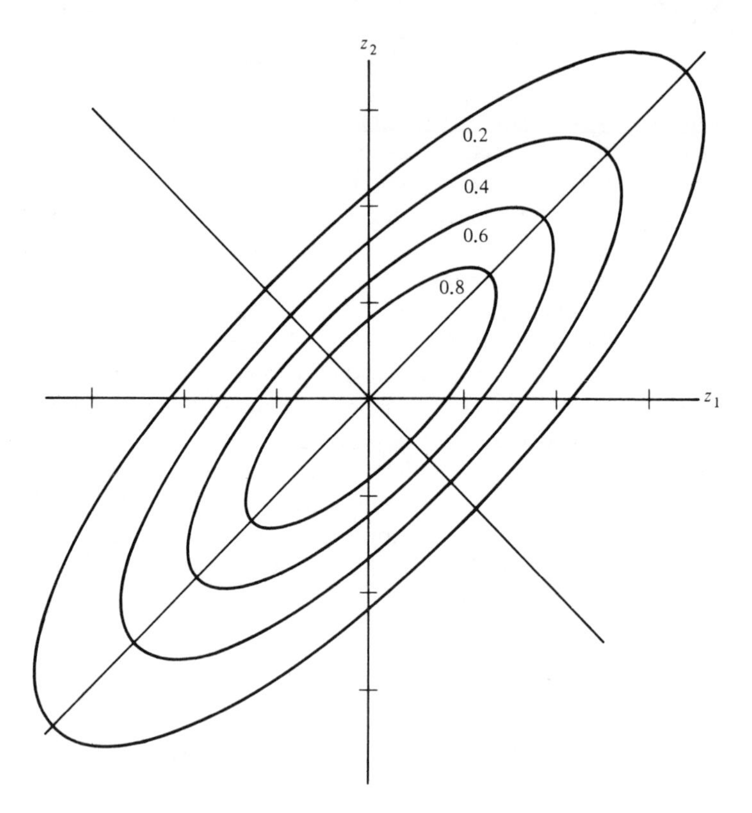

Figure 1-7 Contours of a bivariate normal density.

fixed at some particular value x. As we have indicated for Figure 1-6, those conditional densities are also normal and have means that are linearly related to x. The densities for four values of x have the appearance of Figure 1-8: Their means fall on a straight line, and their variances are the same for each x. The conditional density is given by the formula

$$f(y \mid x) = \frac{f(x, y)}{f_1(x)} \tag{7}$$

To find the conditional density we rewrite the exponent of $f(x, y)$ by "completing the square":

$$\begin{aligned} f(x, y) \\ &= \frac{1}{2\pi\sigma_1\sigma_2\sqrt{1-\rho^2}} \exp\left\{-\frac{1}{2}\frac{1}{1-\rho^2}\left[\left(\frac{x-\mu_1}{\sigma_1}\right)^2\right.\right. \\ &\quad \left.\left. + \frac{1}{\sigma_2^2}\left[y - \mu_2 - \rho\frac{\sigma_2}{\sigma_1}(x-\mu_1)\right]^2 - \frac{\rho^2}{\sigma_1^2}(x-\mu_1)^2\right]\right\} \\ &= \frac{1}{2\pi\sigma_1\sigma_2\sqrt{1-\rho^2}} \exp\left\{-\frac{1}{2\sigma_2^2(1-\rho^2)}\left[y - \mu_2 - \rho\frac{\sigma_2}{\sigma_1}(x-\mu_1)\right]^2\right. \\ &\quad \left. - \frac{1}{2\sigma_1^2}(x-\mu_1)^2\right\} \end{aligned} \tag{8}$$

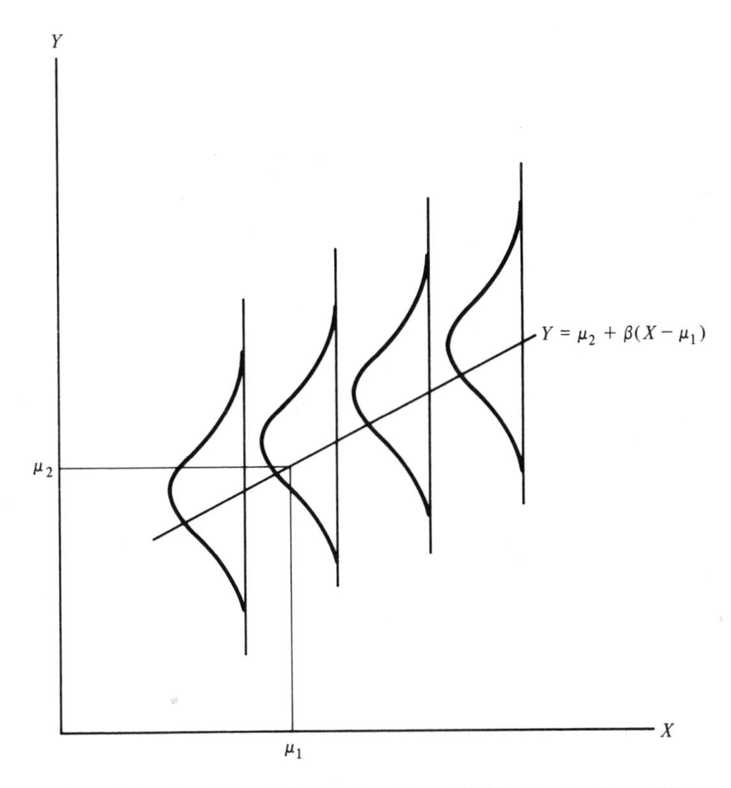

Figure 1-8 Conditional density functions of Y for fixed values of X.

Then

$$f(y \mid x) = \frac{1}{\sqrt{2\pi}\sigma_2\sqrt{1-\rho^2}} \exp\left\{-\frac{1}{2\sigma_2^2(1-\rho^2)}\left(y - \left[\mu_2 + \rho\frac{\sigma_2}{\sigma_1}(x - \mu_1)\right]\right)^2\right\} \tag{9}$$

or a normal density function with mean

$$E(Y \mid x) = \mu_2 + \rho\frac{\sigma_2}{\sigma_1}(x - \mu_1) \tag{10}$$

and variance

$$\text{Var}\,(Y \mid x) = \sigma_2^2(1 - \rho^2) \tag{11}$$

Some properties of the conditional density in Figure 1-8 are evident from these moments:

1. The conditional mean of Y is a *linear* function of the values of the fixed variate.
2. The conditional variance is the same for all values of x.
3. The conditional mean of X for a fixed value of $Y = y$ follows from the notational symmetry of (9) to be

$$E(X \mid y) = \mu_1 + \rho\frac{\sigma_1}{\sigma_2}(y - \mu_2) \tag{12}$$

The conditional variance is

$$\text{Var}(X \mid y) = \sigma_1^2(1 - \rho^2) \tag{13}$$

The mean functions (10) and (12) are known as the *regression function of Y on X* and the *regression function of X on Y*, respectively. The usage "regression" is a curious, yet commonplace, synonym for the method of least squares estimation. Its origin will be explained and illustrated in our subsequent example of correlation analysis.

The conditional distribution in the bivariate normal population provides us with a straight-line model for the case when both the y_i and x_i are observations on random variables. The analogy with the linear model of Section 1-3 is clear: α has been replaced by μ_2, β is now the parametric function $\rho(\sigma_2/\sigma_1)$, and the average $\bar{x}$ has been replaced by the population mean μ_1 of the independent variable. We may fit the straight line of the conditional means by replacing the population parameters by estimates computed from the sample data. The usual estimators of the means are

$$\begin{aligned} \hat{\mu}_1 &= \frac{1}{N} \sum x_i = \bar{x} \\ \hat{\mu}_2 &= \frac{1}{N} \sum y_i = \bar{y} \end{aligned} \tag{14}$$

The unbiased estimators of the variances are

$$\begin{aligned} \hat{\sigma}_1^2 &= \frac{1}{N-1} \sum (x_i - \bar{x})^2 \\ \hat{\sigma}_2^2 &= \frac{1}{N-1} \sum (y_i - \bar{y})^2 \end{aligned} \tag{15}$$

and the estimator of the correlation coefficient is

$$\begin{aligned} \hat{\rho} &= r \\ &= \frac{\sum (y_i - \bar{y})(x_i - \bar{x})}{\sqrt{[\sum (y_i - \bar{y})^2][\sum (x_i - \bar{x})^2]}} \end{aligned} \tag{16}$$

All summations extend over the subscripts $i = 1, \ldots, N$. We note, of course, that hand calculation of the sums of squares and products can be expedited by the shortcut formulas

$$\begin{aligned} \sum (x_i - \bar{x})^2 &= \sum x_i^2 - \frac{(\sum x_i)^2}{N} \\ \sum (y_i - \bar{y})^2 &= \sum y_i^2 - \frac{(\sum y_i)^2}{N} \\ \sum (y_i - \bar{y})(x_i - \bar{x}) &= \sum x_i y_i - \frac{(\sum x_i)(\sum y_i)}{N} \end{aligned} \tag{17}$$

r, like ρ, is always between -1 and 1. It is a pure number index unaffected by the units or scale origins of the original variates. Proof of that invariance property is left as an exercise.

The estimator of the conditional mean, or slope, parameter is

$$\hat{\beta} = \hat{\rho}\left(\frac{\hat{\sigma}_2}{\hat{\sigma}_1}\right)$$
$$= \frac{\sum (y_i - \bar{y})(x_i - \bar{x})}{\sum (x_i - \bar{x})^2} \tag{18}$$

or precisely the least squares estimator (5) of Section 1-3. The estimator of the conditional mean (10) for $X = x_i$ is

$$E(Y|x_i) = \bar{y} + \hat{\beta}(x_i - \bar{x})$$
$$= \hat{y}_i \tag{19}$$

or expression (9), Section 1-3, for the fixed independent variable model.

Our analysis of bivariate data will usually be focused on the correlation coefficient rather than the line of conditional means. We begin by assuming that the bivariate normal density is a reasonable model for the population, or at least that the degree of dependence of the variates can be measured approximately by the sample correlation. Of course, we require that the N pairs of observations are independently drawn from the population. We compute r from the data and note whether its sign indicates a positive or negative relationship between X and Y.

With the sample correlation in hand we would next like to make inferences about the population correlation ρ. Let us begin with the case of independent variates, or that of the bivariate normal population with $\rho = 0$. The density function of r is

$$f(r) = \begin{cases} \dfrac{\Gamma[(n+1)/2]}{\sqrt{\pi}\,\Gamma(n/2)}(1 - r^2)^{(n-2)/2} & -1 \leq r \leq 1 \\ 0 & \text{elsewhere} \end{cases} \tag{20}$$

where the degrees of freedom parameter n is equal to $N - 2$ for a single random sample of N observation pairs. $f(r)$ is a well-behaved function symmetric about $r = 0$. It is shown in Figure 1-9 for $n = 10, 20$, and 30. Its 100α upper percentage points $r_{\alpha;n}$ are given in Table 5, Appendix A. The left-hand, or lower, $100(1 - \alpha)$ percentage points follow from symmetry as $-r_{\alpha;n}$. The percentage points provide us with a test of the null hypothesis

$$H_0\colon \quad \rho = 0 \tag{21}$$

of no correlation in the bivariate normal population. For a test against the one-sided alternative

$$H_1\colon \quad \rho > 0 \tag{22}$$

we compute r, select a significance level or Type I error rate α, and reject H_0 in favor of H_1 if

$$r > r_{\alpha;n} \tag{23}$$

Otherwise, for $r \leq r_{\alpha;n}$ the null hypothesis of no correlation is supported by the data. Similarly, we should reject H_0 in favor of the left-hand one-sided alternative $H_1\colon \quad \rho < 0$ if

$$r < -r_{\alpha:n} \tag{24}$$

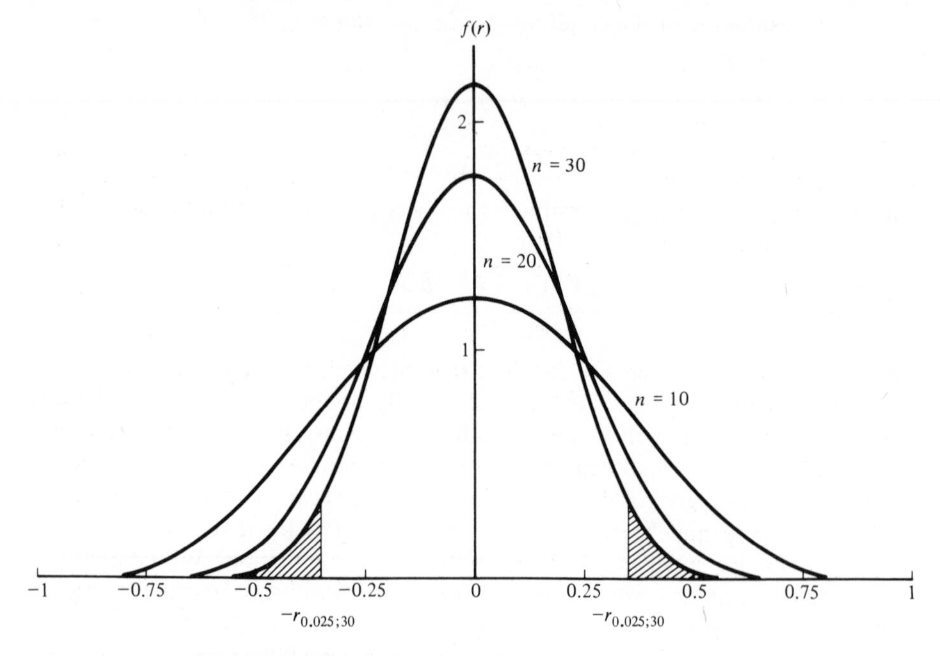

Figure 1-9 Density functions of r for $n = 10, 20,$ and 30.

For the two-sided alternative

$$H_1: \quad \rho \neq 0 \tag{25}$$

of correlation between X and Y of unspecified direction we would reject H_0 at the α level if

$$|r| > r_{\alpha/2;n} \tag{26}$$

If a table of $r_{\alpha;n}$ is not available, it is possible to carry out the preceding tests by computing

$$t = r\sqrt{\frac{n}{1-r^2}} \tag{27}$$

and referring that statistic to the percentage points of the t-distribution with $n = N - 2$ degrees of freedom. The upper 100α percentage points $t_{\alpha;n}$ are given in Table 3, Appendix A. The relationship of r to t is more than an alternative test statistic. Let us evaluate t in terms of the sums of squares and products in r:

$$\begin{aligned} t &= \frac{\sum (y_i - \bar{y})(x_i - \bar{x})}{\sqrt{[\sum (y_i - \bar{y})^2][\sum (x_i - \bar{x})^2]}} \\ &\quad \cdot \sqrt{\frac{n[\sum (y_i - \bar{y})^2][\sum (x_i - \bar{x})^2]}{[\sum (y_i - \bar{y})^2][\sum (x_i - \bar{x})^2] - [\sum (y_i - \bar{y})(x_i - \bar{x})]^2}} \\ &= \left[\frac{\sum (y_i - \bar{y})(x_i - \bar{x})}{\sum (x_i - \bar{x})^2}\right]\sqrt{\frac{\sum (x_i - \bar{x})^2}{\text{SSE}/n}} \\ &= \frac{\hat{\beta}\sqrt{\sum (x_i - \bar{x})^2}}{\sqrt{\hat{\sigma}^2}} \end{aligned} \tag{28}$$

or expression (11) of Section 1-4 with $\beta_0 = 0$. The test for zero correlation in the bivariate normal distribution of X and Y is identical with that for H_0: $\beta = 0$ in the linear model with fixed x_i-values. Thus we are permitted some vagueness in the nature of the independent variable without affecting the test for the absence of a linear relationship.

Example 1-5

The correlation coefficient of the landlord attitude and apartment quality indices given in Table 1-1 was equal to 0.8394. Let us test the hypothesis that the correlation between those variables in a hypothetical population of landlords is zero, as opposed to the alternative of a positive correlation. Since $N = 29$ we must enter Table 5 of Appendix A with $n = 27$ degrees of freedom. The one-sided 0.005 critical value is $r_{0.005;27} = 0.471$. Clearly, the sample correlation is statistically significant at the 0.005 level or any other reasonable one, and we should conclude that the two indices are truly highly correlated.

Example 1-6

The plasma free fatty acid (FFA) levels of Section 1-5 have this correlation with the subjects' ages:

$$\begin{aligned} r &= \frac{-47.3316}{\sqrt{(6052.49)(2.59153)}} \\ &= -0.3779 \end{aligned}$$

If we have no prior information from the physiological nature of FFA or from earlier studies about its relation to age we should adopt the two-sided alternative hypothesis

$$H_1 : \rho \neq 0$$

of a nonzero correlation of either sign. The null hypothesis of zero correlation is rejected if either

$$r > r_{\alpha/2;39} \qquad \text{or} \qquad r < -r_{\alpha/2;39}$$

If $\alpha = 0.02$, the critical value can be found from interpolation in Table 5 to be approximately $r_{0.01;39} = 0.363$. Since the sample correlation is less than -0.363 we should conclude that a negative association is present in the age- and FFA-variate population. Alternatively, if we had evidence that FFA level could only decrease with age, we might choose the one-sided hypothesis H_1: $\rho < 0$. Then the significance level for our test would be halved to 0.01.

The coefficient of determination. In our description of the linear least squares model in Section 1-4 we showed how the total sum of squares about the average of the y_i-values could be decomposed as

$$\sum_{i=1}^{N} (y_i - \bar{y})^2 = \text{SSR} + \text{SSE} \tag{29}$$

where

$$\begin{aligned} \text{SSR} &= \sum_{i=1}^{N} (\hat{y}_i - \bar{y})^2 \\ &= \sum_{i=1}^{N} [\bar{y} + \hat{\beta}(x_i - \bar{x}) - \bar{y}]^2 \\ &= \hat{\beta}^2 \sum_{i=1}^{N} (x_i - \bar{x})^2 \\ &= \frac{[\sum (y_i - \bar{y})(x_i - \bar{x})]^2}{\sum (x_i - \bar{x})^2} \end{aligned} \tag{30}$$

is the sum of squared deviations of the fitted line from $\bar{y}$, and

$$\begin{aligned} \text{SSE} &= \sum_{i=1}^{N} (y_i - \hat{y}_i)^2 \\ &= \sum_{i=1}^{N} (y_i - \bar{y})^2 - \frac{[\sum (y_i - \bar{y})(x_i - \bar{x})]^2}{\sum (x_i - \bar{x})^2} \end{aligned} \tag{31}$$

is the sum of squares of errors about the fitted line introduced in Section 1-2 and (16), Section 1-3. We may write

$$\begin{aligned} \text{SSR} &= \frac{[\sum (y_i - \bar{y})(x_i - \bar{x})]^2}{\sum (y_i - \bar{y})^2 \sum (x_i - \bar{x})^2} \sum (y_i - \bar{y})^2 \\ &= r^2 \sum (y_i - \bar{y})^2 \end{aligned} \tag{32}$$

so that the coefficient of determination

$$r^2 = \frac{\text{SSR}}{\sum (y_i - \bar{y})^2} \tag{33}$$

defined in Sections 1-2 and 1-3 for the least squares model is the proportion of the sum of squares about the mean due to the least squares line. Similarly,

$$\text{SSE} = (1 - r^2) \sum_{i=1}^{N} (y_i - \bar{y})^2 \tag{34}$$

and $1 - r^2$ is the proportion due to error about the line. We see that the coefficient of determination is the meaningful measure of sample correlation, for it is the proportional reduction in the dispersion of the dependent variable explained by fitting a straight line. For example, $r = 0.50$ suggests a moderate correlation between the two variables and is highly significantly different from zero if the sample size is greater than 40 or 50. Nevertheless, only 25% of the variation has been accounted for by the line, and 75% is due to noise about the fitted model. A correlation of -0.70 implies that 51% of the variation is still unexplained by a linear trend. In practice we should always immediately convert correlations to r^2 or $1 - r^2$ and consider whether we have explained sufficient variation to justify prediction by the fitted line. If we apply that rule to the free fatty acid and age data of Section 1-5, we note that only 14.28% of the variation in FFA can be explained by the linear function. The sample line, although statistically significant, has little utility for predicting FFA at a given age.

The expressions (32) and (34) for SSR and SSE can be substituted in the F-ratio (18) of Section 1-4 to give an alternative form of that statistic for testing for zero slope of the least squares line:

$$\begin{aligned} F &= (N - 2) \frac{\text{SSR}}{\text{SSE}} \\ &= \frac{(N - 2) r^2}{1 - r^2} \end{aligned} \tag{35}$$

We would reject $H_0\colon \ \beta = 0$ if F exceeds the critical value $F_{\alpha;1,N-2}$ from Table 4, Appendix A. But the alternative statistic is merely the square of the t-ratio (27) for zero population correlation. We see that the test of zero slope in the fixed-variable least squares model is equivalent to that of no correlation in the bivariate normal model.

Example 1-7

The origin of regression analysis. In the preceding development we have referred to the techniques for fitting linear models as least squares estimation rather than regression analysis. At this point it will be convenient to introduce that name by discussing the original correlation study of Sir Francis Galton in which it first arose. In the wake of the scientific ferment caused by the publication of Darwin's *On the Origin of Species*, Galton (1888, 1889) collected large sets of data on inherited characteristics. Among these were the heights of adult offspring and weighted averages of the heights of their parents shown in Table 1-2. These data require some explanation before proceeding to their analysis. First, we have truncated Galton's original table to eliminate 36 observation pairs merely recorded as "below" or "above" the limits in Table 1-2. For comparability Galton multiplied all female heights by 1.08. The curious class intervals for the offspring heights were chosen to counteract a bias in the measurements in favor of whole inches. Because most parents had several children the sample does *not* contain 892 independent observations.

TABLE 1-2 Truncated Frequency Distribution of the Statures of Parents and Children

Offspring heights [class midpoint (in.)]	Composite parent heights [class midpoint (in.)]									
	64.5	65.5	66.5	67.5	68.5	69.5	70.5	71.5	72.5	Total
62.2	1	—	3	3	—	—	—	—	—	7
63.2	4	9	3	5	7	1	1	—	—	30
64.2	4	5	5	14	11	16	—	—	—	55
65.2	1	7	2	15	16	4	1	1	—	47
66.2	5	11	17	36	25	17	1	3	—	115
67.2	5	11	17	38	31	27	3	4	—	136
68.2	—	7	14	28	34	20	12	3	1	119
69.2	2	7	13	38	48	33	18	5	2	166
70.2	—	5	4	19	21	25	14	10	1	99
71.2	—	2	—	11	18	20	7	4	2	64
72.2	—	1	—	4	4	11	4	9	7	40
73.2	—	—	—	—	3	4	3	2	2	14
Total	22	65	78	211	218	178	64	41	15	892

SOURCE: Data reproduced with the permission of The Macmillan Company.

Let us find the correlation of adult offspring and composite parents' heights and under the bivariate normal model find the estimated line on which the conditional mean heights must lie. To reduce the arithmetic labor we shall translate the class midpoints to smaller numbers. For the composite parent height class midpoints x_j make the transformation

$$u_j = x_j - 68.5 \qquad j = 1, \ldots, 9 \tag{36}$$

Similarly, translate the offspring height class midpoints y_i to

$$v_i = y_i - 61.2 \qquad i = 1, \ldots, 12 \tag{37}$$

Clearly, $v_i = i$, but we shall retain the subscript form for consistency. Introduce the notation

$$f_{ij} = \text{number of offspring in the } (x_j, y_i) \text{ cell of Table 1-2}$$

and the "dot" notation for marginal totals:

$$f_{i.} = \sum_{j=1}^{9} f_{ij}$$

$$= \text{number of children in the } i\text{th offspring height class}$$

$$f_{.j} = \sum_{i=1}^{12} f_{ij}$$

$$= \text{number of offspring in the } j\text{th parent height class}$$

$$\sum_{i=1}^{12} \sum_{j=1}^{9} f_{ij} = \sum_{i=1}^{12} f_{i.} = \sum_{j=1}^{9} f_{.j} = 892$$

These statistics have been computed from the table:

$$\sum_{j=1}^{9} f_{.j}u_j = -161 \qquad \bar{u} = \frac{1}{892} \sum_{j=1}^{9} f_{.j}u_j = -0.18049$$

$$\sum f_{.j}(u_j - \bar{u})^2 = \sum f_{.j}(x_j - \bar{x})^2$$

$$= 2473.9406$$

$$\bar{x} = 68.5 + \bar{u}$$

$$= 68.3195$$

$$\sum_{i=1}^{12} f_{i.}v_i = 6111 \qquad \bar{v} = \frac{1}{892} \sum_{i=1}^{12} f_{i.}v_i = 6.8509$$

$$\sum_{i=1}^{12} f_{i.}(v_i - \bar{v})^2 = \sum_{i=1}^{12} f_{i.}(y_i - \bar{y})^2$$

$$= 5009.1693$$

$$\bar{y} = 61.2 + \bar{v}$$

$$= 68.0509$$

$$\sum_{i=1}^{12} \sum_{j=1}^{9} f_{ij}v_i u_j = 327$$

$$\sum_{i=1}^{12} \sum_{j=1}^{9} f_{ij}(y_i - \bar{y})(x_j - \bar{x}) = \sum_{i=1}^{12} \sum_{j=1}^{9} f_{ij}(v_i - \bar{v})(u_j - \bar{u})$$

$$= 1429.9944$$

The least squares line for predicting offspring height from the composite parent height is

$$\hat{y} = 68.0509 + 0.5780(x - 68.3195) \tag{38}$$

The correlation coefficient of offspring and composite parent heights is $r = 0.4062$. The scatterplot of the data has been summarized in Figure 1-10 by the average offspring heights for each composite parent class. Galton prepared a similar chart from the observation pairs, although with medians of the conditional distributions rather than means. He noted that composite parents of short stature had offspring of higher average height, whereas taller composite parents tended to have children of shorter

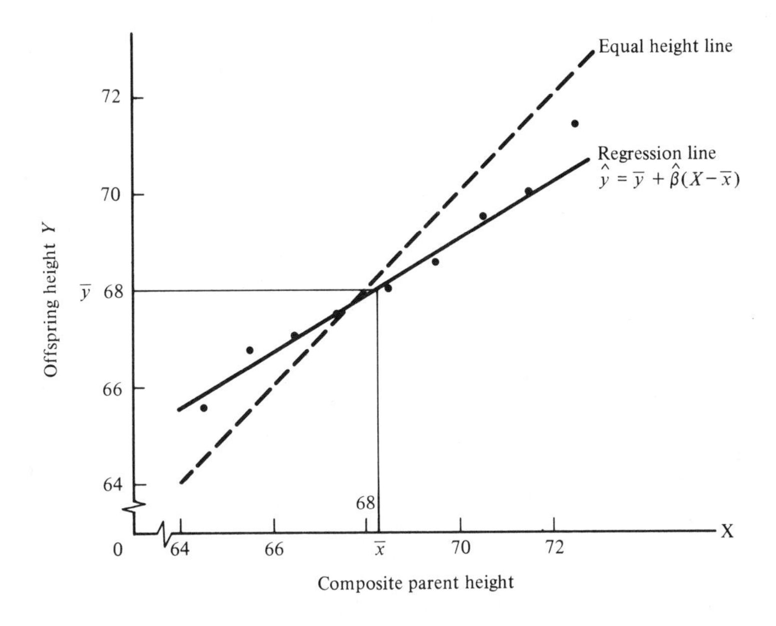

Figure 1-10 Mean adult offspring heights and composite parent heights for the truncated Galton data.

average stature. The dashed line $Y = X$ in Figure 1-10 illustrates this property. The average statures of children with composite parents in the classes 64.5, 65.5, and 66.5 lie well above that equal height line, whereas those of parent classes 68.5–72.5 are below. Galton referred to this shift in stature from one generation to the next as "regression to mediocrity," or in more modern usage, "regression toward the average." Galton termed such studies of pairs of measurements "regression analysis," and that name soon came to mean the fitting of lines and planes by least squares to data sets. Such expressions as (38) will be called the *sample regression* equation. Galton estimated the parameters of his regression equation for the stature data from the properties of the concentration ellipse of the bivariate normal distribution. For that reason the conditional mean function $E(Y|x)$ of (10) is commonly called the *population regression function of Y on X*, whereas $E(X|y)$ of expression (12) is the *population regression function of X on Y*. In general, conditional means are known as regression functions. Henceforth we shall use "regression analysis" synonymously with "least squares estimation," and the study of conditional mean functions.

Inferences about the correlation coefficient when $\rho \neq 0$. When the population correlation coefficient ρ of the bivariate normal distribution is zero we have seen from expression (20) and Figure 1-9 that the density function of the sample correlation is an elementary function symmetric about zero. When ρ is not zero the distribution of r is complicated and cannot be expressed in a simple form. The density function is asymmetric with its longer tail to the left for positive ρ and to the right for ρ negative. The variance of r depends on ρ: When $\rho = 0$ the variance is greatest and decreases to zero for the

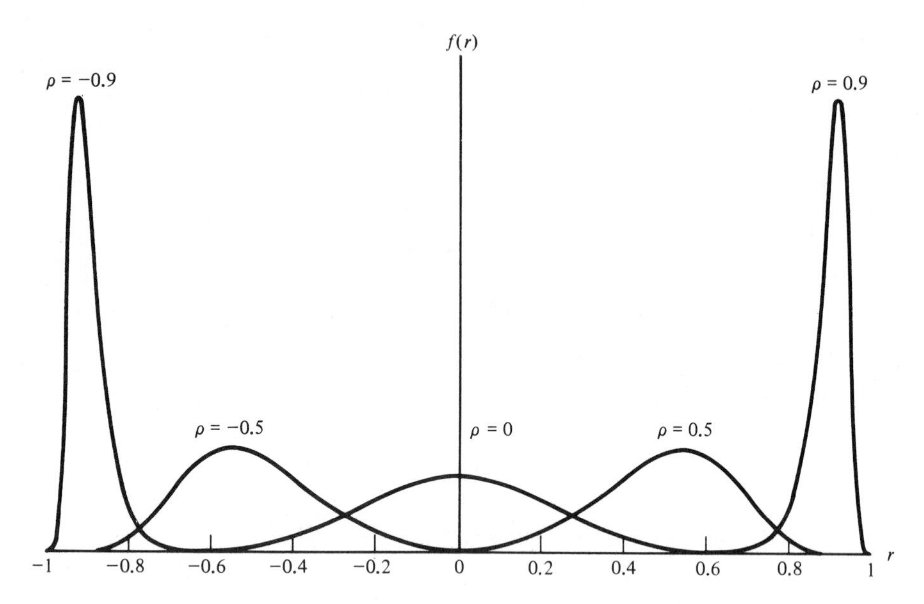

Figure 1-11 Noncentral densities of r for $n = 20$.

extreme cases of $\rho = \pm 1$. These properties of the density are illustrated in Figure 1-11 for the degrees of freedom parameter $n = 20$, or a correlation computed from a simple random sample of $N = 22$ observation pairs.

Because the density of r is a complicated function of the two parameters n and ρ, exact tests of hypotheses and confidence intervals for ρ would require voluminous tables of the lower and upper percentage points of the distribution. Fortunately, an easier approximate method is available. We begin by noting that as the sample size $N = n + 2$ increases, the quantity

$$\frac{\sqrt{N}(r - \rho)}{1 - \rho^2} \tag{39}$$

tends to be distributed as a standard normal variate. However, the large sample variance of r,

$$\frac{(1 - \rho^2)^2}{N} \tag{40}$$

is a function of the population correlation: The density has greatest spread when $\rho = 0$ and approaches the degenerate form of a spike when $\rho = -1$ or 1. The higher moments of r also depend upon ρ if N is small, as Figure 1-11 shows. Unless the sample size is extremely large the limiting normal distribution of (39) is a poor approximation to that of r for extreme values of ρ. R. A. Fisher recognized this dissimilarity of the normal and r densities and sought a transformation of r whose large sample variance and higher moments would be free of ρ. Fisher found that the "variance stabilizing" transformation for r is the inverse hyperbolic tangent function

$$z = \frac{1}{2} \ln \frac{1+r}{1-r}$$

$$= \tanh^{-1} r \tag{41}$$

For even small samples of 15 or more observations z tends to have a normal distribution with a mean

$$\zeta = \frac{1}{2} \ln \frac{1+\rho}{1-\rho}$$

$$= \tanh^{-1} \rho \tag{42}$$

and approximate variance

$$\mathrm{Var}\,(z) = \frac{1}{n-1} \tag{43}$$

We note that

$$\tanh^{-1}(-r) = -z \tag{44}$$

so that the z-transformation has only been given for positive r in Table 6, Appendix A.

The general distribution of the correlation coefficient was found by Fisher (1915). Fisher (1921) derived the z-transformation as the solution of a simple differential equation; the development has been given in detail by Anderson (1958). Hotelling (1953) has discussed the distributions of r and z at length. Winterbottom (1979) has derived the z-transformation for the purpose of normalization, with variance stabilization a secondary consequence.

We may use the z-transformation to test the hypothesis

$$H_0: \quad \rho = \rho_0 \tag{45}$$

against the one-sided alternative

$$H_1^{(1)}: \quad \rho > \rho_0 \tag{46}$$

of a larger correlation. The quantity

$$u = \frac{z - \zeta_0}{\sqrt{\mathrm{Var}\,(z)}}$$

$$\approx -(z - \zeta_0)\sqrt{N-3} \tag{47}$$

tends to be distributed as a normal variate with mean zero and unit variance, where

$$\zeta_0 = \tanh^{-1} \rho_0 \tag{48}$$

The decision rule for carrying out the test at the α level is

$$\begin{aligned} &\text{Accept } H_0 \quad \text{if } u \leq z_\alpha \\ &\text{Accept } H_1^{(1)} \text{ if } u > z_\alpha \end{aligned} \tag{49}$$

z_α is the upper 100α percentage point of the standard normal distribution, or the value for which $P(Z > z_\alpha) = \alpha$.

Example 1-8

In a study of the relation of cognitive performance and education a random sample of $N = 36$ subjects had a correlation $r = 0.84$ between a certain verbal intelligence test score and years of formal education. Previous studies had shown a correlation in the vicinity of 0.60 between these variables. Do the present data support such a population correlation? That is, we wish to test the hypothesis

$$H_0: \rho = 0.60 \tag{50}$$

against the alternative

$$H_1^{(1)}: \rho > 0.60 \tag{51}$$

that the sample arose from a population with a stronger correlation between the verbal score and education. We begin by finding these z-values from Table 6, Appendix A:

$$\begin{aligned} z &= \tanh^{-1}(0.84) \\ &= 1.2212 \\ \zeta_0 &= \tanh^{-1}(0.60) \\ &= 0.6931 \end{aligned}$$

The test statistic is

$$\begin{aligned} u &= (1.2212 - 0.6931)\sqrt{36 - 3} \\ &= 3.034 \end{aligned}$$

For a test at the 0.01 level the normal distribution critical value is $z_{0.01} = 2.326$, and because u exceeds that point we should reject the hypothesis of a correlation of 0.60 in favor of the conclusion that the correlation is higher in the population whence came the sample. We might also note that the risk of a wrong decision is considerably less than 0.01. The probability that a standard normal variate is greater than 3.034 is approximately 0.0012.

Similarly, the null hypothesis (45) might be tested against the opposite one-sided alternative

$$H_1^{(2)}: \rho < \rho_0$$

that the population correlation is smaller than the specified value ρ_0. Then the same statistic (47) is employed with the decision rule

$$\begin{aligned} &\text{Accept } H_0 \quad \text{if } u \geq -z_\alpha \\ &\text{Accept } H_1^{(2)} \text{ if } u < -z_\alpha \end{aligned} \tag{52}$$

By the symmetry of the normal distribution

$$P(Z < -z_\alpha) = \alpha$$

For example, let us suppose that the previous sample correlation between the verbal test score and years of education in the sample of 36 subjects had only been 0.45. Is the value $\rho = 0.60$ still tenable, or should we conclude that the population value is smaller? The new z-transformation is

$$\begin{aligned} z &= \tanh^{-1}(0.45) \\ &= 0.4847 \end{aligned}$$

and the test statistic is

$$\begin{aligned} u &= (0.4847 - 0.6931)\sqrt{33} \\ &= -1.197 \end{aligned}$$

Since this statistic is not less than the 0.05 critical value -1.645, we can conclude that the null hypothesis (50) should not be rejected in favor of a smaller value of ρ.

Often, as in exploratory studies, the direction of change in the correlation coefficient cannot be specified. For example, an experimental condition might lead to a higher correlation by producing more extreme pairs of observations on the two variables, or a lower correlation by increasing the variances of the variables. In that less determinate case we should test H_0 against the two-sided alternative

$$H_1^{(3)}: \quad \rho \neq \rho_0 \tag{53}$$

by referring the statistic u to the decision rule

$$\begin{aligned} &\text{Accept } H_0: \quad \rho = \rho_0 \text{ if } |u| \leq z_{\alpha/2} \\ &\text{Accept } H_1^{(3)}: \quad \rho \neq \rho_0 \text{ if } |u| > z_{\alpha/2} \end{aligned} \tag{54}$$

Example 1-9

Suppose that the correlation of the reaction times of normal subjects to two visual stimuli has been found to be 0.70 from extensive experimentation. Reaction times were recorded for a new version of the method of displaying the two stimuli to a random sample of $N = 20$ individuals. One purpose of the experiment was to determine if the display changed the reaction time correlation. The sample correlation coefficient was equal to 0.52, so that

$$\begin{aligned} u &= (0.5763 - 0.8673)\sqrt{17} \\ &= -1.20 \end{aligned}$$

Since $|u|$ is less than the 0.05 two-sided critical value 1.96, we should conclude that the hypothesis of no change is tenable.

These tests on the correlation coefficient extend easily to the case of two independent random samples from bivariate normal populations. If the respective correlations are denoted by ρ_1 and ρ_2, the hypothesis

$$H_0: \quad \rho_1 = \rho_2 \tag{55}$$

can be tested against one of the usual alternatives

$$\begin{aligned} &H_1^{(1)}: \quad \rho_1 > \rho_2 \\ &H_2^{(2)}: \quad \rho_1 < \rho_2 \\ &H_2^{(3)}: \quad \rho_1 \neq \rho_2 \end{aligned} \tag{56}$$

by computing

$$u = \frac{z_1 - z_2}{\sqrt{\dfrac{1}{N_1 - 3} + \dfrac{1}{N_2 - 3}}} \tag{57}$$

where z_1 and z_2 are the Fisher z-transformations of the correlations r_1 and r_2 from the first and second samples, and N_1, N_2 are the respective sample sizes. As the sample numbers become large u tends to a standard normal random variable. For a test of level α these decision rules should be employed:

1. H_0 against $H_1^{(1)}$: $\rho_1 > \rho_2$.

$$\begin{aligned} &\text{Accept } H_0 \text{ if } u \leq z_\alpha. \\ &\text{Reject } H_0 \text{ in favor of } H_1^{(1)} \text{ if } u > z_\alpha. \end{aligned} \tag{58}$$

2. H_0 against $H_1^{(2)}$: $\rho_1 < \rho_2$.

$$\begin{aligned} &\text{Accept } H_0 \text{ if } u \geq -z_\alpha. \\ &\text{Reject } H_0 \text{ in favor of } H_1^{(2)} \text{ if } u < -z_\alpha. \end{aligned} \tag{59}$$

3. H_0 against $H_1^{(3)}$: $\rho_1 \neq \rho_2$.

$$\begin{aligned} &\text{Accept } H_0 \text{ if } |u| \leq z_{\alpha/2}. \\ &\text{Reject } H_0 \text{ in favor of } H_1^{(3)} \text{ if } |u| > z_{\alpha/2}. \end{aligned} \tag{60}$$

Example 1-10

Now let us illustrate the test for equal correlations with an application from psychology and medicine. Suppose that a random sample of elderly men in good health has been tested on the Wechsler Adult Intelligence Scale (WAIS). In addition, each subject has been evaluated by a psychiatrist for the presence or absence of senile changes. The $N_1 = 37$ subjects without senile changes had a correlation $r_1 = 0.57$ between their Information and Arithmetic WAIS subtest scores, whereas the $N_2 = 12$ subjects with the senile quality had a correlation $r_2 = 0.82$ between those subtest scores. Are those correlations consistent with the hypothesis of a common population correlation coefficient?

We begin by finding the z-transformations of the correlations from Table 6, Appendix A:

$$z_1 = 0.6475 \qquad z_2 = 1.1568$$

The test statistic is

$$\begin{aligned} u &= \frac{0.6475 - 1.1568}{\sqrt{\frac{1}{34} + \frac{1}{9}}} \\ &= \frac{-0.5093}{0.3749} \\ &= -1.36 \end{aligned}$$

Because we have no substantive basis for predicting how senility will affect the correlation let us choose the third alternative hypothesis $H_1^{(3)}$: $\rho_1 \neq \rho_2$ and make the test at the 0.05 level. Since the absolute value of u does not exceed the critical value $z_{0.025} = 1.96$, we should conclude that a common correlation coefficient may hold for the normal and senile populations. At least with the present sample sizes we have no basis for rejecting that contention.

Confidence intervals for the correlation coefficient. The decision rule for testing H_0: $\rho = \rho_0$ against the two-sided alternative (53) states that H_0 should be accepted at the α level if

$$|\tanh^{-1} r - \tanh^{-1} \rho_0| \sqrt{N-3} \leq z_{\alpha/2} \tag{61}$$

or if

$$\tanh^{-1} r - \frac{z_{\alpha/2}}{\sqrt{N-3}} \leq \tanh^{-1} \rho_0 \leq \tanh^{-1} r + \frac{z_{\alpha/2}}{\sqrt{N-3}} \tag{62}$$

From the relation of a function to its inverse, $\tanh(\tanh^{-1} \rho_0) = \rho_0$, so that the values of ρ_0 for which the null hypothesis is tenable at the α level are given by the interval

$$\tanh\left(z - \frac{z_{\alpha/2}}{\sqrt{N-3}}\right) \le \rho \le \tanh\left(z + \frac{z_{\alpha/2}}{\sqrt{N-3}}\right) \tag{63}$$

This is the $100(1-\alpha)\%$ confidence interval for the population correlation, for it covers ρ with probability $1-\alpha$. Values of the tanh function can be found by interpolation in the body of Table 6, Appendix A, or from its definition

$$\tanh x = \frac{e^x - e^{-x}}{e^x + e^{-x}} \tag{64}$$

and the use of a hand calculator with the exponential function.

The confidence interval allows us to express our uncertainty about the value of ρ in probabilistic terms. For example, let us find the 95% confidence interval for the correlation coefficient of the population of older men with the senile quality cited in the last example. Then

$$\tanh\left(1.1568 - \frac{1.96}{3}\right) = 0.465 \qquad \tanh\left(1.1568 + \frac{1.96}{3}\right) = 0.948$$

and the interval is

$$0.465 \le \rho \le 0.948$$

The width of the interval reflects the uncertainty about ρ inherent in a correlation computed from only 12 observations. However, we also may be concerned that the large-sample normal distribution assumption is inappropriate for so few observations. Use of the Pearson–Hartley chart for the exact 95% confidence intervals on ρ (Pearson and Hartley, 1969, Table 15) indicates that this is not so: The confidence interval based on the exact distribution of r is approximately

$$0.44 \le \rho \le 0.95$$

or virtually that given by the normal approximation.

1-7 SOME HISTORY OF LEAST SQUARES AND CORRELATION

The methods of least squares and correlation were seeded in 1733 with the discovery by De Moivre of the normal distribution. De Moivre found the normal density as the limit of a symmetric binomial probability function as the number of trials N became very large. Later, in 1812, Laplace generalized the limit to a binomial with an arbitrary success probability p. These special cases of the central limit theorem for sums of independent random variables gave importance to the normal distribution, particularly as a model for measurement or experimental error. In the early years of the nineteenth century others extended the normal distribution to several dimensions. Robert Adrain, an Irish mathematician who fled to America and taught at Columbia, Rutgers, and the University of Pennsylvania, published the case of the circular bivariate normal density, or two independent normal variates with a common variance, in 1808. In 1810, Laplace gave the general bivariate normal distribution, and two years later Plana also found the bivariate normal density and its extension to three variables. About 1823, the great German mathematician Gauss gave the p-dimensional multinormal density. In 1846, Bravais reconsidered the two- and three-variate normal densities of Plana and pointed out that the exponential terms represented ellipsoids in the space of the variates. Some generations hence Galton, Pearson, and the other early biometricians would build their theories of correlation and regression lines upon those properties of normal random variables.

The origins of the principle of least squares are indeterminate and fraught with controversy, although credit for its discovery is usually acceded to Gauss. The French mathematician Legendre used the method for determining the paths of comets about 1805, whereas Laplace used least squares estimation in books published in 1812 and 1820. Adrain discovered the least squares criterion in 1809 in the context of navigational, geodesic, and surveying applications, although it is not clear whether he knew of Legendre's paper of 1805 (Stigler, 1978). Adrain also made further use of least squares in geodesy in 1818. Gauss published his method of least squares in a paper in 1821 and used it for the determination of astronomical movements in 1809. He also claimed to have used the principle in 1795. Although the claims for priority led to a bitter dispute with Legendre, later research has ascribed the discovery to Gauss (Plackett, 1949; Stigler, 1981). Gauss also showed in 1821 that least squares estimates enjoyed least variance among all unbiased estimators. This property was found independently by the Russian probabilist Markov and is referred to as the Gauss–Markov theorem—we shall encounter it in Section 2-4 for the linear model with several independent variables.

Whereas the least squares principle arose from the resolution of measurement errors in astronomy and surveying, correlation and its associated linear functions came from the life sciences. In the latter part of the nineteenth century the controversy surrounding Darwin's theory of evolution provoked the collection of large sets of measurements on organisms for the quantitative study of life and its inherited characteristics. One of the English scientists so engaged was Sir Francis Galton (1822–1911), a cousin of Charles Darwin. Galton sought a means of measuring the degree of relationship of two variables in the period 1877–1889 and arrived at a concept of the "correlation" of the variables. He noted that the fitted contours of equal frequency in two-way tables of certain measurements seemed to follow concentric ellipses, and he asked the mathematician J. D. H. Dickson to find the family of bivariate density functions with that property. Dickson showed directly that the family would be the bivariate normal density. Galton's response (1889) should be noted (if not memorized) by any unconvinced of the importance of mathematics outside the physical sciences:

> The problem may not be difficult to an accomplished mathematician, but I certainly never felt such a glow of loyalty and respect towards the sovereignty and wide sway of mathematical analysis as when the answer arrived, confirming by purely mathematical reasoning my various and laborious statistical conclu-

sions with more minuteness than I had dared to hope, because the data ran somewhat roughly, and I had to smooth them with tender caution.

Galton chose the symbol r for the correlation coefficient in reference to his "reversion" (later "regression") function, or the linear relationship between the independent variable X and the conditional means of Y for given values of X. Galton noticed that the variation in those conditional frequency tables was about the same for each value of X. Later Karl Pearson coined the term *homoscedasticity* for that property of the normal distribution and the usual linear model of least squares.

Although Galton had defined correlation in terms of the properties of the bivariate normal density surface in the mid-1880s, the actual product moment correlation coefficient was introduced by Karl Pearson in 1896 through what amounted to maximum likelihood estimation. Various workers attempted to find its distribution and moments in the next few years. In 1908, W. S. Gosset, constrained to write under the pseudonym "Student" by the proprietary rules of his employer, Guinness brewers, conjectured from sampling experiments that the distribution of r had the form (20) of Section 1-6 for a bivariate normal population with zero correlation. Fisher found the distribution of r for general ρ in 1915 and gave his z-transformation for large-sample normality in 1921.

Accounts of the early history of correlation and regression have been given by Galton (1908), Walker (1929), Snedecor (1954), and David (1976).

1-8 REFERENCES

ANDERSON, T. W. (1958): *An Introduction to Multivariate Statistical Analysis*, John Wiley & Sons, Inc., New York.

ANSCOMBE, F. J. (1973): Graphs in statistical analysis, *The American Statistician*, vol. 27, no. 1, pp. 17–20.

DAVID, F. N. (1976): Karl Pearson and Correlation, Technical Report No. 30, Department of Statistics, University of California, Riverside, Calif.

FISHER, R. A. (1915): Frequency distribution of the values of the correlation coefficient in samples from an indefinitely large population, *Biometrika*, vol. 10, pp. 507–521.

FISHER, R. A. (1921): On the "probable error" of a coefficient of correlation deduced from a small sample, *Metron*, vol. 1, pp. 1–32.

GALTON, F. (1888): Co-relations and their measurement, chiefly from anthropometric data, *Proceedings of the Royal Society*, vol. 45, pp. 135–140.

GALTON, F. (1889): *Natural Inheritance*, Macmillan & Company, London.

GALTON, F. (1908): *Memories of My Life*, Methuen & Company Ltd., London.

HEALD, F. P., G. ARNOLD, W. SEABOLD, AND D. MORRISON (1967): Plasma levels of free fatty acids in adolescents, *American Journal of Clinical Nutrition*, vol. 20, pp. 1010–1014.

HOHN, F. E. (1964): *Elementary Matrix Algebra*, 2nd ed., Macmillan Publishing Co., Inc., New York.

HOTELLING, H. (1953): New light on the correlation coefficient and its transforms, *Journal of the Royal Statistical Society, Series B*, vol. 15, pp. 193–232.

LANSDELL, H. (1970): Relation of extent of temporal removals to closure and visuomotor factors, *Perceptual and Motor Skills*, vol. 31, pp. 491–498.

PEARSON, E. S., AND H. O. HARTLEY (1969): *Biometrika Tables for Statisticians*, vol. 1, 3rd ed., Cambridge University Press, Cambridge, England.

PITMAN, E. J. G. (1939): A note on normal correlation, *Biometrika*, vol. 31, pp. 9–12.

PLACKETT, R. L. (1949): A historical note on the method of least squares, *Biometrika*, vol. 36, pp. 458–460.

SNEDECOR, G. W. (1954): Biometry, its makers and concepts, in O. Kempthorne et al. (eds.), *Statistics and Mathematics in Biology*, pp. 3–10, Iowa State University Press, Ames, Iowa.

STIGLER, S. M. (1978): Mathematical statistics in the early states, *Annals of Statistics*, vol. 6, pp. 239–265.

STIGLER, S. M. (1981): Gauss and the invention of least squares, *Annals of Statistics*, vol. 9, pp. 465–474.

URBAN, W. D. (1976): *Statistical Analysis of Blood Lead Levels of Children Surveyed in Pittsburgh, Pennsylvania: Analytical Methodology and Summary Results*, NBSIR 76-1024, Institute of Applied Technology, National Bureau of Standards, Washington, D.C.

WALKER, H. M. (1929): *Studies in the History of Statistical Method*, The Williams & Wilkins Company, Baltimore.

WINTERBOTTOM, A. (1979): A note on the derivation of Fisher's transformation of the correlation coefficient, *The American Statistician*, vol. 33, pp. 142–143.

1-9 EXERCISES

1. In a survey of young children's blood lead levels (Urban, 1976) those children whose measures were 40 μg/100 ml or more were subjected to a second micro-assay. These blood levels were obtained for the two finger-puncture samples:

Child	Initial screening	Repeat
1	45	26
2	73	73
3	42	35
4	43	27
5	45	38
6	41	28
7	46	34
8	42	38
9	47	35
10	59	28
11	42	19
12	40	21
13	50	48
14	41	32

(a) Investigate the relation between the second level and the initial measure.
(b) Is the "regression phenomenon" evident in these data? Explain.
(c) To what extent is the regression equation fitted in (a) influenced by any extreme values in the data?

2. The approximate temperatures (°F) of a cup of boiling water were recorded at 1-minute intervals:

Time	0	1	2	3	4	5	6	7	8
Temperature	194	185	177	170	164	157	150	148	144

(a) Plot the temperatures against time. Does a linear model seem reasonable for the data?
(b) Fit a linear model to the data. What proportion of the variation in temperature is explained by the linear function?
(c) Find the 99% confidence interval for the slope parameter of the linear model.
(d) Calculate the 95% prediction interval for the temperature of the water at 4 minutes.

3. The intelligence test scores of $N = 1100$ army inductees were correlated with their levels of a certain chemical compound thought from folklore to be associated with high intellectual achievement. The correlation coefficient was found to be 0.08, a value that would lead to rejection of a zero correlation in the population at the 0.01 level. What "practical" significance might such a finding have?

4. Determination of lead levels in $N = 48$ pairs of 0.25-ml human blood samples was made at the National Bureau of Standards (Urban, 1976) by a mass-spectrometric method and by a spectrophotometric procedure at a clinical laboratory. These values of the NBS (x_i) and clinical laboratory (y_i) blood lead levels (μg/100 ml) were obtained:

x_i	y_i	x_i	y_i	x_i	y_i	x_i	y_i
15.21	10	19.43	15	22.75	30	25.27	32
15.48	11	19.96	20	22.85	27	25.57	19
15.90	16	19.80	25	22.85	15	26.54	17
16.64	20	20.11	21	22.96	21	26.54	19
16.74	11	20.11	21	23.38	24	26.96	26
16.74	16	20.64	21	23.48	20	27.69	31
17.16	15	20.85	26	23.59	28	30.43	37
18.22	15	21.27	24	23.69	19	32.54	35
18.32	14	21.27	17	23.80	16	32.64	32
19.06	17	21.59	19	24.80	21	33.17	32
19.17	15	21.80	23	24.32	18	33.90	32
19.43	15	22.75	21	24.96	28	39.38	59

SOURCE: Data reproduced with the permission of the Institute of Applied Technology, National Bureau of Standards.

(a) Find the estimated regression line for predicting the clinical laboratory blood lead level from the NBS value.
(b) Find the 99% confidence and prediction bands for the regression equation.

(c) Some of the x_i values appear twice. Discuss how this information might be used for a better estimator of the disturbance term variance (for the answer to this question, consult Section 3-4).

(d) Construct the 99% confidence interval for β and use it to test $H_0: \beta = 1$.

(e) The last observation pair contains an extreme value ($y_{48} = 59$) of the clinical laboratory levels. Remove this pair of values and fit a straight line to the remaining pairs. Compare the results with those obtained in (a).

5. For a description of graphical analysis of simple regression results Anscombe (1973) constructed four sets of $N = 11$ (x_i, y_i) observation pairs. The mean of each X series was made to be $\bar{x} = 9$, and the mean of each Y series was $\bar{y} = 7.5$. In addition, for each regression

$$\sum_{i=1}^{11} (x_i - \bar{x})^2 = 110 \qquad \text{SSR} = 27.50 \qquad \text{SSE} = 13.75 \qquad r^2 = 0.6667$$

The first three X series were identical. These are the data:

		Variable					
		X		Y			
Observation	Data set:	1, 2, 3	4	1	2	3	4
1		10	8	8.04	9.14	7.46	6.58
2		8	8	6.95	8.14	6.77	5.76
3		13	8	7.58	8.74	12.74	7.71
4		9	8	8.81	8.77	7.11	8.84
5		11	8	8.33	9.26	7.81	8.47
6		14	8	9.96	8.10	8.84	7.04
7		6	8	7.24	6.13	6.08	5.25
8		4	19	4.26	3.10	5.39	12.50
9		12	8	10.84	9.13	8.15	5.56
10		7	8	4.82	7.26	6.42	7.91
11		5	8	5.68	4.74	5.73	6.89
Mean		9	9	7.5	7.5	7.5	7.5

SOURCE: Data reproduced with the kind permission of Francis J. Anscombe and the American Statistical Association.

Plot the data of each set and discuss the extent to which the Y-values are predicted by the common regression equation

$$\hat{y}_i = 3 + 0.5x_i$$

6. Find the 95% and 99% confidence intervals for the population correlation of the WAIS Information and Arithmetic scores of the $N = 37$ normal aged subjects whose sample correlation for those tests was $r = 0.57$.

7. Lansdell (1970) has investigated the effects of temporal lobe surgery in epileptic patients upon their performance on standard intelligence tests. Three dimensions, or "factors," of cognition have been identified, and their scores correlated with an index of the amount of tissue removed from the temporal lobes of the brain. These correlation coefficients were obtained:

Cognitive factor	Side of removal	
	Left ($N_1 = 7$)	Right ($N_2 = 14$)
Verbal comprehension	−0.35	0.33
Visual construction	0.80	−0.71
Closure	−0.82	−0.18

SOURCE: Reprinted by permission of author and publisher from H. Lansdell: Relation of extent of temporal removals to closure and visuomotor factors. *Perceptual and Motor Skills*, vol. 31 (1970), pp. 491–498, Table 2.

Test the hypotheses of common correlations in the left and right samples for the three factors.

8. The random vector **X** has the bivariate normal distribution with covariance matrix

$$\mathbf{\Sigma} = \begin{bmatrix} 1 & \rho \\ \rho & 1 \end{bmatrix}$$

What are the implications of the orthogonal transformation

$$\mathbf{Y} = \begin{bmatrix} \sqrt{2}/2 & \sqrt{2}/2 \\ -\sqrt{2}/2 & \sqrt{2}/2 \end{bmatrix} \mathbf{X}$$

to a new random vector?

9. The Fisher z-transformation can be found in this way: We expand the function $h(r)$ in a Taylor series about ρ through the linear term, then express the large sample variance of the transformation as

$$\text{Var } h(r) \approx \left[\frac{dh(r)}{dr} \bigg|_{r=\rho} \right]^2 \text{Var } (r)$$

The requirement that the variance of the transformation be the same for all ρ is

$$\left[\frac{dh(\rho)}{d\rho} \right]^2 (1 - \rho^2)^2 = \text{constant}$$

Find $h(r)$ by solving that differential equation (Anderson, 1958, Section 4.2.3).

10. Find the prediction and confidence intervals for the average starting salaries of undergraduate business majors who received their degrees in years 6, 7, and 8. Compare those intervals with these actual averages for the graduating classes:

Year	Average starting salary
6	\$12,920
7	13,814
8	15,062

11. The plot of landlord attitude and apartment quality ratings in Figure 1-4 appears to indicate that the last four pairs of observations may be aberrant or "outlier" observations in some sense. Verify that the following correlations will result as different pairs are omitted from the data set:

Omitted pairs	Correlation
26, 28	0.8771
26, 29	0.7892
27, 28	0.8943
26, 27, 28	0.8874
27, 28, 29	0.8792
26, 27, 28, 29	0.8381

Discuss the effects of the observation points on the correlations in terms of their proximity to the regression line.

12. The average M.B.A. starting salary for the eighth year in the series of Example 1-1 was $24,993. The least squares trend fitted to the eight points has the equation

$$y_t = 18{,}609.12 + 1436.25(t - 4.5)$$

computed from the statistics

$$\sum (y_t - \bar{y})^2 = 91{,}741{,}447 \qquad \sum (y_t - \bar{y})(t - \bar{t}) = 60{,}322.5$$
$$\sum (t - \bar{t})^2 = 42$$

(a) What proportion of variation does the line explain?
(b) Show that the 95% confidence interval for the regression coefficient is

$$1088 \leq \beta \leq 1784$$

13. Show that the prediction interval for the eighth average salary in the continuation of the series in Example 1-1 is

$$20{,}566 \leq Y_8 \leq 24{,}767$$

Compare the prediction interval with the actual average starting salary of $24,993. Does the linear model seem appropriate for the series and for extrapolation from it?

14. Each semester the student association of a graduate school of business publishes faculty evaluation averages obtained from student ratings of courses and their instructors. The individual ratings are on a scale from 1 (best) to 5 (worst). These average evaluations for two scales of the questionnaire were obtained for instructors in two departments of the school:

Accounting			Finance		
	Overall rating			Overall rating	
Instructor	Instructor	Course	Instructor	Instructor	Course
1	1.4	2.8	1	1.4	1.7
2	2.0	2.1	2	2.7	2.3
3	2.0	2.2	3	2.2	2.1
4	1.2	1.4	4	1.7	1.8
5	3.2	3.1	5	1.9	2.1
6	1.8	1.4	6	2.4	2.5
7	1.8	2.7	7	1.6	1.5
8	2.2	2.7	8	2.2	2.4

Accounting			Finance		
	Overall rating			Overall rating	
Instructor	Instructor	Course	Instructor	Instructor	Course
9	1.1	1.2	9	2.5	2.9
10	1.6	1.5	10	1.6	2.7
11	3.6	3.1	11	2.9	2.5
12	1.9	2.2	12	2.2	2.3
13	1.6	2.5	13	2.4	2.7
14	3.6	2.7	14	2.7	3.0
15	3.0	2.8	15	2.0	2.1
16	1.5	1.9	16	1.4	1.7
17	1.8	2.2	17	3.0	3.1
			18	3.5	3.2
			19	3.6	2.9
			20	2.0	1.9
			21	1.7	1.9
			22	1.6	2.1
			23	1.9	2.1
			24	2.9	2.6
			25	2.1	2.2
			26	2.9	2.9
			27	1.5	1.7

(a) Treat the instructor rating as the dependent variable and find the separate regression line for each department.

(b) Test the hypothesis of equal disturbance variances σ_1^2, σ_2^2 in the linear models for the two departments by computing the estimates

$$\hat{\sigma}_1^2 = \frac{\text{SSE}_1}{N_1 - 2} \qquad \hat{\sigma}_2^2 = \frac{\text{SSE}_2}{N_2 - 2}$$

from the separate regression analyses, and referring their ratio

$$F = \frac{\hat{\sigma}_1^2}{\hat{\sigma}_2^2}$$

to critical values of the F-distribution with $N_1 - 2$ and $N_2 - 2$ degrees of freedom. The larger variance should be in the numerator so that the upper percentage points in Table 4, Appendix A can be used directly.

(c) Compute the correlation coefficients for the separate departments and test the hypotheses that their population parameters are zero.

(d) Find the 95% confidence intervals for the population correlations of the two ratings for the individual departments.

(e) Test the hypothesis of equal population correlations of the ratings in the two departments.

15. A statistics department administers a qualifying examination each year for graduate students in other disciplines. The examination consists of three parts, covering linear algebra and probability, random variables and statistical inference, and applied linear models. These scores were obtained for parts I and III by one class of students:

Student	1	2	3	4	5	6	7	8	9	10	11	12	13
Part I	23	22	20	14	21	28	23	32	26	22	21	24	16
Part III	25	23	8	29	25	26	25	27	30	24	25	29	26

(a) What proportion of the variation in one part's scores is explained by a linear relationship with the other score?

(b) Is the correlation between the scores in the two parts significantly different from zero?

(c) Find the least squares equation relating the part I scores to their corresponding part III scores. Repeat the linear regression process with the part III scores as the dependent variable observations. How are the two slope estimates related?

(d) Are the part I–part III score differences correlated with the part I scores?

16. The incumbent Representative in a congressional district received these numbers of votes by municipalities in two successive elections:

Municipality	1978	1980	Municipality	1978	1980
AL	954	1,103	ML	99	143
AS	2,358	3,249	MO	372	545
BH	1,605	2,187	NP	2,804	3,498
CT	674	865	NT	2,104	2,489
CC	5,339	8,543	NO	1,209	1,573
CH	1,587	1,820	PS	388	519
CD	1,565	1,806	PP	1,408	1,627
CW	475	539	RP	1,568	2,037
DB	1,564	2,096	RT	6,476	8,345
DT	1,898	2,702	RV	256	311
EL	654	740	RL	185	250
ED	· 457	524	SH	1,153	1,430
FC	1,361	1,686	SP	6,716	7,381
GL	1,370	1,670	SW	1,650	2,195
LD	2,570	3,438	TI	860	1,073
LC	424	581	TR	292	176
MH	402	485	UP	383	557
MT	5,988	6,665	UD	16,316	19,497
MB	1,202	1,657	YD	2,568	3,123

(a) Find the correlation and coefficient of determination of the votes in the two elections.

(b) Repeat the analysis of (a), but first with the municipality UD eliminated, and then with CC, MT, RT, SP, and UD removed. To what extent is the proportion of explained variation influenced by the townships with very large election turnouts?

(c) Find the equation of the least squares line fitted to the $N = 33$ pairs obtained by removing the five large municipalities listed in (b). Compute the residuals from the line and identify those that appear to be "outliers" from the linear model. What other visual inferences can you make from the residuals?

17. A part-time busboy at a summer resort kept a record of his tips each evening, as well as the number of meals served, weather, and number of waitresses on duty.* The information was used for deciding whether an assignment on a given night would be worthwhile. The data collected for that purpose are shown in the table. Fair weather is indicated by a zero score; we shall discuss such dummy variables at greater length in Section 7-2. At the present time we shall only use the number of meals served as the single independent variable. The other variables were included for the multiple regression model introduced in the next chapter.

Busboy tips	Number of meals served	Weather	Number of waitresses − 7
\$11.50	210	0	1
11.60	190	0	1
9.42	220	1	2
14.90	260	0	2
13.00	240	1	4
17.00	270	0	4
21.93	360	0	4
17.85	290	0	3
16.70	320	1	4
19.90	330	0	3
20.90	350	0	3
18.50	340	1	4
19.30	360	1	4
20.30	360	0	2
19.95	370	1	3
23.20	390	0	5
21.55	400	1	4
23.25	410	0	3
22.40	400	0	2
23.60	450	1	5
18.10	340	0	2
17.25	350	1	2
24.50	340	0	4
18.85	320	0	4
13.30	230	1	4
14.21	240	0	4

(a) Find the sample regression equation relating the dependent variable of busboy tips to the number of meals served.

(b) Does the fitted straight line explain sufficient variation for practical prediction of tips? Are the residuals generally small?

(c) Find the 99% confidence interval for the regression coefficient of the linear model.

*I am indebted to Mitchell Zimmer and Gregory Ahlstrom for this example, and for kindly permitting its use here.

18. In Section 1-3 we gave in expression (7) an alternative form of the straight-line model with its intercept at the origin of the independent variable's scale. The model was

$$y_i = \beta_0 + \beta x_i + e_i$$

and the estimator of its intercept was expressed as

$$\hat{\beta}_0 = \bar{y} - \hat{\beta}\bar{x}$$

If the disturbances e_i in the model are independently and normally distributed with a zero mean and common variance σ^2, $\hat{\beta}_0$ will also have a normal distribution with mean β_0 and variance

$$\begin{aligned} \text{Var}(\hat{\beta}_0) &= \text{Var}(\bar{y} - \hat{\beta}\bar{x}) \\ &= \text{Var}(\bar{y}) + \bar{x}^2 \text{Var}(\hat{\beta}) \\ &= \sigma^2\left[\frac{1}{N} + \frac{\bar{x}^2}{\sum (x_i - \bar{x})^2}\right] \end{aligned}$$

Then

$$t = \frac{\hat{\beta}_0 - \beta_0}{\hat{\sigma}\sqrt{\dfrac{1}{N} + \dfrac{\bar{x}^2}{\sum (x_i - \bar{x})^2}}}$$

has the t-distribution with $N - 2$ degrees of freedom. We may test the hypothesis that β_0 is the true intercept by referring t to Table 3, Appendix A. Or, we may find the confidence interval

$$\hat{\beta}_0 - t_{\alpha/2;N-2}\hat{\sigma}\sqrt{\frac{1}{N} + \frac{\bar{x}^2}{\sum (x_i - \bar{x})^2}} \leq \beta_0 \leq \hat{\beta}_0 + t_{\alpha/2;N-2}\hat{\sigma}\sqrt{\frac{1}{N} + \frac{\bar{x}^2}{\sum (x_i - \bar{x})^2}}$$

and see if it includes the hypothetical value of β_0, for example, zero in the case of a line through the origin.

(a) We would expect the least squares line for busboy tips in Exercise 17 to go through the origin: No dinners served should result in no tips. Compute $\hat{\beta}_0$ for those data and test the hypothesis that β_0 is zero.

(b) Find the 95% confidence interval for β_0. Does the interval include zero and thereby support the null hypothesis?

(c) Use the variance operator described in Section C-1 to verify the expression for $\text{Var}(\hat{\beta}_0)$. Recall that $\bar{y}$ and $\hat{\beta}$ have zero covariance.

19. The covariance of the sum and difference of the random variables X and Y is

$$\begin{aligned} \text{Cov}(X + Y, X - Y) &= \text{Var}(X) - \text{Cov}(X, Y) + \text{Cov}(Y, X) - \text{Var}(Y) \\ &= \text{Var}(X) - \text{Var}(Y) \end{aligned}$$

and is zero only if the variances of X and Y are equal. We see that the hypothesis

$$H_0\colon \text{Var}(X) = \text{Var}(Y)$$

is equivalent to the hypothesis that $X + Y$ and $X - Y$ are uncorrelated. If X and Y are bivariate normal and we have a random sample of N independent observation pairs (x_i, y_i) upon them, we can test the equal-variances hypothesis by computing the correlation r_{uv} of the sums and differences

$$\begin{aligned} u_i &= x_i + y_i \\ v_i &= x_i - y_i \end{aligned} \qquad i = 1, \ldots, N$$

We reject the hypothesis if r_{uv} exceeds an appropriate critical value of Table 5,

Appendix A. The choice of the left, right, or two-sided critical value would depend on the alternative hypothesis. This ingenious way of dealing with the correlation of X and Y was discovered by Pitman (1939).

Test whether the part I and part III variances of the data in Exercise 15 are significantly different at the 0.05 level.

20. A market research firm collected large amounts of information from consumers who agreed to participate in groups called *panels*. Each panel member supplied monthly income and expenditure data via questionnaires or diaries provided by the firm. Among the many statistics computed from the data were these correlations between family income and entertainment expenditures for a winter month for two types of panelists:

	Urban	Suburban
Number of families	48	124
Correlation	0.62	0.35

(a) Test the hypothesis of a common correlation in the urban and suburban populations out of which the panel families came.

(b) Find the 99% confidence intervals for the separate population correlations of the two groups.

21. Following an accident the speed recorder of a freight locomotive was calibrated to determine its accuracy. These true and indicated speeds were given in the published report of the derailment:

Actual	Speedometer reading
25	21
31.4	26
39.3	34

We would like to infer as much as possible about the relationship of the indicated speed to the true velocity of the locomotive.

(a) Plot the three pairs of speeds. Do the points suggest linear or nonlinear response functions?

(b) Fit a straight line to the data with its intercept at the origin. Use the results of Exercise 18 to test the hypothesis that the intercept is zero or that the indicator reads zero when the locomotive is stationary.

(c) Find the 95% confidence interval for the slope of the linear model fitted in (b).

(d) Now let us assume that the straight line passes through the origin, or

$$y_i = \beta x_i + e_i$$

for the ith true speed x_i. The least squares estimator of the slope is

$$\hat{\beta} = \frac{\sum x_i y_i}{\sum x_i^2}$$

The estimator of the variance of the disturbances e_i is

$$\hat{\sigma}^2 = \frac{\sum (y_i - \hat{\beta}x_i)^2}{N-1}$$

$$= \frac{\sum y_i^2 - (\sum x_i y_i)^2 / \sum x_i^2}{2}$$

and the estimated standard deviation of $\hat{\beta}$ is

$$\hat{\sigma}(\hat{\beta}) = \frac{\hat{\sigma}}{\sqrt{\sum x_i^2}}$$

$\hat{\beta}$ and $\hat{\sigma}$ are independently distributed when the disturbances have the usual joint normal distribution, and the ratio

$$t = \frac{\hat{\beta} - \beta}{\hat{\sigma}(\hat{\beta})}$$

has the t-distribution with $N - 1$ degrees of freedom. Compute the estimate of β under the present model with no intercept.

(e) Find the 95% confidence interval for β and compare it with the interval found in (c).

(f) Use the preceding confidence interval to test the hypothesis of a unit slope for the zero-intercept line.

22. The correlation coefficient is unaffected by changes in the origins or units of the scales of its variables. Verify that invariance property by transforming the original observations x_i, y_i to

$$\begin{aligned} u_i &= a + bx_i \\ v_i &= c + dy_i \end{aligned} \qquad i = 1, \ldots, N$$

and evaluating expression (16) of Section 1-6 for the u_i, v_i pairs. Of course, for the sign of the new correlation to remain the same the constants b and d must be of the same sign.

23. Find the effect of the linear transformations in Exercise 22 upon the least squares estimator of the straight line slope. In particular, show that the standard scores

$$u_i = \frac{x_i - \bar{x}}{s_x} \qquad v_i = \frac{y_i - \bar{y}}{s_y}$$

have the correlation coefficient of the (x_i, y_i) pairs for their estimated slope. The divisors s_x and s_y are the usual sample standard deviations equal to the square roots of the variances defined by expression (15), Section 1-6.

24. Seven persons admitted one year to a statistics graduate program received these Graduate Record Examination Aptitude Test scores:

Verbal	Quantitative
780	800
740	760
650	800
650	780
620	730
720	760
490	700

(a) Compute the correlation of the scores and test the hypothesis of zero population correlation.

(b) Find the 95% confidence interval for the correlation.

(c) Test the hypothesis that the population correlation is equal to 0.56.

25. A family checking account had these average numbers (rounded to integers) of checks each month for 13 years:

Year	Average checks per month
1	20
2	19
3	20
4	19
5	20
6	21
7	24
8	27
9	27
10	30
11	33
12	36
13	37

(a) Suppose that a trend line was fitted to the series by a regression analysis packaged program. How would the printout of the residuals show that the linear trend was inappropriate?

(b) Fit a straight line to the averages. What proportion of variation does the line explain?

(c) Find the residuals from the fitted line. Does the model appear to be misspecified?

(d) The averages appear to be nearly constant for the first five or six years, then begin to increase steadily each year. Fit this simple segmental model

$$y_t = \begin{cases} \alpha_0 + e_t & t = 1, \ldots, 5 \\ \alpha + \beta(t - 9.5) + e_t & t = 6, \ldots, 13 \end{cases}$$

to the data, where

$$\hat{\alpha}_0 = \tfrac{1}{5} \sum_{t=1}^{5} y_t = 19.6$$

and the estimates of α and β are computed from the usual least squares estimators based on the last eight observations.

(e) Find the 95% confidence interval for β.

(f) Compare the proportion of variation explained by the segmental model in (d) with that of the naive straight line found in (b).

26. A school district had these birth rates and subsequent kindergarten enrollments five years later:

Year, t	Live births in year t	Kindergarten enrollments in year $(t + 5)$
1	230	276
2	226	237
3	209	213
4	162	177
5	171	181
6	181	182
7	168	180
8	175	—
9	172	—
10	201	—

SOURCE: Data reproduced with the kind permission of the Pennsylvania Economy League, Eastern Division.

(a) Plot the enrollments against the numbers of births. Does the latter appear to be a useful predictor of enrollment?

(b) Fit a linear model to the data and examine its residuals. Compute the standardized residuals

$$\frac{z_i}{\hat{\sigma}} = \frac{y_i - \hat{y}_i}{\hat{\sigma}}$$

Does the first year appear to be an outlier from the fitted line?

(c) Compute the predicted enrollments and 95% prediction intervals for years 8, 9, and 10.

(d) Drop the first year from the data set and fit a straight line to the remaining six points. Compute the predicted enrollments and their 95% prediction intervals for years 8, 9, and 10. Compare the results with those for the complete set of seven years.

chapter 2

Regression and Correlation Models with Several Independent Variables

2-1 INTRODUCTION

In this chapter we extend least squares estimation in the linear model to several independent variables. As in the single-predictor model of Chapter 1 we assume initially that the independent variables are fixed rather than random quantities and that the source of random variation in the dependent variable will be a disturbance term added to the linear function of the predictors. The coefficients of that linear function will be estimated by least squares, although instead of minimizing a sum of squared vertical deviations of points about a line we must find the plane in $(p + 1)$-dimensional space that minimizes such a sum of squares. We find these estimators in Section 2-2 with the aid of matrix notation. The estimators are jointly distributed according to the multivariate normal distribution; the density and properties of that distribution are described in Section 2-3 and applied to the sampling properties of the estimators. We also develop basic tests and confidence intervals for the regression model parameters. The real justification of least squares estimation through the minimum variance properties of its estimators is established in Section 2-4, together with the equality of least squares and maximum likelihood estimators when the disturbance terms are normally distributed.

When the independent variable values are actually observations subject to sampling variation their relationship with the dependent variable should be analyzed by means of the multiple correlation model. We develop multiple correlation in Section 2-5 by means of the multivariate normal conditional distribution. The mean vector of that conditional distribution is the analogue of the original linear multiple regression model with fixed independent variables, whereas the multiple correlation coefficient can be interpreted as the greatest correlation of the dependent variable with any weighted sum of the independent variables, or through its square as a proportion of variance. The hypothesis tests on the regression coefficients will be found to be equivalent with those of Section 2-3 for fixed predictors. The partial correlation of two variables with one or more others held constant is treated in Section 2-6, initially in terms of residuals of regression equations in the fixed variables, and then on the basis of the conditional multinormal distribution.

2-2 THE MULTIPLE REGRESSION MODEL

In the natural and behavioral sciences the study of relationships usually involves more than one independent variable. As in Chapter 1 let us denote the dependent variable by Y and the p independent variables thought to contribute to its variation as $X_1, \ldots, X_p$. The mathematical model for the ith observation on Y has the general form

$$y_i = f(x_{i1}, \ldots, x_{ip}) + e_i \qquad i = 1, \ldots, N \tag{1}$$

where the observations $x_{i1}, \ldots, x_{ip}$ on the independent variables are fixed values under the control of the experimenter or other observer, and the disturbance term e_i is a random variable accounting for the failure of $x_{i1}, \ldots, x_{ip}$ to predict exactly the value y_i of the dependent variable. It is essential to note that the source of random, or unexplained, variation in the y_i is added to the deterministic portion $f(x_{i1}, \ldots, x_{ip})$ of the model. At least the mathematical form of the distribution of the e_i must be specified. As in the simple linear model of Chapter 1, we begin with the assumption of zero expectations, a common variance, and zero correlations, and then vary these conditions as circumstances require.

Now let us turn our attention to the deterministic portion of the model. Two fundamental questions must be resolved about that part:

1. What variables $X_1, \ldots, X_p$ should be included?
2. What mathematical function should be chosen for $f(X_1, \ldots, X_p)$?

The choice of the basic set of p independent variables must be made from the substantive nature of the dependent variable and the hypothesized sources of its variability. In an exploratory search for determinants of Y with no theoretical structure as a guide, one may include all likely independent variables that time and budgeting constraints will permit, but for the final set of predictors Newton's dictum on the simplicity of nature and Occam's razor ("Entities should not be multiplied beyond necessity") must prevail. We consider some statistical methods for arriving at that final set in Chapter 5.

Simplicity and economy also constrain the form of the deterministic portion of the model. From the nature of the variables, from other analyses of similar data, or from plotting the y_i against the observations on each independent variable we should acquire some approximation to the form of the function. We should specify whether the individual relations of Y to the X_i are monotonic and, if so, whether a linear function will suffice. If the plots show a nonlinear response, can this be approximated by a quadratic or other polynomial of low degree? If several oscillations are present, as in seasonal time series, can these be represented by a polynomial, or must more complex harmonic models be used? For variables measuring growth or yields from chemical processes must exponential functions be employed? Do jumps or other discontinuities in the data imply that "dummy" variables indicating the presence or absence of effects must be used? How should the effects of interactions among the X_i be related to Y?

We shall delimit our class of functions to a simple representation that will include many of those relationships as variants or special cases. The basis for our class is a fundamental theorem in the calculus known as the Taylor–Maclaurin expansion in several variables, but we shall prefer to work at a more intuitive and verbal level. We require that $f(X_1, \ldots, X_p)$ represents a smoothly varying surface over the domain of admissible values of the X_i, with no sharp peaks, sudden sinkholes, or precipitous cliffs. Let us further assume for the moment that the surface is a plane, or a linear function of the X_i. Then

$$f(X_1, \ldots, X_p) = \beta_0 + \beta_1 X_1 + \cdots + \beta_p X_p \tag{2}$$

or the Maclaurin expansion of the function through linear terms. However, if the model or data call for a curved surface, we may add product and quadratic terms in the sense of the second-order Maclaurin expansion:

$$\begin{aligned} f(X_1, \ldots, X_p) = \beta_0 + \beta_1 X_1 + \cdots + \beta_p X_p + \beta_{p+1} X_1^2 + \cdots + \beta_{2p} X_p^2 \\ + \beta_{2p+1} X_1 X_2 + \cdots + \beta_{p(p+3)/2} X_{p-1} X_p \end{aligned} \tag{3}$$

Similarly, more complex nonlinear functions can be built up by adding higher powers and products of the X_i or by including various other functions of them. In that way the deterministic function of the multiple regression model may be written as

$$f(X_1, \ldots, X_k) = \beta_0 + \beta_1 X_1 + \cdots + \beta_k X_k \tag{4}$$

or a function linear in the parameters $\beta_0, \beta_1, \ldots, \beta_k$, some of whose $k \geq p$ variables may be nonlinear functions of the original p predictors. If the function is not linear in the β_j, estimation of those parameters and statistical inferences about them may be difficult to accomplish. The models of this text will be restricted to those that are linear in the parameters and with additive random components.

Now let us give the linear model in several independent variables. Its representation for the ith observation on the dependent variable is

$$y_i = \alpha + \beta_1(x_{i1} - \bar{x}_1) + \cdots + \beta_p(x_{ip} - \bar{x}_p) + e_i \qquad i = 1, \ldots, N \tag{5}$$

where the independent variable values have been centered at their respective means

$$\bar{x}_j = \frac{1}{N} \sum_{i=1}^{N} x_{ij} \qquad j = 1, \ldots, p \tag{6}$$

to reduce the computation required for the least squares estimators of the parameters. The e_i are uncorrelated random variables with mean and variance

$$E(e_i) = 0 \qquad \mathrm{Var}\,(e_i) = \sigma^2 \tag{7}$$

for all values of i. α is the intercept parameter of the linear function at the point $(\bar{x}_1, \ldots, \bar{x}_p)$, whereas the β_j will be referred to as the regression or slope parameters of the individual predictor variables.

In this chapter we develop the least squares estimators of the parameters of the model (5), or those that minimize the sum of squared deviations

$$S(\alpha, \beta_1, \ldots, \beta_p) = \sum_{i=1}^{N} [y_i - \alpha - \beta_1(x_{i1} - \bar{x}_1) - \cdots - \beta_p(x_{ip} - \bar{x}_p)]^2 \tag{8}$$

The estimators are found by the usual method of differential calculus: $S(\alpha, \beta_1, \ldots, \beta_p)$ is differentiated with respect to the parameters, the $p + 1$ derivatives are set equal to zero, and the resulting linear equations are solved for the least squares estimators. However, before differentiating it will be more convenient to write the model (5) in vector and matrix form. Not only can the least squares estimators be found more easily, but their statistical properties can be obtained and discussed more directly. Those who are not familiar with matrix notation should turn to Appendix B for some essential definitions and rules of matrix algebra.

The matrix form of the linear model (5) is

$$\mathbf{y} = \mathbf{X}\boldsymbol{\beta} + \mathbf{e} \tag{9}$$

where

$$\mathbf{y} = \begin{bmatrix} y_1 \\ \cdot \\ \cdot \\ \cdot \\ y_N \end{bmatrix} \qquad \mathbf{e} = \begin{bmatrix} e_1 \\ \cdot \\ \cdot \\ \cdot \\ e_N \end{bmatrix} \qquad \boldsymbol{\beta} = \begin{bmatrix} \alpha \\ \beta_1 \\ \cdot \\ \cdot \\ \cdot \\ \beta_p \end{bmatrix} \tag{10}$$

$$\mathbf{X} = \begin{bmatrix} 1 & x_{11} - \bar{x}_1 & \cdots & x_{1p} - \bar{x}_p \\ \cdot & \cdot & & \cdot \\ \cdot & \cdot & & \cdot \\ \cdot & \cdot & & \cdot \\ 1 & x_{N1} - \bar{x}_1 & \cdots & x_{Np} - \bar{x}_p \end{bmatrix} \tag{11}$$

It will be convenient to write $\mathbf{X}$ as the partitioned matrix

$$\mathbf{X} = [\mathbf{j} \quad \mathbf{X}_1] \tag{12}$$

where $\mathbf{j}$ is the column vector of N ones and $\mathbf{X}_1$ is the matrix of centered observations

$$\mathbf{X}_1 = \begin{bmatrix} x_{11} - \bar{x}_1 & \cdots & x_{1p} - \bar{x}_p \\ \cdot & & \cdot \\ \cdot & & \cdot \\ \cdot & & \cdot \\ x_{N1} - \bar{x}_1 & \cdots & x_{Np} - \bar{x}_p \end{bmatrix} \tag{13}$$

The sum of squared deviations (8) about the linear model can be expressed in matrix form as

$$S(\boldsymbol{\beta}) = (\mathbf{y} - \mathbf{X}\boldsymbol{\beta})'(\mathbf{y} - \mathbf{X}\boldsymbol{\beta}) \tag{14}$$

The vector of partial derivatives of $S(\boldsymbol{\beta})$ with respect to the elements of $\boldsymbol{\beta}$ is

$$\frac{\partial S(\boldsymbol{\beta})}{\partial \boldsymbol{\beta}} = -2\mathbf{X}'\mathbf{y} + 2\mathbf{X}'\mathbf{X}\boldsymbol{\beta} \tag{15}$$

A sufficient condition for a stationary minimum of $S(\boldsymbol{\beta})$ is that

$$\frac{\partial S(\boldsymbol{\beta})}{\partial \boldsymbol{\beta}} = \mathbf{0} \tag{16}$$

or

$$\mathbf{X}'\mathbf{X}\boldsymbol{\beta} = \mathbf{X}'\mathbf{y} \tag{17}$$

and if the inverse of $\mathbf{X}'\mathbf{X}$ exists, the least squares estimator of $\boldsymbol{\beta}$ is

$$\hat{\boldsymbol{\beta}} = (\mathbf{X}'\mathbf{X})^{-1}\mathbf{X}'\mathbf{y} \tag{18}$$

Because the values of the independent variables have been expressed as deviations about their means the estimator of $\boldsymbol{\beta}$ can be simplified. Note that

$$\begin{aligned}\mathbf{X}'\mathbf{X} &= [\mathbf{j} \quad \mathbf{X}_1]'[\mathbf{j} \quad \mathbf{X}_1] \\ &= \begin{bmatrix}\mathbf{j}' \\ \mathbf{X}_1'\end{bmatrix}[\mathbf{j} \quad \mathbf{X}_1] \\ &= \begin{bmatrix}\mathbf{j}'\mathbf{j} & \mathbf{j}'\mathbf{X}_1 \\ \mathbf{X}_1'\mathbf{j} & \mathbf{X}_1'\mathbf{X}_1\end{bmatrix} \\ &= \begin{bmatrix}N & \mathbf{0}' \\ \mathbf{0} & \mathbf{X}_1'\mathbf{X}_1\end{bmatrix}\end{aligned} \tag{19}$$

since each column of $\mathbf{X}_1$ must sum to zero. The jth diagonal element of $\mathbf{X}_1'\mathbf{X}_1$ is the sum of squares

$$\sum_{i=1}^{N} (x_{ij} - \bar{x}_j)^2 \tag{20}$$

of the jth independent variable about its mean, whereas the jhth off-diagonal element is the sum of products

$$\sum_{i=1}^{N} (x_{ij} - \bar{x}_j)(x_{ih} - \bar{x}_h) \tag{21}$$

Similarly,

$$\begin{aligned}\mathbf{X}'\mathbf{y} &= \begin{bmatrix}\mathbf{j}' \\ \mathbf{X}_1'\end{bmatrix}\mathbf{y} \\ &= \begin{bmatrix}N\bar{y} \\ \mathbf{X}_1'\mathbf{y}\end{bmatrix}\end{aligned} \tag{22}$$

where $\bar{y}$ is the mean of the dependent variable observations and $\mathbf{X}_1'\mathbf{y}$ has jth element

$$\sum_{i=1}^{N} (x_{ij} - \bar{x}_j)y_i = \sum_{i=1}^{N} (x_{ij} - \bar{x}_j)(y_i - \bar{y}) \tag{23}$$

If we substitute these results into the right-hand side of (18), the estimators simplify as

$$\begin{bmatrix} \hat{\alpha} \\ \hat{\boldsymbol{\beta}}_1 \end{bmatrix} = \begin{bmatrix} N & \mathbf{0}' \\ \mathbf{0} & \mathbf{X}_1'\mathbf{X}_1 \end{bmatrix}^{-1} \begin{bmatrix} N\bar{y} \\ \mathbf{X}_1'\mathbf{y} \end{bmatrix}$$

$$= \begin{bmatrix} \bar{y} \\ (\mathbf{X}_1'\mathbf{X}_1)^{-1}\mathbf{X}_1'\mathbf{y} \end{bmatrix} \tag{24}$$

As in the straight-line model of Chapter 1 the intercept estimator with centered predictor variables is merely the mean of the y_i. It is independent of $\hat{\boldsymbol{\beta}}_1$ in the computational and statistical senses. The solution for $\hat{\boldsymbol{\beta}}_1$ requires the inversion of a $p \times p$ matrix rather than the $(p + 1) \times (p + 1)$ matrix $\mathbf{X}'\mathbf{X}$.

Example 2-1

Let us illustrate least squares estimation with a simple, though real, data set. The variables are ones we shall use on several occasions in the remainder of this text. At the end of each semester students in courses at a university return evaluation forms for each course and its instructor. The newspaper of the graduate business school publishes average ratings for the several scales of the form for each instructor. We would like to use the linear multiple regression model to relate the averages of the sixth scale, Overall Rating of the Instructor (Y) to the independent variables

$$X_1 = \text{ability to present course material}$$

$$X_2 = \text{ability to stimulate student interest}$$

corresponding to the first and second scales. Values of the three variables for $N = 20$ instructors are given in Table 2-1. The instructors were chosen so that the data set would

TABLE 2-1 Average Ratings

	x_{i1}	x_{i2}	y_i	$\hat{y}_i$	$z_i = y_i - \hat{y}_i$
	1.1	1.2	1.1	1.04	0.06
	1.3	1.7	1.4	1.41	−0.01
	1.4	1.6	1.5	1.38	0.12
	2.0	1.5	1.6	1.49	0.11
	2.0	1.4	1.7	1.43	0.27
	1.6	1.4	1.2	1.31	−0.11
	2.3	2.1	2.0	1.96	0.04
	1.6	2.2	1.6	1.82	−0.22
	1.5	2.3	1.6	1.85	−0.25
	2.5	2.2	2.1	2.09	0.01
	2.6	2.7	2.4	2.43	−0.03
	2.1	2.4	2.0	2.09	−0.09
	3.2	3.1	3.2	2.86	0.34
	2.6	3.0	2.7	2.62	0.08
	2.8	1.8	2.0	1.92	0.08
	3.1	2.4	2.0	2.39	−0.39
	2.2	3.1	2.6	2.56	0.04
	3.1	3.4	3.0	3.02	−0.02
	3.1	3.8	3.5	3.27	0.23
	3.9	3.2	2.9	3.13	−0.23
Mean	2.30	2.325	2.105	2.104	0.0015

encompass a wide range of values of X_1 and X_2. These sums of squares and products were computed about the means of the observations:

$$\mathbf{X}_1'\mathbf{X}_1 = \begin{bmatrix} 10.86 & 8.04 \\ 8.04 & 10.8375 \end{bmatrix}$$

$$\mathbf{X}_1'\mathbf{y} = \begin{bmatrix} 8.31 \\ 9.2375 \end{bmatrix}$$

$$\sum_{i=1}^{20} (y_i - \bar{y})^2 = 8.9295$$

The required inverse matrix is

$$(\mathbf{X}_1'\mathbf{X}_1)^{-1} = \begin{bmatrix} 0.204274 & -0.151545 \\ -0.151545 & 0.204698 \end{bmatrix}$$

and from it we may compute these sample regression coefficients

$$\hat{\beta}_1 = 0.298 \qquad \hat{\beta}_2 = 0.632$$

The estimated regression function is

$$\begin{aligned} \hat{y}_i &= 2.105 + 0.298(x_{i1} - 2.30) + 0.632(x_{i2} - 2.325) \\ &= -0.050 + 0.298x_{i1} + 0.632x_{i2} \end{aligned}$$

Values of the overall evaluation averages predicted by the regression equation are shown in the next-to-last column of Table 2-1.

The scatterplot of the observations is sketched in Figure 2-1. We chose the orientation of the X_1 and X_2 axes to show how the fitted regression plane slopes upward

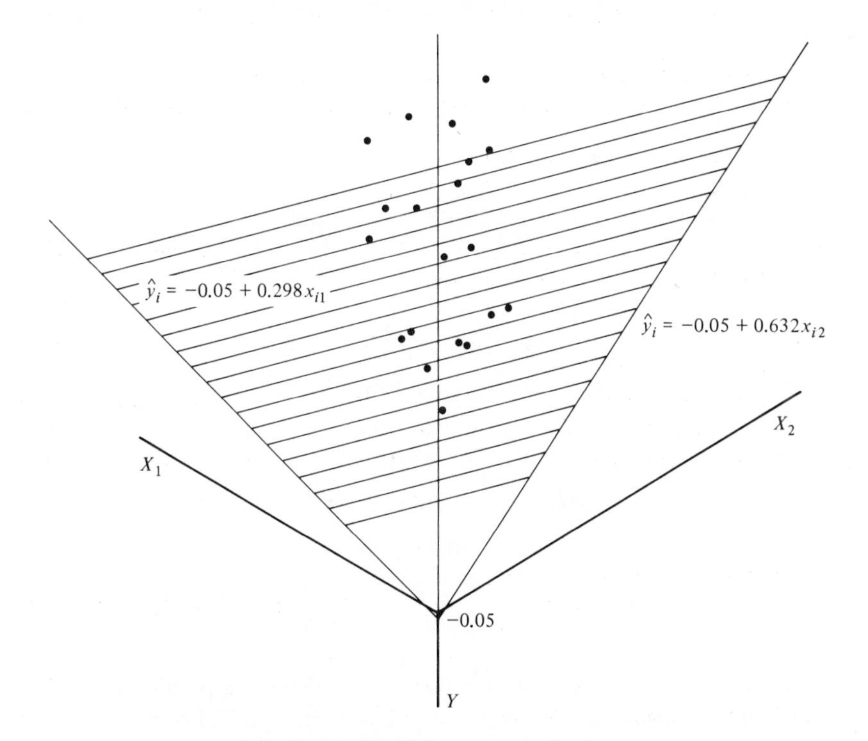

Figure 2-1 Scatterplot of the course evaluation averages.

from the intercept of -0.05. The edges of the pictured segment are the lines given by setting X_1 or X_2 equal to zero in the sample regression function. The predicted values $\hat{y}_i$ are the vertical projections of the y_i onto the plane. The residuals $z_i = y_i - \hat{y}_i$ given in the last column of Table 2-1 are the distances of the y_i above or below the fitted plane. The sum of squared residuals is as small as possible, for the plane was chosen to minimize that criterion. We shall see in Example 2-4 how the residuals can be used to check the assumptions for the multiple regression model and the validity of the fitted model.

Example 2-2

Free fatty acid levels related to age, weight, and skin fold thickness. As a second illustration let us return to the free fatty acid data of Section 1-5. In the original study of prepubescent boys, age, weight, and the thickness of a skin fold were related to free fatty acid (FFA) level. We shall fit a linear regression equation to measure the influence of the three independent variables on FFA. Although the values of the independent variables are what arose in the sample of subjects, we shall treat them as fixed for the least squares model. The data are shown in Table 2-2.

The model we shall fit is

$$y_i = \alpha + \beta_1(x_{i1} - \bar{x}_1) + \beta_2(x_{i2} - \bar{x}_2) + \beta_3(x_{i3} - \bar{x}_3) + e_i$$

and our initial task will be the computation of the least squares estimators of α, β_1, β_2, and β_3. For the data we compute

$$\mathbf{X}_1'\mathbf{X}_1 = \begin{bmatrix} 6052.49 & 2440.22 & 12.6546 \\ 2440.22 & 4134.05 & 58.7966 \\ 12.6546 & 58.7966 & 2.61364 \end{bmatrix}$$

$$\mathbf{X}_1'\mathbf{y} = \begin{bmatrix} -47.3316 \\ -56.1266 \\ -0.388319 \end{bmatrix}$$

The inverse of $\mathbf{X}_1'\mathbf{X}_1$ can be computed to be

$$(\mathbf{X}_1'\mathbf{X}_1)^{-1} = 10^{-4}\begin{bmatrix} 2.304939 & -1.767257 & 28.596330 \\ -1.767257 & 4.912002 & -101.944058 \\ 28.596330 & -101.944058 & 5980.966092 \end{bmatrix}$$

so that the regression coefficients are

$$\hat{\boldsymbol{\beta}}_1 = (\mathbf{X}_1'\mathbf{X}_1)^{-1}\mathbf{X}_1'\mathbf{y} = \begin{bmatrix} -0.002101 \\ -0.015246 \\ 0.204574 \end{bmatrix}$$

The least squares prediction equation for the FFA level of the ith subject is

$$\hat{y}_i = 0.6028 - 0.002101(x_{i1} - 113.71) - 0.015246(x_{i2} - 66.27) + 0.2045(x_{i3} - 0.7312)$$

But how well has the last equation in Example 2-2 predicted FFA as compared to the linear function of age alone in Section 1-5? For an answer let

TABLE 2-2 Values of Free Fatty Acid, Age, Weight, and Skin Fold Thickness

Subject	FFA, Y	Age, X_1 (months)	Weight, X_2	Skin fold thickness, X_3
1	0.759	105	67	0.96
2	0.274	107	70	0.52
3	0.685	100	54	0.62
4	0.526	103	60	0.76
5	0.859	97	61	1.00
6	0.652	101	62	0.74
7	0.349	99	71	0.76
8	1.120	101	48	0.62
9	1.059	107	59	0.56
10	1.035	100	51	0.44
11	0.531	100	80	0.74
12	1.333	101	57	0.58
13	0.674	104	58	1.10
14	0.686	99	58	0.72
15	0.789	101	54	0.72
16	0.641	110	66	0.54
17	0.641	109	59	0.68
18	0.355	109	64	0.44
19	0.256	110	76	0.52
20	0.627	111	50	0.60
21	0.444	112	64	0.70
22	1.016	117	73	0.96
23	0.582	109	68	0.82
24	0.325	112	67	0.52
25	0.368	111	81	1.14
26	0.818	115	74	0.82
27	0.384	115	63	0.56
28	0.509	125	74	0.72
29	0.634	131	70	0.58
30	0.526	121	63	0.90
31	0.337	123	67	0.66
32	0.307	125	82	0.94
33	0.748	122	62	0.62
34	0.401	124	67	0.74
35	0.451	122	60	0.60
36	0.344	129	98	1.86
37	0.545	128	76	0.82
38	0.781	127	63	0.26
39	0.501	140	79	0.74
40	0.524	141	60	0.62
41	0.318	139	81	0.78
Mean	0.6028	113.71	66.27	0.7312
S.D.	0.25454	12.3009	10.1662	0.25562

SOURCE: Data collected by Dr. Felix P. Heald et al. (1967) and reproduced with Dr. Heald's kind permission.

us begin by computing the residuals

$$z_i = y_i - \hat{y}_i \tag{25}$$

of the actual FFA values from those predicted by the least squares equation. The z_i are estimates of the random disturbance terms e_i in the model for FFA. The sum of the residuals, and hence their mean, must always be zero. In matrix notation, the column vector of residuals is

$$\begin{aligned}\mathbf{z} &= \mathbf{y} - \mathbf{X}\hat{\boldsymbol{\beta}} \\ &= \mathbf{y} - \mathbf{X}(\mathbf{X}'\mathbf{X})^{-1}\mathbf{X}'\mathbf{y} \\ &= \mathbf{y} - [\mathbf{j} \quad \mathbf{X}_1]\begin{bmatrix}\bar{y} \\ \hat{\boldsymbol{\beta}}_1\end{bmatrix} \\ &= \mathbf{y} - \mathbf{j}\bar{y} - \mathbf{X}_1\hat{\boldsymbol{\beta}}_1\end{aligned} \tag{26}$$

where $\mathbf{j}$ is the N-component column vector of ones. The sum of the residuals is

$$\mathbf{j}'\mathbf{y} - \mathbf{j}'\mathbf{j}\bar{y} - \mathbf{j}'\mathbf{X}_1\hat{\boldsymbol{\beta}}_1 = N\bar{y} - N\bar{y} = 0 \tag{27}$$

since each column sum of the centered predictor matrix $\mathbf{X}_1$ must be zero. The sum of the squares of the residuals,

$$\begin{aligned}\text{SSE} &= \sum_{i=1}^{N} z_i^2 \\ &= \mathbf{z}'\mathbf{z} \\ &= \mathbf{y}'[\mathbf{I} - \mathbf{X}(\mathbf{X}'\mathbf{X})^{-1}\mathbf{X}'][\mathbf{I} - \mathbf{X}(\mathbf{X}'\mathbf{X})^{-1}\mathbf{X}']\mathbf{y} \\ &= \mathbf{y}'[\mathbf{I} - \mathbf{X}(\mathbf{X}'\mathbf{X})^{-1}\mathbf{X}']\mathbf{y}\end{aligned} \tag{28}$$

might be divided by some number close to N to give a reasonable estimate of the variance σ^2 of the e_i terms. The divisor is in fact $N - p - 1$, or the sample size less the number of parameters $\alpha, \beta_1, \ldots, \beta_p$ in the linear model, and the estimator of the variance is

$$\hat{\sigma}^2 = \frac{1}{N - p - 1}\,\text{SSE} \tag{29}$$

The divisor was chosen so that $\hat{\sigma}^2$ is unbiased, or

$$E(\hat{\sigma}^2) = \sigma^2 \tag{30}$$

The proof of the unbiasedness is given in Appendix C.

Now we rewrite the last line of the definition (28) in a form better suited for computation and by so doing obtain a measure of the efficacy of our estimated regression equation for predicting the dependent variable:

$$\begin{aligned}\text{SSE} &= \mathbf{y}'\left(\mathbf{I} - [\mathbf{j} \quad \mathbf{X}_1]\begin{bmatrix}1/N & \mathbf{0}' \\ \mathbf{0} & (\mathbf{X}_1'\mathbf{X}_1)^{-1}\end{bmatrix}\begin{bmatrix}\mathbf{j}' \\ \mathbf{X}_1'\end{bmatrix}\right)\mathbf{y} \\ &= \mathbf{y}'\left(\mathbf{I} - \frac{1}{N}\mathbf{j}\mathbf{j}'\right)\mathbf{y} - \mathbf{y}'\mathbf{X}_1(\mathbf{X}_1'\mathbf{X}_1)^{-1}\mathbf{X}_1'\mathbf{y} \\ &= \sum_{i=1}^{N}(y_i - \bar{y})^2 - \hat{\boldsymbol{\beta}}_1'\mathbf{X}_1'\mathbf{X}_1\hat{\boldsymbol{\beta}}_1\end{aligned} \tag{31}$$

The first term in the last line is merely the total sum of squares SSY of the deviations of the dependent variables about their mean. The least squares prediction model has decomposed that sum of squares into the component SSE of total variation about the prediction plane and a sum of squares

$$\begin{aligned} \text{SSR} &= \mathbf{y}'\mathbf{X}_1(\mathbf{X}_1'\mathbf{X}_1)^{-1}\mathbf{X}_1'\mathbf{y} \\ &= \hat{\boldsymbol{\beta}}_1'\mathbf{X}_1'\mathbf{X}_1\hat{\boldsymbol{\beta}}_1 \end{aligned} \tag{32}$$

due to the plane itself. The proportion of the total variation in the dependent variable explained by the prediction plane is

$$\begin{aligned} R^2 &= \frac{\text{SSR}}{\text{SSY}} \\ &= 1 - \frac{\text{SSE}}{\text{SSY}} \end{aligned} \tag{33}$$

R^2 is known as the *coefficient of determination*, or the *squared multiple correlation coefficient*, and is treated in detail in Section 2-5. As in the straight-line model, R^2 will provide a convenient measure of the "goodness" of the independent variable set for predicting Y.

The *adjusted coefficient of determination* is sometimes reported in place of R^2, particularly when comparing regression analyses with different numbers of independent variables. If in the definition

$$R^2 = 1 - \frac{\text{SSE}}{\sum (y_i - \bar{y})^2} \tag{34}$$

we replace the sum of squares by unbiased estimators of the corresponding population variances the adjusted coefficient of determination follows as

$$\begin{aligned} R_a^2 &= 1 - \frac{\text{SSE}/(N-p-1)}{\sum (y_i - \bar{y})^2/(N-1)} \\ &= 1 - \frac{N-1}{N-p-1}\frac{\text{SSE}}{\sum (y_i - \bar{y})^2} \\ &= 1 - \frac{N-1}{N-p-1}(1-R^2) \\ &= \frac{-p}{N-p-1} + \frac{N-1}{N-p-1}R^2 \end{aligned} \tag{35}$$

Adjusted R^2 was proposed by Ezekiel (1930); a more recent account of its derivation has been given by Ezekiel and Fox (1959, pp. 300–302).

Example 2-3

In the preceding regression analysis for the prediction of FFA from age, weight, and skin fold thickness,

$$\text{SSY} = 2.5915 \qquad \text{SSE} = 1.7158 \qquad \text{SSR} = 0.8757$$

and $R^2 = 0.3379$. The three variables account for 33.79% of the variation in the free fatty acid levels. 66.21% must be attributed to "noise" or sampling variability about that plane. However, the simple regression model with age described in Section 1-5 accounted for only 14.28% of the variation. The inclusion of weight and skin fold thickness appears to have reduced the uncertainty in our prediction rule. Nevertheless, we shall return to the question of the "goodness" of the three-variable model in Section 2-3 in the context of the "statistical significance" of the three predictors.

Graphical analysis of residuals. By examining the residuals from the fitted regression plane we may be able to determine whether the assumptions for ordinary least squares estimation have been satisfied, and at what points the linear model has failed to represent the dependent variable. We can make these analyses visually from certain plots of the residuals, although without any protection provided by significance levels and other measures of formal statistical inference.

The simplest examination of the residuals might proceed from their histogram or merely their one-dimensional scatterplot. If the disturbances in the least squares model are normally distributed, the histogram should be roughly symmetric and generally shaped like a normal density function. The histogram or plot should be free of large gaps or outlying extreme values of the residuals. It is often convenient to divide the residuals by their standard deviation $\hat{\sigma}$ so that they are expressed in units of $\hat{\sigma}$. A crude rule might be to consider a residual a probable outlier if it is more than 1.5 or 2 standard deviations away from the zero mean. For a more formal test of normality we might plot the residuals on normal probability paper, or their absolute values on half-normal paper: If the sample is nearly normally distributed, the plots should be close to straight lines. Outliers would plot as stragglers away from the ends of the line. Statistical tests for outliers and normality have been developed but are beyond the scope and level of this text.

Two-dimensional plots of the residuals are more informative. In the simplest case we might plot the residuals against their subscripts. Shifts in the width of the plot for different indices might indicate a failure of the constant-variance assumption. Time trends in the data might show through a steady upward or downward movement when the residuals are plotted against time. Next we might plot the residuals with the values of the dependent variable and each of the independent variables to see if the residuals' scatter changes in shape, dispersion, or trend with any of the observed variables. For example, a parabolic shape of the plot against X_1 might indicate that Y has a quadratic rather than linear relationship with X_1. In that way the plots provide a check on the validity of the regression model we have selected and may suggest that some variables should be transformed before inclusion in the model. A steady increase in the width of the residuals' scatter swarm with changing X_1 would imply that the disturbance variance is probably not constant for all values of X_1.

Anscombe (1961, 1973) and Anscombe and Tukey (1963) have proposed plotting the residuals against the predicted values $\hat{y}_i$ of the dependent variable. If the model is correctly specified $\hat{y}_i$ and its residual z_i are uncorrelated. The sample correlation of the $(\hat{y}_i, z_i)$ pairs is also zero. If all assumptions for the model hold we should expect to see an untilted elliptical scatterplot without gaps or empty spots. We can examine the plot for the same patterns we have described earlier: outliers, changing variance, nonlinear trends, or other indications of systematic variation in the residuals unexplained by the regression model.

Example 2-4

We shall use some residual plots to study the multiple regression model for the course evaluation averages given in Example 2-1. Recall that the dependent variable Y was the average score on the overall evaluation scale, and X_1 and X_2 were the average scores for the scales measuring the abilities to present course material and to stimulate student interest. Let us begin with the plots of the residuals from the two-variable model against the values of the predictors. These are shown in Figure 2-2, plots (a) and (b). The plot against X_1 appears to consist of a strong linear trend with some outliers diagonally above and below it. The aberrant points may correspond to instructors whose

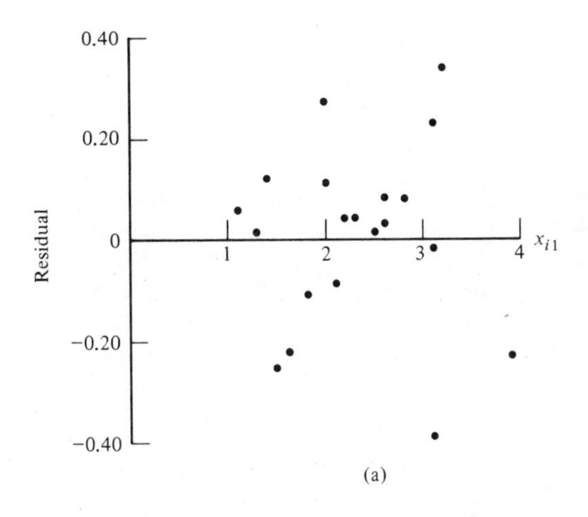

(a)

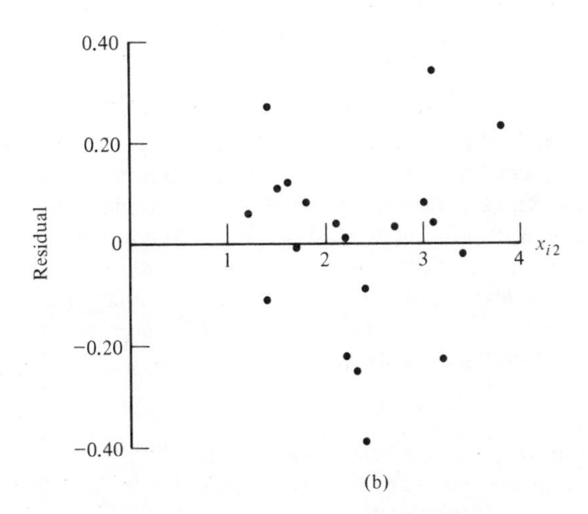

(b)

Figure 2-2 Residual plots for the course evaluation averages.

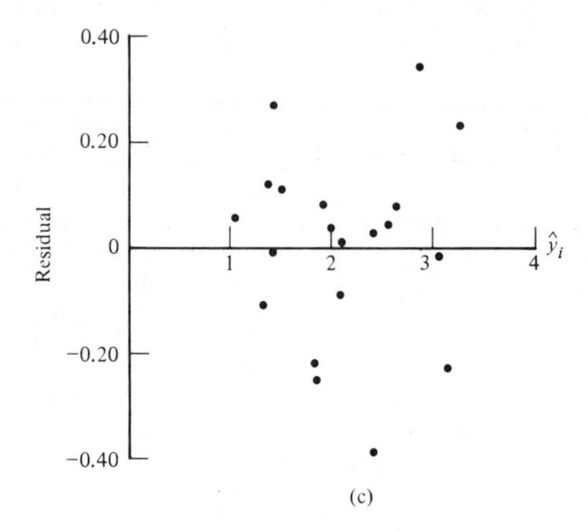

Figure 2-2 (*cont.*)

correlations among the three scales do not fit the general pattern. A dummy variable indicating those courses by a score of one and the remainder by zero might explain additional variation in the dependent variable. The second plot contains a V-shaped trend in the residuals, although one with considerable variation about it. That nonlinear trend appears again in plot (c) of Figure 2-2, where the residuals have been plotted against the predicted values $\hat{y}_i$ of the overall rating. Those plots suggest that the square of X_2 or some centered variant of it should be included in the regression model.

On the basis of the curvature in the last two residual plots we introduced the new variable

$$x_{i3} = (x_{i2} - 2.325)^2 - 0.5419 \qquad i = 1, \ldots, 20$$

That is, the centered X_2 variable was squared and centered again about the mean of the squared terms to realize the computing benefits of centered data. This regression equation was obtained:

$$\hat{y}_i = -0.070 + 0.308x_{i1} + 0.592x_{i2} + 0.168(x_{i2} - 2.325)^2$$

These statistics were computed from the fitted model:

$$\text{SSR} = 8.4591 \qquad \text{SSE} = 0.4704 \qquad \hat{\sigma} = 0.1715 \qquad R^2 = 0.9473$$

By the tests in Section 2-3 we can show that the regression coefficient of the new variable is significantly different from zero. The overall evaluation rating appears to be related in a nonlinear as well as linear way to the ability to stimulate interest. However, the percentage of explained variation of 94.73% is scarcely more than the original percentage 93.03% with X_1 and X_2 alone. The nonlinear relationship may be an artifact of two or three of the observations and might not generalize to a larger data set or to course averages from another semester. Nevertheless, we have seen how the nonlinear trend could be detected from the residual plot.

Example 2-5

The residuals from the three-predictor model for free fatty acid level in Example 2-2 have been plotted against predicted FFA in Figure 2-3. We note that the residuals are closest to zero for the smallest predicted values: The multiple regression model is best for low values of FFA. The dispersion in the residuals is greater for the predicted values

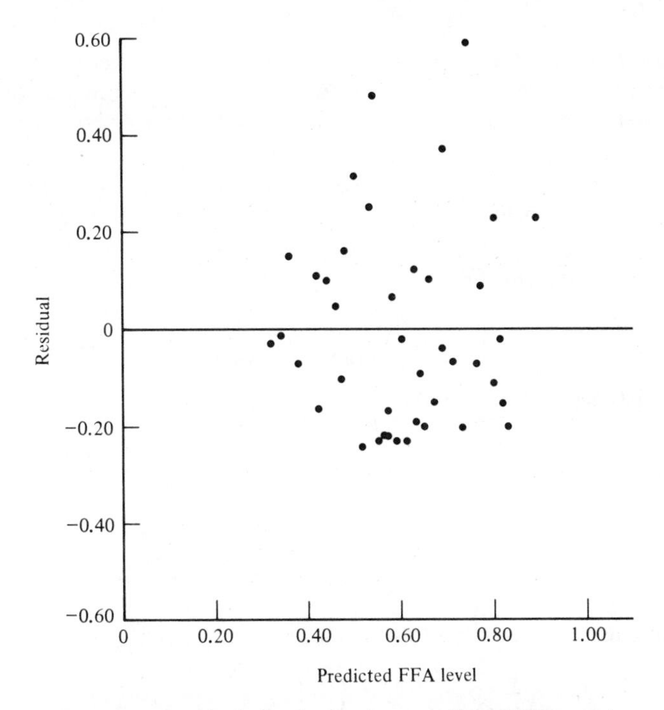

Figure 2-3 Residual plot for the three-predictor FFA model.

in excess of 0.50. The residuals for a given range of predicted values do not appear to be significantly imbalanced by sign, although the more extreme deviations always indicate an underprediction of FFA.

The residual plot for the simple regression of FFA on age described in Section 1-5 is contained in Figure 2-4. The greater amount of unexplained variation in FFA is

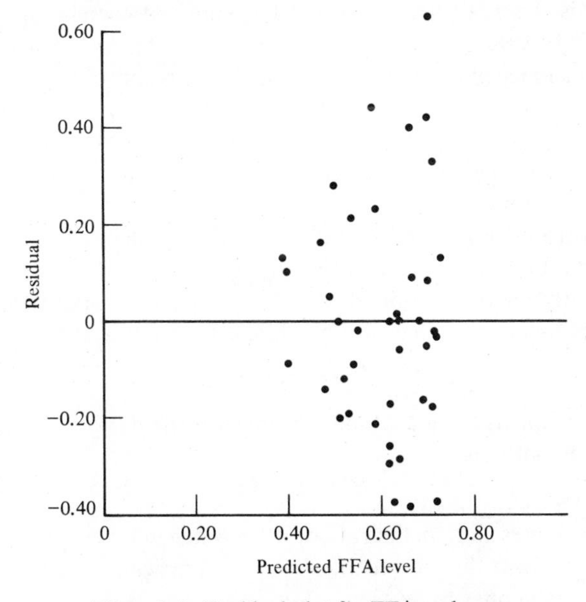

Figure 2-4 Residual plot for FFA and age.

evident here. The predictions of FFA are especially poor for larger values. The trapezoidal shape of the scatterplot shows the increasing residual variance with the predicted value. The residual values are roughly symmetrically distributed about zero.

Scale changes and standardized regression coefficients. Because regression coefficients are rates of change they are unaffected by the choice of the origins of the scales of the dependent and predictor variables. However, the coefficients are measured in units of the ratio

$$\frac{\text{Units of the dependent variable}}{\text{Units of the predictor variable}} \tag{36}$$

The scale changes

$$\begin{aligned} V &= aY \\ U_1 &= c_1 X_1 \\ &\vdots \\ U_p &= c_p X_p \end{aligned} \tag{37}$$

lead to the new parameters and estimators

$$\gamma_j = \left(\frac{a}{c_j}\right)\beta_j \qquad \hat{\gamma}_j = \left(\frac{a}{c_j}\right)\hat{\beta}_j \qquad j = 1, \ldots, p \tag{38}$$

As an example of scale changes in the observations let us take the case of all variables scaled by their standard deviations. Then the sample regression coefficients of the new variables are given by the $p \times 1$ vector

$$\hat{\boldsymbol{\beta}}_s = \mathbf{R}_{xx}^{-1}\mathbf{R}_{xy} \tag{39}$$

in which $\mathbf{R}_{xx}$ is the $p \times p$ matrix of product moment correlations of the independent variable values, and $\mathbf{R}_{xy}$ is the $p \times 1$ vector of correlations of the dependent and independent variables. These regression coefficients are called the *sample beta coefficients*. Because they are the estimators of the regression parameters of the variables expressed as standard scores, and thereby dimensionless, they are sometimes preferred for the interpretation of the relationship of Y to the predictors.

In our introduction of the multiple regression model we have not addressed several important questions. One of the most essential is whether the coefficients β_j are sufficiently removed from zero to justify the inclusion of their variables in the model. To answer that question we should have a test of the hypothesis

$$H_0\colon\ \beta_j = 0 \tag{40}$$

or a method of constructing a confidence interval for β_j. We also require a test of the overall hypothesis

$$H_0\colon\ \boldsymbol{\beta}_1 = \mathbf{0} \tag{41}$$

that the linear regression coefficients of all p independent variables vanish, or its more general form that the coefficients have particular specified values. We develop those techniques in Section 2-3.

2-3 SOME TESTS AND CONFIDENCE INTERVALS FOR REGRESSION PARAMETERS

In this section we develop tests and some confidence statements for the linear model parameters $\alpha, \beta_1, \ldots, \beta_p$, and σ^2. To do this we must know the sampling properties of their estimators: What are the means, variances, and distributions under the assumption of independently and normally distributed disturbances? We begin by giving the joint distribution of the estimators $\hat{\alpha}, \hat{\beta}_1, \ldots, \hat{\beta}_p$ of the regression coefficients. That distribution is the generalization of the normal probability law to several dimensions, the *multivariate normal* or *multinormal* distribution.

The random vector

$$\mathbf{Y}' = [Y_1, \ldots, Y_p] \tag{1}$$

is said to have the multinormal distribution if the joint density function of its elements is

$$f(\mathbf{y}) = \frac{1}{(2\pi)^{1/2p}|\mathbf{\Sigma}|^{1/2}} \exp\left\{-\frac{1}{2}(\mathbf{y}-\boldsymbol{\mu})'\,\mathbf{\Sigma}^{-1}\,(\mathbf{y}-\boldsymbol{\mu})\right\} \tag{2}$$

The two parameters of the multinormal distribution are the mean vector

$$E(\mathbf{Y}) = \begin{bmatrix} \mu_1 \\ \cdot \\ \cdot \\ \cdot \\ \mu_p \end{bmatrix}$$

$$= \boldsymbol{\mu} \tag{3}$$

and the *covariance matrix*

$$\mathrm{Cov}\,(\mathbf{Y}, \mathbf{Y}') = \mathbf{\Sigma}$$

$$= \begin{bmatrix} \sigma_{11} & \sigma_{12} & \cdots & \sigma_{1p} \\ \sigma_{12} & \sigma_{22} & \cdots & \sigma_{2p} \\ \cdot & \cdot & & \cdot \\ \cdot & \cdot & & \cdot \\ \cdot & \cdot & & \cdot \\ \sigma_{1p} & \sigma_{2p} & \cdots & \sigma_{pp} \end{bmatrix} \tag{4}$$

The elements of $\boldsymbol{\mu}$ are the population means of the successive variates in $\mathbf{Y}$. The diagonal positions of $\mathbf{\Sigma}$ contain the variances $\mathrm{Var}\,(Y_i) = \sigma_{ii}$ of the variates, while the *ij*th off-diagonal position contains the covariance σ_{ij} of Y_i and Y_j. If the function (2) is a density $\mathbf{\Sigma}$ must be a symmetric positive definite matrix. Further properties of the multivariate normal distribution are given in Appendix C.

The estimators $\hat{\alpha}, \hat{\beta}_1, \ldots, \hat{\beta}_p$ of the regression parameters have a multinormal distribution. If the disturbances e_i have the usual assumptions of zero means, a common variance, independence, and normality, their joint distribution is multinormal with the null mean vector and the covariance matrix

$$\text{Cov}(\mathbf{e}, \mathbf{e}') = \sigma^2 \begin{bmatrix} 1 & \cdots & 0 \\ \cdots & \cdots & \cdots \\ 0 & \cdots & 1 \end{bmatrix} \tag{5}$$

Since linear functions of multinormal random variables are multinormally distributed, the vector $\mathbf{y} = \mathbf{X}\boldsymbol{\beta} + \mathbf{e}$ is multinormal with mean vector $\mathbf{X}\boldsymbol{\beta}$ and the same covariance matrix (5). Now the least squares estimator $\hat{\boldsymbol{\beta}} = (\mathbf{X}'\mathbf{X})^{-1}\mathbf{X}'\mathbf{y}$ is also a linear function of $\mathbf{y}$ and $\mathbf{e}$, so its distribution is multinormal with mean vector $\boldsymbol{\beta}$ and covariance matrix

$$\text{Cov}(\hat{\boldsymbol{\beta}}, \hat{\boldsymbol{\beta}}') = \sigma^2(\mathbf{X}'\mathbf{X})^{-1} \tag{6}$$

If the independent variable values are centered about their averages as in (13), Section 2-2, the covariance matrix is

$$\text{Cov}(\hat{\boldsymbol{\beta}}, \hat{\boldsymbol{\beta}}') = \sigma^2 \begin{bmatrix} 1/N & \mathbf{0} \\ \mathbf{0} & (\mathbf{X}_1'\mathbf{X}_1)^{-1} \end{bmatrix} \tag{7}$$

The elements of (7) tell us these facts about the distribution of $\hat{\alpha}, \hat{\beta}_1, \ldots, \hat{\beta}_p$:

1. The variance of $\hat{\alpha}$ is σ^2/N, or what we would expect for the variance of $\bar{y}$.
2. $\hat{\alpha}$ and $\hat{\boldsymbol{\beta}}_1$ are independently distributed, for their covariances are zero.
3. $\hat{\alpha}$ is a univariate normal random variable.
4. $\hat{\boldsymbol{\beta}}_1$ has the p-dimensional multivariate normal distribution with mean vector $\boldsymbol{\beta}_1$ and covariance matrix $\sigma^2(\mathbf{X}_1'\mathbf{X}_1)^{-1}$.
5. The jth element of $\hat{\boldsymbol{\beta}}_1$ is a normal random variable with mean β_j and variance $\sigma^2 a_{jj}$, where a_{jj} is the jth diagonal element of
$$\mathbf{A} = (\mathbf{X}_1'\mathbf{X}_1)^{-1} \tag{8}$$
6. The covariance of $\hat{\beta}_i$ and $\hat{\beta}_j$ is $\sigma^2 a_{ij}$. The difference $\hat{\beta}_i - \hat{\beta}_j$ is a normal random variable with mean $\beta_i - \beta_j$ and variance
$$\text{Var}(\hat{\beta}_i - \hat{\beta}_j) = \sigma^2(a_{ii} + a_{jj} - 2a_{ij}) \tag{9}$$

These results follow from the normality of linear functions of multinormal random variables.

Next we need the distribution of the estimator of σ^2. Recall that the estimator was defined by (29), Section 2-2, as

$$\begin{aligned} \hat{\sigma}^2 &= \frac{\mathbf{y}'[\mathbf{I} - \mathbf{X}(\mathbf{X}'\mathbf{X})^{-1}\mathbf{X}']\mathbf{y}}{N - p - 1} \\ &= \frac{\text{SSE}}{N - p - 1} \end{aligned} \tag{10}$$

From the properties of quadratic forms given in Appendices B and C we can show that SSE$/\sigma^2$ or $(N - p - 1)\hat{\sigma}^2/\sigma^2$ has the chi-squared distribution with $N - p - 1$ degrees of freedom. Furthermore, SSE and the variance estimator are distributed independently of $\hat{\alpha}$ and $\hat{\boldsymbol{\beta}}_1$. Those results and the normal distributions of the regression coefficient estimators will enable us to test hypotheses and construct confidence intervals for the parameters of the linear model.

Example 2-6

In Example 1-1 and Exercise 12 of Chapter 1 we gave average starting salaries for graduates of an MBA program for eight years. Let us fit a second-degree polynomial trend to the data by the model

$$y_t = \alpha + \beta_1 t + \beta_2 t^2 + e_t$$

Note that we have not centered the independent variables

$$X_{t1} = t \qquad X_{t2} = t^2$$

about their averages. The data for the analysis had the following values:

y_t	t	t^2
14,200	1	1
15,645	2	4
16,452	3	9
16,884	4	16
18,159	5	25
20,650	6	36
21,890	7	49
24,993	8	64

$$\mathbf{X'X} = \begin{bmatrix} 8 & 36 & 204 \\ 36 & 204 & 1296 \\ 204 & 1296 & 8772 \end{bmatrix}$$

$$(\mathbf{X'X})^{-1} = \begin{bmatrix} 1.9464 & -0.9107 & 0.08939 \\ -0.9107 & 0.5506 & -0.05357 \\ 0.08939 & -0.05357 & 0.005952 \end{bmatrix}$$

$$\text{SSE} = 1{,}274{,}330 \qquad \hat{\sigma}^2 = 254{,}866 \qquad \hat{\sigma} = 504.84$$

The estimates and their estimated standard deviations had these values:

Parameter	Estimate	Standard deviation
α	14,410.72	704.32
β_1	77.41	374.60
β_2	150.98	38.948

We note that the linear coefficient $\hat{\beta}_1$ has a standard deviation nearly five times its value. The linear trend component seems to be well within sampling variation of a zero population value.

It will be of interest to find the correlation coefficients of the three estimates. The correlation matrix is calculated directly from $(\mathbf{X'X})^{-1}$ as shown in (8): It is

$$\begin{bmatrix} 1 & -0.92 & 0.83 \\ -0.92 & 1 & -0.98 \\ 0.83 & -0.98 & 1 \end{bmatrix}$$

The estimates are very highly correlated. The linear coefficient $\hat{\beta}_1$ bears a negative relationship with both the intercept and quadratic component estimates. We shall see presently that these correlations can be removed by centering the linear and quadratic components.

Example 2-7

If the columns of the centered predictor matrix $\mathbf{X}_1$ are mutually orthogonal, interpretation of the fitted equation is simplified. The effect and contribution of each variable can be measured directly and independently of the others. Because $\mathbf{X}_1'\mathbf{X}_1$ is a diagonal matrix the regression coefficient estimates will be uncorrelated. Hypothesis tests on the coefficients will be essentially independent.

As an example of orthogonal predictors let us rewrite the polynomial trend model of the preceding example as

$$y_t = \gamma_0 + \gamma_1(t - 4.5) + \gamma_2[(t - 4.5)^2 - 5.25] + e_t$$

The linear term has been centered about its average $\bar{t} = 4.5$, whereas the quadratic variable is also centered. Then

$$\mathbf{X}_1 = \begin{bmatrix} -3.5 & 7 \\ -2.5 & 1 \\ -1.5 & -3 \\ -0.5 & -5 \\ 0.5 & -5 \\ 1.5 & -3 \\ 2.5 & 1 \\ 3.5 & 7 \end{bmatrix} \qquad \mathbf{X}_1'\mathbf{X}_1 = \begin{bmatrix} 42 & 0 \\ 0 & 168 \end{bmatrix}$$

These estimates and standard deviations were computed:

Parameter	Estimate	Standard deviation
γ_0	18,609.12	178.93
γ_1	1,436.25	77.90
γ_2	150.98	38.95

Now the linear and quadratic coefficients $\hat{\gamma}_1$ and $\hat{\gamma}_2$ are uncorrelated. The linear coefficient measures a straight line trend that is unaffected by the quadratic curvature measured by $\hat{\gamma}_2$. The standard deviation of the linear coefficient is very small compared to $\hat{\gamma}_1$ and indicates that a linear trend in the model is not an artifact of random variation.

The model with orthogonalized linear and quadratic components is an example of *orthogonal polynomials*. We treat those polynomials in detail in Chapter 4.

Hypothesis tests and confidence intervals for the regression coefficients. The results on the distributions of the least square estimators and $\hat{\sigma}^2$ provide us with the means for inferences about the parameters of the multiple regression model. Our statistics for tests on the individual components will have this form:

$$t = \frac{\text{estimator} - \text{parameter}}{\text{standard deviation of the estimator}} \tag{11}$$

The actual statistics are

$$t = \frac{(\bar{y} - \alpha)\sqrt{N}}{\hat{\sigma}} \tag{12}$$

$$\begin{aligned} t_j &= \frac{\hat{\beta}_j - \beta_j}{\sqrt{\widehat{\operatorname{Var}}(\hat{\beta}_j)}} \\ &= \frac{(\hat{\beta}_j - \beta_j)}{\hat{\sigma}\sqrt{a_{jj}}} \qquad j = 1, \ldots, p \end{aligned} \tag{13}$$

where a_{jj} is the jth diagonal element of the matrix $\mathbf{A}$ defined by (8). Each of the statistics $t, t_1, \ldots, t_p$ has the t-distribution with $N - p - 1$ degrees of freedom. To test the hypothesis

$$H_0\colon \quad \alpha = \alpha_0 \tag{14}$$

that the intercept parameter has the specific value α_0 we compute

$$t = \frac{(\bar{y} - \alpha_0)\sqrt{N}}{\hat{\sigma}} \tag{15}$$

and reject H_0 in favor of the alternative

$$H_1^{(1)}\colon \quad \alpha \neq \alpha_0 \tag{16}$$

of a different intercept if

$$|t| > t_{\alpha/2;N-p-1} \tag{17}$$

Similarly, the hypothesis (14) can be tested against the one-sided alternative

$$H_1^{(2)}\colon \quad \alpha < \alpha_0 \tag{18}$$

by rejecting H_0 if $t < -t_{\alpha:N-p-1}$. The test of (14) against

$$H_1^{(3)}\colon \quad \alpha > \alpha_0 \tag{19}$$

is made by rejecting H_0 if

$$t > t_{\alpha;N-p-1} \tag{20}$$

The tests of the hypotheses

$$H_0\colon \quad \beta_j = \beta_{j0} \tag{21}$$

on the individual regression coefficients are based on the statistics t_j defined by expression (13). The most common null hypothesis is that of $\beta_{j0} = 0$ or that the jth predictor has no influence in the least squares model. For the jth predictor the null hypothesis can be tested against the usual three alternatives by the decision rules:

1. Reject H_0 in favor of $H_1^{(1)}$: $\beta_j \neq \beta_{j0}$ if $|t| > t_{\alpha/2;N-p-1}$.
2. Reject H_0 in favor of $H_1^{(2)}$: $\beta_j < \beta_{j0}$ if $t < -t_{\alpha;N-p-1}$. (22)
3. Reject H_0 in favor of $H_1^{(3)}$: $\beta_j > \beta_{j0}$ if $t > t_{\alpha;N-p-1}$.

The probability statements

$$\begin{aligned} P(|t| \leq t_{\alpha/2;N-p-1}) &= 1 - \alpha \\ p(|t_j| \leq t_{\alpha/2;N-p-1}) &= 1 - \alpha \end{aligned} \tag{23}$$

formed from the random variables defined by (12) and (13) lead to the $100(1-\alpha)\%$ confidence intervals

$$\bar{y} - t_{\alpha/2;N-p-1}\frac{\hat{\sigma}}{\sqrt{N}} \leq \alpha \leq \bar{y} + t_{\alpha/2;N-p-1}\frac{\hat{\sigma}}{\sqrt{N}} \tag{24}$$

$$\hat{\beta}_j - t_{\alpha/2;N-p-1}\hat{\sigma}\sqrt{a_{jj}} \leq \beta_j \leq \hat{\beta}_j + t_{\alpha/2;N-p-1}\hat{\sigma}\sqrt{a_{jj}} \qquad j = 1, \ldots, p \tag{25}$$

In addition to providing measures of the "uncertainty" in the estimates the confidence intervals permit us to test at once an infinity of hypotheses. Every value within an interval would be a null hypothesis acceptable at the α level in comparison with the two-sided alternative.

It is essential to remember that the confidence coefficient $1-\alpha$ applies only to a single confidence interval. If a large number of intervals is computed, the probability that one or more do not cover the true value of the parameter will be much less than $1-\alpha$. For moderate values of p and for extensive "data snooping" to determine which predictors or linear combinations of them might be dropped we need *families* of confidence statements with an *experiment* or *family confidence coefficient* or *error rate*. Such tests and intervals are provided by various simultaneous inference procedures. We examine those methods in detail in Chapter 3.

Example 2-8

The free fatty acid data. Let us test the hypotheses of zero population regression coefficients for the predictors age, skin fold thickness, and weight in the linear model for free fatty acid level. The statistics are shown in Table 2-3. If we adopt a two-sided alternative

$$H_1: \beta_j \neq 0 \qquad j = 1, 2, 3$$

more from conservatism than physiological considerations the 0.025 critical value is approximately $t_{0.025;37} = 2.026$. Only weight appears to have a nonzero regression coefficient in the population of normal preadolescent boys. The 99% confidence interval for that coefficient is

$$-0.02820 \leq \beta_3 \leq -0.00229$$

The other coefficients are too close to zero to warrant similar intervals.

TABLE 2-3 Test Statistics for the Free Fatty Acid Regression Coefficients

Predictor	$\hat{\beta}_j$	$\sqrt{\widehat{\text{Var}}(\hat{\beta}_j)}$	t_j
Age	−0.002101	0.003269	−0.64
Skin fold thickness	0.204574	0.166541	1.23
Weight	−0.015246	0.004773	−3.19

Testing the hypothesis that all β_j are zero. The previous results of this section provided a means for individual tests and confidence intervals for the separate regression parameters. In those methods we were afforded protection against Type I errors, or falsely declaring a coefficient "significant," only with respect to the individual hypothesis, although we promised "family rate" protection would be produced in Chapter 3. In addition to the tests on the separate coefficients we would like a test of the hypothesis

$$H_0: \begin{bmatrix} \beta_1 \\ \cdot \\ \cdot \\ \cdot \\ \beta_p \end{bmatrix} = \begin{bmatrix} 0 \\ \cdot \\ \cdot \\ \cdot \\ 0 \end{bmatrix} \tag{26}$$

that the dependent variable is not linearly related to the predictors. For the test we begin with the sum of squares

$$\begin{aligned} \text{SSR} &= \mathbf{y}'\mathbf{X}_1(\mathbf{X}_1'\mathbf{X}_1)^{-1}\mathbf{X}_1'\mathbf{y} \\ &= \hat{\boldsymbol{\beta}}_1'\mathbf{X}_1'\mathbf{X}_1\hat{\boldsymbol{\beta}}_1 \end{aligned} \tag{27}$$

attributable to the regression plane which we introduced in expression (32) of Section 2-2. Under the null hypothesis of zero coefficients SSR/σ^2 has a chi-squared distribution with p degrees of freedom. SSR is also distributed independently of SSE, so that

$$F = \frac{\text{SSR}}{p} \bigg/ \frac{\text{SSE}}{N-p-1} = \frac{N-p-1}{p}\,\frac{\text{SSR}}{\text{SSE}} \tag{28}$$

is an F random variable with p and $N-p-1$ degrees of freedom if $H_0: \boldsymbol{\beta}_1 = \mathbf{0}$ is true. That hypothesis is tenable at the α level if

$$F < F_{\alpha;\,p,N-p-1} \tag{29}$$

Otherwise the hypothesis should be rejected. We can also express the statistic in terms of the coefficient of determination:

$$F = \frac{N-p-1}{p}\,\frac{R^2}{1-R^2} \tag{30}$$

As an example let us test the hypothesis that the free fatty acid levels are unrelated to the variables age, skin fold thickness, and weight. From the values $\text{SSR} = 0.8757$ and $\text{SSE} = 1.7158$ given in Section 2-2 we find $F = 6.30$ with 3 and 37 degrees of freedom. Since the nearest smaller tabulated critical value is $F_{0.005;3,30} = 5.24$, we conclude that the hypothesis of no linear relationship with the three predictors should be rejected.

In practice we usually test the overall hypothesis (26) of no regression relationship, and if it is rejected, proceed to tests and confidence intervals for the individual coefficients. Olshen (1973) has shown that the true confidence coefficient for intervals found only when the hypothesis (26) was rejected is always smaller than the nominal $1 - \alpha$ value. The true confidence coefficients are probabilities conditional on the rejection of the overall F-statistic. Olshen has computed those probabilities for a variety of cases of the conventional least squares model.

Our treatment of hypothesis tests and confidence intervals has been brief and elementary, in the sense of introducing the most important and basic tests for the significance of the regression parameters. We have avoided such questions as these:

1. How can the more general hypothesis $H_0: \boldsymbol{\beta}_1 = \boldsymbol{\beta}_{10}$ be tested?
2. What are the distributions of the quadratic forms SSR and the F-statistics when the null hypothesis is not true? How can these distributions be used to calculate the power, or its complement the Type II error probability, of the test?
3. How does one test hypotheses on subsets of the regression parameters?

These methods and several other tests and confidence statements are treated in Chapter 3. Meanwhile let us return to some other properties of the regression model and its extensions.

2-4 THE MINIMUM VARIANCE AND MAXIMUM LIKELIHOOD PROPERTIES OF LEAST SQUARES ESTIMATORS

Our choice of the minimum sum of squared deviations as a criterion for fitting lines and planes to data points requires some defense beyond the simplicity of its estimates, its connection with the maximum likelihood estimates of the bivariate normal parameters, and the analogy of the sum of squares to the error variance. We might have chosen instead the sum of the absolute deviations or some arbitrary power of them as a minimand. With present computing facilities such criteria are indeed feasible. However, the least squares estimators enjoy a powerful property which serves as a justification of their use. It is known as the Gauss–Markov theorem, and has the following statement:

> The least squares estimators have minimum variance among all unbiased estimators that are linear functions of the observations $\mathbf{y}$.

The proof of the theorem in matrix notation is simple and instructive. It is due to Plackett (1949); a version with further discussion and extensions has also been given by Kendall and Stuart (1961, pp. 79–82). We begin by defining the vector of q linear functions of the parameters, $\mathbf{C}\boldsymbol{\beta}$, where $\mathbf{C}$ is a given $q \times (p+1)$ matrix of constants. The least squares estimator of $\mathbf{C}\boldsymbol{\beta}$,

$$\mathbf{C}\hat{\boldsymbol{\beta}} = \mathbf{C}(\mathbf{X}'\mathbf{X})^{-1}\mathbf{X}'\mathbf{y} \tag{1}$$

is unbiased:

$$\begin{aligned} E(\mathbf{C}\hat{\boldsymbol{\beta}}) &= \mathbf{C}(\mathbf{X}'\mathbf{X})^{-1}\mathbf{X}'\mathbf{E}(\mathbf{y}) \\ &= \mathbf{C}(\mathbf{X}'\mathbf{X})^{-1}\mathbf{X}'\mathbf{X}\boldsymbol{\beta} \\ &= \mathbf{C}\boldsymbol{\beta} \end{aligned} \tag{2}$$

Now let us show that its elements have smallest variances among all unbiased linear estimators. Such an estimator is a linear function

$$\mathbf{t} = \mathbf{L}\mathbf{y} \tag{3}$$

where $\mathbf{L}$ is a $q \times N$ matrix of coefficients to be determined. By the unbiasedness condition

$$\begin{aligned} \mathbf{C}\boldsymbol{\beta} &= E(\mathbf{t}) \\ &= \mathbf{L}E(\mathbf{y}) \\ &= \mathbf{L}\mathbf{X}\boldsymbol{\beta} \end{aligned} \tag{4}$$

so that

$$\mathbf{C} = \mathbf{L}\mathbf{X} \tag{5}$$

if the estimator is to be unbiased. The covariance matrix of the estimators $\mathbf{t}$ is the $q \times q$ matrix

$$\begin{aligned} \text{Cov}\,(\mathbf{t}, \mathbf{t}') &= \text{Cov}\,(\mathbf{L}\mathbf{y}, \mathbf{y}'L') \\ &= \mathbf{L}[\text{Cov}\,(\mathbf{y}, \mathbf{y}')]\mathbf{L}' \\ &= \sigma^2\mathbf{L}\mathbf{L}' \end{aligned} \tag{6}$$

Introduce the identity

$$\begin{aligned} \mathbf{L}\mathbf{L}' = &[\mathbf{L}\mathbf{X}(\mathbf{X}'\mathbf{X})^{-1}\mathbf{X}'][\mathbf{L}\mathbf{X}(\mathbf{X}'\mathbf{X})^{-1}\mathbf{X}']' \\ &+ [\mathbf{L} - \mathbf{L}\mathbf{X}(\mathbf{X}'\mathbf{X})^{-1}\mathbf{X}'][\mathbf{L} - \mathbf{L}\mathbf{X}(\mathbf{X}'\mathbf{X})^{-1}\mathbf{X}']' \end{aligned} \tag{7}$$

and then replace $\mathbf{L}\mathbf{X}$ by the matrix $\mathbf{C}$ as given by the unbiasedness condition (5). The identity is then

$$\begin{aligned} \mathbf{L}\mathbf{L}' = &[\mathbf{C}(\mathbf{X}'\mathbf{X})^{-1}\mathbf{X}'][\mathbf{C}(\mathbf{X}'\mathbf{X})^{-1}\mathbf{X}']' \\ &+ [\mathbf{L} - \mathbf{C}(\mathbf{X}'\mathbf{X})^{-1}\mathbf{X}'][\mathbf{L} - \mathbf{C}(\mathbf{X}'\mathbf{X})^{-1}\mathbf{X}']' \end{aligned} \tag{8}$$

What matrix $\mathbf{L}$ will minimize the diagonal elements of $\mathbf{L}\mathbf{L}'$, and thereby give least variances for the estimators $\mathbf{t}$? The first term in the right-hand side of (8) is a constant fixed by the values in $\mathbf{C}$ and $\mathbf{X}$. The second term is a matrix whose diagonal terms are sums of squares and as such have zero as a minimum value. That value will be attained for all $\mathbf{C}$ if

$$\mathbf{L} = \mathbf{C}(\mathbf{X}'\mathbf{X})^{-1}\mathbf{X}' \tag{9}$$

or if the estimator of $\mathbf{C}\boldsymbol{\beta}$ is

$$\begin{aligned} \mathbf{t} &= \mathbf{C}(\mathbf{X}'\mathbf{X})^{-1}\mathbf{X}'\mathbf{y} \\ &= \mathbf{C}\hat{\boldsymbol{\beta}} \end{aligned} \tag{10}$$

the least squares estimator (1). We have established that the least squares estimators have minimum variance among all unbiased estimators that are linear functions of the dependent variable observations. The covariance matrix of the least squares estimators is

$$\text{Cov}\,[\mathbf{C}\hat{\boldsymbol{\beta}}, (\mathbf{C}\hat{\boldsymbol{\beta}})'] = \sigma^2\mathbf{C}(\mathbf{X}'\mathbf{X})^{-1}\mathbf{C}' \tag{11}$$

or the same matrix obtained in Section 2-3 under the assumption of normally distributed error terms.

The equivalence of least squares and maximum likelihood estimators for normally distributed errors. In our introduction of the least squares criterion by intuitive appeal to goodness of fit and its subsequent rational justification by the minimum variance property we did not impose a particular type of distribution function on the independent error terms e_i. A common method of estimating parameters which begins with the assumption of a particular form of distribution function is that of *maximum likelihood.* Estimates of the parameters of the distribution are found by finding the functions of the observations which maximize the *likelihood* of the sample or the joint density of the random variables corresponding to the observations. In that way we have chosen estimated values for the parameters which are consistent with the greatest probability of observing the given data set. Application of the maximum likelihood technique to random samples from the normal population yields the usual sample mean and (biased) variance estimates of the normal parameters. Similar results consistent with estimators found from sample moments or intuition hold for the Poisson, binomial, and scale-parameter exponential distributions.

The maximum likelihood estimators of the parameters of the regression model with normally and independently distributed error variates are identical with those found by the least squares criterion. The likelihood function of the dependent variable observation vector $\mathbf{y}$ is

$$\mathbf{L}(\boldsymbol{\beta}, \sigma^2) = f(y; \boldsymbol{\beta}, \sigma^2)$$

$$= \frac{1}{(2\pi)^{N/2}(\sigma^2)^{N/2}} \exp\left\{\frac{-\frac{1}{2}[(\mathbf{y} - \mathbf{X}\boldsymbol{\beta})'(\mathbf{y} - \mathbf{X}\boldsymbol{\beta})]}{\sigma^2}\right\} \tag{12}$$

Maximization of $L(\boldsymbol{\beta}, \sigma^2)$ is equivalent to maximization of its natural logarithm

$$l(\boldsymbol{\beta}, \sigma^2) = -(N/2) \ln 2\pi - (N/2) \ln \sigma^2 - \frac{\frac{1}{2}[(\mathbf{y} - \mathbf{X}\boldsymbol{\beta})'(\mathbf{y} - \mathbf{X}\boldsymbol{\beta})]}{\sigma^2} \tag{13}$$

but maximization of this function with respect to $\boldsymbol{\beta}$ is identical with minimization of the least squares criterion in the last term: The two types of estimators of $\boldsymbol{\beta}$ are equivalent. Finally, we note that calculation of the partial derivative of $l(\boldsymbol{\beta}, \sigma^2)$ with respect to the variance and solution of the equation

$$\frac{\partial l(\boldsymbol{\beta}, \sigma^2)}{\partial \sigma^2} = 0 \tag{14}$$

leads to the maximum likelihood estimator

$$\hat{\sigma}^2 = \frac{1}{N}(\mathbf{y} - \mathbf{X}\hat{\boldsymbol{\beta}})'(\mathbf{y} - \mathbf{X}\hat{\boldsymbol{\beta}}) \tag{15}$$

of σ^2. Replacement of N by $N - p - 1$ gives the unbiased estimator (29) of Section 2-2.

2-5 THE MULTIPLE CORRELATION MODEL

Our treatment of the multiple regression model by least squares estimation has been predicated upon the assumption that the values of the independent variables in the matrix $\mathbf{X}$ are fixed, rather than observations on random variables or the results of sampling experiments. As we emphasized in Chapter 1, many collections of data do not arise this way: A random sample of N subjects, forest tracts, or laboratory rats may be chosen, and a dependent variable and p independent variables of interest observed on each sampling unit. The values of the predictors are then as much subject to random and measurement fluctuations as the dependent variable. Such data with several independent variables should be treated by the methods of *multiple correlation* rather than the standard regression analysis of the preceding sections. Our development of multiple correlation will be based on the model of a multivariate normal population.

Recall from Section 2-3 that the $(p + 1)$-dimensional multinormal density function of the variate $\mathbf{W}$ is

$$f(\mathbf{w}) = \frac{1}{(2\pi)^{1/2(p+1)}|\mathbf{\Sigma}|^{1/2}} \exp\{-\tfrac{1}{2}(\mathbf{w} - \boldsymbol{\mu})'\mathbf{\Sigma}^{-1}(\mathbf{w} - \boldsymbol{\mu})\} \tag{1}$$

$\mathbf{W}$ is a partitioned column vector

$$\mathbf{W} = \begin{bmatrix} Y \\ \mathbf{X} \end{bmatrix} \tag{2}$$

consisting of the scalar Y and the $p \times 1$ vector $\mathbf{X}$. Y will assume the role of the dependent variable, whereas the elements of $\mathbf{X}$ will represent the p independent variables.* The mean vector and covariance matrix in partitioned form are

$$\begin{aligned} \boldsymbol{\mu} = E(\mathbf{W}) &= \begin{bmatrix} \mu_1 \\ \boldsymbol{\mu}_2 \end{bmatrix} \\ \mathbf{\Sigma} = \text{Cov}\,(\mathbf{W}, \mathbf{W}') &= \begin{bmatrix} \sigma_{11} & \boldsymbol{\sigma}'_{12} \\ \boldsymbol{\sigma}_{12} & \mathbf{\Sigma}_{22} \end{bmatrix} \end{aligned} \tag{3}$$

The $p \times 1$ vector $\boldsymbol{\sigma}_{12}$ contains the covariances of Y with the elements of $\mathbf{X}$, and $\mathbf{\Sigma}_{22}$ is the $p \times p$ covariance matrix of $\mathbf{X}$. We shall develop the multiple correlation analysis of Y and $\mathbf{X}$ in these steps:

1. We shall find the conditional density of Y for $\mathbf{X}$ fixed at the value $\mathbf{x}$.
2. The mean of the conditional density of Y is a linear function of $\mathbf{x}$ and a population analogue of the least squares model for the dependent variable.
3. The variance of the conditional density of Y will provide a basis for a measure of the multiple correlation of Y with the p variates in $\mathbf{X}$.
4. The parameters of the conditional density will be estimated from a random sample of N observations on the partitioned variate $\mathbf{W}' = [Y, \mathbf{X}']$, and estimates of the multiple correlation and regression parameters will follow from them.

The conditional density of Y for $\mathbf{X} = \mathbf{x}$ is given formally by

$$g(y|\mathbf{x}) = \frac{f(y, \mathbf{x})}{f_1(\mathbf{x})} \tag{4}$$

*The random vector $\mathbf{X}$ of this section should not be confused with the $N \times (p + 1)$ matrix $\mathbf{X}$ of values of the predictors used in the earlier development of the least squares model.

$f(y, \mathbf{x})$ is the density (1) with the partitioned variable $\mathbf{w}' = [y, \mathbf{x}']$ replaced by its two elements for clarity, and $f_1(\mathbf{x})$ is the $N(\boldsymbol{\mu}_2, \boldsymbol{\Sigma}_{22})$ marginal density of the variate $\mathbf{X}$. To obtain $g(y|\mathbf{x})$ we must replace $|\boldsymbol{\Sigma}|$ and $\boldsymbol{\Sigma}^{-1}$ in (1) by their partitioned matrix representations, then carry out the division by $f_1(\mathbf{x})$ and simplify. The details can be found in most texts on multivariate analysis, for example, Morrison (1976, Section 3.4). The density is

$$g(y|\mathbf{x}) = \frac{1}{(2\pi)^{1/2}\sigma_{11.2}{}^{1/2}} \exp\left\{\frac{-\frac{1}{2}[y - \mu_1 - \boldsymbol{\beta}_1'(\mathbf{x} - \boldsymbol{\mu}_2)]^2}{\sigma_{11.2}}\right\} \tag{5}$$

where $\boldsymbol{\beta}_1 = \boldsymbol{\Sigma}_{22}^{-1}\boldsymbol{\sigma}_{12}$ and $\sigma_{11.2} = \sigma_{11} - \boldsymbol{\sigma}_{12}'\boldsymbol{\Sigma}_{22}^{-1}\boldsymbol{\sigma}_{12}$. The subscript 1 has been appended to $\boldsymbol{\beta}$ only for consistency with the symbol for the regression coefficients in Section 2-2. This is a normal density with mean

$$\begin{aligned} E(Y|\mathbf{x}) &= \mu_1 + \boldsymbol{\beta}_1'(\mathbf{x} - \boldsymbol{\mu}_2) \\ &= \mu_1 + \boldsymbol{\sigma}_{12}'\boldsymbol{\Sigma}_{22}^{-1}(\mathbf{x} - \boldsymbol{\mu}_2) \end{aligned} \tag{6}$$

and variance

$$\operatorname{Var}(Y|\mathbf{x}) = \sigma_{11.2} = \sigma_{11} - \boldsymbol{\sigma}_{12}'\boldsymbol{\Sigma}_{22}^{-1}\boldsymbol{\sigma}_{12} \tag{7}$$

The mean is a linear function of the value $\mathbf{x}$ of the set of independent variables, whereas the conditional variance is the same for all vectors $\mathbf{x}$. The conditional mean is called the *regression function of Y on the vector variate* $\mathbf{X}$.

The parallelism of the multivariate normal model and the multiple regression model of Section 2-2 is clear. It will be helpful for an understanding of the correlation approach to compare the parametrization of the two models. For simplicity assume that we have a single value Y of the dependent variable related to the $p \times 1$ vector $\mathbf{x}$ of predictor values:

$$Y = \alpha + \boldsymbol{\beta}_1'(\mathbf{x} - \bar{\mathbf{x}}) + e \tag{8}$$

and

$$E(Y) = \alpha + \boldsymbol{\beta}_1'(\mathbf{x} - \bar{\mathbf{x}}) \tag{9}$$

The parameters of the models are compared in Table 2-4.

TABLE 2-4 Comparison of the Least Squares and Conditional Mean Linear Models

	Model	
Parameter	Least squares	Conditional mean
Predictor mean	$\bar{\mathbf{x}}$	$\boldsymbol{\mu}_2$
Intercept	α	μ_1
Regression coefficients	$\boldsymbol{\beta}_1$	$\boldsymbol{\beta}_1 = \boldsymbol{\Sigma}_{22}^{-1}\boldsymbol{\sigma}_{12}$
Error variance	σ^2	$\sigma_{11.2} = \sigma_{11} - \boldsymbol{\sigma}_{12}'\boldsymbol{\Sigma}_{22}^{-1}\boldsymbol{\sigma}_{12}$

The multiple correlation of the variates Y and X. The correlations

$$\rho_{1j} = \frac{\sigma_{1j}}{\sqrt{\sigma_{11}\sigma_{jj}}} \qquad j = 2, \ldots, p + 1 \tag{10}$$

computed from σ_{11}, the elements of $\boldsymbol{\sigma}_{12}$, and the diagonal elements σ_{jj} of $\boldsymbol{\Sigma}_{22}$ measure the strength of the relationship of Y with the p individual predictors.

Y and $\mathbf{X}$ are independent random variables if every element of $\boldsymbol{\sigma}_{12}$, and hence every correlation ρ_{1j}, is zero. We would prefer a single measure of the correlation between Y and the predictor variates, and we propose for that purpose the *population coefficient of determination*

$$\begin{aligned}\rho^2 &= \frac{\boldsymbol{\sigma}'_{12}\boldsymbol{\Sigma}_{22}^{-1}\boldsymbol{\sigma}_{12}}{\sigma_{11}} \\ &= \frac{\boldsymbol{\beta}'_1\boldsymbol{\Sigma}_{22}\boldsymbol{\beta}_1}{\sigma_{11}} \\ &= 1 - \frac{\sigma_{11.2}}{\sigma_{11}} \end{aligned} \tag{11}$$

or its positive square root, ρ, the *multiple correlation coefficient.*

When more than one dependent variable or several sets of predictors require multiple correlations or determination coefficients the variables are indicated by subscripts. In the present content,

$$\rho^2_{1.2\ldots(p+1)} = \rho^2$$

denotes the coefficient of determination of the first variable, Y, with the next p variables, the predictors $X_1, \ldots, X_p$.

The interpretations of the coefficient of determination as a proportion of variance and ρ as a product moment correlation follow from these properties:

Property 1. ρ^2 is the proportion of the variance of the dependent variable due to the conditional mean function.

If we apply the variance operator to that linear function with the particular value $\mathbf{x}$ of the predictor variates replaced by its random variable $\mathbf{X}$,

$$\begin{aligned}\operatorname{Var}[\mu_1 + \boldsymbol{\beta}'_1(\mathbf{X} - \boldsymbol{\mu}_2)] &= \boldsymbol{\beta}'_1\boldsymbol{\Sigma}_{22}\boldsymbol{\beta}_1 \\ &= \boldsymbol{\sigma}'_{12}\boldsymbol{\Sigma}_{22}^{-1}\boldsymbol{\sigma}_{12} \\ &= \rho^2\sigma_{11} \end{aligned} \tag{12}$$

as the property states.

Property 2. $1 - \rho^2$ is the proportion of the variance of the dependent variable about the conditional mean function.

The property follows from the definition (7) of the conditional distribution variance and (11) or by applying the variance operator to the residual variate

$$Z = Y - \mu_1 - \boldsymbol{\beta}'_1(\mathbf{X} - \boldsymbol{\mu}_2) \tag{13}$$

giving the variation about the conditional mean function. Then

$$\begin{aligned}\operatorname{Var}(Z) &= \operatorname{Var}(Y - \mu_1) + \operatorname{Var}[\boldsymbol{\beta}'_1(\mathbf{X} - \boldsymbol{\mu}_2)] \\ &\quad - 2\operatorname{Cov}[Y - \mu_1, \boldsymbol{\beta}'_1(\mathbf{X} - \boldsymbol{\mu}_2)] \\ &= \sigma_{11} + \boldsymbol{\beta}'_1\boldsymbol{\Sigma}_{22}\boldsymbol{\beta}_1 - 2\boldsymbol{\beta}'_1\sigma_{12} \\ &= \sigma_{11}(1 - \rho^2) \end{aligned} \tag{14}$$

as we wished to show.

Property 3. The predictor variates are uncorrelated with the residual variate Z.

That is,

$$\operatorname{Cov}[Y - \mu_1 - \boldsymbol{\beta}_1'(\mathbf{X} - \boldsymbol{\mu}_2), \mathbf{X}' - \boldsymbol{\mu}_2'] = \boldsymbol{\sigma}_{12}' - \boldsymbol{\beta}_1'\boldsymbol{\Sigma}_{22} = \mathbf{0}$$

Property 4. ρ^2 is invariant under transformations

$$U = aY + b \qquad \mathbf{V} = \mathbf{CX} + \mathbf{d} \tag{15}$$

within the Y and $\mathbf{X}$ variates. $a \neq 0$, and the $p \times p$ matrix $\mathbf{C}$ consists of constants such that $|\mathbf{C}| \neq 0$.

Since ρ^2 is only a function of the central second moments, it is unaffected by the translation constants b and $\mathbf{d}$. The coefficient of determination of U with respect to the new predictor variate vector $\mathbf{V}$ is

$$\begin{aligned} \rho_{u,v}^2 &= \frac{a\boldsymbol{\sigma}_{12}'\mathbf{C}'(\mathbf{C}\boldsymbol{\Sigma}_{22}\mathbf{C}')^{-1}\mathbf{C}\boldsymbol{\sigma}_{12}a}{a^2\sigma_{11}} \\ &= \frac{a^2\boldsymbol{\sigma}_{12}'\mathbf{C}'(\mathbf{C}')^{-1}\boldsymbol{\Sigma}_{22}^{-1}\mathbf{C}^{-1}\mathbf{C}\boldsymbol{\sigma}_{12}}{a^2\sigma_{11}} \\ &= \frac{\boldsymbol{\sigma}_{12}'\boldsymbol{\Sigma}_{22}^{-1}\boldsymbol{\sigma}_{12}}{\sigma_{11}} \\ &= \rho^2 \end{aligned} \tag{16}$$

It is essential that the transformations be made separately on the dependent and predictor variates. Otherwise one could change the value of the coefficient of determination at will by an appropriate choice of transformation matrix.

Property 5. The multiple correlation ρ is interpretable as the correlation of Y and the variate $V = \mu_1 + \boldsymbol{\beta}_1'(\mathbf{X} - \boldsymbol{\mu}_2)$.

$$\begin{aligned} \operatorname{Cov}[Y, \mu_1 + \boldsymbol{\beta}_1'(\mathbf{X} - \boldsymbol{\mu}_2)] &= \boldsymbol{\beta}_1'[\operatorname{Cov}(Y, \mathbf{X})] \\ &= \boldsymbol{\beta}_1'\boldsymbol{\sigma}_{12} \\ &= \boldsymbol{\sigma}_{12}'\boldsymbol{\Sigma}_{22}^{-1}\boldsymbol{\sigma}_{12} \end{aligned} \tag{17}$$

and

$$\begin{aligned} \operatorname{Var}[\mu_1 + \boldsymbol{\beta}_1'(\mathbf{X} - \boldsymbol{\mu}_2)] &= \boldsymbol{\beta}_1'[\operatorname{Cov}(\mathbf{X}, \mathbf{X}')]\boldsymbol{\beta}_1 \\ &= \boldsymbol{\beta}_1'\boldsymbol{\Sigma}_{22}\boldsymbol{\beta}_1 \\ &= \boldsymbol{\sigma}_{12}'\boldsymbol{\Sigma}_{22}^{-1}\boldsymbol{\sigma}_{12} \end{aligned} \tag{18}$$

The correlation is

$$\rho_{YV} = \frac{\boldsymbol{\sigma}_{12}'\boldsymbol{\Sigma}_{22}^{-1}\boldsymbol{\sigma}_{12}}{\sqrt{\sigma_{11}\boldsymbol{\sigma}_{12}'\boldsymbol{\Sigma}_{22}^{-1}\boldsymbol{\sigma}_{12}}} \tag{19}$$

or the positive square root of ρ^2. Consequently, $0 \leq \rho \leq 1$ for all multinormal distributions.

Property 6. The multiple correlation coefficient ρ is the maximum product moment correlation between Y and any nonnull linear compound $\mathbf{a}'\mathbf{X}$ of the predictor variates.

The proof of Property 6 can be found in many texts on multivariate statistical analysis, for example, Morrison (1976).

Property 7. The vector $\mathbf{a}$ such that the residual variance

$$\text{Var}\,(\mathbf{Y} - \mathbf{a}'\mathbf{X}) \tag{20}$$

is a minimum is the regression coefficient vector $\boldsymbol{\beta}_1 = \boldsymbol{\Sigma}_{22}^{-1}\boldsymbol{\sigma}_{12}$.

If we calculate

$$V = \text{Var}\,(Y - \mathbf{a}'\mathbf{X}) = \sigma_{11} - 2\mathbf{a}'\boldsymbol{\sigma}_{12} + \mathbf{a}'\boldsymbol{\Sigma}_{22}\mathbf{a} \tag{21}$$

set its partial derivative

$$\frac{\partial V}{\partial \mathbf{a}} = -\,2\boldsymbol{\sigma}_{12} + 2\boldsymbol{\Sigma}_{22}\mathbf{a} \tag{22}$$

equal to the $p \times 1$ null vector, $\mathbf{a} = \boldsymbol{\Sigma}_{22}^{-1}\boldsymbol{\sigma}_{12} = \boldsymbol{\beta}_1$, as the property states.

Estimation of the multiple correlation parameters from sample data. The preceding results have described the multiple correlation, the coefficient of determination, and the regression coefficient vector as parameters of a multinormal distribution of $p + 1$ variates. Apart from motivating and parametrizing the multiple correlation model, these properties are useful only in the rare cases where the multinormal parameters are known. Usually, we must estimate the parameters from a finite sample of observations, and we address that technique now on the basis of the multinormal population model.

The estimators we propose for $\boldsymbol{\beta}_1$ and ρ^2 are those found from the maximum likelihood principle introduced in Section 2-4. We begin by assuming that N independent observation vectors consisting of a value for the dependent variable Y and values for each of the p predictor variables have been drawn from the $(p + 1)$-dimensional multinormal distribution with density function (1). Let the data be arranged in matrix form* as

$$\mathbf{W} = \begin{bmatrix} y_1 & x_{11} & \cdots & x_{1p} \\ \cdot & \cdot & & \cdot \\ \cdot & \cdot & & \cdot \\ \cdot & \cdot & & \cdot \\ y_N & x_{N1} & \cdots & x_{Np} \end{bmatrix} = \begin{bmatrix} y_1 & \mathbf{x}_1' \\ \cdot & \cdot \\ \cdot & \cdot \\ \cdot & \cdot \\ y_N & \mathbf{x}_N' \end{bmatrix} \tag{23}$$

where the rows denote the independent sampling units. The maximum likelihood estimator of the mean vector is

*The matrix $\mathbf{W}$, the vectors $\mathbf{x}_1', \ldots, \mathbf{x}_N'$, and their elements will denote *observed* values of random variables introduced at the beginning of this section.

$$\begin{bmatrix} \hat{\mu}_1 \\ \hat{\boldsymbol{\mu}}_2 \end{bmatrix} = \begin{bmatrix} \bar{y} \\ \bar{\mathbf{x}} \end{bmatrix} \tag{24}$$

where

$$\bar{y} = \frac{1}{N} \sum_{i=1}^{N} y_i$$

$$\bar{\mathbf{x}} = \begin{bmatrix} \bar{\mathbf{x}}_1 \\ \cdot \\ \cdot \\ \cdot \\ \bar{\mathbf{x}}_p \end{bmatrix} = \frac{1}{N} \sum_{i=1}^{N} \mathbf{x}_i \tag{25}$$

As in the univariate normal population, the estimators of the mean parameters are the sample averages of the observations on the respective variates. The estimators of the elements of the partitioned covariance matrix (3) are

$$\hat{\sigma}_{11} = s_{11} = \frac{1}{N-1} \sum_{i=1}^{N} (y_i - \bar{y})^2$$

$$\boldsymbol{\sigma}_{12} = \mathbf{s}_{12} = \frac{1}{N-1} \sum_{i=1}^{N} (y_i - \bar{y})(\mathbf{x}_i - \bar{\mathbf{x}})$$

$$= \frac{1}{N-1} \begin{bmatrix} \sum_{i=1}^{N} (y_i - \bar{y})(x_{i1} - \bar{x}_1) \\ \cdot \\ \cdot \\ \cdot \\ \sum_{i=1}^{N} (y_i - \bar{y})(x_{ip} - \bar{x}_p) \end{bmatrix} \tag{26}$$

$$\boldsymbol{\Sigma}_{22} = \mathbf{S}_{22} = \frac{1}{N-1} \sum_{i=1}^{N} (\mathbf{x}_i - \bar{\mathbf{x}})(\mathbf{x}_i - \bar{\mathbf{x}})'$$

$$= \frac{1}{N-1} \begin{bmatrix} \sum_{i=1}^{N} (x_{i1} - x_1)^2 & \cdots & \sum_{i=1}^{N} (x_{i1} - \bar{x}_1)(x_{ip} - \bar{x}_p) \\ \cdot & & \cdot \\ \cdot & & \cdot \\ \cdot & & \cdot \\ \sum_{i=1}^{N} (x_{i1} - \bar{x}_1)(x_{ip} - \bar{x}_p) & \cdots & \sum_{i=1}^{N} (x_{ip} - \bar{x}_p)^2 \end{bmatrix}$$

These estimators have been obtained by replacing the divisor N of the maximum likelihood estimators by $N - 1$ so that they are unbiased:

$$E(\hat{\sigma}_{11}) = \sigma_{11} \qquad E(\hat{\boldsymbol{\sigma}}_{12}) = \boldsymbol{\sigma}_{12} \qquad E(\hat{\boldsymbol{\Sigma}}_{22}) = \boldsymbol{\Sigma}_{22} \tag{27}$$

The details of their derivation can be found in many texts, for example, Anderson (1958, Chapter 3) or Morrison (1976, Chapter 3).

We can find the estimators of ρ^2, $\boldsymbol{\beta}_1$ and the regression function by replacing the parameters in those quantities by their estimators (24) and (26). The estimator of the coefficient of determination is the sample statistic

$$R^2 = \frac{\mathbf{s}_{12}' \mathbf{S}_{22}^{-1} \mathbf{s}_{12}}{s_{11}} \tag{28}$$

and the estimator of $\boldsymbol{\beta}_1$ is

$$\hat{\boldsymbol{\beta}}_1 = \mathbf{S}_{22}^{-1}\mathbf{s}_{12} \tag{29}$$

In these two estimators the divisors of the sums of squares and products cancel, making the choice of N or $N-1$ moot. The estimator of the conditional mean function is the sample regression equation

$$y = \bar{y} + \hat{\boldsymbol{\beta}}_1'(\mathbf{x} - \bar{\mathbf{x}}) \tag{30}$$

In arriving at these results we have made use of an important property: Under very general conditions the maximum likelihood estimator of a function of parameters is merely the function with the parameters replaced by their maximum likelihood estimators.

The estimators of ρ^2 and $\hat{\boldsymbol{\beta}}_1$ are identical with those for the coefficient of determination and the regression vector in expressions (33) and (24) of Section 2-2. Recall that

$$\begin{aligned} \mathbf{X}_1'\mathbf{X}_1 &= \sum_{i=1}^{N} (\mathbf{x}_i - \bar{\mathbf{x}})(\mathbf{x}_i - \bar{\mathbf{x}})' \\ \mathbf{X}_1'\mathbf{y} &= \sum_{i=1}^{N} (\mathbf{x}_i - \bar{\mathbf{x}})(y_i - \bar{y}) \\ \text{SSR} &= (N-1)\mathbf{s}_{12}'\mathbf{S}_{22}^{-1}\mathbf{s}_{12} \\ \text{SSY} &= (N-1)s_{11} \end{aligned} \tag{31}$$

so that

$$\begin{aligned} R^2 &= \frac{\text{SSR}}{\text{SSY}} \\ \hat{\boldsymbol{\beta}}_1 &= (\mathbf{X}_1'\mathbf{X}_1)^{-1}\mathbf{X}_1'\mathbf{y} \end{aligned} \tag{32}$$

as before. Other results on hypothesis tests will also carry over from the fixed predictor model. Before considering them it will be helpful to illustrate multiple correlation with some test scores from the psychology of personality.

Example 2-9

Psychological tests for personality. In an interdisciplinary study of human aging Singer (1963) administered a battery of projective psychological tests to a group of healthy men of ages 65 and over. The scores for $N = 46$ subjects on five of the tests are shown in columns 2 through 6 of Table 2-5. We have designated the score on the test Draw a Person as the dependent variable Y for our regression analysis. Essentially, we wish to determine how well the scores on that "performance" test can be predicted by the scores

X_1 = Level of Aspiration test total score

X_2 = Proverbs test interpretations rating

X_3 = number of statements made in the Homonyms test

X_4 = total of all performance ratings made of the tests in the battery

The Level of Aspiration test is also considered to be a performance instrument, whereas Proverbs and Homonyms are cognitive tests measuring verbal manipulation. The score X_1 is the number of X's the subject can draw in a square in a limited time, whereas X_2 takes on values on an eight-point scale ranging from good abstract interpretation (rating one) to a lack of comprehension (rating eight).

TABLE 2-5 Projective Psychological Test Scores

Subject	Y	X_1	X_2	X_3	X_4	$\hat{Y}$	$Z = Y - \hat{Y}$
1	12	61	6	17	46	16.55	−4.55
2	14	67	6	17	47	15.39	−1.39
3	12	81	5	27	33	22.93	−10.93
4	26	67	1	36	16	30.76	−4.76
5	7	49	6	19	57	12.12	−5.12
6	24	80	2	26	33	18.81	5.19
7	17	81	7	21	35	22.98	−5.98
8	22	70	7	15	51	13.68	8.32
9	12	24	8	10	74	5.59	6.41
10	43	58	6	24	31	26.96	16.04
11	7	34	8	9	76	3.22	3.78
12	19	83	4	24	28	23.45	−4.45
13	12	84	7	18	36	21.38	−9.38
14	26	81	5	21	34	20.89	5.11
15	8	83	6	17	43	16.00	−8.00
16	42	90	4	28	28	23.73	18.27
17	19	70	2	24	25	23.79	−4.79
18	16	82	3	29	35	19.56	−3.56
19	14	71	7	21	52	14.51	−0.51
20	35	92	3	35	18	29.52	5.48
21	22	45	5	24	49	16.91	5.09
22	15	55	8	19	47	19.75	−4.75
23	23	77	2	26	24	24.13	−1.13
24	12	63	7	23	46	19.16	−7.16
25	11	37	3	19	51	12.71	−1.71
26	14	54	8	12	68	6.40	7.60
27	8	64	6	18	53	12.59	−4.59
28	13	36	7	20	59	13.90	−0.90
29	8	82	3	23	47	11.37	−3.37
30	11	58	1	27	37	17.73	−6.73
31	11	38	8	11	66	8.89	2.11
32	27	77	5	30	25	28.54	−1.54
33	26	84	7	25	34	24.23	1.77
34	22	59	5	19	36	21.51	0.49
35	45	83	1	38	13	31.31	13.69
36	20	77	1	19	25	20.52	−0.52
37	17	61	1	26	35	18.29	−1.29
38	27	48	2	23	34	20.75	6.25
39	29	83	2	35	32	21.29	7.71
40	26	70	1	27	29	20.97	5.03
41	18	61	1	41	24	28.14	−10.14
42	35	79	1	43	14	32.40	2.60
43	29	67	2	28	32	21.18	7.82
44	10	60	3	24	39	18.32	−8.32
45	14	63	2	33	28	25.05	−11.05
46	26	59	4	31	27	28.17	−2.17
Mean	19.70	66.26	4.33	23.96	38.52	19.70	0
S.D.	9.7202	16.3536	2.4590	7.7458	15.0787	—	—

SOURCE: Data reproduced with the kind permission of Dr. Margaret Thaler Singer.

X_3 consists of the number of meanings given by the subject for 10 common homonyms. The performance on each test was rated independently by another psychologist or research assistant, and the rating scores were summed to give the X_4 measure of each subject. Large values of X_4 indicate many poor ratings of performance.

For our analyses we must proceed under the rather tenuous assumption that the $N = 46$ observation vectors were drawn independently from a five-dimensional multinormal population. The frequency plots of some of the individual scores do not appear to support normality. Y and X_1 have long-tailed histograms and are nearly bimodal. X_2 is clearly bimodal with its least frequency at the middle score of four. X_3 and X_4 have more symmetrical frequency functions. The observations may be mixtures from different distributions and may contain some outliers.

The sample means and standard deviations of the scores are shown at the bottom of Table 2-5. The covariance and correlation matrices of the variables in the order Y, $X_1, \ldots, X_4$ are

$$\mathbf{S} = \begin{bmatrix} 94.4831 & 65.7256 & -9.3874 & 44.5865 & -97.5932 \\ & 267.4415 & -13.2647 & 61.8783 & -170.3836 \\ & & 6.0469 & -14.0522 & 27.3594 \\ & & & 59.9981 & -97.8879 \\ & & & & 227.3662 \end{bmatrix}$$

$$\mathbf{R} = \begin{bmatrix} 1 & 0.4135 & -0.3927 & 0.5922 & -0.6659 \\ & 1 & -0.3299 & 0.4885 & -0.6910 \\ & & 1 & -0.7378 & 0.7379 \\ & & & 1 & -0.8381 \\ & & & & 1 \end{bmatrix}$$

We note immediately that the absolute values of the correlations in the first row of **R** between Y and the predictor variables each exceed the two-sided 0.01 critical value 0.376. Because those correlations are significantly different from zero the multiple regression analysis would appear to be a productive use of the four predictor variables. The vector of regression coefficients is

$$\hat{\boldsymbol{\beta}}_1 = \begin{bmatrix} -0.1013 \\ 1.3259 \\ 0.2476 \\ -0.5581 \end{bmatrix} \tag{34}$$

and the estimated regression equation for predicting the Draw a Person scores from the other four tests is

$$\begin{aligned} \hat{y}_i = 19.70 &- 0.1013(x_{i1} - 66.26) + 1.3259(x_{i2} - 4.33) \\ &+ 0.2476(x_{i3} - 23.96) - 0.5581(x_{i4} - 38.52) \end{aligned} \tag{35}$$

The predicted values for the subjects are shown in Table 2-5, together with the residuals $z_i = y_i - \hat{y}_i$ of the actual observations from those predicted by the four independent variables.

We plotted the residuals against the predictors and the predicted values $\hat{y}_i$ of the dependent variable. The four predictor plots contained no informative patterns and will not be given here. An earlier scatterplot of Y and X_4 suggested a nonlinear relationship of those variables; that model was not supported by the residual plot with X_4. The plot with the $\hat{y}_i$ was more interesting and is shown in Figure 2-5. Subjects 10, 16,

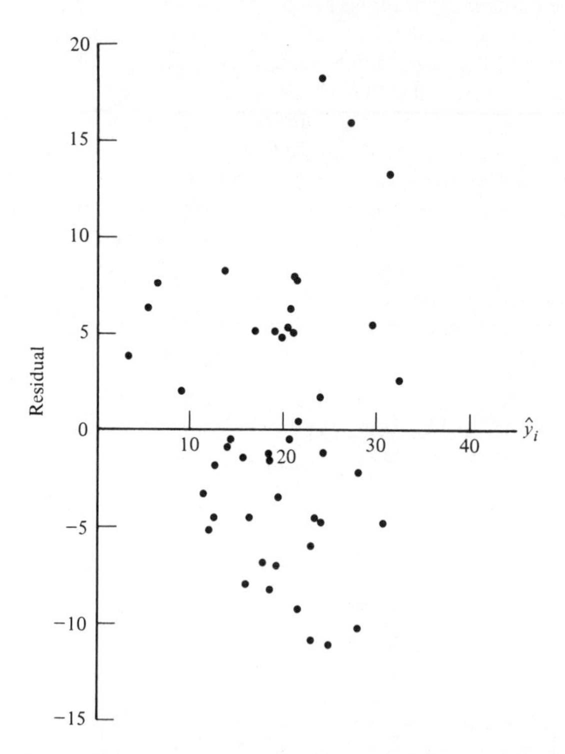

Figure 2-5 Residuals plotted against the predicted Draw a Person scores.

and 35 stand alone as outliers 2.21, 2.51, and 1.88 standard deviations, respectively, above the zero mean. Since high scores on the Draw a Person test are indicative of intellectual deterioration, the multiple regression model failed for subjects with pathologically poor scores. Without the extreme points the residuals seem slightly negatively correlated with the predicted values. If the outliers are omitted, the residuals appear to have greatest variance in the middle range of the predicted values.

We can compute the coefficient of determination from the residual values in Table 2-5. The sum of the squared residuals is

$$\text{SSE} = \mathbf{z}'\mathbf{z} = \sum_{i=1}^{46} z_i^2 = 2163.6727$$

The sum of squares of the dependent variable scores about their mean is SSY = 4251.7391, so that the sum of squares due to the regression equation (35) is SSR = SSY − SSE = 2088.0664. The coefficient of determination is

$$R^2 = \frac{\text{SSR}}{\text{SSY}} = 0.4911$$

We have accounted for almost one-half the variation in the Draw a Person scores by their linear relationship with the other four predictors. The multiple correlation coefficient of Y with the other scores is 0.7008. Whereas those values are moderately large, we still must interpret them in the context of the multinormal model and its correlation and regression parameters. We introduce those hypothesis tests next and apply them to the present data set.

Hypothesis tests on $\boldsymbol{\beta}_1$ and $\boldsymbol{\rho}^2$. The test of the hypothesis

$$H_0: \quad \boldsymbol{\beta}_1 = \begin{bmatrix} \beta_1 \\ \cdot \\ \cdot \\ \cdot \\ \beta_p \end{bmatrix} = \begin{bmatrix} 0 \\ \cdot \\ \cdot \\ \cdot \\ 0 \end{bmatrix} \tag{36}$$

of no regression relation of Y to the p predictors is equivalent to the test of independence

$$H_0: \quad \boldsymbol{\sigma}_{12} = \mathbf{0} \tag{37}$$

of Y and $\mathbf{X}$, since $\boldsymbol{\beta}_1 = \boldsymbol{\Sigma}_{22}^{-1}\boldsymbol{\sigma}_{12}$. It is also equivalent to the test of a zero coefficient of determination, for

$$\rho^2 = \frac{\boldsymbol{\sigma}_{12}'\boldsymbol{\Sigma}_{22}^{-1}\boldsymbol{\sigma}_{12}}{\sigma_{11}} = \frac{\boldsymbol{\beta}_1'\boldsymbol{\Sigma}_{22}\boldsymbol{\beta}_1}{\sigma_{11}}$$

is zero if and only if $\boldsymbol{\beta}_1$ is the null vector. When H_0 is true

$$F = \frac{N-p-1}{p}\frac{R^2}{1-R^2} \tag{38}$$

has the F-distribution with p and $N-p-1$ degrees of freedom. This is the same statistic given by expression (30) of Section 2-3 for the fixed-predictor case, and has the same distribution. We accept H_0 at the α level if

$$F < F_{\alpha;\, p, N-p-1}$$

and otherwise reject it in favor of a nonnull regression coefficient vector.

When the population multiple correlation is zero R^2 has the beta distribution (Mood et al., 1974, pp. 115–116) on the interval from zero to one. Its mean and variance are

$$\begin{aligned} E(R^2) &= \frac{p}{N-1} \\ \operatorname{Var}(R^2) &= \frac{2(N-p-1)p}{(N^2-1)(N-1)} \end{aligned} \tag{39}$$

Because R and R^2 are always positive their expected values will be biased upward from the population ρ and ρ^2 of zero. In fact, if the number of predictors is large relative to the sample size the expected value of R^2 will approach 1—an absurd example of a nonsense correlation generated by the dimensions of the data matrix.

When ρ^2 is not zero the distribution of R^2 is a complicated function. It was found by Fisher (1928) and has been discussed by Anderson (1958, Chapter 3) and by Kendall and Stuart (1961, Chapter 27). Kendall and Stuart give these moments for R^2 through terms of order N^{-1}:

$$\begin{aligned} E(R^2) &\approx \rho^2 + \frac{p}{N-1}(1-\rho^2) - \frac{2(N-p-1)}{N^2-1}\rho^2(1-\rho^2) \\ \operatorname{Var}(R^2) &\approx \frac{4\rho^2(1-\rho^2)^2(N-p-1)^2}{(N^2-1)(N+3)} \end{aligned} \tag{40}$$

Note that the variance of R^2 is not equal to that of (39) with $\rho^2 = 0$. As Kendall and Stuart point out, the distribution of R^2 tends to normality as N increases when ρ^2 is not zero. However, when $\rho^2 = 0$ R^2 converges to the value zero at the lower limit of its range as N becomes large—hardly a normal density function.

Tables of certain percentage points of the distribution of R have been prepared by Kramer (1963) and Lee (1972); the Kramer table has been reproduced by Pearson and Hartley (1972) in volume 2 of their *Biometrika Tables for Statisticians*. Through those tables confidence intervals for ρ or ρ^2 can be constructed.

Example 2-10

To test the hypothesis that the Draw a Person score is independent of the four predictor variates in the population of subjects we compute the F-statistic (38):

$$F = \frac{46 - 4 - 1}{4} \frac{0.4911}{0.5089} = 9.89$$

That value exceeds any common upper percentage point of the F-distribution with 4 and 41 degrees of freedom, and we should conclude that the variates are dependent.

Tests on the individual regression coefficients. In the multinormal regression model we may test the hypotheses

$$H_0: \quad \beta_j = \beta_{j0} \tag{41}$$

on the individual regression coefficients by the same t-statistic and decision rules of Section 2-3 for the fixed-predictor model. The statistic (13) of that section,

$$t_j = \frac{(\hat{\beta}_j - \beta_{j0})}{\hat{\sigma}\sqrt{a_{jj}}} \qquad j = 1, \ldots, p \tag{42}$$

is computed from the jth diagonal element a_{jj} of the inverse

$$\begin{aligned} \mathbf{A} &= (\mathbf{X}_1'\mathbf{X}_1)^{-1} \\ &= [(N-1)\mathbf{S}_{22}]^{-1} \end{aligned} \tag{43}$$

of the $p \times p$ matrix of the sums of squares and products of the independent variable values about their means. $\hat{\sigma}$ is defined by

$$\begin{aligned} \hat{\sigma} &= \sqrt{\frac{\text{SSE}}{N-p-1}} \\ &= \sqrt{\frac{(1-R^2)\sum(y_i - \bar{y})^2}{N-p-1}} \end{aligned} \tag{44}$$

As in Section 2-3 we would reject H_0 for values of t in excess of the appropriate critical value of the Student t-distribution with $N - p - 1$ degrees of freedom. The confidence intervals (25) of that section can also be used for individual statements on the β_j.

Example 2-11

Let us test the hypotheses

$$H_0: \beta_j = 0 \qquad j = 1, 2, 3, 4$$

that the respective predictor variates in the projective psychological test battery are unrelated to the Draw a Person score. The test statistics are given in Table 2-6. Only the variable X_4, total rating, has a test statistic greater in absolute value than the approximate two-sided critical value $t_{0.025;41} \approx 2.02$. We might consider dropping the other three predictor variables; the study of the predictive efficacy of X_4 alone is left as an exercise. The formal problem of variate subset selection is treated in Chapter 5.

TABLE 2-6 Regression Statistics for the Psychological Tests

Predictor variate	$\hat{\beta}_j$	$a_{jj} \times 10^4$	t_j
X_1: Aspiration score	−0.1013	1.86983	−1.02
X_2: Proverbs rating	1.3259	100.40895	1.82
X_3: Homonyms statements	0.2476	14.129358	0.91
X_4: Ratings sum	−0.5581	6.27338	−3.08

2-6 PARTIAL CORRELATION

The *partial correlation* of two variables is their product moment correlation computed as if one or more other variables were held at some constant values. Let us motivate the notion of partial correlation by considering the sample correlation of the variables X_1 and X_2 with a third variable X_3 held fixed. We represent X_1 and X_2 by the linear regression models

$$\begin{aligned} x_{i1} &= \mu_1 + \beta_{13}(x_{i3} - \bar{x}_3) + e_{i1} \\ x_{i2} &= \mu_2 + \beta_{23}(x_{i3} - \bar{x}_3) + e_{i2} \end{aligned} \qquad i = 1, \ldots, N \tag{1}$$

as if the values $x_{13}, \ldots, x_{N3}$ of X_3 are nonrandom quantities. Then the least squares predicted values of x_{i1} and x_{i2} are

$$\begin{aligned} \hat{x}_{i1} &= \bar{x}_1 + \hat{\beta}_{13}(x_{i3} - \bar{x}_3) \\ \hat{x}_{i2} &= \bar{x}_2 + \hat{\beta}_{23}(x_{i3} - \bar{x}_3) \end{aligned} \tag{2}$$

where

$$\hat{\beta}_{j3} = \frac{\sum_{i=1}^{N} (x_{ij} - \bar{x}_j)(x_{i3} - \bar{x}_3)}{\sum_{i=1}^{N} (x_{i3} - \bar{x}_3)^2} \qquad j = 1, 2 \tag{3}$$

The residuals

$$\begin{aligned} z_{i1} &= x_{i1} - \hat{x}_{i1} \\ &= x_{i1} - \bar{x}_1 - \hat{\beta}_{13}(x_{i3} - \bar{x}_3) \\ z_{i2} &= x_{i2} - \hat{x}_{i2} \\ &= x_{i2} - \bar{x}_2 - \hat{\beta}_{23}(x_{i3} - \bar{x}_3) \end{aligned} \tag{4}$$

represent the variation in X_1 and X_2 *after* their linear relationship with X_3 has been removed or if the effect of X_3 on them had been held constant. Since the residuals sum to zero their correlation coefficient is

$$r_{z_1,z_2} = \frac{\sum z_{i1}z_{i2}}{\sqrt{(\sum z_{i1}^2)(\sum z_{i2}^2)}}$$

$$= \frac{\sum [x_{i1} - \bar{x}_1 - \hat{\beta}_{13}(x_{i3} - \bar{x}_3)][x_{i2} - \bar{x}_2 - \hat{\beta}_{23}(x_{i3} - \bar{x}_3)]}{\sqrt{\{\sum [x_{i1} - \bar{x}_1 - \hat{\beta}_{13}(x_{i3} - \bar{x}_3)]^2\}\{\sum [x_{i2} - \bar{x}_2 - \hat{\beta}_{23}(x_{i3} - \bar{x}_3)]^2\}}} \qquad (5)$$

$$= \frac{\sum (x_{i1} - \bar{x}_1)(x_{i2} - \bar{x}_2) - \hat{\beta}_{12}\hat{\beta}_{23} \sum (x_{i3} - \bar{x}_3)^2}{\sqrt{[\sum (x_{i1} - \bar{x}_1)^2 - \hat{\beta}_{13}^2 \sum (x_{i3} - \bar{x}_3)^2][\sum (x_{i2} - \bar{x}_2)^2 - \hat{\beta}_{23}^2 \sum (x_{i3} - \bar{x}_3)^2]}}$$

If we replace $\hat{\beta}_{12}$, $\hat{\beta}_{23}$ by their expressions in (3) and divide numerator and denominator by

$$\sqrt{[\sum (x_{i1} - \bar{x}_1)^2][\sum (x_{i2} - \bar{x}_2)^2]}$$

we may express the partial correlation as

$$r_{12.3} = r_{z_1,z_2}$$

$$= \frac{r_{12} - r_{13}r_{23}}{\sqrt{(1 - r_{13}^2)(1 - r_{23}^2)}} \qquad (6)$$

in terms of the product moment correlations r_{j} of the original three variables. The notation $r_{kl.m}$ will denote the sample partial correlation of the kth and lth variables with the mth variable held constant.

Let us illustrate partial correlation by some contrived data. Let the variables X_1, X_2, X_3 have the following observations:

X_1	X_2	X_3
3	2	3
3	4	2
4	6	6
5	8	7
8	8	10
7	10	11
9	14	13
13	13	16
10	16	18
11	15	20

A glance at the data tells us that the three variables are highly positively correlated. The values of X_1 and X_2 predicted from X_3 by least squares are

$$\hat{x}_{i1} = 1.69 + 0.529x_{i3}$$

$$\hat{x}_{i2} = 1.71 + 0.744x_{i3}$$

The scatterplot of their residuals $z_{i1} = x_{i1} - \hat{x}_{i1}$, $z_{i2} = x_{i2} - \hat{x}_{i2}$ is shown in Figure 2-6. The points show no trend; by removing the effect of X_3 we have eliminated the correlation between the first two variables.

Now we compute the actual partial correlation coefficient by the formula

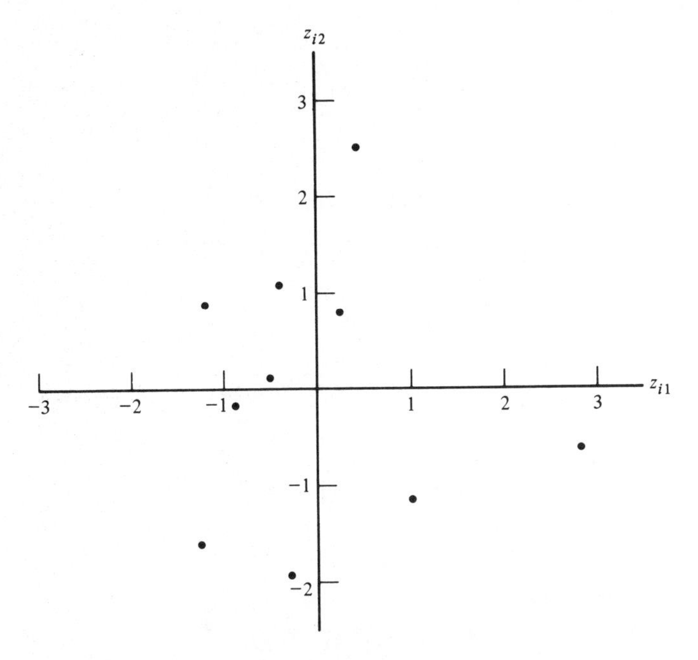

Figure 2-6 Residual scatterplot.

(6). The variables X_1, X_2, X_3 have this correlation matrix:

$$\mathbf{R} = \begin{bmatrix} 1 & 0.8926 & 0.9357 \\ & 1 & 0.9571 \\ & & 1 \end{bmatrix} \tag{7}$$

The partial correlation of X_1 and X_2 with X_3 held constant is

$$\begin{aligned} r_{12.3} &= \frac{0.8926 - (0.9357)(0.9571)}{\sqrt{[1 - (0.9357)^2][1 - (0.9571)^2]}} \\ &= -0.0289 \end{aligned}$$

The high correlation between the values of X_1 and X_2 disappears when the third variable is fixed. Any relationship of X_1 and X_2 may be due to the effect of the third variable on each. Of course, any causal relationships of X_1 and X_2 with X_3 that are imputed must be consistent with the substantive natures of the series. In the present example the values of X_1 and X_2 were in fact generated from those of X_3 plus a random disturbance, and the X_3 observations were perturbations from a linear trend.

Next we develop the sample partial correlation of two variables with several other variables held constant. We shall do this in terms of the correlation of the residuals of the first two variables from their least squares predictions by the remaining fixed variables. As before let the free variables be X_1 and X_2 and the constant variables be $X_3, \ldots, X_k$. Denote the N observed values of X_1 and X_2 by

$$\mathbf{x}_1 = \begin{bmatrix} x_{11} \\ \cdot \\ \cdot \\ \cdot \\ x_{N1} \end{bmatrix} \qquad \mathbf{x}_2 = \begin{bmatrix} x_{12} \\ \cdot \\ \cdot \\ \cdot \\ x_{N2} \end{bmatrix} \tag{8}$$

and the centered values of the fixed variables by

$$\mathbf{X}_3 = \begin{bmatrix} x_{13} - \bar{x}_3 & \cdots & x_{1k} - \bar{x}_k \\ \cdot & & \cdot \\ \cdot & & \cdot \\ \cdot & & \cdot \\ x_{N3} - \bar{x}_3 & \cdots & x_{Nk} - \bar{x}_k \end{bmatrix}$$

Then the residual vectors are

$$\begin{aligned} \mathbf{z}_1 &= \mathbf{x}_1 - \bar{x}_1\mathbf{j} - \mathbf{X}_3\hat{\boldsymbol{\beta}}_{13} \\ \mathbf{z}_2 &= \mathbf{x}_2 - \bar{x}_2\mathbf{j} - \mathbf{X}_3\hat{\boldsymbol{\beta}}_{23} \end{aligned} \tag{9}$$

where

$$\hat{\boldsymbol{\beta}}_{i3} = (\mathbf{X}_3'\mathbf{X}_3)^{-1}\mathbf{X}_3'\mathbf{x}_i \qquad i = 1, 2 \tag{10}$$

and in matrix notation,

$$\begin{aligned} \mathbf{z}_1 &= [\mathbf{I} - \mathbf{X}_3(\mathbf{X}_3'\mathbf{X}_3)^{-1}\mathbf{X}_3'](\mathbf{x}_1 - \bar{x}_1\mathbf{j}) \\ \mathbf{z}_2 &= [\mathbf{I} - \mathbf{X}_3(\mathbf{X}_3'\mathbf{X}_3)^{-1}\mathbf{X}_3'][\mathbf{x}_2 - \bar{x}_2\mathbf{j}) \end{aligned} \tag{11}$$

$\mathbf{j}$ is of course the column vector of N ones. The partial correlation is the ordinary product moment correlation of the corresponding elements of $\mathbf{z}_1$ and $\mathbf{z}_2$ or

$$\begin{aligned} r_{12.3\ldots k} &= \frac{\mathbf{z}_1'\mathbf{z}_2}{\sqrt{(\mathbf{z}_1'\mathbf{z}_1)(\mathbf{z}_2'\mathbf{z}_2)}} \\ &= \frac{(\mathbf{x}_1 - \bar{x}_1\mathbf{j})'[\mathbf{I} - \mathbf{X}_3(\mathbf{X}_3'\mathbf{X}_3)^{-1}\mathbf{X}_3'](\mathbf{x}_2 - \bar{x}_2\mathbf{j})}{\sqrt{\begin{aligned}&\{(\mathbf{x}_1 - \bar{x}_1\mathbf{j})'[\mathbf{I} - \mathbf{X}_3(\mathbf{X}_3'\mathbf{X}_3)^{-1}\mathbf{X}_3'](\mathbf{x}_1 - \bar{x}_1\mathbf{j})\} \\ &\cdot\{(\mathbf{x}_2 - \bar{x}_2\mathbf{j})'[\mathbf{I} - \mathbf{X}_3(\mathbf{X}_3'\mathbf{X}_3)^{-1}\mathbf{X}_3'](\mathbf{x}_2 - \bar{x}_2\mathbf{j})\}\end{aligned}}} \end{aligned} \tag{12}$$

Let us put the partial correlation in a form more suitable for computation and manipulation. Write the sum of squares and products in the second line of (12) as

$$\begin{aligned} a_{11} &= (\mathbf{x}_1 - \bar{x}_1\mathbf{j})'(\bar{\mathbf{x}}_1 - \bar{x}_1\mathbf{j}) = \sum (x_{i1} - \bar{x}_1)^2 \\ a_{22} &= (\mathbf{x}_2 - \bar{x}_2\mathbf{j})'(\bar{\mathbf{x}}_2 - \bar{x}_2\mathbf{j}) = \sum (x_{i2} - \bar{x}_2)^2 \\ a_{12} &= (\mathbf{x}_1 - \bar{x}_1\mathbf{j})'(\mathbf{x}_2 - \bar{x}_2\mathbf{j}) = \sum (x_{i1} - \bar{x}_1)(x_{i2} - \bar{x}_2) \\ \mathbf{a}_{13} &= \mathbf{X}_3'(\mathbf{x}_1 - \bar{x}_1\mathbf{j}) \\ \mathbf{a}_{23} &= \mathbf{X}_3'(\mathbf{x}_2 - \bar{x}_2\mathbf{j}) \\ \mathbf{A}_{33} &= \mathbf{X}_3'\mathbf{X}_3 \end{aligned} \tag{13}$$

These can be gathered into the symmetric partitioned sums of squares and products matrix

$$\mathbf{A} = \begin{bmatrix} a_{11} & a_{12} & \mathbf{a}_{13}' \\ a_{12} & a_{22} & \mathbf{a}_{23}' \\ \mathbf{a}_{13} & \mathbf{a}_{23} & \mathbf{A}_{33} \end{bmatrix} \tag{14}$$

The partial correlation coefficient is

$$r_{12.3\ldots k} = \frac{a_{12} - \mathbf{a}'_{13}\mathbf{A}_{33}^{-1}\mathbf{a}_{23}}{\sqrt{(a_{11} - \mathbf{a}'_{13}\mathbf{A}_{33}^{-1}\mathbf{a}_{13})(a_{22} - \mathbf{a}'_{23}\mathbf{A}_{33}^{-1}\mathbf{a}_{23})}} \tag{15}$$

Similarly, if we scale the elements of $\mathbf{A}$ to give the usual correlation matrix

$$R = \begin{bmatrix} 1 & r_{12} & \mathbf{r}'_{13} \\ r_{12} & 1 & \mathbf{r}'_{23} \\ \mathbf{r}_{13} & \mathbf{r}_{23} & \mathbf{R}_{33} \end{bmatrix} \tag{16}$$

of all k variables the partial correlation can be written in terms of the simple correlations as

$$r_{12.3\ldots k} = \frac{\mathbf{r}_{12} - \mathbf{r}'_{13}\mathbf{R}_{33}^{-1}\mathbf{r}_{23}}{\sqrt{(1 - \mathbf{r}'_{13}\mathbf{R}_{33}^{-1}\mathbf{r}_{13})(1 - \mathbf{r}'_{23}\mathbf{R}_{33}^{-1}\mathbf{r}_{23})}} \tag{17}$$

Example 2-12

Let us find the correlation of the variables Y (Draw a Person score) and X_1 (Level of Aspiration score) given in Table 2-5 with the remaining variables X_2 (Proverbs score), X_3 (number of homonyms), and X_4 (sum of the ratings in the test battery) held constant. The necessary simple correlations are given in the second matrix of (33), Section 2-5. In the present subscript notation

$$r_{12} = -0.6659 \qquad \mathbf{r}_{13} = \begin{bmatrix} 0.4135 \\ -0.3927 \\ 0.5922 \end{bmatrix} \qquad \mathbf{r}_{23} = \begin{bmatrix} -0.6910 \\ 0.7379 \\ -0.8381 \end{bmatrix} \tag{18}$$

$$\mathbf{R}_{33} = \begin{bmatrix} 1 & -0.3299 & 0.4885 \\ & 1 & -0.7378 \\ & & 1 \end{bmatrix}$$

and

$$\mathbf{r}'_{13}\mathbf{R}_{33}^{-1}\mathbf{r}_{23} = -0.5310 \qquad \mathbf{r}'_{13}\mathbf{R}_{33}^{-1}\mathbf{r}_{13} = 0.3744 \qquad \mathbf{r}'_{23}\mathbf{R}_{33}^{-1}\mathbf{r}_{23} = 0.8442 \tag{19}$$

The partial correlation is

$$r_{12.345} = \frac{-0.6659 - (-0.5310)}{\sqrt{(1 - 0.3744)(1 - 0.8442)}} = -0.4321 \tag{20}$$

The correlation between the performance tests Draw a Person and Level of Aspiration is less when the two verbal tests and the sum of the ratings are held at constant values.

Estimation of partial correlation. Our introduction of the partial correlation coefficient in terms of residuals from fitted regression equations has only provided a descriptive measure of correlation. To obtain hypothesis tests and confidence intervals for the partial correlation coefficient it will be necessary to begin with a population model for the original observations. Let us suppose that some $p + q$ variables of interest can be described by a random vector $\mathbf{X}$ partitioned as

$$\mathbf{X} = \begin{bmatrix} \mathbf{X}_1 \\ \mathbf{X}_2 \end{bmatrix} \tag{21}$$

where $\mathbf{X}_1$ is a $p \times 1$ vector and $\mathbf{X}_2$ is $q \times 1$. $\mathbf{X}$ has the $(p + q)$-dimensional multivariate normal distribution with density function $f(\mathbf{x}_1, \mathbf{x}_2)$ given by expression (2) of Section 2-3. We shall develop the partial correlations of the elements of $\mathbf{X}_1$ with those of $\mathbf{X}_2$ held fixed as the correlations in the conditional distribution of $\mathbf{X}_1$ for fixed $\mathbf{X}_2$. That conditional distribution has the density

$$h(\mathbf{x}_1 \mid \mathbf{x}_2) = \frac{f(\mathbf{x}_1, \mathbf{x}_2)}{f_2(\mathbf{x}_2)} \tag{22}$$

where $f_2(\mathbf{x}_2)$ is the density function of the elements of $\mathbf{X}_2$ alone. To find the conditional density we begin by partitioning the mean vector and covariance matrix of $\mathbf{X}$ in accordance with the dimensions of its subvectors $\mathbf{X}_1$ and $\mathbf{X}_2$. The partitioned parameters are

$$\boldsymbol{\mu} = \begin{bmatrix} \boldsymbol{\mu}_1 \\ \boldsymbol{\mu}_2 \end{bmatrix} \qquad \boldsymbol{\Sigma} = \begin{bmatrix} \boldsymbol{\Sigma}_{11} & \boldsymbol{\Sigma}_{12} \\ \boldsymbol{\Sigma}'_{12} & \boldsymbol{\Sigma}_{22} \end{bmatrix} \tag{23}$$

where $\boldsymbol{\mu}_1$ is $p \times 1$, $\boldsymbol{\mu}_2$ $q \times 1$, $\boldsymbol{\Sigma}_{11}$ $p \times p$, $\boldsymbol{\Sigma}_{12}$ $p \times q$, and $\boldsymbol{\Sigma}_{22}$ $q \times q$. $f_2(\mathbf{x}_2)$ is a multinormal density with mean vector $\boldsymbol{\mu}_2$ and covariance matrix $\boldsymbol{\Sigma}_{22}$. After some algebraic manipulation [see, for example, Morrison (1976, Section 3.4)] we find the conditional density of $\mathbf{X}_1$ for $\mathbf{X}_2$ fixed at the value $\mathbf{x}_2$ to be

$$\begin{aligned} h(\mathbf{x}_1 \mid \mathbf{x}_2) = {} & \frac{1}{(2\pi)^{p/2} \mid \boldsymbol{\Sigma}_{11.2} \mid^{1/2}} \\ & \cdot \exp\{-\tfrac{1}{2}[\mathbf{x}_1 - \boldsymbol{\mu}_1 - \boldsymbol{\Sigma}_{12}\boldsymbol{\Sigma}_{22}^{-1}(\mathbf{x}_2 - \boldsymbol{\mu}_2)]'\boldsymbol{\Sigma}_{11.2}^{-1} \\ & \cdot [\mathbf{x}_1 - \boldsymbol{\mu}_1 - \boldsymbol{\Sigma}_{12}\boldsymbol{\Sigma}_{22}^{-1}(\mathbf{x}_2 - \boldsymbol{\mu}_2)]\} \end{aligned} \tag{24}$$

or the density of a multinormal random variable with mean vector

$$\boldsymbol{\mu}_1 + \boldsymbol{\Sigma}_{12}\boldsymbol{\Sigma}_{22}^{-1}(\mathbf{x}_2 - \boldsymbol{\mu}_2) \tag{25}$$

and covariance matrix

$$\boldsymbol{\Sigma}_{11.2} = \boldsymbol{\Sigma}_{11} - \boldsymbol{\Sigma}_{12}\boldsymbol{\Sigma}_{22}^{-1}\boldsymbol{\Sigma}'_{12} \tag{26}$$

Because the elements of $\boldsymbol{\Sigma}_{11.2}$ are the variances and covariances of the variates in $\mathbf{X}_1$ with those of $\mathbf{X}_2$ constrained to be constant the correlations found from $\boldsymbol{\Sigma}_{11.2}$ are the partial correlations of $\mathbf{X}_1$ with $\mathbf{X}_2$ fixed. Their matrix is given by

$$\begin{aligned} \mathbf{P}_{11.2} &= \mathbf{D}^{-1/2}(\boldsymbol{\Sigma}_{11} - \boldsymbol{\Sigma}_{12}\boldsymbol{\Sigma}_{22}^{-1}\boldsymbol{\Sigma}'_{12})\mathbf{D}^{-1/2} \\ &= \mathbf{D}^{-1/2}\boldsymbol{\Sigma}_{11.2}\mathbf{D}^{-1/2} \end{aligned} \tag{27}$$

where

$$\mathbf{D} = \operatorname{diag}(\boldsymbol{\Sigma}_{11} - \boldsymbol{\Sigma}_{12}\boldsymbol{\Sigma}_{22}^{-1}\boldsymbol{\Sigma}'_{12})$$

is the diagonal matrix consisting of the diagonal elements of $\boldsymbol{\Sigma}_{11.2}$, and $\mathbf{D}^{-1/2}$ contains the reciprocals of the square roots of its elements. Pre- and post-multiplication by $\mathbf{D}^{-1/2}$ has the effect of scaling the conditional covariances to product moment correlations.

Since our motivation is the application of statistical theory to data, we must find a way to estimate $\mathbf{P}_{11.2}$. As in our development of multiple correlation in Section 2-5 suppose that we have drawn N independent observation vectors $\mathbf{x}_i$ from the $(p + q)$-dimensional multinormal population. The observations have been partitioned as

$$\begin{aligned} \mathbf{x}'_i &= [\mathbf{x}'_{i1}, \mathbf{x}'_{i2}] \\ &= [x_{i1}, \ldots, x_{ip}, x_{i,p+1}, \ldots, x_{i,p+q}] \qquad i = 1, \ldots, N \end{aligned} \tag{28}$$

as in the partitioning of $\boldsymbol{\mu}$ in (23). From these we compute the matrices of sums of squares and products

$$\mathbf{A}_{rs} = \sum_{i=1}^{N} (\mathbf{x}_{ir} - \bar{\mathbf{x}}_r)(\mathbf{x}_{is} - \bar{\mathbf{x}}_s)' \qquad s = 1, 2; \quad r = 1, 2 \tag{29}$$

where $\bar{\mathbf{x}}_1$ and $\bar{\mathbf{x}}_2$ are the mean vectors of the variates in the first and second groups, respectively. The $\mathbf{A}_{rs}$ are the submatrices of the matrix

$$\mathbf{A} = \begin{bmatrix} \mathbf{A}_{11} & \mathbf{A}_{12} \\ \mathbf{A}'_{12} & \mathbf{A}_{22} \end{bmatrix} \tag{30}$$

corresponding to the partitioned form of $\boldsymbol{\Sigma}$ in (23). The estimator of those parameters is the sample covariance matrix

$$\begin{aligned} \mathbf{S} &= \left[\frac{1}{N-1}\right]\mathbf{A} \\ &= \begin{bmatrix} \mathbf{S}_{11} & \mathbf{S}_{12} \\ \mathbf{S}'_{12} & \mathbf{S}_{22} \end{bmatrix} \end{aligned} \tag{31}$$

and the estimator of the similarly partitioned matrix of simple correlations of the $p + q$ variates is

$$\mathbf{R} = \begin{bmatrix} \mathbf{R}_{11} & \mathbf{R}_{12} \\ \mathbf{R}'_{12} & \mathbf{R}_{22} \end{bmatrix} \tag{32}$$

Then the estimator of the partial correlation matrix (27) is

$$\begin{aligned} \mathbf{R}_{11.2} &= \mathbf{D}_S^{-1/2}(\mathbf{S}_{11} - \mathbf{S}_{12}\mathbf{S}_{22}^{-1}\mathbf{S}'_{12})\mathbf{D}_S^{-1/2} \\ &= \mathbf{D}_R^{-1/2}(\mathbf{R}_{11} - \mathbf{R}_{12}\mathbf{R}_{22}^{-1}\mathbf{R}'_{12})\mathbf{D}_R^{-1/2} \end{aligned} \tag{33}$$

in which

$$\begin{aligned} \mathbf{D}_S &= \operatorname{diag}(\mathbf{S}_{11} - \mathbf{S}_{12}\mathbf{S}_{22}^{-1}\mathbf{S}'_{12}) \\ \mathbf{D}_R &= \operatorname{diag}(\mathbf{R}_{11} - \mathbf{R}_{12}\mathbf{R}_{22}^{-1}\mathbf{R}'_{12}) \end{aligned} \tag{34}$$

$\mathbf{R}_{11.2}$ is a symmetric positive semidefinite matrix with all the properties of a matrix of simple correlations.

Example 2-13

More projective test scores. Let us examine a larger number of psychological test scores in the battery administered by Singer (1963) that we have used in previous examples. This time we have chosen nine tests classified as "projective," "performance," or "verbal." These are the test score variables:

Group	Test score
Projective	1. Sentence Completion
	2. Emotional Projection
	3. Family Scene
	4. Problem Situation
	5. Thematic Apperception Test
Performance	6. Draw a Person
	7. Level of Aspiration (score)
Verbal	8. Proverbs
	9. Homonyms

We shall partition the 9×9 correlation matrix of the variables according to the three types of tests:

$$\mathbf{R} = \begin{bmatrix} \mathbf{R}_{11} & \mathbf{R}_{12} & \mathbf{R}_{13} \\ \mathbf{R}'_{12} & \mathbf{R}_{22} & \mathbf{R}_{23} \\ \mathbf{R}'_{13} & \mathbf{R}'_{23} & \mathbf{R}_{33} \end{bmatrix} \tag{35}$$

These are the submatrices:

$$\mathbf{R}_{11} = \begin{bmatrix} 1 & 0.39 & 0.46 & 0.49 & 0.63 \\ & 1 & 0.53 & 0.72 & 0.73 \\ & & 1 & 0.63 & 0.74 \\ & & & 1 & 0.73 \\ & & & & 1 \end{bmatrix}$$

$$\mathbf{R}_{12} = \begin{bmatrix} -0.30 & -0.49 \\ -0.36 & -0.52 \\ -0.18 & -0.50 \\ -0.37 & -0.55 \\ -0.36 & 0.53 \end{bmatrix} \qquad \mathbf{R}_{13} = \begin{bmatrix} 0.39 & -0.40 \\ 0.45 & -0.54 \\ 0.36 & -0.51 \\ 0.35 & -0.41 \\ 0.60 & -0.65 \end{bmatrix} \tag{36}$$

$$\mathbf{R}_{22} = \begin{bmatrix} 1 & 0.41 \\ 0.41 & 1 \end{bmatrix} \qquad \mathbf{R}_{33} = \begin{bmatrix} 1 & -0.74 \\ -0.74 & 1 \end{bmatrix} \qquad \mathbf{R}_{23} = \begin{bmatrix} -0.39 & 0.59 \\ -0.33 & 0.49 \end{bmatrix}$$

The correlations have been rounded to two places to save space, although six decimal places were used in the subsequent computations.

Presumably the subject's responses to the projective tests will depend on his ability to function intellectually as measured by the performance cognitive tests and his communication skills as reflected by the verbal tests. Let us begin by calculating the correlations among the projective tests when the performance test scores are held constant. That matrix of partial correlations is

$$\mathbf{R}_{11.2} = \begin{bmatrix} 1 & 0.16 & 0.29 & 0.29 & 0.49 \\ & 1 & 0.38 & 0.59 & 0.60 \\ & & 1 & 0.50 & 0.67 \\ & & & 1 & 0.61 \\ & & & & 1 \end{bmatrix} \tag{37}$$

The partial correlations are lower than the original correlations in $\mathbf{R}_{11}$. The intellectual qualities measured by the performance tests appear to stretch the ranges of the projective test scores and by so doing contribute to their intercorrelations. It is interesting to note that the reduction in the simple correlations has not been uniform for all pairs of projective tests.

Now we shall determine the effect on the projective test interrelations with the two verbal tests fixed. That partial correlation matrix is

$$\mathbf{R}_{11.3} = \begin{bmatrix} 1 & 0.21 & 0.33 & 0.39 & 0.52 \\ & 1 & 0.35 & 0.64 & 0.58 \\ & & 1 & 0.54 & 0.65 \\ & & & 1 & 0.68 \\ & & & & 1 \end{bmatrix} \tag{38}$$

The correlations among the projective tests are lower, although generally not as much as in the case with fixed performance scores. Finally, let us hold both performance and verbal test scores constant. The partial correlations are

$$\mathbf{R}_{11.2,3} = \begin{bmatrix} 1 & 0.09 & 0.25 & 0.26 & 0.45 \\ & 1 & 0.28 & 0.58 & 0.52 \\ & & 1 & 0.51 & 0.60 \\ & & & 1 & 0.64 \\ & & & & 1 \end{bmatrix} \tag{39}$$

With only two exceptions these partial correlations are the lowest. The correlations among the five projective tests apparently have strong verbal and performance cognitive components.

Inferences about partial correlations. The distribution of a single sample partial correlation computed from N independent observations from the multinormal population and with k other variables held constant is merely that of a simple product moment correlation calculated from $N - k$ observation pairs. When $\rho_{ij\cdot 1\ldots k} = 0$, the density of $r \equiv r_{ij\cdot 1\ldots k}$ is that of (20), Section 1-6, with $n = N - 2 - k$. We may test the hypothesis

$$H_0\colon \quad \rho_{ij\cdot 1\ldots k} = 0 \tag{40}$$

of no correlation between the ith and jth variates with k other variates held fixed by referring the sample correlation to the appropriate critical value. As for simple correlation we would reject H_0 in favor of

$$H_1\colon \quad \rho_{ij\cdot 1\ldots k} > 0 \tag{41}$$

at the α level if

$$r_{ij\cdot 1\ldots k} > r_{\alpha;N-2-k} \tag{42}$$

Rules for the one- and two-sided alternatives

$$H_1\colon \quad \rho_{ij\cdot 1\ldots k} < 0 \qquad \text{and} \qquad H_1\colon \quad \rho_{ij\cdot 1\ldots k} \neq 0 \tag{43}$$

identical with those of (24) and (26), Section 1-6, hold, with $n = N - 2 - k$ in the critical values.

When the population partial correlation is not equal to zero, its sample estimator has the same noncentral distribution as the simple correlations but with parameters and degrees of freedom $n - k = N - 2 - k$ for the usual single random sample of N observation vectors. Similarly,

$$z = \tanh^{-1}(r_{ij\cdot 1\ldots k}) \tag{44}$$

has a large-sample normal distribution with mean

$$\zeta = \tanh^{-1}(\rho_{ij\cdot 1\ldots k}) \tag{45}$$

and approximate variance

$$\begin{aligned} \operatorname{Var}(z) &= \frac{1}{n - 1 - k} \\ &= \frac{1}{N - 3 - k} \end{aligned} \tag{46}$$

for a single random sample. We may test the hypothesis that the partial correlation has a particular value by computing its z-value and applying the decision rule (49), (52), or (54) of Section 1-6 for the appropriate alternative. The $100(1-\alpha)\%$ confidence interval for a single partial correlation is

$$\tanh\left(z - \frac{z_{\alpha/2}}{\sqrt{N-3-k}}\right) \leq \rho \leq \tanh\left(z + \frac{z_{\alpha/2}}{\sqrt{N-3-k}}\right) \tag{47}$$

where the values of tanh can be found from Table 6, Appendix A, or its definition (64) of Section 1-6.

Example 2-14

Let us see what inferences can be made about the population counterpart of the psychological test partial correlations in the matrices (37)–(39). We shall begin by testing the hypotheses

$$H_0: \rho_{ij\cdot rs} = 0$$

against the one-sided alternative of positive population values for each partial correlation. Admittedly, we have chosen the more powerful one-sided alternative because all of the sample partial correlations are positive. In a more general situation two-sided tests would be more conservative. We shall choose an error rate of $\alpha = 0.01$ for each test and ignore the problem of error rates for multiple testing.* The original simple correlations were computed from $N = 46$ independent subjects, so that linear interpolation in Table 5, Appendix A, gives the critical value $r_{0.01;42} = 0.35$ for the correlations of the matrices (37) and (38) in which two other test scores were held constant. The critical value for the correlations in (39) for which the four verbal and performance tests were fixed is $r_{0.01;40} = 0.358$. When the performance tests are held constant, only the correlations of the Sentence Completion test with the Emotional Projection, Family Scene, and Problem Situation scores do not exceed the critical value. At the 0.01 level their population correlations might well be taken as zero. When the verbal test scores are fixed only the partial correlations of Sentence Completion with Emotional Projection and Family Scene, and equivocally, Emotional Projection with Family Scene, do not exceed the critical value: It is doubtful that those tests have even low to moderate population partial correlations. Similar conclusions follow from comparing the correlations of (39) with the critical value 0.358: Zero partial correlations would be tenable for Sentence Completion with Emotional Projection, Family Scene, and Problem Situation, and Emotional Projection with Family Scene. The remaining correlations probably describe covariation which could not have been removed by holding the performance and verbal tests at fixed scores.

2-7 PACKAGED STATISTICAL PROGRAMS FOR REGRESSION ANALYSIS

The computations for most of the examples in this text were carried out by short functions or ad hoc expressions in the APL programming language. Some analyses were run on local programs written for course use. We have assumed that the reader will have access to a computer and, at the least, subroutines for the matrix operations that will give the estimates and statistics for regression analysis. Alternatively, one might work the exercises with current statistical

*The multiple comparison or simultaneous testing problem will be discussed in Chapter 3; for its treatment with regard to tests on many correlations, see Morrison (1976, Section 3.8).

packages. Standard results can be obtained more quickly and directly, albeit without checks provided by intermediate steps in the computations, and within the rigidity of the output the package designer has chosen to prepare or print. We describe the regression capabilities of three available statistical packages in this section.

BMDP biomedical computer programs—P series 1977. The BMDP P series (1977) statistical program package contains two basic programs for carrying out the regression and correlation analyses of Chapters 1 and 2: P6D for bivariate scatterplots and simple regression, and P1R for multiple linear regression.

P6D *Bivariate (Scatter) Plots* prints scatterplots for pairs of variables or all possible pairs of a set of variables. The program will also print for each pair of variables their correlation coefficient, means, standard deviations, regression equations of Y against X and X against Y, and the residual mean square σ^2 for each fitted regression line.

P1R *Multiple Linear Regression* will fit the regression model by least squares to a data set consisting of N complete observations on a dependent variable and p-predictors. The output lists the mean, standard deviation, coefficient of variation, minimum, and maximum for each variable's observations. For the multiple regression the program prints out the coefficient of determination, the multiple correlation, and the residual standard deviation $\hat{\sigma}$. Next the program prints the regression and error sums of squares SSR and SSE, their mean squares SSR/p and SSE/$(N - p - 1)$, the F-ratio (28) of Section 2-3, and the significance probability of that F-statistic. Finally, the fitted regression model is specified in a table containing the intercept at the origin of the independent variable scales, the regression coefficients, their standard deviations, the beta coefficients for a regression based on standardized variables, the t-statistics (13) of Section 2-3, and the two-tail significance probabilities of those t-values. Optional outputs include the covariance and correlation matrices for the $p + 1$ variables, the correlation matrix for the regression coefficients, the residuals, predicted values $\hat{y}_i$, variable values for each observation vector, and various scatter and normal probability plots of the residuals.

IDA (Interactive Data Analysis). The IDA language has commands for executing least squares regression analysis. These are described in Chapter 9 of the *User's Manual for IDA* (Ling and Roberts, 1980). The IDA system was constructed to apply automatically many tests for model validity on the residuals from the fitted regression model. The program prints out warnings when gross violations of the normality, constant-variance, or independence assumptions on the residuals have occurred.

The command for a multiple regression analysis does not print out the results of fitting the model. Instead, the user selects a new command to print a particular subset of the results. These include commands for the overall F-test, one for the table of regression coefficients, beta coefficients, standard deviations, and t-statistics, another for $\hat{\sigma}$ and adjusted and unadjusted R and R^2, and commands for printing the covariance or correlation matrices of the regression coefficients. Means, standard deviations, variances, and covariances of the original variables would have to be obtained from commands for some other summary and single-sample statistics.

IDA provides a command for printing the residual, predicted value, and y_i observation for each sampling unit. A command will print the normal probability plot of the residuals or any other variable. The residuals can be plotted against the $\hat{y}_i$ or any of the original variables. The independence assumption on the disturbances can be checked by the Durbin–Watson test we shall meet in Chapter 6. Commands are available for many of the regression analyses and tests we shall describe in the next three chapters. In all of them the essential feature of the IDA language is *interactive* analysis of the data: Techniques and their commands are chosen as the results unfold from the previous steps.

SAS Statistical Analysis System. Three regression programs from the SAS Statistical Analysis System (1979) can be used for the computations of Chapters 1 and 2. A fourth procedure, STEPWISE, can be employed to form optimal sets of independent variables by some of the methods described in Chapter 5.

1. GLM *General Linear Model Analysis.* The GLM procedure has been designed to analyze very general linear models for the analysis of variance methods in Chapters 8 through 10, but it will provide estimates and tests for the special case of multiple regression. The output of the GLM procedure begins with the breakdown of the total sum of squares of the centered dependent variable into SSR and SSE, and the F-statistic for testing the overall hypothesis of no regression relationship. Next come R^2, $\hat{\sigma}/\bar{y}$, $\hat{\sigma}$, and $\bar{y}$. The program then prints two kinds of F-statistics: The second set provides tests of the individual hypotheses of zero regression coefficients. Finally, a table is printed whose columns contain the intercept and regression coefficients, their statistics t_j for the individual zero-coefficient hypotheses, the two-sided significance probabilities, and the estimated standard deviations of the coefficients. This portion of the SAS regression procedure does not print means and other statistics for the independent variables in the regression model.

2. CORR *Correlation Analysis Procedure.* The CORR procedure program computes correlations and other descriptive statistics for sets of observations. The correlations are the ordinary Pearson product moment coefficients introduced in Chapter 1, although Spearman or Kendall correlations can be specified. The printout begins with a table giving the number of observations, mean, standard deviation, median, minimum, and maximum for each variable. Next comes the complete correlation matrix, with its rows and columns labeled with the variable names. Under each correlation is the two-sided significance probability for the test of zero population correlation. When the option COV is specified in the CORR procedure statement the covariance matrix is also printed. The program contains options for dealing with data matrices with missing observations: Each correlation is computed only from its complete pairs of observations, although elements of the incomplete pairs may be used for

computing the other correlations. For that reason the correlation, covariance, and sums of squares and products matrices are not necessarily positive semidefinite. Multiple and partial correlations computed from them may be undefined. Alternatively, an option NOMISS will compute the correlation matrix only from the complete sampling units in the original data.

3. RSQUARE *Procedure for All Possible Regressions.* The RSQUARE procedure will provide all possible R^2 values for a dependent variable and a set of independent variables. The procedure begins with the p simple regressions, then computes the $p(p-1)/2$ regressions with two independent variables, and so on until all 2^p models (including that with only an intercept) have been fitted. The regression coefficients and other statistics are not given in the printout. Presumably, one would choose those variable combinations with large values of R^2 and make the complete regression analysis by the GLM procedure.

2-8 REFERENCES

ANDERSON, T. W. (1958): *An Introduction to Multivariate Statistical Analysis*, John Wiley & Sons, Inc., New York.

ANSCOMBE, F. J. (1961): Examination of residuals, *Proceedings of the Fourth Berkeley Symposium on Mathematical Statistics and Probability*, pp. 1–36, University of California Press, Berkeley.

ANSCOMBE, F. J. (1973): Graphs in statistical analysis, *The American Statistician*, vol. 27, no. 1, pp. 17–20.

ANSCOMBE, F. J., and J. W. TUKEY (1963): The examination and analysis of residuals, *Technometrics*, vol. 5, pp. 141–160.

BMDP Biomedical Computer Programs—P-Series (1977), Health Sciences Computing Facility, University of California, University of California Press, Los Angeles.

DURAND, D. (1954): Joint confidence regions for multiple regression coefficients, *Journal of the American Statistical Association*, vol. 49, pp. 130–146.

EZEKIEL, M. (1930): *Methods of Correlation Analysis*, John Wiley & Sons, Inc., New York.

EZEKIEL, M., and K. A. FOX (1959): *Methods of Correlation and Regression Analysis*, 3rd ed., John Wiley & Sons, Inc., New York.

FISHER, R. A. (1928): The general sampling distribution of the multiple correlation coefficient, *Proceedings of the Royal Society of London, Series A*, vol. 121, pp. 654–673.

HEALD, F. P., G. ARNOLD, W. SEABOLD, and D. MORRISON (1967): Plasma levels of free fatty acids in adolescents, *American Journal of Clinical Nutrition*, vol. 20, pp. 1010–1014.

KENDALL, M. G., and A. STUART (1961): *The Advanced Theory of Statistics*, vol. 2, Charles Griffin & Company Ltd., London.

KRAMER, K. H. (1963): Tables for constructing confidence limits on the multiple correlation coefficient, *Journal of the American Statistical Association*, vol. 58, pp. 1082–1085.

LEE, Y.-S. (1972): Tables of upper percentage points of the multiple correlation coefficient, *Biometrika*, vol. 59, pp. 175–189.

LING, R. F., and H. V. ROBERTS (1980): *User's Manual for IDA*, The Scientific Press, Palo Alto, Calif.

LUCAS, J. M. (1973): Large differences between partial correlation coefficients, *The American Statistician*, vol. 27, pp. 77–78.

MOOD, A. M., F. A. GRAYBILL, and D. C. BOES (1974): *Introduction to the Theory of Statistics*, 3rd ed., McGraw-Hill Book Company, New York.

MORRISON, D. F. (1976): *Multivariate Statistical Methods*, 2nd ed., McGraw-Hill Book Company, New York.

OLSHEN, R. A. (1973): The conditional level of the *F*-test, *Journal of the American Statistical Association*, vol. 68, pp. 692–698.

PEARSON, E. S., and H. O. HARTLEY (1972): *Biometrika Tables for Statisticians*, vol. 2, Cambridge University Press, Cambridge, England.

Philadelphia Inquirer, Philadelphia, Penn., July 26, 1972.

PLACKETT, R. L. (1949): A historical note on the method of least squares, *Biometrika*, vol. 36, pp. 458–460.

SAS INSTITUTE (1979): *SAS User's Guide, 1979 Edition*, SAS Institute, Inc., Raleigh, N.C.

SINGER, M. T. (1963): Personality measurements in the aged, in J. E. Birren et al. (eds.), *Human Aging: A Biological and Behavioral Study*, pp. 217–249, U.S. Government Printing Office, Washington, D.C.

2-9 EXERCISES

1. The three quantities

Y: argon-40 concentration in mica in (mol/g $\times 10^9$)

X_1: weight percent of potassium in rock sample

X_2: (potassium/rubidium) weight ratio of rock specimen

were measured in $N = 8$ samples of mica from granite specimens.* These values were obtained:

Y	X_1	X_2
4.392	4.17	349
1.677	1.79	124
1.659	1.76	632
0.783	1.09	415
0.613	0.49	304
4.114	3.32	292
1.800	2.25	506
2.500	3.87	501

SOURCE: Data reproduced with the kind permission of Kenneth A. Foland.

*I am indebted to Simon George for this example.

(a) Fit a linear regression equation to the data to determine the effects of X_1 and X_2 on the argon-40 concentration.

(b) What is the coefficient of determination in the multiple regression equation?

(c) Test the individual hypotheses that the population regression coefficients of X_1 and X_2 are zero and decide whether the regression equation can be reduced to a single independent variable.

(d) The geochemical hypothesis

$$H_1: \text{Ar}^{40} \text{ and } K \text{ were simultaneously absorbed by the forming mica}$$

would be associated with a high correlation of Y and X_2 and a significant multiple regression coefficient of Y with X_2. Do the data appear to support H_1?

2. From the data in Table 2-4 find the simple regression equation for predicting the Draw a Person score (Y) from performance ratings total (X_4). Compute the predicted values of Y for the subjects and compare their residuals with those of the multiple regression equation in Table 2-4.

3. Data were collected on $N = 56$ employees with the same job classification in a certain company to determine the effect of gender on salary. The dependent variable Y was annual salary in dollars at the date of the study; these were the independent variables:

$$X_1 = \begin{cases} 1 & \text{if male} \\ 0 & \text{if female} \end{cases}$$

X_2 = length of service in years

X_3 = educational level
(1 = high school, 2 = some college, 3 = Bachelor's degree,
4 = Master's degree, 5 = doctorate)

These statistics were obtained from the regression analysis of the data:

Variable	Mean	S.D.	$\hat{\beta}_j$	$\sqrt{\widehat{\text{Var}}(\hat{\beta}_j)}$	t_j
Y	20,911	3,054	—	—	—
X_1	0.6786	0.22208	1770	806.4	2.20
X_2	5.34	2.8913	260.9	126.8	2.06
X_3	3.57	1.2772	−305.58	276.7	−1.10

The correlation matrix of the four variables had these values:

$$\begin{bmatrix} 1 & 0.3509 & 0.4466 & -0.3005 \\ & 1 & 0.2667 & 0.0086 \\ & & 1 & -0.5043 \\ & & & 1 \end{bmatrix}$$

The analysis of variance for testing the hypothesis of no relation of Y to X_1, X_2, X_3 had these values:

Source	Sum of squares $\times 10^{-8}$	d.f.	Mean square $\times 10^{-6}$	F
Regression	1.40694	3	46.898	6.55
Error	3.72258	52	7.1588	
Total	5.12952	55	—	

(a) State the estimated multiple regression equation for predicting salary from the three independent variables.

(b) Test the individual hypotheses of zero population regression coefficients against two-sided alternatives. Use $\alpha = 0.05$.

(c) What proportion of the variation in salaries do the three independent variables explain?

(d) Use the correlation matrix and the standard deviations of the variables to construct the vector $\mathbf{X}_1'\mathbf{y}$ and the matrix $\mathbf{X}_1'\mathbf{X}_1$ for Y and the first two independent variables. Find the regression coefficients for X_1 and X_2 alone and their coefficient of determination. Does the sex of the employee still have a significant effect on salary?

(e) What other variables should have been included in the predictor set?

4. The Environmental Protection Agency has published gas mileage measurements for 1978-model cars and trucks sold in the United States. Data for $N = 35$ models selected from the lists of middle-size and large passenger cars have been arranged as these independent variables:

$$X_1 = \begin{cases} 0 & \text{middle-size} \\ 1 & \text{large} \end{cases}$$

$$X_2 = \text{number of cylinders}$$

$$X_3 = \text{engine volume (cubic inches)}$$

The dependent variables are

$$Y_1 = \text{city mileage (miles per gallon)}$$

$$Y_2 = \text{highway mileage}$$

$$Y_3 = \text{combined city and highway mileage}$$

All models have automatic transmissions and standard carburetors. The data are shown in Table 2-7.

(a) Estimate the regression coefficients of X_1, X_2, and X_3 when Y_1, Y_2, and Y_3 are used successively as the dependent variables.

(b) Compute the coefficients of determination for each of the separate regression analyses of (a).

(c) If the 35 auto models are treated as independent sampling units, carry out the usual tests of hypotheses on the set of regression coefficients and the individual coefficients for each of the three analyses of (a).

(d) Find the predicted mileages and their residuals from each of the regression models.

TABLE 2-7 Fuel Consumption Characteristics

Model	X_1	X_2	X_3	Miles per gallon Y_1 (city)	Y_2 (highway)	Y_3 (combined)
1	0	4	140	23	33	26
2	0	6	231	19	27	22
3	0	6	200	19	26	22
4	0	6	200	19	25	21
5	0	8	260	19	27	22
6	0	8	305	17	25	20
7	0	6	225	17	22	19
8	0	8	302	16	23	19
9	0	6	231	16	28	19
10	0	8	318	15	22	18
11	0	8	302	15	22	17
12	0	8	318	14	21	16
13	0	8	351	14	20	16
14	0	8	400	13	20	15
15	0	8	400	13	18	15
16	0	8	400	13	19	15
17	0	8	360	12	17	15
18	0	8	360	10	17	13
19	0	8	425	10	15	11
20	0	8	440	10	14	11
21	1	8	260	18	25	21
22	1	6	231	17	25	20
23	1	8	301	17	24	20
24	1	6	231	17	25	20
25	1	8	305	16	22	19
26	1	6	250	17	24	19
27	1	8	350	16	23	19
28	1	8	350	15	22	18
29	1	8	403	14	20	17
30	1	8	351	13	21	16
31	1	8	425	13	19	15
32	1	8	360	12	17	14
33	1	8	400	11	18	14
34	1	8	460	12	17	14
35	1	8	440	10	16	12

(e) What other variables might have been included in the regression models? What use might be made of transformations?

(f) Find the estimates of the parameters of the linear regression equation relating Y_3 to Y_2 and Y_1.

(g) Now assume that the combined mileage Y_3 was formed by the equation

$$Y_{i3} = (1 - p) Y_{i1} + p Y_{i2} + e_i$$

where e_i represents round-off error for the ith car model, and p is an unknown proportion parameter. Find an estimator of p with "good" properties of the sort described in this chapter.

5. From the correlation matrices of (36), Section 2-6, carry out the following analyses:
   (a) Find the partial correlation of the performance tests with the verbal tests held constant and vice versa.
   (b) Test the hypothesis of zero population correlation for the two partial correlations in (a) and find the 95% confidence intervals for the correlation parameters.
6. Compute the 99% confidence intervals for the partial correlation of the Problem Situation and Thematic Apperception Test scores of Section 2-6 with (a) the performance test scores, (b) the verbal test scores, and (c) both verbal and performance tests held constant.
7. Lucas (1973) has given an engineering example of a partial correlation with sign opposite that of its original simple correlation. The three variables considered by Lucas are the characteristics of artificial fibers given by

$$X_1 = \text{tenacity} = \frac{\text{force applied to fiber}}{\text{cross-sectional area of fiber}}$$

$$X_2 = \text{elongation} = \frac{\text{increase in length}}{\text{original length}}$$

$$X_3 = \text{draw ratio} = \frac{\text{stretched length}}{\text{original length}}$$

In experiments the draw ratio, or the amount of stretching of the fiber during its production, is often varied; Lucas gives the correlation matrix of X_1, X_2, X_3 from such an experiment as

$$\begin{bmatrix} 1 & -0.55 & 0.90 \\ & 1 & -0.80 \\ & & 1 \end{bmatrix}$$

   (a) Find the correlation of tenacity and elongation when the draw ratio is held constant.
   (b) What is the smallest sample size in which $H_0: \rho_{12.3} = 0$ will be rejected at the 0.01 level in favor of the alternative of a nonzero partial correlation?
8. The following 10 independent sets of observations on a variable Y and three predictor variables H, L, and W were collected on certain sampling units:

Y	H	L	W
0.8	3	12	4
1.3	3	18	4
0.9	3	9	9
1.8	3	18	9
3.0	3	27	9
2.4	3	18	14
2.2	3	15	14
1.9	6	9	10
3.9	6	18	10
7.7	6	36	10

   (a) Fit the usual linear regression function in the three predictor variables for the values of Y.

(b) Can the regression function in (a) be improved by adding squared or product terms in the predictors?

(c) Now let us consider regression functions within the context of the four variables: H, L, and W are nominal dimensions in quarter inches of blocks of pine wood, whereas Y is the weight of the block in ounces on a crude scale. Fit the regression plane to the natural logarithms of the weights with the dimensions also transformed to logs.

(d) Fit the linear model

$$y_i = \beta_0 + \beta(h_i l_i w_i) + e_i$$

relating weight to volume and compare its estimates and properties with those found for the model fitted in (c).

9. The observations on the three fictitious variables X_1, X_2, X_3 used to illustrate partial correlation in Section 2-6 contain linear trends. Their correlations with the trend variable $X_4 = t$, $t = 1, \ldots, 10$, are

$$r_{14} = 0.9286 \qquad r_{24} = 0.9762 \qquad r_{34} = 0.9907$$

Find the partial correlations of X_1, X_2, and X_3 with X_4 held constant. Note how the very large correlations of those variables with X_4 have increased the magnitudes of the partial correlations.

10. Compare the adjusted coefficient of determination R_a^2 with

$$R^2 - E(R^2)$$

where the expectation has been computed for the case of $\rho^2 = 0$ in the multinormal population.

11. Compute $E(R^2)$ under the assumption that the values of the independent variables in the data matrix $\mathbf{X}_1$ are *fixed* rather than observations on random variables and compare the expectation with that obtained in the multinormal model.

12. Durand (1954) found the following matrix of sums of squares and products about the means of the variables log shares (X_1), log dividends (X_2), log prices (X_3), and log capital (Y) of $N = 17$ New York bank stocks:

$$\begin{bmatrix} 3.25538 & 3.51281 & 3.42601 & -0.30838 \\ & 7.25942 & 3.55375 & -3.65961 \\ & & 3.69840 & -0.16441 \\ & & & 3.23758 \end{bmatrix}$$

(a) Compute the regression coefficients of the three predictors.

(b) Test the hypotheses of zero population regression coefficients and the hypothesis that all three coefficients are simultaneously zero.

(c) Compute R^2 and the adjusted coefficient of determination R_a^2 and compare their values.

(d) Find the correlations among the four variables.

(e) Compute the partial correlations of Y with each of the predictors, holding the other two predictors constant.

13. For an article on local restaurants the Food Editor of the *Philadelphia Inquirer* (1972) collected data on hamburgers sold at $N = 18$ sandwich shops, fast-food outlets, and other restaurants. These values of Y = price, X_1 = weight of meat patty (ounces), and X_2 = amount of side items (0 = none, . . . , 3 = French fries or salad, etc.) were obtained for hamburgers purchased at the restaurants:

Restaurant	Y	X_1	X_2
1	0.20	1.50	0
2	0.60	3.25	0
3	0.22	1.00	0
4	0.50	2.25	0
5	0.55	1.50	0
6	0.25	2.00	0
7	0.70	3.50	0
8	0.80	4.00	0
9	0.85	4.00	1
10	1.35	4.00	1
11	0.65	3.50	1
12	0.80	3.75	1
13	0.85	3.88	1
14	1.55	3.75	1
15	1.75	5.00	2
16	1.15	3.00	3
17	1.55	4.00	3
18	1.75	5.88	3

(a) Plot Y against X_1 with the points indicated by the values 0, 1, 2, and 3 of X_2.
(b) Find the linear regression equation of Y as a function of X_1 and X_2. What proportion of the variation in Y is explained by that relationship?
(c) Test the hypotheses of zero regression coefficients of X_1 and X_2.

14. The three variables Y, X, and W have been observed in each of N independent sampling units. The observations are represented by the $N \times 1$ centered vectors $\mathbf{y}_1$, $\mathbf{x}_1$, and $\mathbf{w}_1$. If the simple linear regression coefficients of Y with X and Y with W are estimated, the residuals of those fitted equations are

$$\mathbf{z}_{yw} = \mathbf{y}_1 - \mathbf{w}_1\left(\frac{\mathbf{y}_1'\mathbf{w}_1}{\mathbf{w}_1'\mathbf{w}_1}\right)$$

$$\mathbf{z}_{xw} = \mathbf{x}_1 - \mathbf{w}_1\left(\frac{\mathbf{x}_1'\mathbf{w}_1}{\mathbf{w}_1'\mathbf{w}_1}\right)$$

Show that the regression coefficient

$$\frac{\mathbf{z}_{yw}'\mathbf{z}_{xw}}{\mathbf{z}_{xw}'\mathbf{z}_{xw}}$$

of $\mathbf{z}_{yw}$ related to $\mathbf{z}_{xw}$ is equivalent to the multiple regression coefficient of $\mathbf{y}_1$ and $\mathbf{x}_1$ with $\mathbf{w}_1$ held constant.

15. Recall the checking account data of Exercise 25, Section 1-9: The average numbers of checks each month remained fairly constant for the first six years, then increased steadily in the remaining seven years. We might represent that trend by the second-degree polynomial model

$$y_t = \beta_0 + \beta_1 t + \beta_2 t^2 + e_t \qquad t = 1, \ldots, 13$$

This is merely the multiple regression model with the index t as the first variable and t^2 as the second, so that

$$\mathbf{X'X} = \begin{bmatrix} 13 & 91 & 819 \\ & 819 & 8{,}281 \\ & & 89{,}271 \end{bmatrix} \qquad \mathbf{X'y} = \begin{bmatrix} 333 \\ 2{,}623 \\ 25{,}357 \end{bmatrix}$$

(a) Find the least squares estimates of β_0, β_1, and β_2.

(b) Find the residuals

$$z_t = y_t - \hat{y}_t$$

of the predicted and actual numbers of checks. Calculate the error sum of squares

$$\text{SSE} = \sum_{1}^{13} z_t^2$$

and from it the estimate of the disturbance variance.

(c) Plot the residuals against the index t of the years. Does the assumption of a common variance appear to be valid? Does the second-degree model appear to fit the observations, or should another model be tried for a better fit?

(d) What proportion of variation does the quadratic model explain?

(e) Plot the tth residual z_t against its predecessor z_{t-1} for $t = 2, \ldots, 13$. Do the successive residuals appear to be uncorrelated? A formal test of that assumption is given in Chapter 6.

16. Repeat the analysis of Exercise 15 but with the linear and quadratic terms centered to give orthogonal independent variables. The model is

$$y_t = \gamma_0 + \gamma_1(t - 7) + \gamma_2[(t - 7)^2 - 14] + e_t \qquad t = 1, \ldots, 13$$

and the necessary statistics are

$$\mathbf{X'X} = \begin{bmatrix} 13 & 0 & 0 \\ & 182 & 0 \\ & & 2002 \end{bmatrix}$$

$$\mathbf{X'y} = \begin{bmatrix} 333 \\ 292 \\ 290 \end{bmatrix}$$

$$\sum (y_i - \bar{y})^2 = 521.0769$$

17. For the course evaluation average data of Example 2-1, carry out the following analyses:

(a) Test the individual hypotheses of zero regression coefficients for the independent variables X_1 and X_2.

(b) Find the individual 99% confidence intervals for the regression coefficients.

(c) Test the hypothesis of no overall relationship of Y to X_1 and X_2 by the F-statistic.

(d) Compute the partial correlations of Y and X_1 with X_2 fixed, and Y and X_2 with X_1 fixed.

18. Fit a linear multiple regression function to the complete data set of Exercise 17, Section 1-9. Compare the proportion of variation explained by all three predictors with that accounted for by the number of meals served.

19. Plot the residuals from the fitted plane in the preceding example against the indices of the rows in the data matrix and compare the plot with that for the simple regression function of tips against meals served. Have any systematic patterns of variation in the latter plot disappeared with the inclusion of the other two predictors?

20. Calculate the three correlation coefficients of the variables in Example 2-1 and from them the partial correlation of X_1 and X_2 with Y held constant.

21. The linear regression equations for predicting X_1 and X_2 from Y in Example 2-1 are, respectively,

$$\hat{x}_{i1} = 0.3410 + 0.9306y_i$$

$$\hat{x}_{i2} = 0.1474 + 1.0345y_i$$

Find the residuals from these fitted lines and verify that their correlation is equal to the partial correlation calculated in Exercise 20.

22. Plot the residuals from each line in Exercise 21 against their respective predicted values $\hat{y}_i$ of the overall evaluation variable. Which instructors appear to be outliers from the regression lines?

23. Data* were gathered on $N = 27$ baseball players who had played for eight or more years and in one or more World Series games to determine whether the player's total number of home runs was related to other characteristics of his career. These variables were employed for a multiple regression analysis:

Y = total number of home runs to date

X_1 = batting average

X_2 = average playing weight

X_3 = number of games played to date

X_4 = years played to date

These statistics were computed from the data:

Variable	Mean	Standard deviation
Y	138.56	181.99
X_1	.269	0.0360
X_2	183.93	15.56
X_3	1476.81	714.57
X_4	15.70	3.95

Correlation matrix:

$$\begin{matrix} Y \\ X_1 \\ X_2 \\ X_3 \\ X_4 \end{matrix} \begin{bmatrix} 1 & 0.6286 & 0.5480 & 0.5596 & 0.3260 \\ & 1 & 0.1004 & 0.7590 & 0.4675 \\ & & 1 & -0.0584 & 0.0266 \\ & & & 1 & 0.6605 \\ & & & & 1 \end{bmatrix}$$

$$\text{SSR} = 598{,}946 \qquad \text{SSE} = 262{,}210 \qquad \hat{\sigma}^2 = 11{,}919$$

*I am indebted to Rebecca Saeger and Robin Wallach for this example, and for kindly permitting its use here.

Variable	Regression coefficient	t
X_1	1432.38	1.52
X_2	6.43	4.52
X_3	0.117	2.09
X_4	−5.71	−0.78

Regression equation:

$$\hat{y}_i = -1512.4 + 1432.38x_{i1} + 6.43x_{i2} + 0.117x_{i3} - 5.71x_{i4}$$

(a) What proportion of the variation in the number of home runs is explained by a linear function of the four predictors?

(b) Which predictors might be eliminated from the model?

(c) What number of home runs would you predict for a player with a batting average of .275, playing weight of 200, 1600 games played, and a playing career of 17 years?

(d) Test the hypothesis that the four variables have zero regression coefficients.

(e) If the first and fourth independent variables were eliminated, these regression statistics would be obtained:

Variable	Regression coefficient	t
X_2	6.82	4.83
X_3	0.151	4.92

$$\text{SSR} = 561{,}056 \qquad \text{SSE} = 300{,}100 \qquad \hat{\sigma}^2 = 12{,}504$$

Give the regression equation and compute the predicted number of home runs for the player described in (c).

(f) What proportion of variation does the reduced model in (e) explain?

(g) When taken together do variables X_2 and X_3 provide more information for predicting Y than separately? Why or why not?

24. Let us consider another baseball application, although one involving team performance rather than individual players.* Originally, these variables were employed for an analysis of the $N = 26$ major league teams in the 1980 season:

Y = number of wins

X_1 = team batting average

X_2 = number of runs scored

X_3 = number of home runs hit

X_4 = team earned run average (ERA)

X_5 = hits allowed

X_6 = allowed bases on balls

The six predictors explained 85.8% of the variation in the number of wins, but the predictors X_1, X_3, and X_6 had regression coefficients whose $|t|$ statistics were very small. The second multiple regression analysis with those variables omitted gave

*This example is due to James Ramenda and is used here with his kind permission.

these results:

Predictor	Mean	Standard deviation	Regression coefficient	t
X_2	694.3	79.57	0.1094	8.93
X_4	3.84	0.391	−11.10	−2.79
X_5	1467	64.33	−0.0636	−2.62

$$\bar{y} = 80.81 \qquad \text{SSR} = 2885.56 \qquad \text{SSE} = 506.47$$

(a) Test the hypothesis of no overall linear relationship between number of wins and the predictors X_2, X_4, and X_5.

(b) What is the estimate of the disturbance variance in the model?

(c) What proportion of the variation in Y do the three predictors explain?

(d) Figure 2-7 contains a plot of the residuals $y_i - \hat{y}_i$ against the values $\hat{y}_i$ predicted

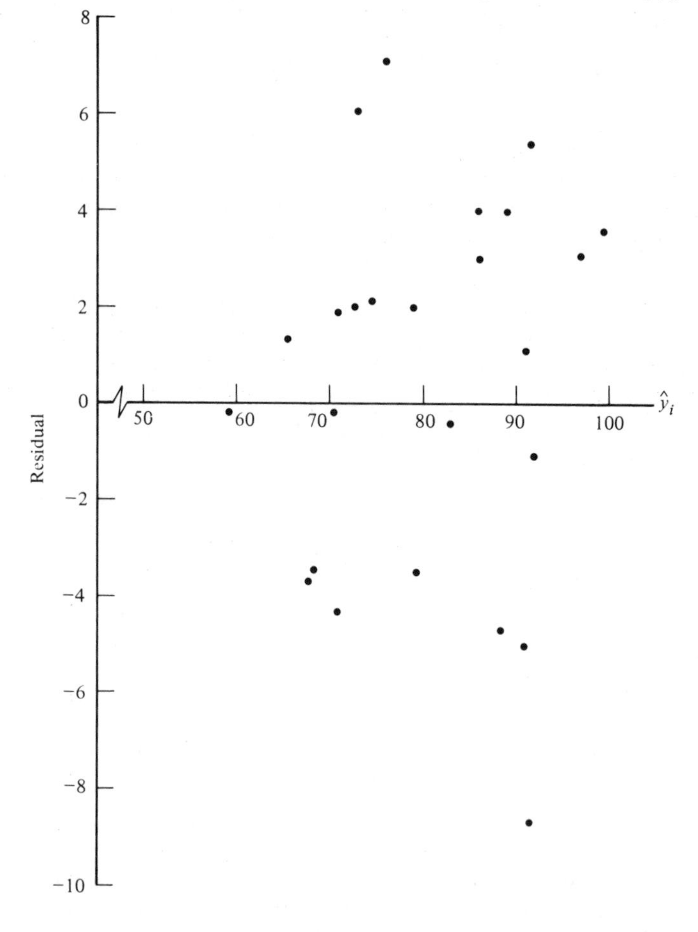

Figure 2-7 Plot of the residuals against predicted number of wins.

by the three-variable model. What can you infer about the appropriateness of the fitted model from this plot? Do certain teams appear to be outliers? Does the assumption of a constant variance seem to hold?

25. Use the covariance operator to show that the $N \times 1$ vector of predicted values $\hat{\mathbf{y}} = \mathbf{X}\hat{\boldsymbol{\beta}}$ for the conventional linear model $\mathbf{y} = \mathbf{X}\boldsymbol{\beta} + \mathbf{e}$ is uncorrelated with the $N \times 1$ vector of residuals $\mathbf{z} = \mathbf{y} - \hat{\mathbf{y}}$, that is,

$$\text{Cov}\,(\hat{\mathbf{y}}, \mathbf{z}') = 0 \quad (N \times N)$$

26. Show that the residuals and predicted values in the preceding exercise have zero sample covariance, or

$$\hat{\mathbf{y}}'\mathbf{z} = \sum_{i=1}^{N} \hat{y}_i z_i = 0$$

Hence the sample correlation of the $(\hat{y}_i, z_i)$ pairs is zero.

27. The residuals from the three-variable quadratic model for the course evaluation averages described in Example 2-4 have been plotted against the predicted dependent variable values in Figure 2-8. Compare the plot with Figure 2-2(c). What effect has the third variable had on the fit of the least squares predictions of overall evaluation averages?

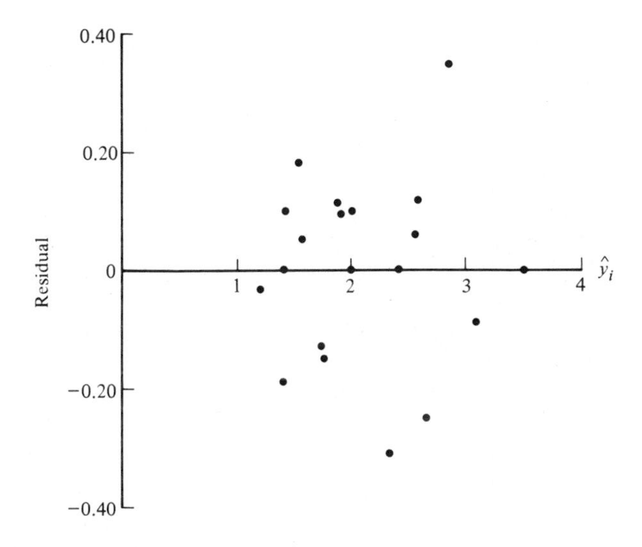

Figure 2-8 Course evaluation residual plot.

chapter 3

Further Inference for Regression Models

3-1 INTRODUCTION

As we have seen in Chapters 1 and 2, statistical inference is concerned with ways for studying the properties of an unknown population of random variables, its parameters, and perhaps mathematical models built about this population. In the present context the mathematical model is that of linear multiple regression, whereas the population is the distribution of the random disturbance component in the model. We have already seen how some basic inferences could be made by tests of hypotheses or confidence intervals on the parameters when the disturbance terms are assumed to be independently and normally distributed. In this chapter we develop a theoretical basis for some of those tests and give several extensions of them and their related confidence intervals.

Most of the new tests of this chapter deal with hypotheses on several regression coefficients and depend on statistics that are sums of squares. Under the present assumptions of independence and normality these sums of squares, or *quadratic forms*, will be distributed as chi-squared random variables. When the null hypothesis is true the distribution will be a *central*, or ordinary, chi-squared. When some alternative hypothesis holds instead the probability mass of the distribution will be squeezed toward the right, and will have a more complicated form known as the *noncentral chi-squared* distribution. In Section 3-2 we give some properties of quadratic forms which imply chi-squared distributions. We also show conditions under which quadratic forms are independently distributed and a powerful theorem for breaking a sum of squares into independently distributed variates with chi-squared distributions. We shall find the noncentral chi-squared and F-distributions and used them to determine the power probabilities of our hypothesis tests.

In Section 3-3 we develop tests of hypotheses on predetermined subsets of the regression coefficients. These tests are generalizations of the familiar hypotheses

$$H_0: \ \beta_j = 0$$

for individual regression coefficients considered in Section 2-4. The test follows directly from the results on quadratic forms in normal variates presented in Section 3-2: A sum of squares is partitioned out of the regression sum of squares and is shown to have a central chi-squared distribution only if the hypothesis on the subset of parameters is true. This test also leads to joint confidence regions on sets of the parameters or linear functions of them.

If several values of the dependent variable Y are observed for the same vector of independent variables, we may gain some very useful information. First, the variation in the Y values within the same dependent variable set provides an estimator of the error variance which does not depend on the kind of regression model specified for the data. Second, the difference of that within-set sum of squares and the usual error sum of squares about the fitted regression function provides a measure of the "goodness of fit" of the particular model we have chosen to represent the data. These methods are developed in Section 3-4. As in Section 3-3, we propose estimates and tests that seem to be intuitively reasonable, then justify their choice by determining their distributions and properties from the general results on quadratic forms introduced in Section 3-2.

The tests and confidence statements of Sections 2-4 and 3-3 were made with Type I error rates and confidence coefficients calculated with respect to each individual hypothesis. In practice one usually scans the output of regression coefficients and their t-statistics for large "significant" values. If the number p of independent variables is large, and if perhaps the variables are highly correlated, values of t in excess of the 0.05 or 0.01 critical values may appear merely by chance. If many tests are to be carried out on the correlated regression coefficients, or if the hypotheses are to be generated a posteriori from an inspection of the estimates, we should prefer some kind of protection against Type I errors with a fixed error probability for the entire "family" of tests. Similarly, a set of several confidence statements should have a coefficient that refers to the entire set. We describe such simultaneous tests and confidence intervals in Section 3-5.

In Section 3-6 we extend the notion of a prediction interval for a future Y observation and the confidence interval for a conditional mean of Y to the multiple regression model.

Frequently, one wishes to determine whether a single regression plane or perhaps a family of several parallel planes will suffice as models for k independent samples of observations on the same variables. We shall give an analysis of variance for testing the equality of k independent regression coefficient vectors and the equality of the k intercepts. Simultaneous tests and confidence intervals will be provided for determining which coefficients and intercepts are in fact different when the hypotheses of overall equality have been rejected. Once again the tests of this section will follow from the general theorems on quadratic forms in normal variates we describe in Section 3-2. Let us hasten to those results without further delay.

3-2 FURTHER INFERENCES ABOUT REGRESSION PARAMETERS

In Section 2-3 we saw how the hypothesis of zero regression coefficients could be tested by means of the statistic (28):

$$F = \frac{N - p - 1}{p} \frac{\text{SSR}}{\text{SSE}} \tag{1}$$

The F-distribution of the statistic followed from the independent chi-squared distributions for SSR/σ^2 and SSE/σ^2 when the hypothesis was true. In the remainder of this text we shall meet many other test statistics that are ratios of sums of squares, and it will be useful to have some rules and properties of them that determine their distributions. In particular, we wish to know when the sums of squares, or quadratic forms, are proportional to chi-squared random variables and when they are independent, for it is those conditions that lead to F-distributions of their ratios.

Let us begin by defining a chi-squared variate. If $Z_1, \ldots, Z_\nu$ are independently distributed normal random variables with means zero and variances one, their sum of squares

$$\chi^2 = Z_1^2 + \cdots + Z_\nu^2 \tag{2}$$

has the chi-squared distribution with the degrees of freedom parameter ν. Upper critical values of that distribution are given in Table 2 of Appendix A; the density function and further properties are discussed in Appendix C. The variate defined by (2) is implicitly a *central* chi-squared, for the Z_i terms are all centered at a zero mean. Conversely, if Z_i' are independent normal random variables with respective means $\mu_1, \ldots, \mu_\nu$ and variances one, the variate

$$\chi'^2 = Z_1'^2 + \cdots + Z_\nu'^2 \tag{3}$$

has the *noncentral* chi-squared distribution with degrees of freedom ν and noncentrality parameter

$$\lambda = \sum_{i=1}^{\nu} \mu_i^2 \tag{4}$$

If $\lambda = 0$ the distribution of χ'^2 is central chi-squared. The noncentral chi-squared density and distribution functions are a bit more complicated than the central versions; some of their properties are shown in Appendix C. The formal concepts of the two kinds of variates will suffice for our purposes here.

Almost all the sums of squares, or equivalently, quadratic forms, that we shall need in the sequel will be proportional to central or noncentral chi-squared variates. Generally, the distribution will be central if the null hypothesis is true and noncentral if some alternative hypothesis holds. We may summarize the conditions under which the sum of squares has the distribution by the following theorem.

Theorem 3-1. Let the $N \times 1$ random vector $\mathbf{y}$ be multinormally distributed with mean $E(\mathbf{y}) = \boldsymbol{\mu}$ and covariance matrix $\boldsymbol{\Sigma}$. The quadratic form

$$Q = \mathbf{y}'\mathbf{A}\mathbf{y}$$

with the symmetric matrix $\mathbf{A}$ of rank r has the noncentral chi-squared distribution with r degrees of freedom and noncentrality parameter

$$\mu = \boldsymbol{\mu}'\mathbf{A}\boldsymbol{\mu}$$

if and only if $\mathbf{A\Sigma}$ is an idempotent matrix.

A proof of the theorem has been given by Searle (1971, pp. 57–58); its origin is in a fundamental result by Cochran (1934).

Let us use the theorem to find the distribution of the quadratic form

$$Q_1 = \frac{\text{SSR}}{\sigma^2} = \frac{\mathbf{y}'\mathbf{X}_1(\mathbf{X}_1'\mathbf{X}_1)^{-1}\mathbf{X}_1'\mathbf{y}}{\sigma^2} \tag{5}$$

appearing in the test statistic (1). In the present case $\mathbf{y}$ has the mean vector

$$E(\mathbf{y}) = \mathbf{X}\boldsymbol{\beta} = [\mathbf{j} \quad \mathbf{X}_1]\begin{bmatrix}\alpha \\ \boldsymbol{\beta}_1\end{bmatrix} = \alpha\mathbf{j} + \mathbf{X}_1\boldsymbol{\beta}_1 \tag{6}$$

and covariance matrix $\boldsymbol{\Sigma} = \sigma^2\mathbf{I}$ under the usual ordinary least squares model. Then the matrix

$$(\mathbf{A\Sigma})^2 = \left\{\left[\frac{\mathbf{X}_1(\mathbf{X}_1'\mathbf{X}_1)^{-1}\mathbf{X}_1'}{\sigma^2}\right]\sigma^2\mathbf{I}\right\}^2 = \mathbf{X}_1(\mathbf{X}_1'\mathbf{X}_1)^{-1}\mathbf{X}_1' \tag{7}$$

is idempotent, so Q_1 must have a chi-squared distribution with p degress of freedom. The noncentrality parameter is

$$\begin{aligned}\lambda &= \frac{(\alpha\mathbf{j} + \mathbf{X}_1\boldsymbol{\beta}_1)'\mathbf{X}_1(\mathbf{X}_1'\mathbf{X}_1)^{-1}\mathbf{X}_1'(\alpha\mathbf{j} + \mathbf{X}_1\boldsymbol{\beta}_1)}{\sigma^2} \\ &= \frac{\boldsymbol{\beta}_1'(\mathbf{X}_1'\mathbf{X}_1)^{-1}\boldsymbol{\beta}_1}{\sigma^2}\end{aligned} \tag{8}$$

The terms involving α vanished because $\mathbf{j}'\mathbf{X}_1 = \mathbf{0}$ from the centered nature of the elements in each column of $\mathbf{X}_1$. We note that $\lambda = 0$; the distribution of SSR/σ^2 is central chi-squared only if the hypothesis

$$H_0: \quad \boldsymbol{\beta}_1 = \mathbf{0} \tag{9}$$

is true.

The theorem can also be used to establish that

$$Q_2 = \frac{\text{SSE}}{\sigma^2} \tag{10}$$

has the central chi-squared distribution with $N - p - 1$ degrees of freedom regardless of the validity of the hypothesis (9) on $\boldsymbol{\beta}_1$. Write

$$Q_2 = \frac{\mathbf{y}'\left[\mathbf{I} - \dfrac{1}{N}\mathbf{J} - \mathbf{X}_1(\mathbf{X}_1'\mathbf{X}_1)^{-1}\mathbf{X}_1'\right]\mathbf{y}}{\sigma^2} \tag{11}$$

as in (31), Section 2-2. $\mathbf{J}$ is the $N \times N$ matrix of ones. Then

$$(\mathbf{A\Sigma})^2 = \left[\mathbf{I} - \frac{1}{N}\mathbf{J} - \mathbf{X}_1(\mathbf{X}_1'\mathbf{X}_1)^{-1}\mathbf{X}_1'\right]^2$$

$$= \mathbf{I} - \frac{1}{N}\mathbf{J} - \mathbf{X}_1(\mathbf{X}_1'\mathbf{X}_1)^{-1}\mathbf{X}_1' \tag{12}$$

so that the idempotency requirement is satisfied. The noncentrality parameter is

$$\begin{aligned}
\lambda &= (\alpha\mathbf{j} + \mathbf{X}_1\boldsymbol{\beta}_1)'\left[\mathbf{I} - \frac{1}{N}\mathbf{J} - \mathbf{X}_1(\mathbf{X}_1'\mathbf{X}_1)^{-1}\mathbf{X}_1'\right](\alpha\mathbf{j} + \mathbf{X}_1\boldsymbol{\beta}_1) \\
&= \boldsymbol{\beta}_1'\mathbf{X}_1'\left[\mathbf{I} - \frac{1}{N}\mathbf{J} - \mathbf{X}_1(\mathbf{X}_1'\mathbf{X}_1)^{-1}\mathbf{X}_1'\right]\mathbf{X}_1\boldsymbol{\beta}_1 \\
&= \boldsymbol{\beta}_1'\mathbf{X}_1'\mathbf{X}_1\boldsymbol{\beta}_1 - \boldsymbol{\beta}_1'\mathbf{X}_1'\mathbf{X}_1\boldsymbol{\beta}_1 \\
&= 0
\end{aligned} \tag{13}$$

Since λ is zero, SSE/σ^2 always has the central chi-squared distribution.

Sums of chi-squared variates also have the chi-squared distribution. We may describe that property for quadratic forms in this theorem:

Theorem 3-2. If the quadratic forms $Q_1, \ldots, Q_k$ are independently distributed as noncentral chi-squared random variables with respective degrees of freedom $\nu_1, \ldots, \nu_k$ and noncentrality parameters $\lambda_1, \ldots, \lambda_k$, their sum

$$Q = Q_1 + \cdots + Q_k \tag{14}$$

has the noncentral chi-squared distribution with respective degrees of freedom and noncentrality parameters

$$\nu = \sum_{i=1}^{k} \nu_i \qquad \lambda = \sum_{i=1}^{k} \lambda_i \tag{15}$$

The proof follows easily from the moment generating functions given in Appendix C.

Because the quadratic forms in our test statistics must be independently distributed, this theorem due to Aitken (1950) will frequently be useful:

Theorem 3-3. If $\mathbf{y}$ has the multinormal distribution with covariance matrix $\mathbf{\Sigma}$ the quadratic forms $Q_1 = \mathbf{y}'\mathbf{A}\mathbf{y}$ and $Q_2 = \mathbf{y}'\mathbf{B}\mathbf{y}$ are independently distributed if and only if

$$\mathbf{A\Sigma B} = \mathbf{0} \tag{16}$$

The theorem has a useful corollary: The linear compound $w = \mathbf{b}'\mathbf{y}$ in the same random vector $\mathbf{y}$ is independently distributed of the quadratic form Q_1 if and only if

$$\mathbf{A\Sigma b} = \mathbf{0} \tag{17}$$

We may use Theorem 3-3 to verify that SSR and SSE are independently distributed. The product (16) of their matrices is

$$\sigma^2 \mathbf{X}_1(\mathbf{X}_1'\mathbf{X}_1)^{-1}\mathbf{X}_1'\left(\mathbf{I} - \frac{1}{N}\mathbf{J} - \mathbf{X}_1(\mathbf{X}_1'\mathbf{X}_1)^{-1}\mathbf{X}_1'\right) = \mathbf{0}$$

and independence follows.

Cochran's theorem. The previous results for independence and chi-squared distributions of quadratic forms in independent unit normal variates are subsumed under a remarkable theorem in linear algebra and statistics due to Cochran (1934). We shall give a version of the theorem as extended by James (1952). The presentation is in terms of central chi-squared variates; the case of noncentral variates can be handled with appropriate attention to the necessary noncentrality parameters.

Theorem 3-4. Let $\mathbf{y}$ be an N-dimensional random vector distributed according to the $N(\mathbf{0}, \sigma^2\mathbf{I})$ distribution. The $N \times N$ real symmetric matrices $\mathbf{A}_1, \ldots, \mathbf{A}_k$ have respective ranks $r_1, \ldots, r_k$ and obey the property

$$\mathbf{I} = \sum_{i=1}^{k} \mathbf{A}_i \tag{18}$$

so that the k quadratic forms $Q_i = \mathbf{y}'\mathbf{A}_i\mathbf{y}$ have as sum

$$\sum_{i=1}^{k} Q_i = \mathbf{y}'\mathbf{y} \tag{19}$$

or the sum of squares of the original variates. Then any one of the following three conditions implies that the other two must hold:

1. $r_1 + \cdots + r_k = N$
2. The quadratic form Q_i/σ^2 has the chi-squared distribution with r_i degrees of freedom, for $i = 1, \ldots, k$.
3. $Q_1, \ldots, Q_k$ are mutually independently distributed.

Proofs of the theorem have been given by James (1952), Lancaster (1954), Kendall and Stuart (1958), and Searle (1971). We note that condition 2 is equivalent to

$$\mathbf{A}_i^2 = \mathbf{A}_i \qquad i = 1, \ldots, k \tag{20}$$

and that condition 3 is the same as

$$\mathbf{A}_i\mathbf{A}_j = \mathbf{0} \qquad \text{all } i \neq j \tag{21}$$

Cochran's theorem provides a useful connection between the ranks, or degrees of freedom, of sums of squares and the joint distribution of the sums. We shall see this connection in the forthcoming *analysis of variance* tables for testing hypotheses on regression models.

A joint confidence region for $\boldsymbol{\beta}_1$. We have seen from Theorem 3-1 that the scaled regression sum of squares SSR$/\sigma^2$ has a noncentral chi-squared distribution. By "centering" $\hat{\boldsymbol{\beta}}_1$ about its population mean $\boldsymbol{\beta}_1$ we can obtain a joint confidence region for the regression parameters. We begin by noting that $\hat{\boldsymbol{\beta}}_1$ has the multinormal distribution with mean vector $\boldsymbol{\beta}_1$ and covariance matrix

$$\text{Cov}\,(\hat{\boldsymbol{\beta}}_1, \hat{\boldsymbol{\beta}}_1') = \sigma^2(\mathbf{X}_1'\mathbf{X}_1)^{-1} \tag{22}$$

Hence the quadratic form

$$Q = \frac{(\hat{\boldsymbol{\beta}}_1 - \boldsymbol{\beta}_1)'\mathbf{X}_1'\mathbf{X}_1(\hat{\boldsymbol{\beta}}_1 - \boldsymbol{\beta}_1)}{\sigma^2} \tag{23}$$

will be a central chi-squared variate with p degrees of freedom because

$$E(\hat{\boldsymbol{\beta}}_1 - \boldsymbol{\beta}_1) = \mathbf{0}$$

and

$$\left(\frac{\mathbf{X}_1'\mathbf{X}_1}{\sigma^2}\right)\sigma^2(\mathbf{X}_1'\mathbf{X}_1)^{-1} = \mathbf{I}$$

satisfies the idempotency requirement of Theorem 3-1. Then

$$F = \frac{N-p-1}{p\,\text{SSE}}(\hat{\boldsymbol{\beta}}_1 - \boldsymbol{\beta}_1)'\mathbf{X}_1'\mathbf{X}_1(\hat{\boldsymbol{\beta}}_1 - \hat{\boldsymbol{\beta}}_1) \tag{24}$$

has the F-distribution with p and $N - p - 1$ degrees of freedom, and the $100(1 - \alpha)\%$ joint confidence region is the p-dimensional ellipsoid

$$(\boldsymbol{\beta}_1 - \hat{\boldsymbol{\beta}})'\mathbf{X}_1'\mathbf{X}_1(\boldsymbol{\beta}_1 - \hat{\boldsymbol{\beta}}_1) \leq \frac{p\,\text{SSE}}{N-p-1}F_{\alpha;\,p,N-p-1} \tag{25}$$

in the space of the elements of $\boldsymbol{\beta}_1$. The ellipsoid is centered at the sample estimate $\hat{\boldsymbol{\beta}}_1$ of the parameters.

The joint confidence region can be used to test hypotheses. If the proposed vector $\boldsymbol{\beta}_{10}$ lies within the ellipsoid defined by (25), or algebraically, if the inequality still holds when $\boldsymbol{\beta}_{10}$ replaces $\boldsymbol{\beta}_1$ in the left-hand side, the hypothesis $H_0\colon \boldsymbol{\beta}_1 = \boldsymbol{\beta}_{10}$ is tenable at the $100\alpha\%$ level. If the inequality no longer holds, so that $\boldsymbol{\beta}_{10}$ lies outside the ellipsoid, the alternative $H_1\colon \boldsymbol{\beta}_1 \neq \boldsymbol{\beta}_{10}$ should be accepted.

Example 3-1

We shall find the 95% joint confidence region for the regression coefficients of the free fatty acid data of Section 2-2 when X_1 = age and X_2 = weight are used as the independent variables. Then

$$\mathbf{X}_1'\mathbf{X}_1 = \begin{bmatrix} 6052.488 & 2440.219 \\ 2440.219 & 4134.049 \end{bmatrix} \qquad \hat{\boldsymbol{\beta}}_1 = \begin{bmatrix} -0.00308 \\ -0.01176 \end{bmatrix} \qquad \hat{\sigma}^2 = 0.046995$$

The F critical value for 2 and 38 degrees of freedom can be found by linear interpolation in the reciprocals of the second degrees of freedom to be approximately 3.25. The equation of the joint confidence ellipse is

$$\begin{aligned} 6052.488(\beta_1 + 0.00308)^2 &+ 4880.438(\beta_1 + 0.00308)(\beta_2 + 0.01176) \\ &+ 4134.049(\beta_2 + 0.01176)^2 \leq 0.3055 \end{aligned}$$

The ellipse is shown in Figure 3-1. The region does not contain zero values for β_2 regardless of the β_1 value, so that we could conclude that a real relationship exists

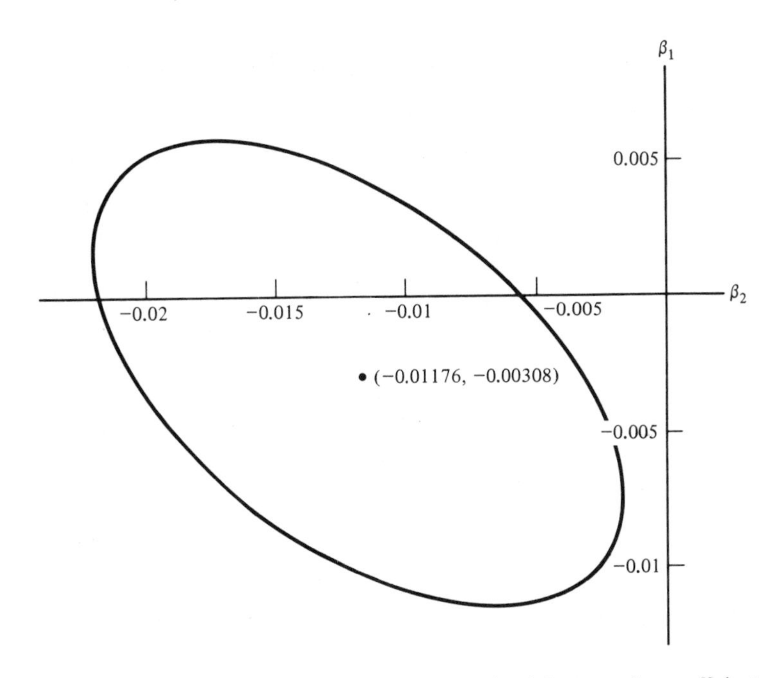

Figure 3-1 Joint confidence region for the age and weight regression coefficients.

between FFA and weight. Such inferences and infinitely many others can be made about the pairs of the regression parameters with an overall error rate of 0.05 or a joint confidence coefficient of 0.95. We shall say more about such families of tests and intervals in our discussion of simultaneous inference in Section 3-4.

The noncentral *F*-distribution. The test statistics in our treatment of regression and the analysis of variance will usually involve ratios of quadratic forms. When the forms $Q_1 = \mathbf{x}'\mathbf{A}_1\mathbf{x}$ and $Q_2 = \mathbf{x}'\mathbf{A}_2\mathbf{x}$ are independently distributed as chi-squared variates with respective degrees of freedom m and n, the ratio

$$F = \frac{Q_1}{m}\bigg/\frac{Q_2}{n} = \frac{n}{m}\frac{Q_1}{Q_2} \tag{26}$$

has the F, or variance ratio, distribution with degrees of freedom m and n. If the chi-squared distributions of Q_1 and Q_2 are both central, the distribution of the ratio is also a central, or ordinary, F-distribution. Upper critical values of the F-distribution are given in Table 4 of Appendix A; some further properties are discussed in Appendix C.

When the numerator quadratic form of the ratio (26) has the noncentral chi-squared distribution with m degrees of freedom and noncentrality parameter λ, the ratio has the noncentral F-distribution with parameters m, n, and λ. The density function of that distribution is shown in Appendix C. The noncentral F-distribution provides power or Type II error probabilities for hypotheses tested by F. In the present notation the power of the test is

$$\begin{aligned}\text{Power} &= P(\text{reject } H_0 \mid H_1 \text{ true}) \\ &= 1 - P(\text{accept } H_0 \mid H_1 \text{ true}) \\ &= P(F > F_{\alpha; m, n} \mid \lambda, m, n) \end{aligned} \tag{27}$$

Tables and charts of the power probability for given α, m, n and various values of the noncentrality parameter have been prepared by different workers. The Pearson and Hartley (1951, 1972) charts of power probabilities have been reproduced as Charts 1 through 10 of Appendix A. The 10 charts correspond to the values $m \equiv \nu_1 = 1, 2, \ldots, 8, 12, 24$ of the first degrees of freedom parameter. Each chart contains two families of power curves for the values $\alpha = 0.05$ and 0.01 of the Type I error probability. The individual curves have been computed for different values of the second degrees of freedom parameter $n \equiv \nu_2$ from $n = 6$ to $n = 60$ and $n = \infty$. The noncentrality parameter used for the horizontal axis of the charts is

$$\phi = \sqrt{\frac{\lambda}{\nu_1 + 1}} \tag{28}$$

To compute the probability (27) that an F-statistic with degrees of freedom m and n will exceed the $100\alpha\%$ critical value we choose the chart for the value of m and the family of curves for $\alpha = 0.05$ or 0.01. Next we compute ϕ from λ and m and locate its value on the ϕ scale. We read up to the curve corresponding to n or its interpolated value, then read across to the vertical scale to obtain the power probability.

As an example of the use of the Pearson–Hartley charts let us calculate the probability that an F'-variate with $m = 3$ and $n = 20$ degrees of freedom exceeds the $\alpha = 0.01$ critical value $F_{0.01;3,20} = 4.94$ when $\lambda = 20$, 30, and 36. We use Chart 3, the right-hand family of curves, and the lower ϕ scale on the horizontal axis. The three values of ϕ are 2.23, 2.74, and 3, respectively. For the first value of the noncentrality parameter the power probability is approximately 0.82; it is 0.95 for the second, and 0.98 for the third.

Extensive tables of power probabilities from the noncentral F-distribution have also been computed by Tiku (1967). Their use avoids the visual interpolation required by many of the Pearson–Hartley charts.

Let us apply these results for the noncentral F-distribution to find the power of the test of H_0: $\boldsymbol{\beta}_1 = \mathbf{0}$ against the alternative of a general nonzero regression coefficient vector $\boldsymbol{\beta}_1$. Then the test statistic

$$F = \frac{\hat{\boldsymbol{\beta}}_1' \mathbf{X}_1' \mathbf{X}_1 \hat{\boldsymbol{\beta}}_1}{p\hat{\sigma}^2} \tag{29}$$

has the noncentral F-distribution with parameters p, $N - p - 1$, and

$$\lambda = \frac{\boldsymbol{\beta}_1' \mathbf{X}_1' \mathbf{X}_1 \boldsymbol{\beta}_1}{\sigma^2} \tag{30}$$

Let the sample size be $N = 24$, the number of independent variables be $p = 3$, and the significance level of the test be $\alpha = 0.01$. Assume further that

$$\mathbf{X}_1'\mathbf{X}_1 = \begin{bmatrix} 20 & 8 & 0 \\ 8 & 30 & 12 \\ 0 & 12 & 15 \end{bmatrix} \tag{31}$$

and the population regression coefficients scaled by the disturbance term standard deviation have the values given by

$$\frac{\boldsymbol{\beta}_1'}{\sigma} = [0.5, \quad 0.25, \quad 0.2] \tag{32}$$

Then $\lambda = 10.675$ and $\phi = 1.634$. Since $\nu_1 = p = 3$, we find Chart 3, locate the value of ϕ on the lower abscissa scale, and read up to the $\nu_2 = N - p - 1 = 20$ curve of the lower family. The power probability of correctly rejecting H_0: $\boldsymbol{\beta}_1 = \mathbf{0}$ may be read from the vertical scale as approximately 0.63. The same process might have been used for $\alpha = 0.05$; in that case the power increases slightly to about 0.72.

Example 3-2

Let us use the noncentral F power charts to determine the sample size for a hypothetical investigation whose results will be analyzed by a linear regression model. It has been conjectured by a psychologist that a certain measure of the ability to solve complex spatial problems is related to two other cognitive tests measuring the performance of the subject's central nervous system. The latter tests require elaborate measurement and recording equipment in the laboratory, whereas the spatial problems can be administered with paper, pencil, and a stopwatch. As a first order of approximation we shall use the linear model

$$y_i = \alpha + \beta_1(x_{i1} - \bar{x}_1) + \beta_2(x_{i2} - \bar{x}_2) + e_i \qquad i = 1, \ldots, N \tag{33}$$

to relate the spatial score y_i for the ith subject to his or her performance scores x_{i1} and x_{i2}. The performance scores will be regarded as fixed quantities, whereas the random variation in the y_i will be attributed to the independently normally distributed e_i disturbance terms. A major purpose of the experiment will be the testing of the hypothesis

$$H_0\colon \beta_1 = \beta_2 = 0 \tag{34}$$

that the spatial problem score is unrelated to the performance measurements. The alternative hypothesis is that of positive β_1, β_2 coefficients. The test will be carried out at the $\alpha = 0.05$ level. We wish to choose the number of subjects N such that the power of rejecting H_0 in favor of some given pair of values of the β_j will be at least 0.95.

To determine N we not only need the vector $\boldsymbol{\beta}_1$ of the alternative hypothesis but σ^2 and $\mathbf{X}_1'\mathbf{X}_1$ as well. We wish a test sensitive to rather small values of β_1 and β_2, perhaps ones smaller than the variation in the disturbance term. Let us state that the test should detect β_1 and β_2 as small as $\sigma/2$, so that

$$\frac{\boldsymbol{\beta}_1}{\sigma} = \begin{bmatrix} \beta_1/\sigma \\ \beta_2/\sigma \end{bmatrix} = \begin{bmatrix} \frac{1}{2} \\ \frac{1}{2} \end{bmatrix} \tag{35}$$

We must still choose the matrix $\mathbf{X}_1'\mathbf{X}_1$ for the noncentrality parameter, and it is here that a large degree of arbitrariness and subjectivity enters the power calculations. The choice of $\mathbf{X}_1'\mathbf{X}_1$ requires some knowledge of the variation of each independent variable as expressed in its variance or sum of squared deviations about its average and the correlations among the independent variables. Those measures must be obtained from earlier experiments, published accounts, or at worst, conjectures about the behavior of unobserved variables. Let us suppose that we are able to assemble such a matrix $\mathbf{S}_0$ of variances and covariances of the independent variables by one of those means. If it is treated as an unbiased sample estimator in the sense of Section 2-5, it is equal to

$$\mathbf{S}_0 = \frac{1}{N-1}\mathbf{X}_1'\mathbf{X}_1 \tag{36}$$

so that $\mathbf{X}_1'\mathbf{X}_1 = (N-1)\mathbf{S}_0$. Division by $N-1$ or N is somewhat ambiguous because of the conjectured or hypothetical nature of the elements of $\mathbf{S}_0$; we chose $N-1$ for consistency with the earlier sample results in Chapter 2. Given $\boldsymbol{\beta}_1/\sigma$ and $\mathbf{S}_0$ we may now write the noncentrality parameter as

$$\lambda = (N-1)\left(\frac{\boldsymbol{\beta}_1'}{\sigma}\right)\mathbf{S}_0\left(\frac{\boldsymbol{\beta}_1}{\sigma}\right) \tag{37}$$

and the argument for the Pearson–Hartley charts as

$$\phi = \sqrt{(N-1)\left(\frac{\boldsymbol{\beta}_1'}{\sigma}\right)\mathbf{S}_0\left(\frac{\boldsymbol{\beta}_1}{\sigma}\right)\Big/(p+1)} \tag{38}$$

In the present example let us take the conjectured covariance matrix to be

$$\mathbf{S}_0 = \begin{bmatrix} 1 & 0 \\ 0 & 2 \end{bmatrix} \tag{39}$$

We have postulated that the variance of the first independent variable relative to β_1/σ is one, and that the variance of the second is twice as great. We have also specified that the variables are uncorrelated. We know from their nature that they cannot have negative correlations. If we erroneously assume that their correlations are positive, the larger values of the noncentrality parameter will lead to an inadequate sample size and a consequent drop in the power probability.

Now we are ready to determine N from the power charts. We compute from (37) $\lambda = \frac{3}{4}(N-1)$ and $\phi = \frac{1}{2}\sqrt{N-1}$. Rather then attempting to guess N directly from the charts, we shall start by assuming that the second degrees of freedom parameter $\nu_2 = N - p - 1 = N - 3$ is infinite. Locate 0.95 on the vertical scale, read across to the first, or $\nu_2 = \infty$, line and drop a line perpendicularly to the upper ϕ axis. The value of ϕ is about 2.25. Set

$$2.25 = \tfrac{1}{2}\sqrt{N-1}$$

and solve for $N - 1 = 20.25$. Because $\nu_2 = 20$ has a curve in the family we shall take $N = 23$ for the next sample size. The value of ϕ corresponding to a power of 0.95 and $\nu_2 = 20$ is approximately 2.45, so that N must satisfy

$$2.45 = \tfrac{1}{2}\sqrt{N-1}$$

or $N - 1 = 24.01$. If we take $N = 25$ as our sample size and interpolate visually between the $\nu_2 = 30$ and $\nu_2 = 20$ curves for $\nu_2 = 22$, $\phi = 2.45$, the power will be 0.95. Any larger sample size will only increase the power.

Because the sample size depends so strongly on the assumed values of the parameters it will be well to try some other reasonable values of $\mathbf{S}_0$. Since we wish our sample size to be conservative in the sense of being larger than necessary for a power of 0.95, we shall still assume a zero correlation; negative correlations will be ruled out from the nature of the central nervous system performance variables. Let us try

$$\mathbf{S}_0 = \begin{bmatrix} 1 & 0 \\ 0 & 1 \end{bmatrix} \tag{40}$$

and the same vector (35) of scaled regression coefficients. Then $\lambda = (N - 1)/2$ and $\phi = \sqrt{(N - 1)/6}$. The same interative process in the third Pearson–Hartley chart leads to a sample size of $N = 35$. If

$$\mathbf{S}_0 = \begin{bmatrix} \frac{1}{2} & 0 \\ 0 & 1 \end{bmatrix} \tag{41}$$

and β_1/σ remains the same, the minimum sample size assuring a power of 0.95 would be about $N = 46$ subjects. We note that visual interpolation in that region of the chart is difficult, and one should not be concerned with finding the exact number of sampling units leading exactly to a power of 0.95.

We also note that this example of sample size determination has been greatly simplified. Such an actual experiment would probably involve multiple determinations of the three variables for each subject, if similar, or "parallel," versions of the tests were available. Then the mean or median score for each might be used, or a more complicated regression model involving all the data might be employed. More than two independent variables would probably be involved. If more than one dependent variable were measured, the problem would be that of multivariate regression analysis (Morrison, 1976, Section 5-3). It is also unlikely that such a costly experiment would be run for a "yea or nay" answer about the relations of spatial problem solving to central nervous system performance. The investigators would undoubtedly wish some measure of the strength and nature of the relationship in terms of confidence intervals and regions.

The analysis of variance table. The sums of squares and F-statistics for testing the hypotheses $H_0: \alpha = 0$ and $H_0: \boldsymbol{\beta}_1 = \mathbf{0}$ in the multiple regression model can be summarized conveniently in an *analysis of variance table.* Such tables will be used extensively in presenting the tests of the later sections of this chapter, particularly in the coming chapters on the analysis of data from designed experiments. The analysis of variance table is a way of organizing the decomposition of the sum of squares of the y_i observations by Cochran's theorem into components due to sampling error and the hypotheses under test. Table 3-1 gives the analysis of variance for the basic hypotheses considered to this point for the regression model. Each line of the table displays an independent component of the total sum of squares, its degrees of freedom, mean square, and where appropriate, an expected mean square and F-statistic for the relevant hypothesis. We note that the expected mean squares are related to the noncentrality parameters of the F-tests in this way:

$$\begin{aligned} &\text{1. Intercept:} && \lambda = \frac{E(\text{mean square}) - \sigma^2}{\sigma^2} \\ &\text{2. Regression } (\hat{\boldsymbol{\beta}}_1): && \lambda = \frac{pE(\text{mean square}) - \sigma^2}{\sigma^2} \end{aligned} \tag{42}$$

In the next sections we consider analyses of variance for hypotheses on subsets of the regression coefficients, tests for equality of several regression vectors, and many other hypotheses in the full-rank model.

TABLE 3-1 Analysis of Variance for the Multiple Regression Model

Source	Sum of squares	d.f.	Mean square	E(MS)	F
Intercept (mean)	$\text{SSM} = N\bar{y}^2$	1	SSM	$\sigma^2 + N\alpha^2$	$(N-p-1)\dfrac{\text{SSM}}{\text{SSE}}$
Regression coefficients ($\hat{\boldsymbol{\beta}}_1$)	$\text{SSR} = \mathbf{y}'\mathbf{X}_1(\mathbf{X}_1'\mathbf{X}_1)^{-1}\mathbf{X}_1'\mathbf{y}$	p	$\dfrac{\text{SSR}}{p}$	$\sigma^2 + \dfrac{\boldsymbol{\beta}_1'\mathbf{X}_1'\mathbf{X}_1\boldsymbol{\beta}_1}{p}$	$\dfrac{N-p-1}{p}\dfrac{\text{SSR}}{\text{SSE}}$
Error	$\text{SSE} = \mathbf{y}'\mathbf{y} - \text{SSM} - \text{SSR}$	$N-p-1$	$\dfrac{\text{SSE}}{N-p-1}$	σ^2	—
Total	$\mathbf{y}'\mathbf{y} = \sum y_i^2$	N	—	—	—

3-3 HYPOTHESIS TESTS FOR SETS OF REGRESSION COEFFICIENTS

In the preceding section we considered tests of hypotheses on the entire vector of p regression coefficients in the linear regression model. The most important of these was the test that all coefficients were simultaneously equal to zero, or the dependent variable was not linearly related to the predictor variables. Not only was the set of predictors considered immutable, but inferences about the coefficients were addressed to the entire vector. In Chapter 2 we saw how individual hypotheses H_0: $\beta_j = 0$ might be tested on the separate coefficients to determine if their predictor variables should be retained in the model. Similarly, in deciding which variables should be omitted we might wish to test the hypothesis that the coefficients of some subset of the p predictors are simultaneously zero, or perhaps equal to a specified vector. In this section we develop such tests and confidence statements for sets of elements of $\boldsymbol{\beta}_1$.

Let us begin with the simplest case and work up to more complicated and general hypotheses. Suppose that we wish to test the hypothesis

$$H_0: \quad \beta_1 = \cdots = \beta_r = 0 \tag{1}$$

that the first r regression coefficients of $\boldsymbol{\beta}_1$ are zero, as opposed to the alternative hypothesis of general values not all simultaneously zero. The r predictors were chosen on some prior basis rather than a posterior one of selecting the r variables with largest $\hat{\beta}_j$ or test statistics t_j, so that we need not be concerned with the effects of their selection upon the error rates of the test. Finally, we note that the *first* r predictors were chosen for notational convenience; the subsequent test will work for any set of r variables and their regression coefficients.

The hypothesis (1) can be represented in terms of the complete vector $\boldsymbol{\beta}_1$ as

$$H_0: \quad \mathbf{C}\boldsymbol{\beta}_1 = \mathbf{0} \tag{2}$$

where $\mathbf{C}$ is the $r \times p$ matrix

$$\mathbf{C} = \begin{bmatrix} 1 & 0 & \cdots & 0 & 0 & \cdots & 0 \\ 0 & 1 & \cdots & 0 & 0 & \cdots & 0 \\ \cdot & \cdot & & \cdot & \cdot & & \cdot \\ \cdot & \cdot & & \cdot & \cdot & & \cdot \\ \cdot & \cdot & & \cdot & \cdot & & \cdot \\ 0 & 0 & \cdots & 1 & 0 & \cdots & 0 \end{bmatrix} = [\mathbf{I} \quad \mathbf{0}] \tag{3}$$

consisting of ones in its first r diagonal positions and zeros elsewhere. The estimators of the β_j in the set are given by $\mathbf{C}\hat{\boldsymbol{\beta}}_1$, where $\hat{\boldsymbol{\beta}}_1$ is the usual least squares estimator of $\boldsymbol{\beta}_1$ in the full p-predictor model. Under the ordinary assumptions of independent and normally distributed disturbance terms $\mathbf{C}\hat{\boldsymbol{\beta}}_1$ has the multinormal distribution with mean vector $\mathbf{C}\boldsymbol{\beta}_1$ and covariance matrix $\sigma^2\mathbf{C}(\mathbf{X}_1'\mathbf{X}_1)^{-1}\mathbf{C}'$. Then, by the results on quadratic forms in Section 3-2,

$$\chi^2 = \frac{\hat{\boldsymbol{\beta}}_1'\mathbf{C}'[\mathbf{C}(\mathbf{X}_1'\mathbf{X}_1)^{-1}\mathbf{C}']^{-1}\mathbf{C}\hat{\boldsymbol{\beta}}_1}{\sigma^2} \tag{4}$$

has the chi-squared distribution with r degrees of freedom when the hypothesis (2) under test is true. Because σ^2 is nearly always unknown we must test the hypothesis by referring the statistic

$$F = \frac{[(N-p-1)/r]\hat{\boldsymbol{\beta}}_1' \mathbf{C}'[\mathbf{C}(\mathbf{X}_1'\mathbf{X}_1)^{-1}\mathbf{C}']^{-1}\mathbf{C}\hat{\boldsymbol{\beta}}_1}{\text{SSE}} \tag{5}$$

to the critical value $F_{\alpha;r,N-p-1}$ of the F-distribution with r and $N-p-1$ degrees of freedom. If

$$F > F_{\alpha;r,N-p-1} \tag{6}$$

we should reject the null hypothesis (2) at the α level of significance.

Let us examine the statistic (5). Just as the matrix $\mathbf{C}$ picked out the first r elements of $\boldsymbol{\beta}_1$, the square symmetric matrix $\mathbf{C}(\mathbf{X}_1'\mathbf{X}_1)^{-1}\mathbf{C}'$ consists of the first r rows and columns of $(\mathbf{X}_1'\mathbf{X}_1)^{-1}$. That submatrix is proportional to the variances and covariances of the r estimators selected from $\hat{\boldsymbol{\beta}}_1$. The symmetry of the statistic (5) indicates that it is unnecessary to regroup the coefficients and their estimators into the first r positions: $\mathbf{C}$ may be any $r \times p$ matrix whose rows successively select the elements of $\boldsymbol{\beta}_1$ comprising the set to be tested. For example, if the hypothesis is

$$H_0: \quad \beta_2 = \beta_3 = \beta_5 = 0$$

the matrix would be

$$\mathbf{C} = \begin{bmatrix} 0 & 1 & 0 & 0 & 0 & 0 & \cdots & 0 \\ 0 & 0 & 1 & 0 & 0 & 0 & \cdots & 0 \\ 0 & 0 & 0 & 0 & 1 & 0 & \cdots & 0 \end{bmatrix}$$

However, we may take this generality even further. Let $\mathbf{C}$ be any $r \times p$ matrix of real numbers with rank $r \leq p$. Then the hypothesis

$$H_0: \quad \mathbf{C}\boldsymbol{\beta}_1 = \mathbf{0} \tag{7}$$

specified by $\mathbf{C}$ on the regression coefficients can be tested by the F-statistic (5) with the same decision rule described by the inequality (6). In that way any set of r linearly independent linear hypotheses on the β_j can be specified and tested.

As an example of the general hypothesis suppose that the regression model consists of five predictors. The first four variables are of the same type and are measured in commensurate units. One hypothesis of interest to the analyst is that of whether those four variables have the same effect upon the dependent variable, or whether

$$H_0: \quad \beta_1 = \beta_2 = \beta_3 = \beta_4 \tag{8}$$

of equal regression coefficients is true. This hypothesis is equivalent to

$$H_0: \quad \beta_1 - \beta_2 = \beta_2 - \beta_3 = \beta_3 - \beta_4 = 0 \tag{9}$$

and can be represented in matrix form (7) by

$$\mathbf{C} = \begin{bmatrix} 1 & -1 & 0 & 0 & 0 \\ 0 & 1 & -1 & 0 & 0 \\ 0 & 0 & 1 & -1 & 0 \end{bmatrix} \tag{10}$$

We note that the common value of the four coefficients under H_0 need not be specified.

A second example provides a connection with our earlier hypothesis tests in Section 2-3 on single regression coefficients. The hypothesis

$$H_0: \quad a_1\beta_1 + \cdots + a_p\beta_p = \mathbf{a}'\boldsymbol{\beta}_1 = 0$$

that a single linear compound of the regression coefficients is equal to zero would be tested by

$$F = (N - p - 1)\frac{(\mathbf{a}'\hat{\boldsymbol{\beta}}_1)^2}{[\mathbf{a}'(\mathbf{X}_1'\mathbf{X}_1)^{-1}\mathbf{a}]\,\mathrm{SSE}} \tag{11}$$

or an F-statistic with 1 and $N - p - 1$ degrees of freedom. However, that statistic is the square of

$$t = \frac{\mathbf{a}'\hat{\boldsymbol{\beta}}_1}{\hat{\sigma}\sqrt{\mathbf{a}'(\mathbf{X}_1'\mathbf{X}_1)^{-1}\mathbf{a}}} \tag{12}$$

where $\hat{\sigma}^2 = \mathrm{SSE}/(N - p - 1)$. By a little generalization of the argument used to derive the t-statistic in Section 2-3 for hypotheses on single regression coefficients it follows that (12) has the Student t-distribution with $N - p - 1$ degrees of freedom, and a test based on the F-statistic (11) would be equivalent to a two-sided t-test made by referring the absolute value of (12) to the critical value $t_{\alpha/2;N-p-1}$. In particular, if $\mathbf{a}$ contains one in the jth position and zeros elsewhere, (12) becomes

$$t_j = \frac{\hat{\beta}_j}{\hat{\sigma}\sqrt{a_{jj}}} \tag{13}$$

in which a_{jj} is the jth diagonal element of $(\mathbf{X}_1'\mathbf{X}_1)^{-1}$, or expression (13) of Section 2-3 with $\beta_j = 0$. Similarly, for a test of

$$H_0: \quad \beta_i = \beta_j \qquad i \neq j \tag{14}$$

$\mathbf{a}' = [0, \ldots, 0, 1, 0, \ldots, -1, 0, \ldots, 0]$, where the nonzero elements are in the ith and jth positions, respectively. Then, in the same notation as (13),

$$F = (N - p - 1)\frac{(\hat{\beta}_i - \hat{\beta}_j)^2}{(a_{ii} + a_{jj} - 2a_{ij})\,\mathrm{SSE}} \tag{15}$$

and

$$t = \frac{\hat{\beta}_i - \hat{\beta}_j}{\hat{\sigma}\sqrt{a_{ii} + a_{jj} - 2a_{ij}}} \tag{16}$$

The F-tests on single linear hypotheses are identical to t-tests of the same hypotheses against two-sided alternatives.

Example 3-3

Let us apply these tests to a regression analysis involving psychological test scores obtained from a sample of $N = 46$ male subjects. The data were collected and discussed by Singer (1963), and portions of the scores were used in Examples 2-9 and 2-13 of Sections 2-5 and 2-6. In those examples of multiple and partial correlation we properly considered each subject's vector of scores as an observation from a multinormal distribution. In the present example we must treat the independent variables as fixed quantities in keeping with the usual least squares model. The dependent variable will be the score on the Draw a Person test. The $p = 11$ independent variables are listed in Table

TABLE 3-2 Regression Analysis of the Psychological Test Scores

Variable	Mean	S.D.	$\hat{\beta}_j$	t_j
Dependent				
Y: Draw a Person	19.70	9.72	—	—
Independent				
1. Aspiration Level	66.26	16.35	−0.14	−1.66
2. Proverbs	4.33	2.46	1.95	3.08
3. Homonyms	23.96	7.75	−0.03	−0.12
4. Sum of ratings	38.52	15.08	−1.28	−5.38
5. Aspiration Index	4.65	5.08	−0.35	−1.75
6. Weigl	2.74	1.48	1.13	1.55
7. Sentence Completion	1.74	1.10	−0.07	−0.06
8. Emotional Projection	2.78	1.74	1.29	1.48
9. Family Scene	2.17	1.37	1.48	1.40
10. Problem Situations	1.65	0.85	−0.35	−0.18
11. Thematic Apperception Test	2.30	1.17	4.17	2.19

3-2 as the scores from verbal, performance, or projective tests. Variables 1, 5, and 6 are performance test scores, variables 2 and 3 are verbal test scores, variable 4 is the total of the "rating" scores of the performance on all tests in the battery, and variables 7 through 11 are "projective" test measures. Their means and standard deviations are given to indicate the magnitudes of the scores. The regression coefficients of the Draw a Person scores on the independent variables for the multiple regression model and the test statistics for the coefficients are shown in the last two columns of the table. The squared multiple correlation for the fitted regression model was 0.7159, so that the 11 independent variables explained 71.59% of the variation in the Draw a Person scores. The F-statistic for testing the hypothesis that $\boldsymbol{\beta}_1 = \mathbf{0}$ is equal to $F = (34/11)(0.7159)/0.2841 = 7.79$; the hypothesis should be rejected at any reasonable significance level.

Only variables 2, 4, and 11 have regression coefficients with $|t|$-statistics in excess of the two-sided approximate critical value $t_{0.025;34} \approx 2.03$. Nevertheless, let us focus attention on the performance test scores given by variables 1, 5, and 6 and test the hypothesis

$$H_0\colon \beta_1 = \beta_5 = \beta_6 = 0$$

that their regression coefficients in the complete regression model are simultaneously equal to zero. Our $\mathbf{C}$ matrix is

$$\mathbf{C} = \begin{bmatrix} 1 & 0 & 0 & 0 & 0 & 0 & \cdots & 0 \\ 0 & 0 & 0 & 0 & 1 & 0 & \cdots & 0 \\ 0 & 0 & 0 & 0 & 0 & 1 & \cdots & 0 \end{bmatrix} \tag{17}$$

in which columns 7 through 11 contain zeros. Then

$$\hat{\boldsymbol{\beta}}_1'\mathbf{C}' = [-0.1427, \quad -0.3492, \quad 1.1314]$$

$$\mathbf{C}(\mathbf{X}_1'\mathbf{X}_1)^{-1}\mathbf{C}' = 10^{-4}\begin{bmatrix} 2.07567 & -0.18170 & -2.99099 \\ & 11.24444 & 2.51987 \\ & & 149.63006 \end{bmatrix}$$

$$[\mathbf{C}(\mathbf{X}_1'\mathbf{X}_1)^{-1}\mathbf{C}']^{-1} = \begin{bmatrix} 4964.384 & 58.202 & 98.254 \\ & 893.380 & -13.882 \\ & & 69.029 \end{bmatrix}$$

$$\hat{\sigma}^2 = 35.5268 \qquad \hat{\sigma} = 5.9604$$

The F-statistic (5) for testing H_0 is equal to 2.66. Since $F_{0.05;3,34}$ is approximately 2.89 the hypothesis cannot be rejected at the 0.05 level. The evidence is not strong that the Draw a Person score is related to the three performance test scores in the battery.

A more general hypothesis: H_0: $\mathbf{C}\boldsymbol{\beta}_1 = \mathbf{a}$. Suppose that an investigator wishes to test that one or more linear compounds of the regression coefficients are equal to specific constants rather than zero. For example, an economist constructing a regression model for the selling prices of homes as functions of the independent variables X_1, X_2, X_3, and X_4 might wish to test the hypothesis that the relations

$$\beta_1 - \beta_2 = 2 \qquad \beta_2 = \beta_3 \qquad \beta_4 = 5 \tag{18}$$

among the regression coefficients were simultaneously true. Similarly, a chemist conducting a series of experiments measuring the effect of industrial air pollution on the life and wearing quality of a new paint might wish to test the hypothesis

$$H_0: \quad \begin{bmatrix} \beta_1 \\ \beta_2 \\ \beta_3 \end{bmatrix} = \begin{bmatrix} -2 \\ -5 \\ -12 \end{bmatrix} \tag{19}$$

on the regression coefficients of the concentrations of three chemical compounds in the atmosphere surrounding the paint applications. The particular values of the β_j were suggested by previous studies of the relationship of an index of life and quality of the painted surface with the three concentrations. In the present experiment the chemist wishes to determine if the same coefficients still obtain.

Let the generalization of the hypothesis (2) be

$$H_0: \quad \mathbf{C}\boldsymbol{\beta}_1 = \mathbf{a} \tag{20}$$

where $\mathbf{C}$ is a given $r \times p$ matrix of constants, and $\mathbf{a}$ is an $r \times 1$ vector also specified by the investigator. Then the statistic

$$F = \frac{[(N-p-1)/r](\mathbf{C}\hat{\boldsymbol{\beta}}_1 - \mathbf{a})'[\mathbf{C}(\mathbf{X}_1'\mathbf{X}_1)^{-1}\mathbf{C}']^{-1}(\mathbf{C}\hat{\boldsymbol{\beta}}_1 - \mathbf{a})}{\text{SSE}} \tag{21}$$

has the central F-distribution with degrees of freedom r and $N - p - 1$ when H_0 is true. Clearly, r cannot exceed p if the matrix $\mathbf{C}(\mathbf{X}_1'\mathbf{X}_1)^{-1}\mathbf{C}'$ is to have an inverse. We reject H_0 at the α level if

$$F > F_{\alpha;r,N-p-1}$$

and otherwise consider that H_0 is tenable in light of the available data.

Conversely, we may test an infinity of such hypotheses by means of the $100(1-\alpha)\%$ joint confidence region

$$(\mathbf{C}\boldsymbol{\beta}_1 - \mathbf{C}\hat{\boldsymbol{\beta}}_1)'[\mathbf{C}(\mathbf{X}_1'\mathbf{X}_1)^{-1}\mathbf{C}']^{-1}(\mathbf{C}\boldsymbol{\beta}_1 - \mathbf{C}\hat{\boldsymbol{\beta}}_1) \leq \frac{r\,\text{SSE}}{N-p-1} F_{\alpha;r,N-p-1} \tag{22}$$

In practice, though, such joint confidence regions are tedious to evaluate and difficult to visualize for p greater than three or even two. A more convenient representation based on linear confidence intervals is treated in our development of multiple comparisons and simultaneous inference in Section 3-5.

The reduced model analysis of variance for hypotheses on subsets of regression coefficients. Frequently, the test of the hypothesis that r of the regression coefficients are zero is given in a different manner through an analysis of variance table. Let us assume that the hypothesis refers to the first r coefficients, as in (1), and that its matrix representation is in terms of the $r \times p$ matrix $\mathbf{C}$ defined by (3). We begin by partitioning the components of the regression model as

$$\begin{aligned}\mathbf{y} &= [\mathbf{j} \quad \mathbf{X}_r \quad \mathbf{X}_c]\begin{bmatrix}\alpha \\ \boldsymbol{\beta}_r \\ \boldsymbol{\beta}_c\end{bmatrix} + \mathbf{e} \\ &= \alpha\mathbf{j} + \mathbf{X}_r\boldsymbol{\beta}_r + \mathbf{X}_c\boldsymbol{\beta}_c + \mathbf{e} \qquad (23)\end{aligned}$$

where the $N \times r$ matrix $\mathbf{X}_r$ consists of the first r columns of $\mathbf{X}_1$ and the $N \times (p - r)$ matrix $\mathbf{X}_c$ contains the remaining values of the independent variables. The subscripts r and c refer to the r variables and their complement, respectively. Then the least squares estimators of the regression parameters can be computed from the formula for the inverse of a partitioned matrix (Morrison, 1976, Section 2.11) as

$$\begin{aligned}\begin{bmatrix}\hat{\boldsymbol{\beta}}_r \\ \hat{\boldsymbol{\beta}}_c\end{bmatrix} &= \begin{bmatrix}\mathbf{X}_r'\mathbf{X}_r & \mathbf{X}_r'\mathbf{X}_c \\ \mathbf{X}_c'\mathbf{X}_r & \mathbf{X}_c'\mathbf{X}_c\end{bmatrix}^{-1}\begin{bmatrix}\mathbf{X}_r'\mathbf{y} \\ \mathbf{X}_c'\mathbf{y}\end{bmatrix} \\ &= \begin{bmatrix}\mathbf{A}_{rr.c}^{-1} & -\mathbf{A}_{rr.c}^{-1}\mathbf{X}_r'\mathbf{X}_c(\mathbf{X}_c'\mathbf{X}_c)^{-1} \\ -(\mathbf{X}_c'\mathbf{X}_c)^{-1}\mathbf{X}_c'\mathbf{X}_r\mathbf{A}_{rr.c}^{-1} & (\mathbf{X}_c'\mathbf{X}_c)^{-1} + (\mathbf{X}_c'\mathbf{X}_c)^{-1}\mathbf{X}_c'\mathbf{X}_r\mathbf{A}_{rr.c}^{-1}(\mathbf{X}_r'\mathbf{X}_c)(\mathbf{X}_c'\mathbf{X}_c)^{-1}\end{bmatrix} \\ &\quad \cdot \begin{bmatrix}\mathbf{X}_r'\mathbf{y} \\ \mathbf{X}_c'\mathbf{y}\end{bmatrix} \qquad (24)\end{aligned}$$

where

$$\mathbf{A}_{rr.c} = \mathbf{X}_r'\mathbf{X}_r - \mathbf{X}_r'\mathbf{X}_c(\mathbf{X}_c'\mathbf{X}_c)^{-1}\mathbf{X}_c'\mathbf{X}_r \qquad (25)$$

Then the sum of squares due to the p independent variables in the full regression model may be written as

$$\begin{aligned}\mathrm{SSR}(\boldsymbol{\beta}_1) &= \hat{\boldsymbol{\beta}}_1'\mathbf{X}_1'\mathbf{X}_1\hat{\boldsymbol{\beta}}_1 \\ &= \mathbf{y}'\mathbf{X}_r\mathbf{A}_{rr.c}^{-1}\mathbf{X}_r'\mathbf{y} - 2\mathbf{y}'\mathbf{X}_r\mathbf{A}_{rr.c}^{-1}\mathbf{X}_r'\mathbf{X}_c(\mathbf{X}_c'\mathbf{X}_c)^{-1}\mathbf{X}_c'\mathbf{y} \\ &\quad + \mathbf{y}'\mathbf{X}_r(\mathbf{X}_c'\mathbf{X}_c)^{-1}\mathbf{X}_c'\mathbf{X}_r\mathbf{A}_{rr.c}^{-1}\mathbf{X}_r'\mathbf{X}_c(\mathbf{X}_c'\mathbf{X}_c)^{-1}\mathbf{X}_c'\mathbf{y} \\ &\quad + \mathbf{y}'\mathbf{X}_c(\mathbf{X}_c'\mathbf{X}_c)^{-1}\mathbf{X}_c'\mathbf{y} \qquad (26)\end{aligned}$$

Now let us consider a new model involving only the variables in the complementary set. It is

$$\mathbf{y} = \alpha_c \mathbf{j} + \mathbf{X}_c \boldsymbol{\beta}_c^* + \mathbf{e} \tag{27}$$

The different intercept α_c has been noted; the disturbance variates are also different, although we have used the same symbol $\mathbf{e}$ for them. The least squares estimator of $\boldsymbol{\beta}_c^*$ is of course

$$\hat{\boldsymbol{\beta}}_c^* = (\mathbf{X}_c'\mathbf{X}_c)^{-1}\mathbf{X}_c'\mathbf{y} \tag{28}$$

and the sum of squares due to its fitted regression model is

$$\text{SSR}(\boldsymbol{\beta}_c^*)_{\text{unadj.}} = \mathbf{y}'\mathbf{X}_c(\mathbf{X}_c'\mathbf{X}_c)^{-1}\mathbf{X}_c\mathbf{y} \tag{29}$$

The fact that the estimator $\hat{\boldsymbol{\beta}}_c^*$ is computed only from the independent variables of $\mathbf{X}_c$ without taking into account the values of the variables in the first set is indicated by the qualification "unadjusted" in the symbol for the regression sum of squares (29). That sum of squares is the last term of $\text{SSR}(\boldsymbol{\beta}_1)$, the sum of squares for the full regression model. Hence

$$\begin{aligned}
\text{SSR}(\boldsymbol{\beta}_1) - \text{SSR}(\boldsymbol{\beta}_c^*)_{\text{unadj.}} &= \mathbf{y}'\mathbf{X}_r\mathbf{A}_{rr.c}^{-1}\mathbf{X}_r'\mathbf{y} \\
&\quad -2\mathbf{y}'\mathbf{X}_r\mathbf{A}_{rr.c}^{-1}\mathbf{X}_r'\mathbf{X}_c(\mathbf{X}_c'\mathbf{X}_c)^{-1}\mathbf{X}_c'\mathbf{y} \\
&\quad +\mathbf{y}'\mathbf{X}_c(\mathbf{X}_c'\mathbf{X}_c)^{-1}\mathbf{X}_c'\mathbf{X}_r\mathbf{A}_{rr.c}^{-1}\mathbf{X}_r'\mathbf{X}_c(\mathbf{X}_c'\mathbf{X}_c)^{-1}\mathbf{X}_c'\mathbf{y} \\
&= \mathbf{y}'[\mathbf{X}_r - \mathbf{X}_c(\mathbf{X}_c'\mathbf{X}_c)^{-1}\mathbf{X}_c'\mathbf{X}_r]\mathbf{A}_{rr.c}^{-1}\mathbf{A}_{rr.c} \\
&\quad \cdot \mathbf{A}_{rr.c}^{-1}[\mathbf{X}_r' - \mathbf{X}_r'\mathbf{X}_c(\mathbf{X}_c'\mathbf{X}_c)^{-1}\mathbf{X}_c']\mathbf{y} \\
&= \hat{\boldsymbol{\beta}}_r'\mathbf{A}_{rr.c}\hat{\boldsymbol{\beta}}_r \\
&= (\mathbf{C}\hat{\boldsymbol{\beta}}_1)'[\mathbf{C}(\mathbf{X}_1'\mathbf{X}_1)^{-1}\mathbf{C}']^{-1}\mathbf{C}\hat{\boldsymbol{\beta}}_1 \\
&= \text{SSR}(\boldsymbol{\beta}_r)
\end{aligned} \tag{30}$$

or the sum of squares given by (4) for testing the hypothesis (1). We may summarize the decomposition of the total regression sum of squares in the analysis of variance of Table 3-3. To test H_0 we would compute

TABLE 3-3 Analysis of Variance for $H_0: \boldsymbol{\beta}_1 = \cdots = \boldsymbol{\beta}_r = 0$

Source	Sum of squares	d.f.	Mean square
Intercept (mean)	$N\bar{y}^2$	1	$N\bar{y}^2$
Subset $\mathbf{X}_r$ (adjusted)	$\text{SSR}(\boldsymbol{\beta}_r) = \text{SSR}(\boldsymbol{\beta}_1) - \text{SSR}(\boldsymbol{\beta}_c^*)_{\text{unadj.}}$	r	$\frac{\text{SSR}(\boldsymbol{\beta}_r)}{r}$
$\mathbf{X}_c$ (unadjusted)	$\text{SSR}(\boldsymbol{\beta}_c^*)_{\text{unadj.}}$	$p - r$	—
Full regression model	$\text{SSR}(\boldsymbol{\beta}_1)$	p	$\frac{\text{SSR}(\boldsymbol{\beta}_1)}{p}$
Error	SSE	$N - p - 1$	$\frac{\text{SSE}}{N - p - 1}$
Total	$\mathbf{y}'\mathbf{y}$	N	—

$$F = \frac{N-p-1}{r} \frac{\text{SSR}(\boldsymbol{\beta}_r)}{\text{SSE}} \tag{31}$$

as in (5) and refer it to the same F critical value with r and $N - p - 1$ degrees of freedom.

The unadjusted sum of squares $\text{SSR}(\boldsymbol{\beta}_c^*)_{\text{unadj.}}$ has no use for a hypothesis test unless $\mathbf{X}_r'\mathbf{X}_c = 0$, or orthogonality of the two sets of variables. Then $\text{SSR}(\boldsymbol{\beta}_c^*)_{\text{unadj.}}$ is distributed independently of $\text{SSR}(\boldsymbol{\beta}_r)$ and can be used to test the hypothesis H_0: $\beta_{r+1} = \cdots = \beta_p = 0$ of zero regression coefficients for the variables in the complementary set. If the orthogonality property does not hold, the sum of squares for $\boldsymbol{\beta}_c$ is not properly adjusted for the effects of the variables in $\mathbf{X}_r$ and cannot be used for the test.

Example 3-4

We repeat the hypothesis test of Example 3-3 for zero regression coefficients of variables 1, 5, and 6 by the reduced model analysis of variance. The sums of products and regression coefficient vectors for the remaining eight independent variables are

$$\mathbf{X}_c'\mathbf{y} = \begin{bmatrix} -422.43 \\ 2006.39 \\ -4391.70 \\ -142.65 \\ -270.04 \\ -106.57 \\ -136.90 \\ -182.74 \end{bmatrix} \qquad \hat{\boldsymbol{\beta}}_c^* = \begin{bmatrix} 1.39 \\ 0.28 \\ -0.85 \\ 0.35 \\ 1.37 \\ 2.30 \\ -1.91 \\ 2.83 \end{bmatrix}$$

and the necessary sums of squares are

$$\text{SSR}(\boldsymbol{\beta}_1) = 3043.83 \qquad \text{SSR}(\boldsymbol{\beta}_c^*)_{\text{unadj.}} = 2760.36$$

$$\sum_{i=1}^{46} y_i^2 = 22{,}096 \qquad N\bar{y}^2 = 17{,}844.26$$

The analysis of variance for H_0: $\beta_1 = \beta_5 = \beta_6 = 0$ is given in Table 3-4.

TABLE 3-4 Analysis of Variance for Testing H_0: $\beta_1 = \beta_5 = \beta_6 = 0$ in the Psychological Test Data

Source	Sum of squares	d.f.	Mean square	F
Mean	17,844.26	1	17,844.26	—
Variables 1, 5, 6	283.47	3	94.49	2.66
$\mathbf{X}_c$ (unadjusted)	2,760.36	8	—	—
Full regression model	3,043.83	11	276.712	7.99
Error	1,207.91	34	35.5268	—
Total	22,096	46	—	—

3-4 MULTIPLE OBSERVATIONS AT THE SAME PREDICTOR VALUES

In the preceding development of linear regression and correlation we have assumed general values of the dependent and independent variables. No provision was made for the case in which several observations were obtained on the dependent variable for a single value of the independent variable or its vector of values in the multiple regression model. Although the estimators and tests are still valid in that case, we will have failed to make use of some other inferential procedures that are possible with the multiple Y-values. In particular these include:

1. Estimation of the disturbance term variance σ^2 from the groups of multiple observations, so that the estimator is unaffected by the validity of the specified model
2. A test for the "goodness of fit" of the model to the observations

In this section we develop the multiple observation model and an analysis of variance for the estimator and the fit of the model to the data.

Let us begin with an illustration. A large laboratory possesses equipment for very accurate assays of the concentrations of contaminants in water, vegetable matter, or such body fluids as blood. Six specimens of a fluid are prepared with different levels of a contaminant, for example, mercury or lead, and then assayed. The respective concentrations were determined to be 9, 22, 30, 39, 50, and 60 micrograms per 100 milliliters. Each specimen was divided into five equal parts, and the 30 samples were sent in identical coded containers to another laboratory for assay of the contaminant's levels by a less elaborate and precise device. The results of the two types of assays might have the appearance of Figure 3-2. The spread of the second assay values for a given level of the first gives an immediate sense of the error variance without fitting a particular model. Comparison of the spreads for the six initial levels provides a visual test of the common-variance assumption. The plot of the means denoted by the open points suggests the kind of relationship between the two kinds of assay. If the common-variance assumption seems tenable the sums of squared deviations about the means can be pooled for an estimate of the variance.

For the development of the least squares estimators and their statistical properties in the general case of p independent variables we start with the usual linear model

$$\mathbf{y} = [\mathbf{j} \quad \mathbf{X}_1]\begin{bmatrix} \alpha \\ \boldsymbol{\beta} \end{bmatrix} + \mathbf{e} \tag{1}$$

in which the matrix of centered independent variable values consists of the vector $\mathbf{x}_1'$ in the first N_1 rows, the vector $\mathbf{x}_2'$ in the next N_2 rows, and so on, until the last N_k rows containing the vector $\mathbf{x}_k'$. The jth such vector

$$\mathbf{x}_j' = [x_{j1} - \bar{x}_1, \ldots, x_{jp} - \bar{x}_p] \tag{2}$$

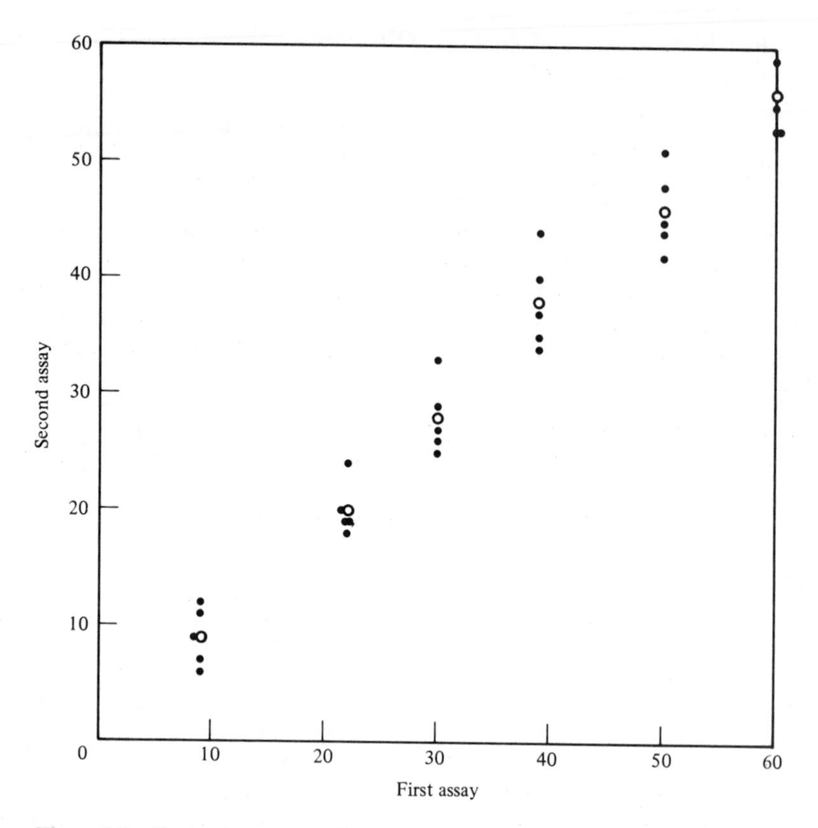

Figure 3-2 Contaminant assays from the two laboratories. Open points denote the group means.

consists of the jth set of values of the p independent variables expressed as deviations about their grand means

$$\bar{x}_h = \frac{1}{N} \sum_{j=1}^{k} N_j x_{jh} \qquad h = 1, \ldots, p \tag{3}$$

where $N = N_1 + \cdots + N_k$ is the total number of independent observations on the dependent variable Y. The dependent variable observations can be written as the partitioned vector

$$\mathbf{y} = \begin{bmatrix} \mathbf{y}_1 \\ \cdot \\ \cdot \\ \cdot \\ \mathbf{y}_k \end{bmatrix} \tag{4}$$

where

$$\mathbf{y}_j = \begin{bmatrix} y_{1j} \\ \cdot \\ \cdot \\ \cdot \\ y_{N_j j} \end{bmatrix} \tag{5}$$

Let the mean of those multiple observations be

$$\bar{y}_j = \frac{1}{N_j} \sum_{i=1}^{N_j} y_{ij} \tag{6}$$

We shall also need the $k \times p$ matrix

$$\mathbf{X}_1^* = \begin{bmatrix} \mathbf{x}_1' \\ \cdot \\ \cdot \\ \cdot \\ \mathbf{x}_k' \end{bmatrix} \tag{7}$$

of the centered distinct independent variable values, the $k \times 1$ vector

$$\bar{\mathbf{y}} = \begin{bmatrix} \bar{y}_1 \\ \cdot \\ \cdot \\ \cdot \\ \bar{y}_k \end{bmatrix} \tag{8}$$

of the means of the dependent variable values for the k independent variable groups, and a diagonal matrix

$$\mathbf{D} = \begin{bmatrix} N_1 & \cdots & 0 \\ \cdot & & \cdot \\ \cdot & & \cdot \\ \cdot & & \cdot \\ 0 & \cdots & N_k \end{bmatrix} \tag{9}$$

of the group sample sizes. Then

$$\begin{aligned} \mathbf{X}_1'\mathbf{X}_1 &= \sum_{j=1}^{k} N_j \mathbf{x}_j \mathbf{x}_j' = \mathbf{X}_1^{*\prime}\mathbf{D}\mathbf{X}_1^* \\ \mathbf{X}_1'\mathbf{y} &= \sum_{j=1}^{k} N_j \mathbf{x}_j \bar{y}_j = \mathbf{X}_1^{*\prime}\mathbf{D}\bar{\mathbf{y}} \end{aligned} \tag{10}$$

and the least squares estimator of the regression coefficients is

$$\hat{\boldsymbol{\beta}}_1 = [\mathbf{X}_1^{*\prime}\mathbf{D}\mathbf{X}_1^*]^{-1}\mathbf{X}_1^{*\prime}\mathbf{D}\bar{\mathbf{y}} \tag{11}$$

The estimator of the intercept parameter is of course merely the grand mean of all the y_i, or

$$\hat{\alpha} = \bar{y} = \frac{1}{N} \sum_{j=1}^{k} N_j \bar{y}_j \tag{12}$$

We note that the estimators (11) and (12) are *equivalent to fitting the regression model to the means $\bar{y}_1, \ldots, \bar{y}_k$ with those values and the independent variable observations weighted inversely by the variances $\sigma^2/N_1, \ldots, \sigma^2/N_k$ of the means.* This is an example of *weighted least squares*, or least squares estimation applied to a linear model whose disturbance terms have unequal variances $\sigma^2/N_1, \ldots,$ σ^2/N_k. We shall derive and justify such estimators in our treatment of generalized least squares in Chapter 6.

Now let us see what will come of manipulating the error sum of squares in the present case of multiple observations. It is

$$\begin{aligned} \text{SSE} &= (\mathbf{y} - \hat{\mathbf{y}})'(\mathbf{y} - \hat{\mathbf{y}}) \\ &= (\mathbf{y} - \bar{y}\mathbf{j} - \mathbf{X}_1\hat{\boldsymbol{\beta}}_1)'(\mathbf{y} - \bar{y}\mathbf{j} - \mathbf{X}_1\hat{\boldsymbol{\beta}}_1) \\ &= \sum_{j=1}^{k}\sum_{i=1}^{N_j}(y_{ij} - \bar{y} - \mathbf{x}_j'\hat{\boldsymbol{\beta}}_1)^2 \end{aligned} \tag{13}$$

Introduce $-\bar{y}_j + \bar{y}_j$ within the parentheses of the last term. Then

$$\begin{aligned} \text{SSE} &= \sum_{j=1}^{k}\sum_{i=1}^{N_j}[y_{ij} - \bar{y}_j - (\bar{y} + \mathbf{x}_j'\hat{\boldsymbol{\beta}}_1 - \bar{y}_j)]^2 \\ &= \sum_{j=1}^{k}\sum_{i=1}^{N_j}(y_{ij} - \bar{y}_j)^2 + \sum_{j=1}^{k} N_j(\bar{y}_j - \bar{y} - \mathbf{x}_j'\hat{\boldsymbol{\beta}}_1)^2 \end{aligned} \tag{14}$$

since the crossproduct term vanishes upon evaluating the sum $\sum_i (y_{ij} - \bar{y}_j)$ contained in it. The first term in (14) is the sum of squared deviations about the means of the dependent variable in each of the k common-value groups:

$$\text{SSW} = \sum_{j=1}^{k}\sum_{i=1}^{N_j}(y_{ij} - \bar{y}_j)^2 \tag{15}$$

is the sum of k independent $\chi^2\sigma^2$ variates with respective degrees of freedom $N_1 - 1, \ldots, N_k - 1$. Hence SSW/σ^2 is a chi-squared random variable with $N - k$ degrees of freedom. This distribution depends only on the independence, normal distributions, and common variances of the disturbance terms in the linear regression model, not on the validity of the model chosen for the analysis of the data. For that reason the estimator

$$\hat{\sigma}^2 = \frac{\text{SSW}}{N - k} \tag{16}$$

is sometimes referred to as the estimator based on "pure error." A useful discussion of its properties vis-à-vis those of the estimator based upon SSE has been given by Draper and Smith (1981).

The second term comprising SSE in (14) is

$$\begin{aligned} \text{SSF} &= \sum_{j=1}^{k} N_j(\bar{y}_j - \bar{y} - \mathbf{x}_j'\hat{\boldsymbol{\beta}}_1)^2 \\ &= \sum_{j=1}^{k} N_j(\bar{y}_j - \hat{y}_j)^2 \end{aligned} \tag{17}$$

or the *sum of squares due to the fit of the model* (1) *to the data*. We see from the second line of (17) that SSF is the weighted sum of squared deviations of the group means from their values predicted by the linear regression model. SSF is distributed independently of SSW, a fact that follows from an application of Cochran's theorem to the decomposition

$$\sum_{j=1}^{k}\sum_{i=1}^{N_j} y_{ij}^2 = N\bar{y}^2 + \text{SSR} + \text{SSW} + \text{SSF} \tag{18}$$

of the total sum of squares of the dependent variable observations. Cochran's theorem also implies that SSF/σ^2 has a chi-squared distribution with $k - p - 1$ degrees of freedom. If the model (1) is the correct one for the data, $E(\mathbf{y}) = \alpha\mathbf{j} + \mathbf{X}_1\boldsymbol{\beta}_1$, and SSF/σ^2 will have a *central* chi-squared distribution. We may test the goodness of fit of the model to the data, or

$$H_0: \quad E(\bar{y}_j) = \mu_j = \alpha + \mathbf{x}_j'\boldsymbol{\beta}_1 \qquad j = 1, \ldots, k \tag{19}$$

against the alternative of general μ_j values by referring

$$F = \frac{[(N-k)/(k-p-1)]\mathrm{SSF}}{\mathrm{SSW}} \tag{20}$$

to the $100\alpha\%$ critical value of the F-distribution with $k - p - 1$ and $N - k$ degrees of freedom.

The usual hypothesis $H_0: \boldsymbol{\beta}_1 = \mathbf{0}$ of no relation between the dependent and independent variables may be tested by computing

$$F = \frac{[(N-k)/p]\mathrm{SSR}}{\mathrm{SSW}} \tag{21}$$

where

$$\mathrm{SSR} = \hat{\boldsymbol{\beta}}_1'\mathbf{X}_1^{*\prime}\mathbf{D}\mathbf{X}_1^*\hat{\boldsymbol{\beta}}_1 \tag{22}$$

and rejecting H_0 at the α level if

$$F > F_{\alpha;\,p,N-k}$$

This test and that for the fit of the model may be summarized in the analysis of variance of Table 3-5. If the goodness-of-fit statistic (20) does not exceed its chosen critical value, a more sensitive test of the hypothesis of zero regression coefficients can be obtained by replacing (21) by the statistic

$$F = \frac{[(N-p-1)/p]\mathrm{SSR}}{\mathrm{SSE}} \tag{23}$$

and referring it to the critical value $F_{\alpha;\,p,N-p-1}$.

Example 3-5

As part of an investigation of blood lead levels in children an experiment was conducted to compare determinations of blood lead level by an isotopic dilution mass-spectrometric technique used by the National Bureau of Standards and a microanalytic spectrophotometric procedure used by a commercial clinical laboratory (Urban, 1976). Blood samples were obtained from pigs that had been fed lead-bearing foods, and samples with different levels of lead were made by blending blood from pigs with different lead levels. From each of those samples with accurately measured lead levels x_j several small specimens were taken and sent to the commercial clinical laboratory for assay. The data for $k = 6$ NBS levels $x_1, \ldots, x_6$ and a total of $N = 57$ clinical laboratory assays are shown in Table 3-6 and Figure 3-3. The statistics computed from those values are contained in columns 2 through 5 of Table 3-7. The means and sums of squares and products from the total sample of 57 pairs of determinations are

$$\bar{x} = 30.679 \qquad \bar{y} = 29.193$$

$$\sum N_j(x_j - \bar{x})^2 = 38{,}594.49 \qquad \sum\sum (y_{ij} - \bar{y})^2 = 27{,}470.88$$

$$\sum N_j(x_j - \bar{x})(\bar{y}_j - \bar{y}) = 31{,}598.73$$

TABLE 3-5 Analysis of Variance for Regression with Multiple Observations

Source	Sum of squares	d.f.	Mean square	F
Linear regression	$SSR = \hat{\boldsymbol{\beta}}_1' \mathbf{X}^{*\prime} \mathbf{D} \mathbf{X}_1^* \hat{\boldsymbol{\beta}}_1$	p	$\frac{SSR}{p}$	$\frac{[(N-k)/p]\,SSR}{SSW}$
Lack of fit	$SSF = \sum_{j=1}^{k} N_j(\bar{y}_j - \hat{y}_j)^2$	$k-p-1$	$\frac{SSF}{k-p-1}$	$\frac{[(N-k)/(k-p-1)]SSF}{SSW}$
Within groups	$SSW = \sum_{j=1}^{k} \sum_{i=1}^{N_j} (y_{ij} - \bar{y}_j)^2$	$N-k$	$\frac{SSW}{N-k}$	—
Error	$SSE = SSF + SSW$	$N-p-1$	$\frac{SSE}{N-p-1}$	—
Total (less mean)	$\sum_{j=1}^{k} \sum_{i=1}^{N_j} (y_{ij} - \bar{y})^2$	$N-1$	—	—

TABLE 3-6 NBS and Clinical Laboratory Porcine Blood Levels (μg Pb/100 ml)

National Bureau of Standards level, x_j									
17.5			20.0		23.2	23.8		33.7	105.3
23	21	12	15	20	29	29	14	31	99
23	29	17	21	25	29	24	29	28	99
18	17	19	12	24	27	28	17	34	85
18	16	15	20	21	22	19	13	31	98
19	17	15	19	30	18	23		33	72
19	19		22	14	22	29		33	89

SOURCE: Data reproduced with the permission of the Institute of Applied Technology, National Bureau of Standards.

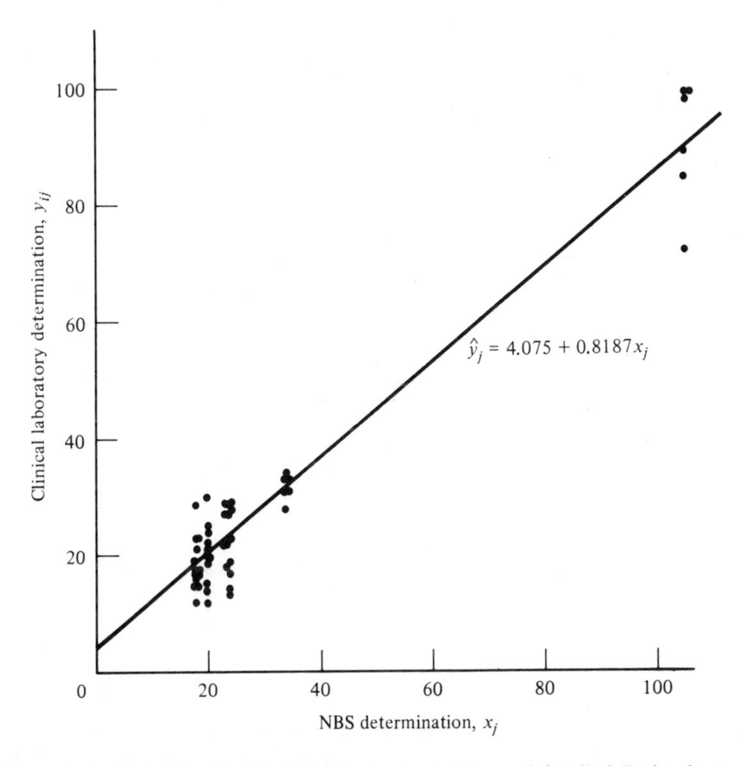

Figure 3-3 Blood lead levels. (Data from the Institute of Applied Technology, National Bureau of Standards.)

TABLE 3-7 Clinical Laboratory Blood Statistics

NBS value, x_j	N_j	$\bar{y}_j$	$\sum (y_{ij} - \bar{y}_j)^2$	SD_j	$\hat{y}_j$	$N_j(\hat{y}_j - \bar{y}_j)^2$
17.5	17	18.65	237.88	3.86	18.40	1.0138
20.0	12	20.25	272.25	4.97	20.45	0.4788
23.2	6	24.50	101.50	4.51	23.07	12.2741
23.8	10	22.50	364.50	6.36	23.56	11.2566
33.7	6	31.67	23.33	2.16	31.67	0
105.3	6	90.33	575.33	10.73	90.29	0.0122
Total	57	—	1574.80	—	—	25.0355

The estimated regression equation is

$$\begin{aligned}\hat{y}_j &= 29.193 + 0.8187(x_j - 30.679) \\ &= 4.075 + 0.8187x_j \qquad j = 1, \ldots, 6\end{aligned} \tag{24}$$

The predicted values of the clinical laboratory blood lead level for the six National Bureau of Standards determinations are shown in column six of Table 3-7, while the weighted squared deviations of the actual and predicted levels are given in the last column. The coefficient of determination based on the simple correlation of the two series of levels is 0.9418.

The analysis of variance is shown in Table 3-8. We see that the estimate of σ^2 obtained from the variation of the clinical laboratory values within the six NBS groups is 30.88. The F-statistic for testing the fit hypothesis is only equal to 0.20, and is not of course statistically significant. For that reason the lack of fit sum of squares probably should be pooled with the pure error sum of squares from the NBS assay groups. Then the additional degrees of freedom give the smaller estimate of σ^2 equal to 29.09.

TABLE 3-8 Analysis of Variance for the Clinical Laboratory Blood Lead Levels

Source	Sum of squares	d.f.	Mean square	F
Linear regression	25,871.04	1	25,871.04	838
Lack of fit	25.04	4	6.26	0.20
Within NBS levels	1,574.80	51	30.88	—
Error	1,599.84	55	29.09	—
Total (less mean)	27,470.88	56	—	—

Several other questions might be asked of these data, and answered in turn by statistical analysis. In particular, we would prefer to have a one-to-one relationship between the two series of blood lead determinations, with a regression coefficient within sampling variation range of unity. Rather than test $H_0\colon \beta = 1$ we shall find the 95% confidence interval for the parameter. The limits of the interval are given by

$$\hat{\beta} \pm t_{0.025;\nu}\hat{\sigma}/\sqrt{\sum N_j(x_j - \bar{x})^2}$$

where we shall use the within-groups estimator of $\hat{\sigma}$ with its $\nu = 51$ degrees of freedom. Then

$$t_{0.025;51} = 2.01 \qquad \hat{\sigma} = 5.557 \qquad \sum N_j(x_j - \bar{x})^2 = 38{,}594.49$$

and the confidence interval is

$$0.7618 \leq \beta \leq 0.8756$$

The hypothesis of a unit regression coefficient is untenable at the 0.05 level, or at any other reasonably smaller one. We must conclude that the clinical laboratory blood lead determinations tend to be lower than those obtained by the National Bureau of Standards as the lead concentration increases.

3-5 SIMULTANEOUS TESTS AND CONFIDENCE INTERVALS FOR REGRESSION COEFFICIENTS

In Chapter 2 and the preceding sections of this chapter we have given hypothesis tests and confidence regions or intervals for the intercept parameter and regression coefficient vector, subsets of the regression coefficients, and individual β_j. In each case we have assumed that the hypothesis or other inference was chosen prior to an examination of the data or the results of the regression analysis, and that the Type I error rate and power probability of the test or confidence statement referred *only* to that test or interval. That is, we came to the analysis with specific questions to ask of its results, rather than generating those questions as we examined a computer program printout. Our significance levels or error rates held only for each *individual* test, with no provision for the effects of making many dependent tests on the same body of observations.

In practice our inferences about the individual regression coefficients or sets of them do not proceed in such an idealized manner. Although we may have prior or substantive reasons for expecting certain predictor variables to be related to the dependent variable, and plan to test $H_0: \beta_j = 0$ for the values of j corresponding to them, it is more likely that we shall read down the column of $\hat{\boldsymbol{\beta}}_j$- and t_j-values with a 0.05 or 0.01 critical value of t in mind and mark asterisks at the lines whose t-statistics are significant at those levels. However, if some independent variables are highly correlated with one another and with the dependent variable, or if the number of independent variables is large, those individual t- or F-tests will tend to overstate the number of significant coefficients: Chance alone would lead to $|t_j|$-values greater than the critical point if a large number of tests is carried out. Clearly, we need some means of controlling the *family error rate* when many hypotheses are to be tested or several confidence intervals contructed. The simultaneous inference methods of this section will provide control on the Type I error rates of families of tests and confidence statements.

It will be helpful to begin with an example of a *family of tests*. Suppose that we are interested in this simplified, if rather artificial, regression analysis: $p = 20$ independent variables have been related to a dependent variable Y on $N = 200$ independent sampling units. The independent variables were truly under the control of the investigator, so that their values were chosen to be mutually orthogonal, with zero sums, and sums of squares equal to unity. Then $\mathbf{X}_1'\mathbf{X}_1 = \mathbf{I}$, or the 20×20 identity matrix. In addition, the disturbance term variance σ^2 will be assumed known from previous information, or at least its estimate based on $N - p - 1 = 179$ degrees of freedom will be regarded as a parameter. We may test each individual hypothesis

$$H_{0j}: \quad \beta_j = 0 \qquad j = 1, \ldots, 20 \tag{1}$$

at the γ level by referring the test statistics $t_j = \hat{\beta}_j/\sigma$ to the $100\gamma\%$ two-sided critical value $z_{\gamma/2}$ of the standard normal distribution. If all 20 hypotheses are in fact true the probability of accepting all of them is

$$\begin{aligned} &P(\text{accept the family of hypotheses} \mid \text{all } H_{0j} \text{ true}) \\ &\quad = \prod_{j=1}^{20} P(\text{accept } H_{0j} \mid H_{0j} \text{ true}) \\ &\quad = (1 - \gamma)^{20} \end{aligned} \tag{2}$$

and the probability that at least one will be rejected when all are true is

$$\alpha = P(\text{reject one or more } H_{0j} \mid \text{all } H_{0j} \text{ true}) = 1 - (1 - \gamma)^{20} \tag{3}$$

We shall use the latter probability as the *error rate of the family of hypotheses* (1), or an example of a general *family* error rate. The family error rate will usually be much larger than the individual rate γ. In general, in the present example of independent test statistics each test is a Bernoulli trial with probability of a correct decision, or "success," $1 - \gamma$. Then the probability of exactly s acceptances, or $p - s$ rejections, is the binomial probability

$$p(s) = \binom{p}{s}(1 - \gamma)^s \gamma^{p-s} \qquad s = 0, \ldots, p \tag{4}$$

and we would expect $p\gamma$ hypotheses to be rejected even when all are true. In the present case of $p = 20$ and an individual test level of $\gamma = 0.05$, the family error rate (3) would be 0.6415, or an unacceptably high value. We should expect, in the average sense, one false rejection in the series of 20 tests. If $\gamma = 0.01$, the family error rate drops to 0.1821—still too large a value. However, we see that we can control the family rate by using a smaller probability of false rejection for the individual tests. In general

$$\gamma = 1 - (1 - \alpha)^{1/p} \tag{5}$$

For a family rate of 0.05, we should make the individual tests at the level $\gamma = 0.00256$. For $\alpha = 0.01$, $\gamma = 0.000502$. The respective decision rules would be

$$\begin{aligned} &1.\ \text{Reject } H_{0j} \quad \text{if} \quad |t_j| \geq z_{\gamma/2} = 3.022. \\ &2.\ \text{Reject } H_{0j} \quad \text{if} \quad |t_j| \geq z_{\gamma/2} = 3.48. \end{aligned} \tag{6}$$

Larger values of the $\hat{\beta}_j$ will be needed to declare their population values different from zero, but the risks of incorrect decisions will be held to less than 0.05 and 0.01 for the first and second rules.

Let us conjure up $p = 20$ hypothetical t-statistics for the preceding motivating example. The first 14 t_j were drawn from a normal distribution with mean zero and variance 1, whereas $t_{15}, \ldots, t_{20}$ came from a normal distribution with mean 2 and unit variance. The values of the t_j and their significance with respect to the ordinary single-test two-sided normal critical values $z_{0.025} = 1.96$ and $z_{0.005} = 2.576$ and the simultaneous test critical values (6) are given in Table 3-9. Three of the first 14 hypotheses would have been rejected at the 0.05

TABLE 3-9 Test Statistics t_j and Their Significance

		Significance	
j	t_j	Single test	Simultaneous tests
1	−1.05		
2	−0.67		
3	2.54	*	
4	0.62		
5	−3.12	**	*
6	0.36		
7	−1.10		
8	0.81		
9	−2.46	*	
10	−0.38		
11	−1.20		
12	−0.85		
13	−0.35		
14	0.97		
15	2.60	**	
16	3.85	**	**
17	2.66	**	*
18	0.77		
19	3.39	**	*
20	2.37	*	

*$|t_j|$ exceeds the 0.05 critical value.
**$|t_j|$ exceeds the 0.01 critical value.

level by the usual single-test rule when in actuality the population regression parameters were zero. With the simultaneous critical value of 3.022 for a family error rate of 0.05 only that for β_5 would be erroneously rejected. The conservatism of the simultaneous test should be apparent here. For the last six regression coefficients, whose population values were each equal to $\beta_j = 2$, the simultaneous test criterion led only to rejection of the null hypotheses for β_{16}, β_{17} and β_{19}: The conservatism of the simultaneous test causes it to be less sensitive to moderate departures from the null hypothesis.

The example of independent, or orthogonal, predictors and a known disturbance variance was contrived to illustrate the problem of multiple testing and to provide an exact way of controlling the Type I error rate for the entire family of tests. In most regression applications we would need a simultaneous inference procedure that allowed for correlated predictors and a general $\mathbf{X}_1'\mathbf{X}_1$

matrix. In some cases we would also prefer a method that allowed tests or confidence statements to be generated indefinitely from linear compounds of the coefficients suggested by the data. In this section we discuss two such methods of multiple tests: the *Bonferroni inequality* family of conservative tests and confidence intervals and the Scheffé method for simultaneous inferences about all linear compounds of the regression parameters. Because the Bonferroni method is an approximate variant of the approach used in the motivating example we begin with that method.

Bonferroni simultaneous tests and confidence statements. We begin with a statement of Bonferroni's inequality for the probability of the union of several general events. Let the K random events be $A_1, \ldots, A_K$; their negations are $\bar{A}_1, \ldots, \bar{A}_K$. The A_j, and hence the $\bar{A}_j$, are neither mutually exclusive nor independent. Their number K must be specified in advance. In the present context of multiple inference the jth event might be

$$\text{Decide that} \quad H_0: \beta_j = \beta_{0j} \text{ is tenable at level } \gamma \tag{7}$$

Then when H_0 is true $P(A_j) = 1 - \gamma$, $P(\bar{A}_j) = \gamma$, and the various joint events of the A_j have the probabilities

$$\begin{aligned} P(A_i \text{ and } A_j \text{ both occur}) &= P(A_i \cap A_j) \\ &\vdots \\ P(A_1, \ldots, A_K \text{ all occur}) &= P(A_1 \cap \cdots \cap A_K) \end{aligned} \tag{8}$$

Or A_j might be the event

$$\begin{aligned} &\text{the } 100(1-\gamma) \text{ confidence interval } [L_j, U_j] \text{ covers} \\ &\text{the value of the parameter } \beta_j \end{aligned} \tag{9}$$

Again $P(A_j) = 1 - \gamma$, $P(\bar{A}_j) = \gamma$.

For K simultaneous hypothesis tests (7) or confidence statements (9) we wish to have

$$P(A_1 \cap \cdots \cap A_K) = 1 - \alpha \tag{10}$$

for a given family error rate α or confidence coefficient $1 - \alpha$. If the joint probability cannot be computed exactly, we would at least wish that it be no less than $1 - \alpha$, or conversely that the family error rate not exceed α. To find such bounds we invoke De Morgan's law to express the joint event of the occurrence of all A_j in terms of the union of the negations of the A_j:

$$\begin{aligned} P(A_1 \cap \cdots \cap A_K) &= P[\overline{(\bar{A}_1 \cup \cdots \cup \bar{A}_K)}] \\ &= 1 - P(\bar{A}_1 \cup \cdots \cup \bar{A}_K) \end{aligned} \tag{11}$$

By the inclusion–exclusion principle (Feller, 1957, Chapter IV),

$$\begin{aligned} P(\bar{A}_1 \cup \cdots \cup \bar{A}_K) = \sum_{j=1}^{K} P(\bar{A}_j) &- \sum\sum_{i<j} P(\bar{A}_i \cap \bar{A}_j) \\ &+ \sum\sum\sum_{h<i<j} P(\bar{A}_h \cap \bar{A}_i \cap \bar{A}_j) \\ &+ \cdots (-1)^{K+1} P(\bar{A}_1 \cap \cdots \cap \bar{A}_K) \end{aligned} \tag{12}$$

and the first Bonferroni inequality (Feller, 1957, pp. 100–101) states that

$$P(\bar{A}_1 \cup \cdots \cup \bar{A}_K) \leq \sum_{j=1}^{K} P(\bar{A}_j) \tag{13}$$

Similarly, the first two terms in the expansion (12) provide a lower bound, or the second Bonferroni inequality:

$$P(\bar{A}_1 \cup \cdots \cup \bar{A}_K) \geq \sum_{j=1}^{K} P(\bar{A}_j) - \sum_{i<j}\sum P(\bar{A}_i \cap \bar{A}_j) \tag{14}$$

Use of (11) and the inequality (13) gives

$$P(A_1 \cap \cdots \cap A_K) \geq 1 - \sum_{j=1}^{K} P(\bar{A}_j) = 1 - K\gamma \tag{15}$$

If we choose the individual test level to be $\gamma = \alpha/K$, the family error rate for K hypothesis tests of the type (7) would be

$$\begin{aligned} &P(\text{one or more hypotheses rejected} \mid \text{all hypotheses true}) \\ &\qquad = 1 - P(A_1 \cap \cdots \cap A_K) \geq K\left(\frac{\alpha}{K}\right) = \alpha \end{aligned} \tag{16}$$

so that the family error rate is no greater than the specified value α. Similarly, if we use the second definition (9) of the A_j as confidence intervals containing the true parameter values, the inequality (15) states that the family confidence coefficient

$$P[(L_1 \leq \beta_1 \leq U_1) \cap \cdots \cap (L_K \leq \beta_K \leq U_K)] \tag{17}$$

must exceed $1 - \alpha$ if each interval in the family has confidence coefficient $1 - \gamma = 1 - \alpha/K$. The Bonferroni method of simultaneous tests and intervals merely requires that the error rate for the individual tests be divided by the number of tests K in the family.

Further discussion of the Bonferroni method has been given by Miller (1981). In particular, the individual test error rates need not be equal, but may be any set of probabilities satisfying the constraint $\sum \gamma_j = \alpha$. If the joint probabilities $P(A_i \cap A_j)$ can be calculated free from unknown, or "nuisance," parameters, the second Bonferroni inequality (14) can be used as a lower bound on the unknown family error rate. Note that the technique can be applied to *any* K inferences about the elements of β_1; those may include the individual regression coefficients, selected differences, or linear compounds of the β_j. The intercept parameter α also may be included in the set, although its orthogonality with respect to the regression coefficients usually leads to separate tests and confidence statements.

Example 3-6

Let us assess the psychological test regression coefficients given in Table 3-2. We shall merely test the $p = 11$ hypotheses $H_0\colon \beta_j = 0$ for the individual hypotheses. A family error rate no greater than 0.05 will be used, so that each individual two-sided t test must be made with probability $0.025/11 = 0.002273$ in the left-hand and right-hand tails. The critical value for a t-distribution with $N - p - 1 = 34$ degrees of freedom and that probability can be found from interpolation in more extensive tables of that distribution [e.g., Pearson and Hartley (1969)] to be $t_{0.002273;34} \approx 3.05$. Only the t-

statistics for variable 2, Proverbs, and variable 4, sum of ratings, exceed that point in absolute value. The coefficient for variable 11, the Thematic Apperception Test, while significant at the 0.025 level in the ordinary testing sense, had a t-statistic well below the Bonferroni critical value.

Alternatively, we could have constructed a family of 11 confidence intervals for the regression parameters with family confidence coefficient 0.95. The intervals would be computed by the formula

$$\hat{\beta}_j - t_{0.025/11;34}\sqrt{\widehat{\text{Var}}(\hat{\beta}_j)} \leq \beta_j \leq \hat{\beta}_j + t_{0.025/11;34}\sqrt{\widehat{\text{Var}}(\hat{\beta}_j)} \tag{18}$$

Only the intervals for variables 2 and 4 would not contain zero, indicating that only those predictors have regression parameters significantly different from zero at the 0.05 level in the simultaneous inference sense.

Scheffé simultaneous tests and confidence intervals. The Bonferroni tests required that we specify the number of inferences K in advance. The actual error rate for the family of inferences was unknown, but a tight upper bound for it was provided by the Bonferroni inequality. A more general and exact method of multiple inferences has been found by Scheffé (1953). The Scheffé approach provides tests of hypotheses

$$H_0: \quad \mathbf{a}'\boldsymbol{\beta}_1 = \mathbf{a}'\boldsymbol{\beta}_{10} \tag{19}$$

on all linear compounds of the regression coefficients, or a family of simultaneous confidence intervals on all parametric functions. The error rate for the family of tests is exactly some specified value α, and the confidence coefficient for the collection of intervals is $1 - \alpha$. By this method we may generate hypotheses ad infinitum on interesting linear functions of the parameters, with the assurance that our error rate will never exceed the chosen value α. The price we must pay for this freedom is that of slightly less sensitive tests or longer confidence intervals.

For the Scheffé confidence intervals we begin with the joint confidence region for $\boldsymbol{\beta}_1$ given by expression (25) of Section 3-2:

$$(\boldsymbol{\beta}_1 - \hat{\boldsymbol{\beta}}_1)'\mathbf{X}_1'\mathbf{X}_1(\boldsymbol{\beta} - \hat{\boldsymbol{\beta}}_1) \leq \hat{\sigma}^2 pF_{\alpha;p,N-p-1} \tag{20}$$

By definition the probability that this ellipsoid includes the true population regression vector $\boldsymbol{\beta}_1$ is $1 - \alpha$. Now the ellipsoid is of course a figure with a curved surface, and we wish to obtain probability statements on linear functions of the β_j. However, the ellipsoid can be generated by an infinity of parallel planes tangent to the surface of the ellipsoid at opposite points. The intersection of the sets of points bounded by all such *supporting hyperplanes*

$$\mathbf{a}'\boldsymbol{\beta}_1 = \mathbf{a}'\hat{\boldsymbol{\beta}}_1 \pm \sqrt{\hat{\sigma}^2\mathbf{a}'(\mathbf{X}_1'\mathbf{X}_1)^{-1}\mathbf{a}pF_{\alpha;\,p,N-p-1}} \tag{21}$$

is the ellipsoid and its interior. Because a point $\boldsymbol{\beta}_1$ will be in the ellipsoid if and only if it is within that intersection the confidence region (25) of Section 3-2 is identical with

$$\bigcap_{\mathbf{a}} [|\mathbf{a}'(\boldsymbol{\beta}_1 - \hat{\boldsymbol{\beta}}_1)| \leq \sqrt{pF_{\alpha;\,p,N-p-1}\hat{\sigma}^2\mathbf{a}'(\mathbf{X}_1\mathbf{X}_1)^{-1}\mathbf{a}}] \tag{22}$$

in which the intersection is taken over all nonnull p-component vectors $\mathbf{a}$. The inequalities defining the infinitely many pairs of planes constitute the family of confidence intervals whose general member is

$$\mathbf{a}'\hat{\boldsymbol{\beta}}_1 - \sqrt{pF_{\alpha;\,p,N-p-1}\hat{\sigma}^2\mathbf{a}'(\mathbf{X}_1'\mathbf{X}_1)^{-1}\mathbf{a}} \le \mathbf{a}'\boldsymbol{\beta}_1 \le \mathbf{a}'\hat{\boldsymbol{\beta}}_1 + \sqrt{pF_{\alpha;\,p,N-p-1}\hat{\sigma}^2\mathbf{a}'(\mathbf{X}_1'\mathbf{X}_1)^{-1}\mathbf{a}} \qquad (23)$$

Perhaps the simplest derivation of the equations (21) and the subsequent confidence intervals is one given by Miller (1981) based on the Cauchy–Schwartz inequality, or the mathematical property of sums of squares and products that guarantees the product moment correlation to be between -1 and 1. An instructive proof based on the geometry of supporting planes is due to Scheffé (1959). The intervals also follow from Roy's union–intersection development of hypothesis tests on the regression parameters (Roy and Bose, 1953).

We may test a family of hypotheses of the sort (19) via the Scheffé simultaneous confidence intervals by observing whether an interval contains the value of the parametric function $\mathbf{a}'\boldsymbol{\beta}_{10}$ specified by its hypothesis. If so, the hypothesis is tenable; otherwise it should be rejected at the α level of significance. The most common value of the function is zero, as in simultaneous tests in the families of hypotheses

$$H_0\colon\ \beta_j = 0 \qquad j = 1, \ldots, p$$

and

$$H_0\colon\ \beta_i - \beta_j = 0 \qquad i < j \qquad (24)$$

Alternatively, we may compute the statistic

$$t = \frac{\mathbf{a}'\hat{\boldsymbol{\beta}}_1 - \mathbf{a}'\boldsymbol{\beta}_{10}}{\sqrt{\hat{\sigma}^2\mathbf{a}'(\mathbf{X}_1'\mathbf{X}_1)^{-1}\mathbf{a}}} \qquad (25)$$

and reject the particular hypothesis (19) if

$$|t| > \sqrt{pF_{\alpha;\,p,N-p-1}} \qquad (26)$$

We shall refer to the right-hand side of that inequality as the *Scheffé simultaneous test critical value.* The hypothesis test approach is probably most efficient for the examination of a set of regression coefficients from a computer printout. Conversely, if a regression program is being written anew, the incorporation of a routine for computing a given family of the confidence intervals (23) would save time and effort in the future applications of the program.

If multiple observations have been obtained on Y for the same set of predictor values, the "pure error" estimator (16) of Section 3-4 may be substituted for $\hat{\sigma}^2$ in the preceding Bonferroni and Scheffé formulas and the error degrees of freedom changed from $N - p - 1$ to $N - k$. Of course, if the new estimator of the variance is much larger, or if $N - k$ is considerably smaller than $N - p - 1$, the usual estimate of σ^2 based on SSE should be used.

Example 3-7

We shall examine the regression coefficients of Table 3-2, for the psychological test scores by means of the Scheffé method. We begin by computing the critical value of the rule (26); after some double linear interpolation in the reciprocals of the F degrees of freedom we have

$$F_{0.05;\,11,34} \approx 2.0816 \qquad \sqrt{11F_{0.05;\,11,34}} = 4.785$$

In the tests of the 11 hypotheses $H_0\colon \beta_j = 0$ only the statistic for variable 4, sum-of-ratings, exceeds that critical value. Because the Scheffé method must protect against Type I, or false positive, errors for an uncountably infinitely large family of hypotheses its critical value will usually exceed that of the Bonferroni t-test. In the present example the Scheffé method seems excessively conservative for deciding which variables should remain in the regression model, for according to it all but one would be eliminated.

Let us test the hypothesis

$$H_0\colon \beta_9 = \beta_{11}$$

or the hypothesis that the Family Scene and Thematic Apperception Test regression coefficients are equal. The hypothesis may be written in matrix form (19) with $\mathbf{a}'\boldsymbol{\beta}_{10} = 0$ and $\mathbf{a}$ an 11×1 vector with -1 in the ninth position, one in the eleventh, and zeros elsewhere. Then the submatrix of $(\mathbf{X}_1'\mathbf{X}_1)^{-1}$ given by the intersections of the ninth and eleventh rows and columns is

$$\begin{bmatrix} 0.031456 & -0.026926 \\ -0.026926 & 0.102201 \end{bmatrix}$$

and

$$\mathbf{a}'(\mathbf{X}_1'\mathbf{X}_1)^{-1}\mathbf{a} = 0.187509$$

The standard deviation of $\hat{\beta}_{11} - \hat{\beta}_9$ is

$$\sqrt{\hat{\sigma}^2\mathbf{a}'(\mathbf{X}_1'\mathbf{X}_1)^{-1}\mathbf{a}} = 2.3439$$

The 95% Scheffé confidence interval is

$$-8.53 \le \beta_{11} - \beta_9 \le 13.91$$

and it is evident that the hypothesis of equal parameters cannot be rejected. Alternatively, we might have computed the value of the statistic (25) to be 1.15; this is well below the Scheffé critical value 4.785, and rejection should not follow.

Some further references. A lucid survey of the multiple comparisons problem is given by the exemplary monograph of Miller (1981). Miller (1977) has also provided a survey of work done from 1966 through 1976, together with an extensive bibliography for that period. Several new methods for simultaneous tests, confidence intervals, and confidence bands in regression are described in the articles of the bibliography. Survey papers cited include a general one by O'Neill and Wetherill (1971), and others directed at particular disciplines, for example, Chew (1976).

3-6 PREDICTION AND CONFIDENCE INTERVALS

In Section 1-5 we constructed probability intervals for a future value of the dependent variable when the predictor variable score was given. Similarly, we found a confidence interval for the population mean of the dependent variable for the given predictor score. The limits of the intervals formed the branches of hyperbolas about the estimated regression line and illustrated the usefulness of the line for predicting individual values or averages of Y from the independent variable. We now extend those methods to the multiple regression model with p independent variables.

Let us start with the prediction interval. Suppose that a new test of a certain aspect of verbal cognition has been developed. Its critics claim that its total score can be predicted very closely from a battery of p readily available standardized psychological tests. For a trial of that contention $N = 100$ high school juniors considered to be a random sample from a homogeneous and well-defined population are given the $p + 1$ tests as part of a biennial evaluation program. The coefficient of determination of Y, the new verbal cognitive test, with the p standard tests is large, but the investigators are still left with the question of how well the new scores would be predicted for various sets of scores of the other tests. One measure of closeness would be the prediction interval $\hat{y}_L \leq Y_F \leq \hat{y}_U$ within which a future score Y_F would lie with a specified probability $1 - \alpha$.

The future score is related to the vector $\mathbf{x}_0$ of the p standard test scores by the usual model

$$Y_F = \alpha + \boldsymbol{\beta}_1'\mathbf{x}_0 + e_0 \tag{1}$$

in which $\mathbf{x}_0$ is expressed in centered form as the deviations of the scores from their averages $\bar{\mathbf{x}}' = (\bar{x}_1, \ldots, \bar{x}_p)$ computed over the 100 subjects in the sample. $\mathbf{x}_0$ may or may not be equivalent to a row in the independent variable data matrix $\mathbf{x}_1$, but e_0 *must* be a random variable distributed independently of the disturbance vector $\mathbf{e}$ in the original model for the regression analysis. Similarly, Y_F must be a future, or unobserved, score on the new verbal test, and not a score from the 100×1 data vector $\mathbf{y}$ used in the regression. Under these assumptions the least squares prediction of Y_F is

$$\hat{y}_F = \bar{y} + \hat{\boldsymbol{\beta}}_1'\mathbf{x}_0 \tag{2}$$

where of course $\hat{\boldsymbol{\beta}}_1 = (\mathbf{X}_1'\mathbf{X}_1)^{-1}\mathbf{X}_1'\mathbf{y}$. Now Y_F has the normal distribution with mean

$$E(Y_F) = \alpha + \boldsymbol{\beta}_1'\mathbf{x}_0 \tag{3}$$

and variance σ^2, while its estimate, $\hat{y}_F$, viewed as a random variable is also normally distributed with the same mean (3), but variance

$$\begin{aligned} \operatorname{Var}(\hat{y}_F) &= \operatorname{Var}(\bar{y} + \hat{\boldsymbol{\beta}}_1'\mathbf{x}_0) \\ &= \operatorname{Var}(\bar{y}) + \operatorname{Cov}(\mathbf{x}_0'\hat{\boldsymbol{\beta}}_1, \hat{\boldsymbol{\beta}}_1'\mathbf{x}_0) \\ &= \sigma^2\left(\frac{1}{N} + \mathbf{x}_0'(\mathbf{X}_1'\mathbf{X}_1)^{-1}\mathbf{x}_0\right) \end{aligned} \tag{4}$$

Because Y_F is a future observation it is distributed independently of $\hat{y}_F$, and the difference $Y_F - \hat{y}_F$ has the normal distribution with mean zero and variance

$$\operatorname{Var}(Y_F - \hat{y}_F) = \sigma^2\left[1 + \frac{1}{N} + \mathbf{x}_0'(\mathbf{X}_1'\mathbf{X}_1)^{-1}\mathbf{x}_0\right] \tag{5}$$

The estimator of that variance is merely

$$\hat{\sigma}^2\left[1 + \frac{1}{N} + \mathbf{x}_0'(\mathbf{X}_1'\mathbf{X}_1)^{-1}\mathbf{x}_0\right] \tag{6}$$

In the usual model without multiple observations at certain combinations of the predictor values $\hat{\sigma}^2 = \text{SSE}/(N - p - 1)$. Since Y_F is a future observation and $\hat{\sigma}^2$ is distributed independently of $\bar{y}$, $\hat{\boldsymbol{\beta}}_1$, and $\hat{y}_F$,

$$t = \frac{Y_F - \hat{y}_F}{\hat{\sigma}\sqrt{1 + 1/N + \mathbf{x}_0'(\mathbf{X}_1'\mathbf{X}_1)^{-1}\mathbf{x}_0}} \tag{7}$$

has the t-distribution with $N - p - 1$ degrees of freedom, and by definition

$$P\left(\frac{|Y_F - \hat{y}_F|}{\hat{\sigma}\sqrt{1 + 1/N + \mathbf{x}_0'(\mathbf{X}_1'\mathbf{X}_1)^{-1}\mathbf{x}_0}} \leq t_{\alpha/2;N-p-1}\right) = 1 - \alpha \tag{8}$$

Expansion of the terms within the probability statement gives the $100(1 - \alpha)\%$ prediction interval for Y_F:

$$\bar{y} + \hat{\boldsymbol{\beta}}_1'\mathbf{x}_0 - t_{\alpha/2;N-p-1}\hat{\sigma}\sqrt{1 + \frac{1}{N} + \mathbf{x}_0'(\mathbf{X}_1'\mathbf{X}_1)^{-1}\mathbf{x}_0}$$
$$\leq Y_F \leq \bar{y} + \hat{\boldsymbol{\beta}}_1'\mathbf{x}_0 + t_{\alpha/2;N-p-1}\hat{\sigma}\sqrt{1 + \frac{1}{N} + \mathbf{x}_0'(\mathbf{X}_1'\mathbf{X}_1)^{-1}\mathbf{x}_0} \tag{9}$$

When $p = 1$ this interval reduces to expression (10) of Section 1-5.

The first term of one within the square root will usually dominate that factor, and if $\hat{\sigma}$ is large because of variation unexplained by the estimated regression equation the prediction interval may be wide and imprecise. The interval is narrowest when $\mathbf{x}_0 = \mathbf{0}$, or the independent variable values are equal to their sample averages. We should keep in mind that prediction intervals measure the uncertainty in the estimate of a *single* future observation. Unless the relationship of Y to the X_j is very close such a prediction must have a large degree of error in it that does not disappear with an increasing size of the sample used to estimate the regression function.

The prediction probability $1 - \alpha$ holds only for a *single* interval based on some given vector $\mathbf{x}_0$. Although in practice we may plot the hyperboloid surfaces defined by the upper and lower prediction limits of (9), or at least evaluate those limits for several predictor sets $\mathbf{x}_0$, the probability is $1 - \alpha$ that one particular interval will contain its associated future value of Y. If we wish to construct simultaneous prediction intervals for several values of Y corresponding to different $\mathbf{x}_0$-vectors, we must use more elaborate or complicated procedures. Some of these have been developed by Lieberman (1961) and Hahn (1972) and will not be discussed here.

Confidence intervals for $E(Y|\mathbf{x}_0)$. In some applications prediction of single values of the dependent variable is less important than that of the mean of Y for a given set of independent variable values $\mathbf{x}_0$. For example, if ceramic objects are to be fired in a kiln at different temperatures and lengths of time, the quality assurance engineer might wish a confidence interval on the mean crushing strength of the population of pieces fired at 700°C for 90 minutes. The interval would be used to make inferences about a parameter of ceramic strength for given firing conditions, rather than the strength of one individual piece that might be fired in the future. Although the independent variables are not random, for convenience we shall denote the mean of Y for given $\mathbf{x}_0$ by the conditional expectation notation

$$E(Y|\mathbf{x}_0) = \alpha + \boldsymbol{\beta}_1'\mathbf{x}_0 \tag{10}$$

$\mathbf{x}_0$ is of course expressed in centered form. The estimate is merely

$$\widehat{E(Y|\mathbf{x}_0)} = \hat{y}_F = \bar{y} + \hat{\boldsymbol{\beta}}_1'\mathbf{x}_0 \tag{11}$$

or the predicted future value (2) used in the development of prediction intervals. Now $\hat{y}_F - E(Y|\mathbf{x}_0)$ is normally distributed with mean zero, but variance

$$\begin{aligned} \operatorname{Var}[\hat{y}_F - E(Y|\mathbf{x}_0)] &= \operatorname{Var}(\hat{y}_F) \\ &= \sigma^2\left[\frac{1}{N} + \mathbf{x}_0'(\mathbf{X}_1'\mathbf{X}_1)^{-1}\mathbf{x}_0\right] \end{aligned} \tag{12}$$

We see that the variance does not contain the component σ^2, for now we are estimating a parameter rather than the future value of a random variable. The estimator of the variance is

$$\hat{\sigma}^2\left[\frac{1}{N} + \mathbf{x}_0'(\mathbf{X}_1'\mathbf{X}_1)^{-1}\mathbf{x}_0\right] \tag{13}$$

and as before

$$t = \frac{\hat{y}_F - E(Y|\mathbf{x}_0)}{\hat{\sigma}\sqrt{1/N + \mathbf{x}_0'(\mathbf{X}_1'\mathbf{X}_1)^{-1}\mathbf{x}_0}} \tag{14}$$

has the t-distribution with $N - p - 1$ degrees of freedom. Use of a probability statement similar to (8) gives the $100(1 - \alpha)\%$ confidence interval on $E(Y|\mathbf{x}_0) = \alpha + \beta_1'\mathbf{x}_0$:

$$\begin{aligned} &\bar{y} + \hat{\boldsymbol{\beta}}_1'\mathbf{x}_0 - t_{\alpha/2;N-p-1}\hat{\sigma}\sqrt{\frac{1}{N} + \mathbf{x}_0'(\mathbf{X}_1'\mathbf{X}_1)^{-1}\mathbf{x}_0} \\ &\qquad \le \alpha + \boldsymbol{\beta}_1'\mathbf{x}_0 \le \bar{y} + \hat{\boldsymbol{\beta}}_1'\mathbf{x}_0 + t_{\alpha/2;N-p-1}\hat{\sigma}\sqrt{\frac{1}{N} + \mathbf{x}_0'(\mathbf{X}_1'\mathbf{X}_1)^{-1}\mathbf{x}_0} \end{aligned} \tag{15}$$

These intervals are considerably shorter than the prediction intervals. As the sample size N increases the width of the interval goes to zero.

As in the case of prediction intervals the confidence coefficient $1 - \alpha$ is the probability that the individual interval (15) covers the conditional mean $\alpha + \boldsymbol{\beta}_1'\mathbf{x}_0$ for the single given vector $\mathbf{x}_0$. However, a family of $100(1 - \alpha)\%$ simultaneous confidence bands enclosing the linear function

$$E(Y|\mathbf{X}) = \alpha + \boldsymbol{\beta}_1'\mathbf{X} \tag{16}$$

for all $\mathbf{X}$ admissible for consideration by the investigator can be obtained by replacing the t critical value $t_{\alpha;N-p-1}$ by

$$\sqrt{(p+1)F_{\alpha;p+1,N-p-1}} \tag{17}$$

Unlike other simultaneous confidence statements in regression analysis the first degrees of freedom parameter is $p + 1$, the total number of parameters in the linear model. Such simultaneous confidence bands were first found by Working and Hotelling (1929); their derivation follows by the same development as that of the Scheffé confidence intervals [see, for example, Miller (1981)]. Other families of intervals (for example, Bonferroni families) have been proposed. They have been described by Miller (1977, 1981), Hahn (1972), and Wynn and Bloomfield (1971).

Example 3-8

As an example of prediction and confidence intervals we shall use the psychological tests regression analysis of Example 3-3. The means and standard deviations of the dependent variable and $p = 11$ independent variables, together with the regression coefficients and their t-statistics, are given in Table 3-2. Recall that $\hat{\sigma} = 5.9604$ for the 11-predictor model. The matrix $\mathbf{X}_1'\mathbf{X}_1$ will be omitted because of its size. Let us begin with the simplest case of the intervals at the averages of the independent variables, or $\mathbf{x}_0 = \mathbf{0}$. Then the 95% prediction interval for a single future observation has limits

$$y \pm t_{0.025;34}\hat{\sigma}\sqrt{1 + \tfrac{1}{46}}$$

and the interval is

$$7.44 \leq Y_F \leq 31.95$$

Since the observed y_i ranged from 7 to 45 the prediction interval is rather wide. The linear regression equation based on the 11 predictors does not give a precise estimate of the Draw a Person test score for any one individual. The 95% confidence limits for the mean Draw a Person score for the population of subjects with independent variable values exactly at the sample averages is

$$\bar{y} \pm \frac{t_{0.025;34}\hat{\sigma}}{\sqrt{46}}$$

and the confidence interval is

$$17.91 \leq E(Y|\mathbf{x}_0) \leq 21.48$$

where $E(Y|\mathbf{x}_0) = \alpha$ when $\mathbf{x}_0 = \mathbf{0}$. The difference in the intervals is striking: The confidence interval has a width of only 3.57 points on the test scale and locates the population mean fairly precisely.

Now we shall calculate the 95% prediction and confidence intervals for three independent variable vectors. The first two vectors are hypothetical scores on the independent variables, while the third consists of the observations on subject 10 in the original data matrix for the regression analysis. We have chosen the third vector in the interests of realism rather than inferences about the predictability of the observed y_i. The use of the regression equation and prediction intervals in that way is not an unbiased and fair way of appraising its predictive usefulness. The score vectors are

$$\mathbf{x}_1' = [70, 3, 24, 50, 10, 4, 5, 6, 5, 4, 5]$$

$$\mathbf{x}_2' = [50, 1, 30, 20, 0, 1, 1, 1, 1, 1, 1]$$

$$\mathbf{x}_3' = [58, 6, 24, 31, 4, 1, 1, 2, 4, 1, 2]$$

From these we calculate the centered vectors $\mathbf{x}_{0j} = \mathbf{x}_j - \bar{\mathbf{x}}$ for use in the interval formulas. The predicted values of Y, the quadratic forms $\mathbf{x}_{0j}'(\mathbf{X}_1'\mathbf{X}_1)^{-1}\mathbf{x}_{0j}$, and the three intervals are shown in Table 3-10. The independent variable values have little effect on the widths of the prediction intervals, although their effect on the confidence interval lengths can be seen to be considerable. The prediction intervals seem too wide to be of use in practice, while even the confidence intervals indicate much uncertainty in the estimates of the conditional means. The predictive ability of the regression equation implied by its R^2 of 0.7159 should be greatly discounted.

TABLE 3-10 **Prediction and Confidence Intervals**

Independent variable vector	$\hat{y}_j$	$\mathbf{x}'_{0j}(\mathbf{X}'_1\mathbf{X}_1)^{-1}\mathbf{x}_{0j}$	Confidence interval	Prediction interval
$\mathbf{x}_1$	19.93	0.5646	$10.65 \le E(Y \mid \mathbf{x}_{01}) \le 29.21$	$4.66 \le y_1 \le 35.20$
$\mathbf{x}_2$	29.55	0.2594	$23.12 \le E(Y \mid \mathbf{x}_{02}) \le 35.98$	$15.83 \le y_2 \le 43.27$
$\mathbf{x}_3$	32.73	0.316	$25.68 \le E(Y \mid \mathbf{x}_{03}) \le 39.78$	$18.71 \le y_3 \le 46.75$

3-7 TESTING THE EQUALITY OF SEVERAL REGRESSION MODELS

Frequently, the dependent and independent variables of a regression model are observed under different experimental or social conditions. For example, suppose that an investigation is being conducted into the relationship of blood cholesterol with age and occupation for adult males of normal health. Occupations A, B, and C have been singled out for their degrees of stress on the subjects, and homogeneous populations of workers in A, B, or C meeting certain medical criteria have been defined. Cholesterol levels tend to increase with age, and one part of the statistical analysis of the data obtained in the investigation will consist of determining whether that relationship is the same for each of the three occupations. If we denote the age of the ith man in occupation j by x_{ij} and his cholesterol level by y_{ij} we might begin by postulating the simple linear model

$$y_{ij} = \alpha_j + \beta_j(x_{ij} - \bar{x}_j) + e_{ij} \tag{1}$$

estimating the parameters for each group, and testing the hypothesis that the α_j, β_j are the same for each occupation. We should also examine a host of other assumptions and hypotheses about variables that might influence cholesterol level, but for the moment we shall consider only age and occupational status. In this section we give an analysis of variance for the equality of several regression models applied to independent random samples.

We begin with the models and some notation. A dependent variable Y and p independent variables $X_1, \ldots, X_p$ are observed on the $N_1, \ldots, N_k$ independent subjects or other sampling units in k independent and disjoint groups. The ith observation on Y in the jth group is related to the independent variable values by the linear regression model

$$y_{ij} = \alpha_j + \beta_{1j}(x_{ij1} - \bar{x}_{j1}) + \cdots + \beta_{pj}(x_{ijp} - \bar{x}_{jp}) + e_{ij}$$
$$i = 1, \ldots, N_j; \quad j = 1, \ldots, k \tag{2}$$

where

$$\bar{x}_{jh} = \frac{1}{N_j}\sum_{i=1}^{N_j} x_{ijh} \qquad h = 1, \ldots, p \tag{3}$$

is the average of the hth independent variable values in the jth group. In matrix notation the model for the jth group is

$$\mathbf{y}_j = \alpha_j \mathbf{j}_j + \mathbf{X}_{1j}\boldsymbol{\beta}_{1j} + \mathbf{e}_j \tag{4}$$

where $\mathbf{y}_j$ and $\mathbf{X}_{1j}$ have respective dimensions $N_j \times 1$ and $N_j \times p$, and $\mathbf{j}_j$ is the $N_j \times 1$ vector of ones. The complete model can be written as

$$\mathbf{y} = \begin{bmatrix} \mathbf{y}_1 \\ \cdot \\ \cdot \\ \cdot \\ \mathbf{y}_k \end{bmatrix} = \begin{bmatrix} \mathbf{j}_1 & \mathbf{0} & \cdots & \mathbf{0} & \mathbf{X}_1 & \cdots & \mathbf{0} \\ \cdot & \cdot & & \cdot & \cdot & & \cdot \\ \cdot & \cdot & & \cdot & \cdot & & \cdot \\ \cdot & \cdot & & \cdot & \cdot & & \cdot \\ \mathbf{0} & \mathbf{0} & \cdots & \mathbf{j}_k & \mathbf{0} & \cdots & \mathbf{X}_k \end{bmatrix} \begin{bmatrix} \alpha_1 \\ \cdot \\ \cdot \\ \cdot \\ \alpha_k \\ \boldsymbol{\beta}_{11} \\ \cdot \\ \cdot \\ \cdot \\ \boldsymbol{\beta}_{1k} \end{bmatrix} + \mathbf{e} \tag{5}$$

We shall require that all $N = N_1 + \cdots + N_k$ disturbance terms e_{ij} have the same variance σ^2. Because the groups are independent the estimators of the parameters follow separately as

$$\begin{aligned} \hat{\alpha}_j &= \bar{y}_j = \frac{1}{N_j}\sum_{i=1}^{N_j} y_{ij} \\ \hat{\boldsymbol{\beta}}_{1j} &= (\mathbf{X}'_{1j}\mathbf{X}_{1j})^{-1}\mathbf{X}'_{1j}\mathbf{y}_j \end{aligned} \qquad j = 1,\ldots,k \tag{6}$$

The estimator of the common variance is

$$\begin{aligned} \hat{\sigma}^2 &= \frac{1}{N - k(p+1)}\left[\sum_{j=1}^{k}\sum_{i=1}^{N_j}(y_{ij} - \bar{y}_j)^2 - \sum_{j=1}^{k}\hat{\boldsymbol{\beta}}'_{1j}(\mathbf{X}'_{1j}\mathbf{X}_{1j})\hat{\boldsymbol{\beta}}_{1j}\right] \\ &= \frac{1}{N - k(p+1)}\left[\sum_{j=1}^{k}\sum_{i=1}^{N_j}(y_{ij} - \bar{y}_j)^2 - \sum_{j=1}^{k}\mathbf{y}'_j\mathbf{X}_{1j}(\mathbf{X}'_{1j}\mathbf{X}_{1j})^{-1}\mathbf{X}'_{1j}\mathbf{y}_j\right] \end{aligned} \tag{7}$$

We arrived at this estimator merely by adding the residual sums of squares for the k separate regressions. Since the expected values of those sums are $(N_j - 1 - p)\sigma^2$, our divisor for (7) follows as

$$N - k(p+1) = \sum_{j=1}^{k}(N_j - 1 - p) \tag{8}$$

Under the assumption of independent and normally distributed e_{ij}, $\hat{\boldsymbol{\beta}}_{1j}$ has the p-dimensional multinormal distribution with mean vector $\boldsymbol{\beta}_{1j}$ and covariance matrix $\sigma^2(\mathbf{X}'_{1j}\mathbf{X}_{1j})^{-1}$. The $\hat{\boldsymbol{\beta}}_{1j}$ are of course independently distributed for the k groups. The $\hat{\alpha}_j$ are distributed independently of the $\hat{\boldsymbol{\beta}}_{1j}$ and one another as $N(\alpha_j, \sigma^2/N_j)$ variates.

The first hypothesis we shall test is that of the parallelism of the regression planes, or

$$H_0\colon\ \boldsymbol{\beta}_{11} = \cdots = \boldsymbol{\beta}_{1k} \tag{9}$$

of equal regression parameter vectors. The test can be developed by the generalized likelihood ratio principle [see, for example, Graybill (1976, Section 8.6)] or by use of the quadratic form results given in Section 3-2. The sum of squares for the parallelism hypothesis (9) is

$$\mathrm{SSP} = \sum_{j=1}^{k} \mathbf{y}_j'\mathbf{X}_{1j}(\mathbf{X}_{1j}'\mathbf{X}_{1j})^{-1}\mathbf{X}_{1j}'\mathbf{y}_j - \left(\sum_{j=1}^{k} \mathbf{y}_j'\mathbf{X}_{1j}\right)\left(\sum_{j=1}^{k} \mathbf{X}_{1j}'\mathbf{X}_{1j}\right)^{-1}\left(\sum_{j=1}^{k} \mathbf{X}_{1j}'\mathbf{y}_j\right) \tag{10}$$

If the hypothesis is true, SSP/σ^2 has the chi-squared distribution with $p(k-1)$ degrees of freedom and is distributed independently of the within-groups error variance $\hat{\sigma}^2$. The test statistic is

$$F = \frac{\mathrm{SSP}/[p(k-1)]}{\hat{\sigma}^2} \tag{11}$$

When the hypothesis is true, F has the F-distribution with $p(k-1)$ and $N - k(p+1)$ degrees of freedom. We should reject the hypothesis of a common regression parameter vector at the α level if

$$F > F_{\alpha;\, p(k-1),\, N-k(p+1)} \tag{12}$$

Let us examine the sum of squares SSP. It is in fact a weighted sum of squared deviations of the group regression vectors $\hat{\boldsymbol{\beta}}_{1j}$ about the *within-groups regression vector*

$$\hat{\boldsymbol{\beta}}_{1W} = \left(\sum_{j=1}^{k} \mathbf{X}_{1j}'\mathbf{X}_{1j}\right)^{-1}\left(\sum_{j=1}^{k} \mathbf{X}_{1j}'\mathbf{y}_j\right) \tag{13}$$

and can be written as

$$\begin{aligned}\mathrm{SSP} &= \sum_{j=1}^{k} \hat{\boldsymbol{\beta}}_{1j}'(\mathbf{X}_{1j}'\mathbf{X}_{1j})\hat{\boldsymbol{\beta}}_{1j} - \hat{\boldsymbol{\beta}}_{1W}'\left(\sum_{j=1}^{k} \mathbf{X}_{1j}'\mathbf{X}_{1j}\right)\hat{\boldsymbol{\beta}}_{1W} \\ &= \sum_{j=1}^{k} (\hat{\boldsymbol{\beta}}_{1j} - \hat{\boldsymbol{\beta}}_{1W})'(\mathbf{X}_{1j}'\mathbf{X}_{1j})(\hat{\boldsymbol{\beta}}_{1j} - \hat{\boldsymbol{\beta}}_{1W})\end{aligned} \tag{14}$$

As the individual within-groups regression coefficients depart from their weighted average $\hat{\boldsymbol{\beta}}_{1W}$, SSP becomes large.

Simultaneous tests and confidence intervals for k regressions. The F-statistic (11) only provides a test of the overall hypothesis of equality of the k regression coefficient vectors. We still must determine which groups are different with respect to which coefficients. We may make such comparisons by the Scheffé or Bonferroni methods introduced in Section 3-5. We shall start with some specific paired comparisons, then build to a general simultaneous test and confidence interval.

Suppose that we wish to test several of the hypotheses

$$H_0:\ \beta_{hj} = \beta_{hl} \qquad j \neq l \tag{15}$$

that the regression coefficient for the hth independent variable is the same in the jth and lth group regression models, where a number of pairs of j and l will be selected after inspecting the group regression estimators. To simplify our notation we shall write

$$\mathbf{A}_j = (\mathbf{X}_{1j}'\mathbf{X}_{1j})^{-1} \qquad j = 1, \ldots, k \tag{16}$$

where $\mathbf{A}_j$ has general element a_{rsj}. Then the statistic

$$t = \frac{\hat{\beta}_{hj} - \hat{\beta}_{hl}}{\hat{\sigma}\sqrt{a_{hhj} + a_{hhl}}} \tag{17}$$

is exactly distributed as a Student t-variate with degrees of freedom $N - k(p+1)$ when the hypothesis (15) is true. Now assume that the indices h, j, and l were not chosen a priori but were selected as an interesting or potentially "significant" comparison from a visual inspection of the coefficients. K comparisons were chosen, where K would be $pk(k-1)/2$ if all unique paired comparisons were to be made separately for each of the p-predictors. The K hypotheses of the sort (15) could then be tested with a family error rate no larger than α by adopting the two-sided decision rule

$$\text{Reject } H_0\text{:} \quad \beta_{hj} = \beta_{hl} \text{ if } |t| > t_{\alpha/2K;\,N-k(p+1)} \tag{18}$$

specified by the Bonferroni inequality of Section 3-5. Alternatively, if K has not been set in advance of the inspection of the data any number of the statistics (17) could be referred to the Scheffé simultaneous test critical value

$$\sqrt{p(k-1)F_{\alpha;\,p(k-1),N-k(p+1)}} \tag{19}$$

by the two-sided decision rule

$$\text{Reject } H_0\text{:} \quad \beta_{hj} = \beta_{hl} \text{ if } |t| > \sqrt{p(k-1)F_{\alpha;\,p(k-1),N-k(p+1)}} \tag{20}$$

Because we are only considering paired comparison hypotheses the error rate for the family of tests will be some probability less than the nominal α of the critical value (19). Similarly, we may construct the $100(1-\alpha)\%$ family of Bonferroni confidence intervals

$$\hat{\beta}_{hj} - \hat{\beta}_{hl} - t_{\alpha/2K;\,N-k(p+1)}\hat{\sigma}\sqrt{a_{hhj} + a_{hhl}} \\ \le \beta_{hj} - \beta_{hl} \le \hat{\beta}_{hj} - \hat{\beta}_{hl} + t_{\alpha/2K;\,N-k(p+1)}\hat{\sigma}\sqrt{a_{hhj} + a_{hhl}} \tag{21}$$

or conservative Scheffé confidence intervals

$$\hat{\beta}_{hj} - \hat{\beta}_{hl} - \hat{\sigma}\sqrt{(a_{hhj} + a_{hhl})p(k-1)F_{\alpha;\,p(k-1),N-k(p+1)}} \\ \le \beta_{hj} - \beta_{hl} \le \hat{\beta}_{hj} - \hat{\beta}_{hl} + \hat{\sigma}\sqrt{(a_{hhj} + a_{hhl})p(k-1)F_{\alpha;\,p(k-1),N-k(p+1)}} \tag{22}$$

If an interval does not contain zero, the hypothesis (15) would be untenable at the α level in the simultaneous testing sense.

Now let us take our hypothesis tests one further step in generality. We may test the family of hypotheses

$$H_0\text{:} \quad \sum_{j=1}^{k} c_j\beta_{hj} = 0 \tag{23}$$

about *contrasts* of $\beta_{h1}, \ldots, \beta_{hk}$ with an error rate no greater than some nominal value α. A contrast is a linear function of parameters whose coefficients sum to zero, or

$$\sum_{j=1}^{k} c_j = 0 \tag{24}$$

We shall encounter contrasts of parameters frequently in our treatment of the analysis of variance in Chapters 8 to 10. The estimator of the contrast is $\sum c_j\hat{\beta}_{hj}$, and if the hypothesis (23) is true,

$$t = \frac{\sum_{j=1}^{k} c_j \hat{\beta}_{hj}}{\hat{\sigma}\sqrt{\sum_{j=1}^{k} a_{hhj} c_j^2}} \tag{25}$$

has the t distribution with $N - k(p + 1)$ degrees of freedom for a given set of contrast coefficients $c_1, \ldots, c_k$. We may test K hypotheses of the type (23) for as many linearly independent contrasts by referring the absolute value of (25) to the Bonferroni critical value $t_{\alpha/2K;N-k(p+1)}$. Any hypothesis for which $|t|$ exceeds the critical value can be rejected with the assurance that the family error rate will still remain below α. If K cannot be determined in advance, or if the contrasts are generated by exploring the data very freely, we may wish to use the Scheffé critical value (19) and the same sort of decision rule for the individual hypotheses in the family.

Finally, we may generalize the hypothesis (23) to one concerning contrasts of a common linear compound $\mathbf{a}'\boldsymbol{\beta}_{1j}$ of the regression coefficients, or

$$H_0: \quad \sum_{h=1}^{p} \sum_{j=1}^{k} a_h c_j \beta_{hj} = 0 \tag{26}$$

The constants a_h do not necessarily have the contrast property, but the c_j still satisfy the constraint $\sum c_j = 0$. For example, suppose that the independent variables X_1, X_2, and X_3 all have commensurable units, so that it would be reasonable to inquire if the differences of the regression coefficients of X_1 and X_2 are the same in groups 1, 2, and 3. We might then simultaneously test hypotheses of the sort

$$\begin{aligned} H_{01}&: \quad \beta_{11} - \beta_{21} = \beta_{12} - \beta_{22} \\ H_{02}&: \quad \beta_{11} - \beta_{21} = \beta_{13} - \beta_{23} \\ H_{03}&: \quad \beta_{12} - \beta_{22} = \beta_{13} - \beta_{23} \end{aligned} \tag{27}$$

The values of the constants in the general bilinear form in the left-hand side of (26) are

$$a_1 = 1 \qquad a_2 = -1 \qquad a_3 = 0$$

and

$$\begin{array}{lll} c_1 = 1 & c_2 = -1 & c_3 = 0 \\ c_1 = 1 & c_2 = 0 & c_3 = -1 \\ c_1 = 0 & c_2 = 1 & c_3 = -1 \end{array}$$

for the respective hypotheses H_{01}, H_{02}, H_{03} of the family. The general test statistic for the hypothesis (26) is

$$t = \frac{\sum_{h=1}^{p} \sum_{j=1}^{k} a_h c_j \hat{\beta}_{hj}}{\hat{\sigma}\sqrt{\sum_{j=1}^{k} \mathbf{a}'\mathbf{A}_j \mathbf{a} c_j^2}} \tag{28}$$

where $\mathbf{a}' = [a_1, \ldots, a_p]$, and $\mathbf{A}_j$ is the matrix defined by (16). For a given set of K hypotheses we would test each member by the decision rule

$$\text{Reject } H_0: \sum_{h=1}^{p} \sum_{j=1}^{k} a_h c_j \beta_{hj} = 0 \text{ if } |t| > t_{\alpha/2K; N-k(p+1)} \tag{29}$$

For the Scheffé family of tests with infinitely many bilinear functions and hypotheses we should use the rule

$$\text{Reject } H_0: \sum_{h=1}^{p} \sum_{j=1}^{k} a_h c_j \beta_{hj} = 0 \text{ if } |t| > \sqrt{p(k-1)F_{\alpha; p(k-1), N-k(p+1}} \tag{30}$$

$100(1 - \alpha)\%$ families of simultaneous confidence intervals similar to (21) and (22) can be found from the test statistic and its Bonferroni or Scheffé critical values. We leave the construction of those intervals as an exercise.

Testing the equality of *k* intercept parameters. In addition to a test for the parallelism of the k linear regression functions it would be useful to have a test for the equality of the intercepts of the planes. Taken together the two tests could be used to determine whether the functions are identical in the population of the k groups. We could begin with a test for the equality of the complete functions or approach the problem via the subset tests of Section 3-3, but we prefer a more direct and explicit method involving only the intercepts and their estimators.

We must begin by defining the intercept parameters properly. The quantities α_j of our original model (2) are the intercepts at the coordinates $(\bar{x}_{j1}, \ldots, \bar{x}_{jp})$ given by the averages of the independent variables. Usually, those means, and the points specified by them, vary from group to group unless the investigator has been able to make each mean vector the same. For that reason the $\alpha_1, \ldots, \alpha_k$ parameters are rarely comparable, and we must define the regression function with a constant term at the same location. We revert to the uncentered model

$$y_{ij} = \gamma_j + \beta_{1j} x_{i1j} + \cdots + \beta_{pj} x_{ipj} + e_{ij} \tag{31}$$

introduced briefly in Chapter 1 for a single predictor. We have replaced β_{0j} by γ_j merely to avoid another subscript. By equating coefficients in the two models we see that

$$\gamma_j = \alpha_j - \sum_{h=1}^{p} \beta_{hj} \bar{x}_{hj} \tag{32}$$

The least squares estimators of the β_{hj} remain the same as in (6), while the estimators of the intercepts at the origin of the predictor variable scales are

$$\begin{aligned} \hat{\gamma}_j &= \hat{\alpha}_j - \sum_{h=1}^{p} \hat{\beta}_{hj} \bar{x}_{hj} \\ &= \bar{y}_j - \sum_{h=1}^{p} \hat{\beta}_{hj} \bar{x}_{hj} \end{aligned} \tag{33}$$

The $\hat{\gamma}_j$ are independently and normally distributed variates with means γ_j and variances

$$\operatorname{Var}(\hat{\gamma}_j) = \sigma^2 \left[\frac{1}{N_j} + \bar{\mathbf{x}}_j' (\mathbf{X}_{1j}' \mathbf{X}_{1j})^{-1} \bar{\mathbf{x}}_j \right] \tag{34}$$

For simplicity in the later expressions we abbreviate the term in the variance as

$$v_j = \frac{1}{N_j} + \bar{\mathbf{x}}_j'(\mathbf{X}_{1j}'\mathbf{X}_{1j})^{-1}\bar{\mathbf{x}}_j \tag{35}$$

and let

$$d = \frac{1}{\sum_{j=1}^{k} 1/v_j} \tag{36}$$

To test the hypothesis

$$H_0: \gamma_1 = \cdots = \gamma_k \tag{37}$$

we compute

$$\begin{aligned} \text{SSC} &= \sum_{j=1}^{k} \frac{\hat{\gamma}_j^2}{v_j} - d\left(\sum_{j=1}^{k} \frac{\hat{\gamma}_j}{v_j}\right)^2 \\ &= \sum_{j=1}^{k} \frac{\left(\hat{\gamma}_j - d\sum_{h=1}^{k} \hat{\gamma}_h/v_h\right)^2}{v_j} \end{aligned} \tag{38}$$

We have adopted the notation SSC rather than SSI to avoid confusion with the ubiquitous interaction sums of squares of the later analysis of variance chapters. When the hypothesis (37) is true, $Q = \text{SSC}/\sigma^2$ is a chi-squared variate with $k - 1$ degrees of freedom. It is distributed independently of the estimator $\hat{\sigma}^2$ defined by expression (7), so that our test statistic for the equal-intercepts hypothesis is

$$F = \frac{\text{SSC}}{(k-1)\hat{\sigma}^2} \tag{39}$$

If

$$F > F_{\alpha; k-1, N-k(p+1)}$$

we should reject the hypothesis at the α level.

When we have only $k = 2$ groups we may test the equal-intercepts hypothesis by the t-statistic

$$\begin{aligned} t &= \frac{\hat{\gamma}_1 - \hat{\gamma}_2}{\sqrt{\widehat{\text{Var}}(\hat{\gamma}_1) + \widehat{\text{Var}}(\hat{\gamma}_2)}} \\ &= \frac{\hat{\gamma}_1 - \hat{\gamma}_2}{\hat{\sigma}\sqrt{v_1 + v_2}} \\ &= \frac{\hat{\gamma}_1 - \hat{\gamma}_2}{\hat{\sigma}\sqrt{1/N_1 + 1/N_2 + \bar{\mathbf{x}}_1'(\mathbf{X}_{11}'\mathbf{X}_{11})^{-1}\bar{\mathbf{x}}_1 + \bar{\mathbf{x}}_2'(\mathbf{X}_{12}'\mathbf{X}_{12})^{-1}\bar{\mathbf{x}}_2}} \end{aligned} \tag{40}$$

We reject the hypothesis in favor of one specifying different intercepts if $|t| > t_{\alpha/2; N_1+N_2-2(p+1)}$.

Simultaneous tests about the intercepts. The Bonferroni and Scheffé methods of multiple comparisons can be used to determine which intercepts or combinations of them led to the rejection of the hypothesis (37) of k equal intercepts. Let us begin with the $k(k-1)/2$ paired comparisons

$$H_0: \quad \hat{\gamma}_i = \hat{\gamma}_j \qquad i \neq j \tag{41}$$

of the intercepts. To test an individual hypothesis of that kind we compute

$$t_{ij} = \frac{\hat{\gamma}_i - \hat{\gamma}_j}{\hat{\sigma}\sqrt{v_i + v_j}} \tag{42}$$

as in the two-group t-statistic (40) and refer t_{ij} to the critical value

$$t_{\alpha/k(k-1);N-k(p+1)} \tag{43}$$

for a Bonferroni test with a family error rate no greater than α. Alternatively, we may use the Scheffé critical value

$$\sqrt{(k-1)F_{\alpha;k-1,N-k(p+1)}} \tag{44}$$

In the latter case we can include in the family hypotheses all contrasts of the intercepts, or test all

$$H_0: \sum_{j=1}^{k} c_j \gamma_j = 0 \tag{45}$$

for any set of constants $c_1, \ldots, c_k$ whose sum is zero. The test statistic is then

$$t = \frac{\sum c_j \hat{\gamma}_j}{\hat{\sigma}\sqrt{\sum c_j^2 v_j}} \tag{46}$$

In most applications the paired comparison hypotheses (41) will suffice, unless contrasts among the groups have a substantive meaning.

Further extensions. The results of Section 3-4 on multiple observations for the same values of the predictor variables can be incorporated into the preceding tests and confidence statements: $\hat{\sigma}^2$ should be replaced by the "pure error" estimate and the error degrees of freedom changed accordingly. The details are left as an exercise.

In our development we have separated the tests of equal intercepts and parallelism of the regression planes. A single overall test of the equality of the k regression functions could have been given instead; such an approach may be found in the text by Graybill (1976). Of course one would still wish to determine by simultaneous inference procedures for the individual hypotheses on the γ_j and β_{1j} whether rejection of the overall hypothesis was due to different intercepts or shapes of the fitted regression surfaces.

Other aspects of the k-group regression hypotheses have been treated by Hald (1952). Some of these results first appeared in an appendix by E. S. Pearson to a paper by Wilsdon (1934) on factors determining the crushing strength of Portland cement.

Example 3-9

The following example of regression functions estimated in two independent samples is based on real, if rather simplistic, data. At a university course instructors are rated confidentially by their students at the end of each term. The rating scales cover several aspects of the teaching and content of the courses and run in unit steps from 1 (outstanding) to 5 (worst). The following scales have been selected from an instructor's rating forms in courses A and B for four years:

Y = overall evaluation of the instructor

X_1 = ability to present course material clearly and orderly

X_2 = ability to stimulate interest in the material

X_3 = overall evaluation of the course

X_4 = year of course (0, 1, 2, 3)

With the exception of X_4, a correlation model would be more appropriate for the data, for the values of X_1, X_2, and X_3 were not fixed by the investigator, but were chosen by each respondent with some random error to describe an aspect of the course or instructor. Nevertheless, we shall proceed with the regression model as if the X_i ratings were fixed. These summary statistics were computed for the $N_1 = 16$ questionnaires received in course A and the $N_2 = 36$ received in course B:

Course A

$$\bar{y}_1 = 1.5625 \qquad \bar{\mathbf{x}}'_1 = [1.5625, 2, 1.875, 1.8125]$$

$$\mathbf{X}'_{11}\mathbf{y}_1 = \begin{bmatrix} 4.9375 \\ 7 \\ 5.125 \\ 1.6875 \end{bmatrix} \qquad \mathbf{X}'_{11}\mathbf{X}_{11} = \begin{bmatrix} 7.9375 & 6 & 6.125 & 0.6875 \\ & 12 & 4 & 5 \\ & & 11.75 & 2.625 \\ & & & 20.4375 \end{bmatrix}$$

$$\sum_1^{16} (y_{i1} - \bar{y}_1)^2 = 5.9375$$

Course B

$$\bar{y}_2 = 1.6944 \qquad \bar{\mathbf{x}}'_2 = [1.8889, 2.2778, 1.6667, 1.4167]$$

$$\mathbf{X}'_{12}\mathbf{y}_2 = \begin{bmatrix} 17.7778 \\ 22.0556 \\ 20.3333 \\ -10.4167 \end{bmatrix} \qquad \mathbf{X}'_{12}\mathbf{X}_{12} = \begin{bmatrix} 25.5556 & 22.1111 & 16.6667 & -15.3333 \\ & 37.2222 & 20.3333 & -17.1667 \\ & & 30 & -13 \\ & & & 42.75 \end{bmatrix}$$

$$\sum_1^{36} (y_{i2} - \bar{y}_2)^2 = 21.6389$$

These regression coefficients and their test statistics were obtained from the individual courses:

	Course			
	A		B	
Variable	$\hat{\beta}_{h1}$	t_{h1}	$\hat{\beta}_{h2}$	t_{h2}
X_1	0.05	0.25	0.26	2.47
X_2	0.50	3.63***	0.26	2.95**
X_3	0.26	1.88†	0.39	4.60***
X_4	−0.08	−0.88	0.07	1.17

†Significant at the 0.10 level.
**Significant at the 0.01 level.
***Significant at the 0.001 level.

From these we may compute the estimates of the intercepts:

$$\hat{\gamma}_1 = 0.133 \qquad \hat{\gamma}_2 = -0.138$$

These statistics were computed from the separate course data:

Course A

$$\text{SSR}_\text{A} = \mathbf{y}'_1\mathbf{X}_{11}(\mathbf{X}'_{11}\mathbf{X}_{11})^{-1}\mathbf{X}'_{11}\mathbf{y}_1 = 4.9602$$

$$\text{SSE}_\text{A} = \sum (y_{i1} - \bar{y}_1)^2 - \text{SSR}_\text{A} = 0.9773$$

$$R^2_\text{A} = 0.8354$$

$$\hat{\sigma}_A^2 = \frac{\text{SSE}_A}{11} = 0.088849$$

Course B

$$\text{SSR}_B = \mathbf{y}_2'\mathbf{X}_{12}(\mathbf{X}_{12}'\mathbf{X}_{12})^{-1}\mathbf{X}_{12}'\mathbf{y}_2 = 17.499008$$

$$\text{SSE}_B = \sum (y_{i2} - \bar{y}_2)^2 - \text{SSR}_B = 4.1399$$

$$R_B^2 = 0.8087$$

$$\hat{\sigma}_B^2 = \frac{\text{SSE}_B}{31} = 0.1335$$

We note in passing that the F-statistic

$$F = \frac{\hat{\sigma}_B^2}{\hat{\sigma}_A^2} = 1.50$$

just barely exceeds the upper 25% critical value of the F-distribution with 31 and 11 degrees of freedom: The usual assumption of a common disturbance term variance in each sample indeed seems tenable. The pooled sum of squares estimate of that common variance follows from (7) as

$$\hat{\sigma}^2 = \frac{\text{SSE}_A + \text{SSE}_B}{42} = 0.1218$$

This estimate and the diagonal elements of the inverses of $\mathbf{X}_{11}'\mathbf{X}_{11}$ and $\mathbf{X}_{12}'\mathbf{X}_{12}$ were used to compute the t-statistics of the regression coefficients for the two samples. The significance levels of these statistics for two-sided alternatives are indicated by the usual dagger and asterisk notation. In particular we note that X_4, the variable indicating the annual trend in the ratings, does not have coefficients significantly different from zero in either course, although the directions of the trends are different.

Now we are ready to test the hypothesis

$$H_0: \boldsymbol{\beta}_{11} = \boldsymbol{\beta}_{12}$$

of parallel regression planes in courses A and B. We begin by computing the within-courses matrices and vectors

$$\sum_{j=1}^{2} \mathbf{X}_{1j}'\mathbf{X}_{1j} \qquad \sum_{j=1}^{2} \mathbf{y}_j'\mathbf{X}_{1j}$$

and from them the within-courses regression coefficients (13):

$$\hat{\boldsymbol{\beta}}_{1W} = \begin{bmatrix} 0.20 \\ 0.33 \\ 0.32 \\ 0.02 \end{bmatrix}$$

Then the sum of squares (10) for testing the parallelism hypotheses can be computed as

$$\begin{aligned} \text{SSP} &= \text{SSR}_A + \text{SSR}_B - \hat{\boldsymbol{\beta}}_W'\left(\sum_{j=1}^{2} \mathbf{X}_{1j}'\mathbf{X}_{1j}\right)\hat{\boldsymbol{\beta}}_{1W} \\ &= 0.6059 \end{aligned}$$

and the statistic (11) for testing H_0 is $F = 1.24$. That value does not even exceed the upper 25% point of the F-distribution with 4 and 42 degrees of freedom, and we should conclude that a common regression vector might hold for the two courses. That lack of significance has also rendered moot the question of multiple comparisons among the coefficients.

Finally, we test the hypothesis

$$H_0: \gamma_1 = \gamma_2$$

of the equality of the intercepts. We do so initially by the F-statistic (39), then by the t-statistic, which compares the difference of $\hat{\gamma}_1$ and $\hat{\gamma}_2$ directly to its standard deviation. We begin by computing

$$v_1 = \tfrac{1}{16} + \bar{\mathbf{x}}_1'(\mathbf{X}_{11}'\mathbf{X}_{11})^{-1}\bar{\mathbf{x}}_1 = 0.58889$$

$$v_2 = \tfrac{1}{36} + \bar{\mathbf{x}}_2'(\mathbf{X}_{12}'\mathbf{X}_{12})^{-1}\bar{\mathbf{x}}_2 = 0.41914$$

$$d = \left(\frac{1}{v_1} + \frac{1}{v_2}\right)^{-1} = 0.24486$$

Then

$$\text{SSC} = \sum_{j=1}^{2} \frac{\hat{\gamma}_j^2}{v_j} - d\left(\sum_{j=1}^{2} \frac{\hat{\gamma}_j}{v_j}\right)^2 = 0.07296$$

The statistic (39) is $F = \text{SSC}/\hat{\sigma}^2 = 0.60$, and the hypothesis of a common intercept value would be tenable at any reasonable significance level. The t-statistic is

$$\begin{aligned} t &= \frac{\hat{\gamma}_1 - \hat{\gamma}_2}{\sqrt{\widehat{\text{Var}}(\hat{\gamma}_1) + \widehat{\text{Var}}(\hat{\gamma}_2)}} \\ &= \frac{\hat{\gamma}_1 - \hat{\gamma}_2}{\hat{\sigma}\sqrt{v_1 + v_2}} \\ &= \frac{0.133 - (-0.138)}{0.3491\sqrt{1.0080}} \\ &= 0.773 \end{aligned}$$

or the square root of the F-statistic.

Now that we have established that the same regression function probably holds for both courses, whither should our data analysis proceed? We might begin by dropping X_4, the variable indicating whether the ratings were obtained in year 0, 1, 2, or 3, from the predictor set. The regression coefficients within the two courses are shown below, together with their t-statistics computed from the estimate of the common disturbance variance and various other within-course statistics.

	Course			
	A		B	
Variable	$\hat{\beta}_{h1}$	t_{h1}	$\hat{\beta}_{h2}$	t_{h2}
X_1	0.10	0.52	0.24	2.27*
X_2	0.46	3.56***	0.24	2.82**
X_3	0.23	1.72†	0.38	4.42***
$\hat{\gamma}_1$	0.064		0.057	
SSR_j	4.8649		17.3298	
SSE_j	1.0726		4.3091	
R_j^2	0.819		0.801	
$\hat{\sigma}_j^2$	0.0894		0.1347	

†Significant at the 0.10 level.
*Significant at the 0.05 level.
**Significant at the 0.01 level.
***Significant at the 0.001 level.

The squared multiple correlations have scarcely changed with the deletion of the time variable. The significance levels of the individual regression coefficients remain about the same. If we should test the parallelism hypothesis of equal regression vectors for the courses the statistic (11) is only equal to $F = 1.01$, and the hypothesis is indeed tenable. The within-courses regression coefficient vector (13) is

$$\hat{\boldsymbol{\beta}}'_{1W} = [0.190, 0.326, 0.317]$$

Its associated coefficient of determination is 0.8048—virtually the same as the individual within-courses coefficients.

What conclusions can we draw about the relationship of the overall rating Y to the three predictor ratings from these several regressions, and which of the estimated regression functions appears to summarize that relationship best? For the answer to the second question we turn to yet another regression equation, that of Y related to X_1, X_2, X_3 for the complete sample of $N = 52$ observations in which the courses are unidentified. This is in a sense the *total regression* equation and has the form

$$\hat{y}_i = 0.048 + 0.198x_{i1} + 0.327x_{i2} + 0.309x_{i3}$$

Its coefficient of determination is equal to 0.7921; once again, that is nearly the same as the other R^2 measures. The coefficients are very close to those of the within-course regression equation, but unlike that function *the single equation will hold for both courses*. We have traded a trifling loss in explained variation for generality, just as we traded some explained variation in the original models for equations with one less variable.

Finally, we turn to the meaning of the total regression equation. Its coefficients are significantly different from zero in the two-sided sense at the levels 0.05, 0.001, and 0.001, respectively. The largest and nearly equal contributors to the size of the instructor's overall rating Y are X_2, ability to stimulate interest, and X_3, overall evaluation of the course. X_1, clarity of the presentation, is less important.

3-8 REFERENCES

AITKEN, A. C. (1950): Statistical independence of quadratic forms in normal variates, *Biometrika*, vol. 37, pp. 93–96.

CHEW, V. (1976): Comparing treatment means: a compendium, *HortScience*, vol. 11, pp. 348–357.

COCHRAN, W. G. (1934): The distribution of quadratic forms in a normal system, with applications to the analysis of covariance, *Proceedings of the Cambridge Philosophical Society*, vol. 30, pp. 178–191.

DRAPER, N. R., and H. SMITH (1981): *Applied Regression Analysis*, 2nd ed., John Wiley & Sons, Inc., New York.

FELLER, W. (1957): *An Introduction to Probability Theory and Its Applications*, vol. I, 2nd ed., John Wiley & Sons, Inc., New York.

GRAYBILL, F. A. (1976): *Theory and Application of the Linear Model*, Duxbury Press, North Scituate, Mass.

HAHN, G. J. (1972): Simultaneous prediction intervals for a regression model, *Technometrics*, vol. 14, pp. 203–214.

Hald, A. (1952): *Statistical Theory with Engineering Applications*, John Wiley & Sons, Inc., New York.

James, G. S. (1952): Notes on a theorem of Cochran, *Proceedings of the Cambridge Philosophical Society*, vol. 48, pp. 443–446.

Kendall, M. G., and A. Stuart (1958): *The Advanced Theory of Statistics*, vol. 1, Charles Griffin & Company Ltd., London.

Lancaster, H. O. (1954): Traces and cumulants of quadratic forms in normal variables, *Journal of the Royal Statistical Society, Series B*, vol. 16, pp. 247–254.

Lieberman, G. J. (1961): Prediction regions for several predictions from a single regression line, *Technometrics*, vol. 3, pp. 21–27.

Miller, R. G., Jr. (1977): Developments in multiple comparisons, *Journal of the American Statistical Association*, vol. 72, pp. 779–788.

Miller, R. G., Jr. (1981): *Simultaneous Statistical Inference*, 2nd ed., Springer-Verlag, New York.

Morrison, D. F. (1976): *Multivariate Statistical Methods*, 2nd ed., McGraw-Hill Book Company, New York.

O'Neill, R., and G. B. Wetherill (1971): The present state of multiple comparison methods, *Journal of the Royal Statistical Society, Series B*, vol. 33, pp. 218–250.

Pearson, E. S., and H. O. Hartley (1951): Charts of the power function of the analysis of variance tests, derived from the non-central *F* distribution, *Biometrika*, vol. 38, pp. 112–130.

Pearson, E. S., and H. O. Hartley (1969): *Biometrika Tables for Statisticians*, vol. 1, 3rd ed., Cambridge University Press, Cambridge, England.

Pearson, E. S., and H. O. Hartley (1972): *Biometrika Tables for Statisticians*, vol. 2, Cambridge University Press, Cambridge, England.

Roy, S. N., and R. C. Bose (1953): Simultaneous confidence interval estimation, *Annals of Mathematical Statistics*, vol. 24, pp. 513–536.

Scheffé, H. (1953): A method for judging all contrasts in the analysis of variance, *Biometrika*, vol. 40, pp. 87–104.

Scheffé, H. (1959): *The Analysis of Variance*, John Wiley & Sons, Inc., New York.

Searle, S. R. (1971): *Linear Models*, John Wiley & Sons, Inc., New York.

Singer, M. T. (1963): Personality measurements in the aged, in J. E. Birren et al. (eds.), *Human Aging: A Biological and Behavioral Study*, pp. 217–249, U.S. Government Printing Office, Washington, D.C.

Tiku, M. L. (1967): Tables of the power of the *F*-test, *Journal of the American Statistical Association*, vol. 62, pp. 525–539.

Urban, W. D. (1976): *Statistical Analysis of Blood Lead Levels of Children Surveyed in Pittsburgh, Pennsylvania: Analytical Methodology and Summary Results*, NBSIR 76-1024, Institute of Applied Technology, National Bureau of Standards, Washington, D.C.

Wilsdon, B. H. (1934): Discrimination by specification statistically considered and illustrated by the standard specification for Portland cement, *Supplement to the Journal of the Royal Statistical Society, Series B*, vol. 1, pp. 152–206.

Working, H., and H. Hotelling (1929): Application of the theory of error to the interpretation of trends, *Journal of the American Statistical Association, Supplement (Proceedings)*, vol. 24, pp. 73–85.

WYNN, H. P., and P. BLOOMFIELD (1971): Simultaneous confidence bands in regression analysis (with discussion), *Journal of the Royal Statistical Society, Series B*, vol. 33, pp. 202–217.

3-9 EXERCISES

1. The data in the Pittsburgh Lead-Based Paint Study (Urban, 1976) included observations on the following variables:

Y = blood lead level (μg/100 ml) of children in the dwellings

X_1 = sex (1 = female, 0 = male)

X_2 = race (1 = black, 0 = white)

X_3 = age of child (months)

X_4 = rank of child in family

X_5 = weight of child

X_6 = months in dwelling

X_7 = score computed from all surface lead measurements

X_8 = surface lead score computed from Type I rooms (bath, kitchen, family)

$N = 407$ children had complete observations on all nine variables. A linear regression analysis was carried out with Y as the dependent variable and $X_1, \ldots, X_8$ as the predictors. These final results were obtained:

Variable	Mean	$\sum (y_i - \bar{y})(x_{ij} - \bar{x}_j)$	r_{Yj}	$\hat{\beta}_j$	t_j
1	0.48	−18.2973	−0.01	−0.1779	−0.28
2	0.29	101.6118	0.25	3.0464	4.17
3	45.10	0.6561	−0.03	0.0521	2.35
4	2.44	−158.2260	0.16	0.4767	2.56
5	30.83	−7.1646	−0.18	−0.0722	−3.03
6	33.59	−202.8084	−0.16	−0.0596	−2.91
7	24.02	−180.0958	0.25	0.0729	2.24
8	28.19	−282.2194	0.22	−0.0081	−0.32

$\bar{y} = 22.03 \qquad \sum_{i=1}^{407} (y_i - \bar{y})^2 = 20{,}142.70 \qquad \text{SSR} = 3477.99 \qquad \text{SSE} = 16{,}664.72$

$\hat{\sigma}^2 = 41.8711 \qquad \hat{\sigma} = 6.4708 \qquad R^2 = 0.1727$

(a) Test the hypothesis that Y is unrelated to the eight predictors.
(b) Test the individual hypotheses of zero regression coefficients for each predictor.
(c) Repeat the tests in (b) with family error rates not greater than 0.05 by using first the Bonferroni, then the Scheffé, methods of simultaneous inference.
(d) Based on the tests of (b) and (c), which variables might be dropped from the regression model?
(e) Compare the multiple regression coefficients with the simple correlations r_{Yj} of blood lead level and the predictors.

(f) Estimate blood lead level for a child with the vector of predictor values

$$\mathbf{x}_0' = [1, 1, 27, 1, 32, 1, 6, 6]$$

2. The variables X_1 and X_8 were dropped from the linear regression model of the preceding exercise, and a new variable

$$X_1^* = X_1 X_2$$

taking on the value one for black female children and zero otherwise was introduced. That variable was suggested by an examination of the blood lead level means for the four race and sex combinations: Black females had a significantly elevated mean. The mean of X_1^* is $\bar{x}_1^* = 57/407 = 0.14$, or the proportion of black females in the sample. These results were obtained:

Variable	$\hat{\beta}_j$	t_j
Race	1.819	1.95
Race × sex	2.537	2.13
Age	0.054	2.45
Rank	0.484	2.63
Weight	−0.072	−3.07
Months in dwelling	−0.061	−3.01
Surface lead score	0.065	4.64

$$\text{SSR} = 3657.70 \qquad \text{SSE} = 16{,}485$$

Compare this regression analysis with that of Exercise 1.

3. Let us form from the blood lead data a new set of predictor variables consisting of race (U_1), race × sex (U_2), and surface lead level (U_3). The matrix of the sums of squares and products of the centered variables Y (blood lead level), U_1, U_2, and U_3 is

$$\begin{bmatrix} 20{,}142.70 & 322.8378 & 209.4595 & 16{,}642.84 \\ & 83.3661 & 40.6143 & 23.88 \\ & & 49.0172 & -167.67 \\ & & & 221{,}093 \end{bmatrix}$$

As in the previous exercises, $N = 407$, and

$$\bar{y} = 22.03 \qquad \bar{u}_1 = 0.2875 \qquad \bar{u}_2 = 0.14 \qquad \bar{u}_3 = 24.02$$

(a) Carry out a regression analysis of Y and the three predictors.

(b) Find the prediction intervals for the future values of Y and confidence intervals for the conditional means of Y when the predictors have these vectors of values:

1. [0, 0, 10]
2. [0, 0, 24.02]
3. [1, 1, 40]

What do these intervals appear to imply about the usefulness of the estimated regression function for predicting blood lead level?

4. These summary statistics were calculated for the variables Y (overall rating of the instructor) and X (year) for course B in Example 3-9:

x_j	N_j	$\bar{y}_j$	$\sum (y_{ij} - \bar{y}_j)^2$
0	9	2.33	10
1	11	1.45	2.7273
2	8	1.5	2
3	8	1.5	2
Sum	36	—	16.7273

$$\bar{y} = 1.69 \qquad \bar{x} = 1.42 \qquad \sum N_j(x_j - \bar{x})^2 = 42.75$$

The analysis of variance table with the error sum of squares partitioned into within-years and lack-of-fit components had this form:

Source	Sum of squares	d.f.	Mean square	F
Linear trend	2.54	1	2.54	4.86
Lack of fit	2.37	2	1.19	2.27
Within years	16.73	32	0.52	—
Error	19.10	34	0.56	—
Total (less grand mean)	21.64	35	—	—

Because the lack-of-fit F-statistic was close to the 0.10 critical value we have elected to use the within-years estimate of the variance for the test of $H_0\colon \beta = 0$; that hypothesis can be rejected at the 0.05 level.

(a) Compute the 95% confidence interval for β using first the usual error mean square and then the within-years mean square.

(b) Predict the value of the mean overall rating for year 4, the next future year.

(c) Find the 95% prediction and confidence intervals for individual and mean ratings in year 4.

5. Course A of Example 3-9 had the following summary statistics for the variables Y (overall evaluation of the instructor) and X (year):

x_j	N_j	$\bar{y}_j$	$\sum (y_{ij} - \bar{y}_j)^2$
0	3	1.33	0.6667
1	3	1.67	0.6667
2	4	1.50	1
3	6	1.67	3.3333
Sum	16	—	5.6667

$$\bar{y} = 1.5625 \qquad \bar{x} = 1.8125 \qquad \sum N_j(x_j - \bar{x})^2 = 20.4375$$

$$\text{SSE} = 5.7982 \qquad \text{SSR} = 0.1393 \qquad \hat{\beta} = 0.0826 \qquad r_{yx} = 0.1532$$

Use these quantities and those of Exercise 4 to test the hypothesis that the linear trend regression coefficients in courses A and B are equal. Carry out the test using (a) the error sums of squares without regard for the effect of the multiple observations in each year and (b) the method described in Exercise 4 using the within-years sums of squares for an estimate of "pure error" variation.

6. The case of multiple dependent-variable observations described in Section 3-4 can be extended to the tests and confidence intervals of k independent regressions of Section 3-7. Only the estimator of the common disturbance variance σ^2 and its degrees of freedom are affected. The "pure error" estimator of σ^2 is

$$\hat{\sigma}^2 = \frac{1}{(N-M)} \sum_{j=1}^{k} \text{SSW}_j$$

where

$$\text{SSW}_j = \sum_{h=1}^{M_j} \sum_{i=1}^{L_{hj}} (y_{ihj} - \bar{y}_{.hj})^2$$

is the sum of squared deviations of the Y-values for each of the M_j sets of multiple observations in the jth sample. The hth set contains L_{hj} observations. Then

$$N_j = \sum_{h=1}^{M_j} L_{hj} \qquad N = \sum_{j=1}^{k} N_j \qquad M = \sum_{j=1}^{k} M_j$$

(a) Use the results of Section 3-4 to show that $\hat{\sigma}^2$ is an unbiased estimator of σ^2 regardless of the validity of the k regression models.

(b) Show that $(N-M)\hat{\sigma}^2/\sigma^2$ has the central chi-squared distribution with $N - M$ degrees of freedom.

(c) Give the distributions of the t- and F-statistics of Section 3-7 when they are computed from $\hat{\sigma}^2$.

(d) Find the lack-of-fit tests for the k individual regression functions under the assumption of a common disturbance variance.

7. N independent observations $\mathbf{x}' = [x_1, \ldots, x_N]$ have been made on the $N(\mu, \sigma^2)$ random variable X. Use the corollary of Theorem 3-3 to show that the mean

$$\bar{x} = \frac{1}{N} \sum_{i=1}^{N} x_i = \mathbf{b}'\mathbf{x}$$

and variance

$$s^2 = \frac{1}{N-1} \sum_{i=1}^{N} (x_i - \bar{x})^2 = \frac{1}{N-1} \mathbf{x}'\mathbf{A}\mathbf{x}$$

are independently distributed, where

$$\mathbf{b} = \frac{1}{N}\begin{bmatrix} 1 \\ \cdot \\ \cdot \\ \cdot \\ 1 \end{bmatrix} \qquad \mathbf{A} = \frac{1}{N}\begin{bmatrix} N-1 & -1 & \cdots & -1 \\ -1 & N-1 & \cdots & -1 \\ \cdot & \cdot & & \cdot \\ \cdot & \cdot & & \cdot \\ \cdot & \cdot & & \cdot \\ -1 & -1 & \cdots & N-1 \end{bmatrix}$$

8. Quantities of porcine blood with three levels of lead were prepared by the National Bureau of Standards (Urban, 1976, pp. 21–25) as reference standards x_j, and samples were drawn from those standards for the determination of blood lead level by the microisotopic dilution method (y_{ij}). The following levels (μg/100 ml) were obtained:

Reference standard, x_j	Microisotopic method, y_{ij}
23.2	30.3, 32.4, 25.9, 24.3, 25.5
33.7	33.6, 34.6, 34.1, 33.5, 34.7, 35.9
105.3	105.8, 109.2, 106.2, 106.4, 112.5, 105.6

(a) Estimate the regression equation for the y_{ij} as a linear function of the reference level x_j.

(b) Carry out the analysis of variance for the regression and the goodness-of-fit test. Is the linear model a reasonable one?

(c) Test the hypothesis that the intercept β_0 of the linear regression model at $x_j = 0$ is zero.

(d) Find the 95% prediction intervals for microisotopic dilution blood lead levels at the three reference standard values. Similarly, find the 95% confidence intervals for the conditional means at those values.

9. Use the data of the preceding Exercise 8, that of Example 3-5, and the observations of Exercise 4, Chapter 1, to test the hypothesis of equality of the three linear regression functions relating the dependent variable blood lead levels to the National Bureau of Standards reference values.

10. From the data given in Example 3-5 test the hypothesis

$$H_0: \beta_0 = 0$$

that the intercept in the uncentered model

$$y_{ij} = \beta_0 + \beta x_j + e_{ij}$$

is zero. The statistic is

$$t = \frac{\hat{\beta}_0}{\sqrt{\widehat{\operatorname{Var}(\hat{\beta}_0)}}}$$

where

$$\widehat{\operatorname{Var}(\hat{\beta}_0)} = \hat{\sigma}^2\left[\frac{1}{N} + \frac{\bar{x}^2}{\sum N_j (x_j - \bar{x})^2}\right]$$

11. Consider the data of Example 3-5, with the six observations at $x_6 = 105.3$ omitted. The statistics for the remaining $N = 51$ pairs are

$$\bar{x} = 21.9 \qquad \bar{y} = 22.0 \qquad \sum N_j(x_j - \bar{x})^2 = 1254.12$$

$$\sum\sum (y_{ij} - \bar{y})^2 = 1828 \qquad \sum\sum (x_j - \bar{x})(y_{ij} - \bar{y}) = 1004.1$$

Fit a straight line to these pairs of blood lead levels and test its fit by the analysis of variance in Table 3-5.

12. Find the $100(1 - \alpha)$% Bonferroni and Scheffé simultaneous confidence intervals for the bilinear parametric functions

$$\sum_{h=1}^{p} \sum_{j=1}^{k} a_h c_j \beta_{hj}$$

given by (26) of Section 3-7, in which β_{hj} is the regression coefficient of the hth independent variable in the jth of k mutually exclusive samples.

13. Show that the quadratic form SSC/σ^2 defined by (38), Section 3-7, has a chi-squared distribution. Show that its matrix and the variances (34) satisfy Theorem 3-1.

14. Example 2-9 contains observations on five psychological test scores for $N = 46$ subjects, and the statistics for a multiple regression analysis of the first test with the remaining four are given in the example. Multiply the elements of the matrix **S**, expression (33) of Section 2-5, by $N - 1 = 45$ to obtain the usual sums of squares and products of deviations from the means and carry out these analyses.

(a) Test the hypothesis that the regression coefficients of X_1, X_2, and X_3 are simultaneously zero, after adjustment for the independent variable X_4.

(b) Find the Bonferroni and Scheffé simultaneous confidence intervals for the four regression coefficients.

(c) Compute the linear regression coefficients for Y related individually to X_1, X_2, X_3, and X_4. Compare the coefficients and their statistical significance probabilities with those for the original multiple regression model.

15. In Exercise 23 of Section 2-9 the total numbers of home runs scored by $N = 27$ baseball players were related to four independent variables describing the players.

(a) Convert the elements of the correlation matrix to $\mathbf{X}_1'\mathbf{X}_1$, $\mathbf{X}_1'\mathbf{y}$, and the sum of squares of the dependent variable about its mean.

(b) Find the Scheffé simultaneous confidence intervals for the four regression coefficients given in the exercise.

(c) Test the hypothesis that X_1 and X_4 have zero regression coefficients in the complete model, with adjustment for X_2 and X_3.

(d) Repeat the subset test in (c) for X_1, X_3, and X_4 after adjustment for X_2.

16. For the course evaluation data of Example 2-1, make these analyses:

(a) Test the hypothesis that the population regression coefficients of the two independent variables are equal.

(b) Find the 95% confidence interval for the difference of the two regression coefficients.

(c) Make 99% Scheffé simultaneous tests for zero regression coefficients of the independent variables.

(d) Find the 95% prediction intervals for the future average overall ratings corresponding to these independent variable values:

Pair	X_1	X_2
1	2.0	1.5
2	2.3	2.325
3	2.8	3.0

17. The independent variables of Example 2-1 each contain some multiple values. Fit individual linear models for the dependent variable and use the within-group estimates of the disturbance variances to test the linearity hypotheses for each line.

18. Exercise 14 of Section 1-9 contains average overall ratings of instructors and their courses in two academic departments.

(a) Treat the course averages as the dependent variable observations and fit individual regression lines for each department.

(b) Test the hypothesis of common intercepts and slopes for the two lines.

19. In Exercise 17, Section 1-9, weather was coded by a dummy variable whose values of 0 and 1 indicated fair and foul weather, respectively. Find the separate multiple regression functions for busboy tips as a function of number of meals served and number of waitresses and test the hypothesis that the coefficient vectors are the same in each case.

20. Find the Scheffé simultaneous test critical value with a family error rate of 0.05 for the test statistics given in the first table of Exercise 3, Section 2-9. Which regression coefficients are significantly different from zero in the Scheffé sense? Is the finding consistent with the overall test given by the F-ratio in the next table of the exercise?

21. Compute the 99% Scheffé simultaneous confidence intervals for the three regression coefficients of Exercise 12, Section 2-9. Which variables might be omitted from the regression model?

22. Make Scheffé simultaneous tests of the hypotheses of zero coefficients in the separate regression equations for the dependent variables Y_1, Y_2, and Y_3 of Exercise 4, Section 2-9.

23. Suppose that an experiment was conducted under conditions A, B, and C to relate a response variable Y to the independent variables X_1 and X_2. The experiment was designed so that X_1 and X_2 had the same values under the three conditions. Only five independent observations could be collected under each condition. The data are shown in the table:

	Condition			Common independent variable values	
	A (Y_1)	B (Y_2)	C (Y_3)	X_1	X_2
	5	4	1	−2	2
	7	6	7	−1	−1
	6	9	15	0	−2
	11	10	21	1	−1
	17	20	32	2	2
Sum	46	49	76	0	0
Sum of squared deviations	96.8	152.8	584.8	10	14

$$\mathbf{X}_1'\mathbf{y}_1 = \begin{bmatrix} 28 \\ 14 \end{bmatrix} \qquad \mathbf{X}_1'\mathbf{y}_2 = \begin{bmatrix} 36 \\ 14 \end{bmatrix} \qquad \mathbf{X}_1'\mathbf{y}_3 = \begin{bmatrix} 76 \\ 8 \end{bmatrix}$$

(a) Test the hypothesis that the regression coefficient vectors obtained under the three conditions are the same.

(b) Use the Scheffé method to determine which coefficients are significantly different.

(c) Test the hypothesis of equal intercepts for the three regression models. If the hypothesis is rejected, use the Scheffé multiple test to determine which intercepts are different.

24. Use the corollary of Theorem 3-3 to show that $\hat{\alpha} = \bar{y}$ is distributed independently of $\hat{\boldsymbol{\beta}}_1$, SSR, and SSE.

25. Express the F-ratio defined by (1), Section 3-2, in terms of the coefficient of determination R^2, and then find the upper critical value of R^2 under the least squares model with zero regression coefficients.

26. The expected value of the quadratic from $\mathbf{x}'\mathbf{A}\mathbf{x}$ is given in expression (5) of Section C-4. Use that result to verify the expected mean squares in Table 3-1.

chapter 4

Polynomial Models for Time Series

4-1 INTRODUCTION

When data have been collected over time, trends, some seasonal fluctuations, and other slowly varying movements might be represented by a polynomial in the time variable plus a random disturbance. If the polynomial accounts for most of the variation in the observations, the series can be described concisely by the least squares estimates of the polynomial coefficients. Parallel time series whose trends or seasonal movements have been removed by a polynomial model can be related without the artifacts of nonsense correlations caused by common trends. At the same time, extrapolation by polynomial models outside the range of the time points is extremely hazardous and should be avoided unless supported by the substantive interpretation of the model. If the observations are equally spaced, the coefficients of the polynomial model can be estimated most easily by the use of *orthogonal* polynomials, or ones constructed so that the columns of the **X** matrix of their values will be mutually orthogonal. This use of orthogonal polynomials in time series was introduced by Fisher (1920, 1924) for the study of the effect of rainfall on wheat yields. The particular class of orthogonal polynomials we shall use is that discovered by the Russian mathematician Chebychev some 50 years earlier than Fisher's application.

In this chapter we develop the fitting of orthogonal polynomials by the ordinary least squares model with uncorrelated disturbances. A test for that assumption is deferred to Chapter 6. In the later chapters on the analysis of variance we also use orthogonal polynomials for describing trends in treatment means when the treatment effects can be scaled on a time, temperature, or other continuum.

4-2 ORTHOGONAL POLYNOMIALS

In some of our earlier examples of regression analysis we fitted linear trend lines to observations collected over time. In this way we estimated the rate of a linear change in the observations, found tests and confidence statements for the parameters of the line, and, in the case of multiple observations at each time point, gave a test for the linearity of the trend model. The residuals from the trend line could be related to another parallel time series to determine the correlation of the two series with the effect of a common trend removed or held constant in the "partial correlation" sense.

Many long-term trends in time series are not linear or even monotonic. Economic series fluctuate according to seasonal patterns, climatic observations may show seasonal variation as well as slower year-to-year fluctuations, and measurements in the natural sciences may attain a maximum or minimum at some point of optimum temperature, pressure, or other experimental condition. As a model for the time series y_t we might use the polynomial representation

$$y_t = \alpha + \gamma_1(t - \bar{t}) + \gamma_2(t - \bar{t})^2 + \cdots + \gamma_p(t - \bar{t})^p + e_t \qquad t = 1, \ldots, N \tag{1}$$

and obtain the usual least squares estimators for its parameters. The distribution theory, tests, and confidence statements of Chapters 2 and 3 could be applied to the estimated model. However, in its form (1) the polynomial model suffers from an unfortunate property. The matrix $\mathbf{X}'\mathbf{X}$ will have the form

$$\begin{bmatrix} N & 0 & \sum(t-\bar{t})^2 & \cdots & \sum(t-\bar{t})^p \\ 0 & \sum(t-\bar{t})^2 & 0 & & \sum(t-\bar{t})^{p+1} \\ \cdots & \cdots & \cdots & \cdots & \cdots \\ \sum(t-\bar{t})^p & \sum(t-\bar{t})^{p+1} & \sum(t-\bar{t})^{p+2} & \cdots & \sum(t-\bar{t})^{2p} \end{bmatrix} \tag{2}$$

Alternate columns of the matrix $\mathbf{X}$ will be highly correlated, so that the effects of the quadratic term $(t - \bar{t})^2$ will be related to that of the constant, the linear term will be associated with the cubic, and so on. Such correlations will also cause the estimators $\hat{\alpha}, \hat{\gamma}_1, \ldots, \hat{\gamma}_p$ to be correlated. A preferable model would be one in which the successive constant, linear, quadratic, cubic, . . . , pth-degree terms explain independent components of the total sum of squares $\sum y^2$. At the same time the matrix $\mathbf{X}'\mathbf{X}$ would have the diagonal pattern, so that the solution for the least squares estimators would be arithmetically trivial. A model with these properties will be provided by reparametrizing (1) into orthogonal polynomials.

The simplest example of an orthogonal polynomial was introduced in Chapter 1 when we elected to center our independent variable observations as the deviations $x_1 - \bar{x}, \ldots, x_N - \bar{x}$. Then the deviations formed a vector orthogonal to the constant term vector $\mathbf{j}' = [1, \ldots, 1]$, so that the intercept and slope parameters were independent of one another. Let us extend this orthogonalization process to the quadratic polynomial model

$$y_t = \alpha + \gamma_1(t - \bar{t}) + \gamma_2(t - \bar{t})^2 + e_t \tag{3}$$

by merely centering the second-degree terms about their mean:

$$y_t = \beta_0 + \beta_1(t - \bar{t}) + \beta_2[(t - \bar{t})^2 - \overline{(t - \bar{t})^2}] + e_t \tag{4}$$

where

$$\overline{(t - \bar{t})^2} = \frac{1}{N} \sum_{t=1}^{N} (t - \bar{t})^2$$
$$= \frac{N^2 - 1}{12} \tag{5}$$

The polynomials

$$\begin{aligned} \phi_0(t) &= 1 \\ \phi_1(t) &= \lambda_1(t - \bar{t}) \\ \phi_2(t) &= \lambda_2\left[(t - \bar{t})^2 - \frac{N^2 - 1}{12}\right] \end{aligned} \tag{6}$$

are said to be orthogonal over the set of values $t = 1, \ldots, N$, for

$$\sum_{t=1}^{N} \phi_0(t)\phi_1(t) = \sum_{t=1}^{N} \phi_0(t)\phi_2(t) = \sum_{t=1}^{N} \phi_1(t)\phi_2(t) = 0 \tag{7}$$

The constants λ_1, λ_2 are chosen so that the values of $\phi_1(t)$ and $\phi_2(t)$ are whole numbers, for example,

$$\lambda_1 = \begin{cases} 1 & N \text{ odd} \\ 2 & N \text{ even} \end{cases} \tag{8}$$

The estimators of the parameters are

$$\begin{aligned} \hat{\beta}_0 &= \bar{y} \\ \hat{\beta}_1 &= \frac{\sum \phi_1(t) y_t}{\sum \phi_1^2(t)} \\ \hat{\beta}_2 &= \frac{\sum \phi_2(t) y_t}{\sum \phi_2^2(t)} \end{aligned} \tag{9}$$

Orthogonal polynomials of third, fourth, . . . , degree may be found by the algebra of finite differences [see, for example, Plackett (1960)]. The cubic, quartic, quintic, and sextic polynomials orthogonal over N equally spaced points are given by the formulas

$$\begin{aligned} \phi_3(t) &= \lambda_3\left[\frac{x^3 - x(3N^2 - 7)}{20}\right] \\ \phi_4(t) &= \lambda_4\left[x^4 - \frac{x^2(3N^2 - 13)}{14} + \frac{3(N^2 - 1)(N^2 - 9)}{560}\right] \\ \phi_5(t) &= \lambda_5\left[x^5 - x^3\left(\frac{5}{18}\right)(N^2 - 7) + \frac{x(15N^4 - 230N^2 + 407)}{1008}\right] \\ \phi_6(t) &= \lambda_6\left[x^6 - x^4\left(\frac{5}{44}\right)(3N^2 - 31) + \frac{x^2(5N^4 - 110N^2 + 329)}{176}\right. \\ &\quad \left. - \left(\frac{5}{14{,}784}\right)(N^2 - 1)(N^2 - 9)(N^2 - 25)\right] \end{aligned} \tag{10}$$

where for convenience $x = t - \bar{t} = t - (N + 1)/2$. In practice with short series of observations we may find the values of these polynomials from tables. Values of the $\phi_j(t)$ for N through 14 and j from 1 to 6 are given in Table 9, Appendix A. Higher-degree orthogonal polynomials can be obtained by the recurrence relation (Plackett, 1960)

$$\phi_{j+1}(t) = \phi_1(t)\phi_j(t) - \frac{j^2(N^2 - j^2)}{4(4j^2 - 1)}\phi_{j-1}(t) \tag{11}$$

Tables of orthogonal polynomials for N through 52 and j through 6 have been compiled by Pearson and Hartley (1966) and for N to 75 and j to 5 by Fisher and Yates (1953). Anderson and Houseman (1942) have evaluated the polynomials of first through fifth degree for N as large as 104 points; some errors in their table have been collected by Greenwood and Hartley (1962).

In applications we may start with the orthogonal polynomial model

$$y_t = \beta_0 + \beta_1\phi_1(t) + \cdots + \beta_p\phi_p(t) + e_t \tag{12}$$

of a degree p chosen from the substantive nature of the time series or its apparent trend movements and fluctuations. From the orthogonality of the terms the coefficients can be estimated separately as

$$\hat{\beta}_j = \frac{\sum \phi_j(t)y_t}{\sum \phi_j^2(t)} \tag{13}$$

The error sum of squares for the pth-degree model is

$$\text{SSE}_p = \sum y_t^2 - \frac{(\sum y_t)^2}{N} - \sum_{j=1}^{p} \frac{[\sum \phi_j(t)y_t]^2}{\sum \phi_j^2(t)} \tag{14}$$

The hypotheses

$$H_0\colon \beta_j = 0 \tag{15}$$

of no jth-degree component in the model are tested against the two-sided alternatives

$$H_1\colon \beta_j \neq 0 \tag{16}$$

by referring their statistics

$$t_j = \hat{\beta}_j\sqrt{\frac{(N - p - 1)\sum \phi_j^2(t)}{\text{SSE}_p}} \tag{17}$$

to the critical value $t_{\alpha/2;N-p-1}$. Orthogonal polynomials for which $|t| < t_{\alpha/2;N-p-1}$ may be dropped from the model and a new error sum of squares (14) calculated for the $k < p + 1$ remaining terms. The hypotheses (15) for those terms are tested again. In that way the degree of an appropriate polynomial model can be determined.

Example 4-1

Let us fit orthogonal polynomials to a short time series to see if its seasonal pattern can be represented adequately with a polynomial of low degree. The data are the average monthly sea surface temperatures at Boothbay Harbor, Maine, for the years 1905–1976. In fitting the polynomials we have put aside the questions of whether a time series analysis of the original 852 monthly averages would be more appropriate, of the meaningfulness of averages computed from data containing strong trends and other sources of heterogeneity, and of the effects of possible month-to-month correlation on the validity of the least squares estimators. The mean temperatures rounded to two decimal places are shown in Table 4-1. February has been chosen as the initial month

TABLE 4-1 Mean Monthly Sea Surface Temperatures (°C)

Month	Mean temperature	Temperature predicted from the fourth-degree model
February	1.03	1.14
March	1.96	1.59
April	4.65	4.83
May	8.69	9.04
June	12.74	12.83
July	15.80	15.26
August	16.07	15.80
September	14.10	14.36
October	10.72	11.29
November	7.52	7.36
December	4.20	3.78
January	1.99	2.19

SOURCE: Data reproduced with the kind permission of Walter R. Welch, Maine Department of Marine Resources.

so that the series has a single turning point. We shall fit the zeroth through sixth-degree orthogonal polynomials to the means. The matrix $\mathbf{X}_1$ is found from the orthogonal polynomial tables to be

$$\mathbf{X}_1 = \begin{bmatrix} -11 & 55 & -33 & 33 & -33 & 11 \\ -9 & 25 & 3 & -27 & 57 & -31 \\ -7 & 1 & 21 & -33 & 21 & 11 \\ -5 & -17 & 25 & -13 & -29 & 25 \\ -3 & -29 & 19 & 12 & -44 & 4 \\ -1 & -35 & 7 & 28 & -20 & -20 \\ 1 & -35 & -7 & 28 & 20 & -20 \\ 3 & -29 & -19 & 12 & 44 & 4 \\ 5 & -17 & -25 & -13 & 29 & 25 \\ 7 & 1 & -21 & -33 & -21 & 11 \\ 9 & 25 & -3 & -27 & -57 & -31 \\ 11 & 55 & 33 & 33 & 33 & 11 \end{bmatrix} \tag{18}$$

$\mathbf{X}_1'\mathbf{X}_1$ then has the diagonal elements

$$\text{diag}\,(\mathbf{X}_1'\mathbf{X}_1) = [572, 12{,}012, 5148, 8008, 15{,}912, 4488] \tag{19}$$

while

$$\mathbf{y}'\mathbf{X}_1 = [65.31, -1891.51, -113.79, 493.84, -32.16, -68.66] \tag{20}$$

and

$$\bar{y} = 8.289 \quad \sum_{t=1}^{12} (y_t - \bar{y})^2 = 339.5891 \tag{21}$$

The regression coefficients of the six orthogonal polynomials, the sums of squares due to them, and the F-statistics for testing the hypotheses

$$H_0: \beta_j = 0 \qquad j = 1, \ldots, 6 \tag{22}$$

are given in Table 4-2. All coefficients except that of the fifth-degree orthogonal polynomial exceed the 0.01 critical value of the F-statistic with one and five degrees of freedom. Nevertheless, a seven-parameter model is not a simple or parsimonious one for a series of only 12 values. Let us drop the fifth- and sixth-degree polynomials, or equivalently, pool their sums of squares with the error sum of squares. The error mean square for the resulting fourth-degree orthogonal polynomial model is

TABLE 4-2 Coefficients and Test Statistics for the Sixth-Degree Model

Polynomial	$\hat{\beta}_j$	Sum of squares	d.f.	F_j
Linear	0.1142	7.4570	1	192
Quadratic	−0.1575	297.8530	1	7669
Cubic	−0.0221	2.5152	1	65
Quartic	0.0617	30.4543	1	784
Quintic	−0.0020	0.0650	1	1.67
Sextic	−0.0153	1.0504	1	27
Error sum of squares	—	0.1942	5	—
Total sum of squares	—	339.5891	11	—

$$\begin{aligned} \text{MSE}_4 &= \frac{0.1942 + 1.0504 + 0.0650}{7} \\ &= 0.1871 \end{aligned}$$

The regression coefficients and sums of squares of the four polynomials of course remain the same because of the orthogonality property. Their F-statistics based on MSE_4 are still highly statistically significant, although smaller than those in Table 4-2. The fourth-degree polynomial model accounts for about 99.61% of the variation in the monthly average temperatures. The means predicted from this model and their residuals from the actual means are shown in Table 4-1.

The fourth-degree polynomial seems a reasonable compromise between simplicity and goodness of fit as measured by its coefficient of determination. Its degree is appropriate for a function with a single maximum and smooth relative minima at each end. Quadratic or cubic polynomials fitted to the temperature means do not have these properties, and their approximations of the data, as evidenced by their coefficients of determination and residuals, are poor.

By starting with a sixth-degree model for the average temperatures we achieved an extremely small estimate of the disturbance variance and large F-ratios for the six coefficients. A more conservative approach consists of starting with a low-degree model suggested by the appearance of the series and adding orthogonal polynomials until we gain a balance between complexity and the proportion of explained variation. For example, we might have begun with a quadratic model and the resulting analysis of variance:

Source	d.f.	Mean square	F
Linear	1	7.457	1.96
Quadratic	1	297.853	78.2
Error	9	3.809	—

The linear component is not significant. If we add a cubic orthogonal polynomial, we have this analysis:

Source	d.f.	Mean square	F
Linear	1	7.4570	1.88
Quadratic	1	297.853	75.02
Cubic	1	2.515	0.63
Error	8	3.970	—

We might be tempted to stop at this point and conclude that a quadratic model without a linear term would suffice to describe the averages. However, it is well in practice to fit at least one more component than is deemed "significant." If we do so, we shall arrive at the fourth-degree model with a small error mean square and large F-ratios. The abrupt drop in the error mean square would lead us to discard the simple quadratic. It is likely that the quartic model "overfits" the data, yet at the same time the quadratic is undoubtedly an oversimplification.

Transformation to an ordinary polynomial model. Occasionally, it may be necessary to express the orthogonal polynomial model as an ordinary polynomial in the time variable t or its centered value $x = t - \bar{t}$. That form is especially useful for predicting values of the time series between observation points or in the future or merely for expressing the polynomial as a direct function of t in a form more familiar to most consumers of the statistical analysis.

The regression coefficients of one polynomial representation are linear functions of those in the other model. To find the equations relating the two sets of estimators we write the jth orthogonal polynomial as

$$\phi_j(t) = c_{0j} + c_{ij}x + c_{2j}x^2 + \cdots + c_{jj}x^j \tag{23}$$

where for convenience $x = t - \bar{t}$. Then the predicted value of the tth observation is

$$\begin{aligned}\hat{y}_t &= \bar{y} + \hat{\beta}_1\phi_1(t) + \cdots + \hat{\beta}_p\phi_p(t) \\ &= \bar{y} + \hat{\beta}_1(c_{01} + c_{11}x) + \hat{\beta}_2(c_{02} + c_{12}x + c_{22}x^2) \\ &\quad + \cdots + \hat{\beta}_p(c_{1p} + c_{0p}x + c_{2p}x^2 + \cdots + c_{pp}x^p)\end{aligned} \tag{24}$$

while its expression in terms of the estimated nonorthogonal polynomial (1) is

$$\hat{y}_t = \hat{\alpha} + \hat{\gamma}_1 x + \hat{\gamma}_2 x^2 + \cdots + \hat{\gamma}_p x^p \tag{25}$$

The estimators are related by equating terms in like powers of x:

$$
\begin{aligned}
\hat{\gamma}_p &= c_{pp}\hat{\beta}_p \\
\hat{\gamma}_{p-1} &= c_{p-1,p-1}\hat{\beta}_{p-1} + c_{p-1,p}\hat{\beta}_p \\
&\;\;\vdots \\
\hat{\gamma}_1 &= c_{11}\hat{\beta}_1 + c_{12}\hat{\beta}_2 + \cdots + c_{1p}\hat{\beta}_p \\
\hat{\alpha} &= \bar{y} + c_{01}\hat{\beta}_1 + c_{02}\hat{\beta}_2 + \cdots + c_{0p}\hat{\beta}_p
\end{aligned} \tag{26}
$$

The coefficients c_{ij} are found from the expressions (6), (10), and (11) for the orthogonal polynomials, and the triangular system of equations is solved successively from $\hat{\gamma}_p$ to $\hat{\alpha}$.

Many of the c_{ij} vanish, because odd orthogonal polynomials contain only odd powers of x and even ones only even powers. The explicit equations for the first seven coefficients are these:

$$
\begin{aligned}
\hat{\alpha} &= \bar{y} - \lambda_2\left(\frac{N^2-1}{12}\right)\hat{\beta}_2 + \lambda_4\left[\frac{3(N^2-1)(N^2-9)}{560}\right]\hat{\beta}_4 \\
&\quad - \lambda_6\left[\frac{5}{14{,}784}(N^2-1)(N^2-9)(N^2-25)\right]\hat{\beta}_6 \\
\hat{\gamma}_1 &= \lambda_1\hat{\beta}_1 - \lambda_3\left(\frac{3N^2-7}{20}\right)\hat{\beta}_3 + \lambda_5\left(\frac{15N^4-230N^2+407}{1008}\right)\hat{\beta}_5 \\
\hat{\gamma}_2 &= \lambda_2\hat{\beta}_2 - \lambda_4\left(\frac{3N^2-13}{14}\right)\hat{\beta}_4 + \lambda_6\left(\frac{5N^4-110N^2+329}{176}\right)\hat{\beta}_6 \\
\hat{\gamma}_3 &= \lambda_3\hat{\beta}_3 - \lambda_5\left(\tfrac{5}{18}\right)(N^2-7)\hat{\beta}_5 \\
\hat{\gamma}_4 &= \lambda_4\hat{\beta}_4 - \lambda_6\left(\tfrac{5}{44}\right)(3N^2-31)\hat{\beta}_6 \\
\hat{\gamma}_5 &= \lambda_5\hat{\beta}_5 \\
\hat{\gamma}_6 &= \lambda_6\hat{\beta}_6
\end{aligned} \tag{27}
$$

The scale factors λ_j included in the polynomials to eliminate fractions must be obtained from Table 9 of Appendix A. We note that $\hat{\gamma}_6$ and $\hat{\beta}_6$ differ only by a constant; conclusions about the degree of the polynomial model based on hypothesis tests on the pth coefficient will be the same for either type of polynomial.

Unequally spaced observations. Methods for fitting orthogonal polynomials to time series observed at unequal intervals have been developed by Wishart and Metakides (1953) and Robson (1959).

4-3 AN EXAMPLE: MULTIPLE REGRESSION ANALYSIS OF TIME SERIES DATA

In this section we describe the use of orthogonal polynomials to remove long-term trends in some concurrent time series so that one series designated as the "dependent" observations can be related to other "predictor" series. The dependent variable Y in this example is the annual catch in metric tons of lobsters (*Homarus Americanus*) by Maine fishermen (Dow, 1976). The predictor variables are

X_1 = number of registered lobster traps (in thousands)

X_2 = number of fishermen

X_3 = mean annual sea surface temperature (°C) at Boothbay Harbor

Values of the four variables for the years 1932–1975 are given in Table 4-3. The annual catch and temperature variables are shown in Figure 4-1. Initially, we fit the linear regression model

$$\begin{aligned} y_t = \alpha &+ \beta_1\phi_1(t) + \beta_2\phi_2(t) + \beta_3\phi_3(t) + \beta_4\phi_4(t) \\ &+ \beta_5(x_{t1} - \bar{x}_1) + \beta_6(x_{t2} - \bar{x}_2) + \beta_7(x_{t3} - \bar{x}_3) \\ &+ e_t \qquad t = 1, \ldots, 44 \end{aligned} \tag{1}$$

of orthogonal polynomials of first through fourth degrees and three predictors. We shall also use the notation of the matrix form of the model,

$$\mathbf{y} = [\mathbf{j} \quad \mathbf{T}_1 \quad \mathbf{W}_1]\beta + \mathbf{e} \tag{2}$$

in which $\mathbf{j}$ is the 44×1 vector of ones, $\mathbf{T}_1$ is the 44×4 matrix of orthogonal polynomial values from Appendix A, and $\mathbf{W}_1$ is the 44×3 matrix of centered values of the three predictor variables.

The general trend in lobster catch appears to have a quadratic or cubic shape, and the quartic term has been included as a slight and inexpensive extension which can be dropped if its coefficient turns out not to be significantly different from zero. One would naturally expect catch to depend on the number of fishermen, or perhaps the number of traps, and those variables have been included in the model. The final predictor, mean annual sea temperature, is the key independent variable in the model. As the sea temperature increases the lobster consumes more food and grows large at a faster rate, thereby increasing the size of the catch population. The purpose of the analysis is to estimate the strength of the relationship of annual catch to the mean sea temperature of that year. The multiple regression model will permit us to do this with the trend components and the fishing intensity variables held constant.

For convenience let us partition the sums of squares and products matrix of the seven independent variables as

$$\mathbf{X}_1'\mathbf{X}_1 = \begin{bmatrix} \mathbf{T}_1'\mathbf{T}_1 & \mathbf{T}_1'\mathbf{W}_1 \\ \mathbf{W}_1'\mathbf{T}_1 & \mathbf{W}_1'\mathbf{W}_1 \end{bmatrix} \tag{3}$$

The 4×4 matrix $\mathbf{T}_1'\mathbf{T}_1$ has the diagonal pattern from the orthogonality of the polynomial values. Its successive diagonal elements are

$$28{,}380, \quad 913{,}836, \quad 1{,}257{,}829{,}980, \quad 1{,}173{,}974{,}648 \tag{4}$$

The matrix of sums of cross products of the polynomials and the deviations of X_1, X_2, X_3 from their means is

$$\mathbf{T}_1'\mathbf{W}_1 = 10^3 \begin{bmatrix} 402.202 & 1{,}544.004 & 0.2631 \\ 774.984 & 116.412 & -2.1459 \\ 16{,}117.454 & 123{,}285.658 & 24.2107 \\ 14{,}724.388 & 120{,}358.915 & 143.1595 \end{bmatrix} \tag{5}$$

TABLE 4-3 **Annual Maine Lobster Yields and Related Variables**

Year	Lobster catch (metric tons)	Traps (thousands)	Fishermen	Mean sea temperature (°C)
1932	2,747	208	2,927	8.2
1933	2,675	180	2,956	8.6
1934	2,439	183	2,925	7.6
1935	3,487	185	3,102	8.2
1936	2,323	185	3,265[a]	7.5
1937	3,333	186	3,429[a]	9.0
1938	3,474	258	3,592	7.3
1939	3,005	260	3,722	6.4
1940	3,467	222	3,717	7.0
1941	4,054	194	3,648	7.8
1942	3,812	187	3,511	8.1
1943	5,202	209	4,239	7.4
1944	6,376	252	4,926	8.0
1945	8,677	378	6,241	8.4
1946	8,517	473	6,574	8.5
1947	8,290	516	5,338	9.2
1948	7,223	439	5,345	8.2
1949	8,742	462	5,424	10.1
1950	8,324	430	5,152	9.6
1951	9,415	383	4,653	10.8
1952	9,087	417	5,032	10.1
1953	10,112	490	5,497	11.1
1954	9,818	488	5,794	10.2
1955	10,302	532	6,051	10.0
1956	9,316	533	5,892	9.2
1957	11,068	565	6,068	9.4
1958	9,665	609	6,236	8.5
1959	10,126	717	6,488	8.3
1960	10,889	745	6,636	8.9
1961	9,485	752	6,472	8.5
1962	10,013	767	5,658	8.1
1963	10,344	731	5,695	8.8
1964	9,713	754	5,803	8.3
1965	8,556	789	5,802	7.7
1966	9,034	776	5,613	7.6
1967	7,479	715	5,425	7.3
1968	9,300	747	5,489	8.1
1969	8,997	805	5,750	8.9
1970	8,243	1,180	6,316	8.9
1971	7,964	1,278	6,702	8.7
1972	7,374	1,448	7,045	8.4
1973	7,731	1,172	7,894	8.8
1974	7,465	1,790	10,523	9.2
1975	7,719	1,750	10,455	9.4
Mean	7,395.05	598.64	5,432.32	8.60

SOURCE: Data reprinted with the kind permission of Walter R. Welch, Robert L. Dow, and the editor, *Marine Technology Society Journal.*
[a]Estimated values.

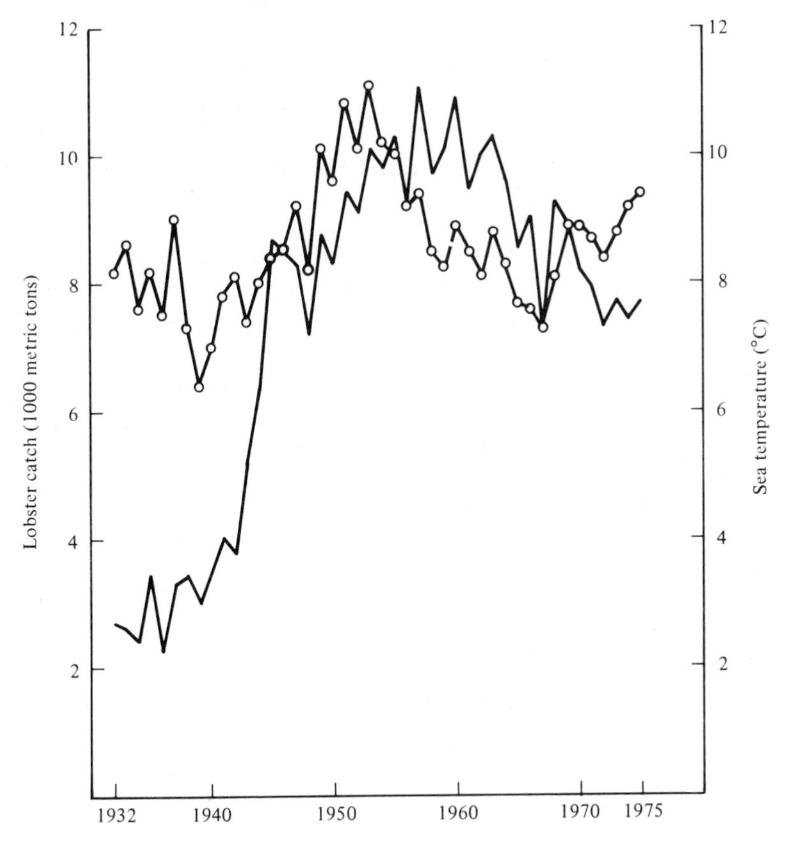

Figure 4-1 Maine lobster catch and mean sea temperature, 1932–1975. Solid line, lobster catch; solid line with circles, sea temperature.

Finally,

$$\mathbf{W}_1'\mathbf{W}_1 = \begin{bmatrix} 7{,}117{,}642 & 26{,}379{,}280 & 3{,}563.6636 \\ & 121{,}616{,}548 & 25{,}419.4318 \\ & & 42.7498 \end{bmatrix} \tag{6}$$

The vector of products of the annual catches and the columns of $\mathbf{X}_1$ is

$$\mathbf{X}_1'\mathbf{y} = 10^4 \begin{bmatrix} 211.9008 \\ -1{,}059.6050 \\ -7{,}555.1104 \\ 13{,}327.8883 \\ 2{,}203.3211 \\ 12{,}613.1313 \\ 6.5876 \end{bmatrix} \tag{7}$$

The estimate of the intercept α of the model is

$$\hat{\alpha} = \bar{y} = 7395.05$$

and the sum of squares of the dependent variable about its mean is

$$\text{SSY} = \sum_{t=1}^{44} (y_t - \bar{y})^2 = 319{,}717{,}456 \tag{8}$$

The correlation matrix of Y, the four orthogonal polynomials treated as observations on random variables, and the three predictors is

$$\mathbf{R} = \begin{bmatrix} 1 & 0.70 & -0.62 & -0.12 & 0.22 & 0.46 & 0.64 & 0.56 \\ & 1 & 0 & 0 & 0 & 0.89 & 0.83 & 0.24 \\ & & 1 & 0 & 0 & 0.30 & 0.01 & -0.34 \\ & & & 1 & 0 & 0.17 & 0.32 & 0.10 \\ & & & & 1 & 0.16 & 0.32 & 0.64 \\ & & & & & 1 & 0.90 & 0.20 \\ & & & & & & 1 & 0.35 \\ & & & & & & & 1 \end{bmatrix} \tag{9}$$

The correlations are sufficiently large and varied to indicate that the multiple regression analysis should yield some significant and informative results.

The regression coefficients, their estimated standard deviations, and their t statistics are given in Table 4-4. The predicted catch for the tth year is

TABLE 4-4 Regression Coefficients and Test Statistics for the First Lobster Catch Model

Variable	$\hat{\beta}_j$	$\sqrt{\widehat{\text{Var}}(\hat{\beta}_j)}$	t_j
$\phi_1(t)$: linear	43.5522	13.388	3.25**
$\phi_2(t)$: quadratic	−8.9130	1.126	−7.91***
$\phi_3(t)$: cubic	−0.1377	0.02256	−6.10***
$\phi_4(t)$: quartic	−0.0324	0.03142	−1.03
X_1: traps	−1.5696	1.0876	−1.44
X_2: fishermen	0.8746	0.1932	4.53***
X_3: sea temperature	622.8	137.72	4.52***

**Significant at the 0.01 level.
***Significant at the 0.001 level.

given by the equation

$$\begin{aligned} \mathbf{y}_t = {} & 7395.05 + 43.5522\,\phi_1(t) - 8.9130\phi_2(t) \\ & -0.1377\phi_3(t) - 0.0324\phi_4(t) \\ & -1.5696(x_{t1} - 598.64) + 0.8746(x_{t2} - 5432.23) \\ & +622.8(x_{t3} - 8.60) \qquad t = 1, \ldots, 44 \end{aligned} \tag{10}$$

The predicted values and the residuals $z_t = y_t - \hat{y}_t$ rounded to integers are given in the first two columns of Table 4-5. The sum of squares due to

TABLE 4-5 Predicted Lobster Catches and Their Residuals for the Two Models

	Model 1 (Fourth-degree polynomial, number of traps, number of fishermen, and sea temperature)		Model 2 (Third-degree polynomial, number of fishermen, and sea temperature)	
Year	Predicted catch, $\hat{y}_t$	Residual, z_t	Predicted catch, $\hat{y}_t$	Residual, z_t
1932	2,313	434	2,504	243
1933	2,805	−130	2,799	−124
1934	2,322	117	2,387	52
1935	3,020	467	2,929	558
1936	2,901	−578	2,837	−514
1937	4,155	−822	3,874	−541
1938	3,309	165	3,325	149
1939	3,047	−42	3,176	−171
1940	3,670	−203	3,717	−250
1941	4,353	−299	4,334	−280
1942	4,638	−826	4,671	−859
1943	5,018	184	5,025	177
1944	6,146	230	6,013	363
1945	7,573	1,104	7,276	1,401
1946	8,009	508	7,808	709
1947	7,532	758	7,713	577
1948	7,275	−52	7,494	−271
1949	8,732	10	8,767	−25
1950	8,473	−149	8,617	−293
1951	9,097	318	9,176	239
1952	9,174	−87	9,284	−197
1953	10,317	−206	10,280	−168
1954	10,240	−422	10,202	−384
1955	10,480	−178	10,430	−128
1956	10,035	−719	10,086	−770
1957	10,440	628	10,416	652
1958	10,113	−448	10,157	−492
1959	10,170	−44	10,268	−142
1960	10,732	157	10,684	205
1961	10,398	−913	10,376	−891
1962	9,446	567	9,643	370
1963	9,962	382	9,924	420
1964	9,653	60	9,598	115
1965	9,117	−561	9,110	−554
1966	8,746	288	8,711	323
1967	8,265	−786	8,160	−681
1968	8,475	825	8,255	1,045
1969	8,743	254	8,408	589
1970	8,202	41	8,279	−36
1971	7,728	236	7,880	84
1972	6,947	427	7,340	34
1973	7,646	85	7,382	349
1974	8,391	−926	8,405	−940
1975	7,572	147	7,661	58

error about the estimated equation is

$$\text{SSE} = \sum_{t=1}^{44} z_t^2 = 10{,}142{,}851$$

so that the coefficient of determination for the model is $R^2 = 0.9683$. The percentage of variation, 96.83, explained by the four trend components and three predictor variables is indeed substantial. The F-statistic computed from it would greatly exceed any reasonable critical value. The estimate of the disturbance term variance is

$$\hat{\sigma}^2 = \frac{\text{SSE}}{44 - 8} = 281{,}745.87$$

and the standard deviation is $\hat{\sigma} = 530.80$.

Now let us examine the test statistics of Table 4-4. Only the fourth-degree polynomial component and the variable X_1 = number of traps are not sufficiently removed from zero to reject the respective hypotheses

$$H_0: \beta_4 = 0 \qquad \text{and} \qquad H_0: \beta_5 = 0$$

The polynomial model for trend should include at least terms through the cubic, while both the number of fishermen and mean sea temperature are important determinants of catch of nearly equal statistical significance. We have established the temperature effect on lobster production after controlling for the other factors.

Finally, let us drop the quartic orthogonal polynomial and the variable X_1, number of traps, from the model and compute the regression coefficients of the remaining variables. The coefficients, their standard deviations, and their t-statistics are shown in Table 4-6. The predicted catches and residuals from this second model are shown in the last two columns of Table 4-5. Other particulars of the fitted regression function are described by these statistics:

$$\text{SSE} = 11{,}062{,}443.17 \qquad \hat{\sigma}^2 = 291{,}116.92 \qquad \hat{\sigma} = 539.55$$
$$R^2 = 0.9654$$

The decrease in R^2 has been slight, yet the number of independent variables in the model has been reduced by two.

TABLE 4-6 Regression Coefficients and Related Statistics for the Second Lobster Catch Model

Variable	$\hat{\beta}_j$	$\sqrt{\widehat{\text{Var}}(\hat{\beta}_j)}$	t_j
Linear	37.8594	6.7033	5.65***
Quadratic	−10.4910	0.6095	−17.21***
Cubic	−0.1277	0.0185	−6.89***
Fishermen	0.5909	0.1121	5.27***
Sea temperature	502.2749	95.9410	5.24***

***Significant at the 0.001 level.

Some further thoughts on the models. The analyses of the lobster catch data could be continued in several directions to study the relationship between yield, sea temperature, and intensity of fishing. These extensions include the following:

1. In all of our preceding least squares or correlation analyses we have assumed that the sampling units and the y_i obtained from them were statistically independent. In the present time series regression we have made that assumption on the catches and the error terms for all years. That part of the model should be tested to assure the appropriateness of the least squares estimators and the validity of the significance tests. In Chapter 6 we give a test based on the residuals z_t from the estimated regression equation. The assumption of independent error terms holds for the models of this section.
2. To what degree do the temperature and perhaps the number of fishermen in past years affect lobster catch? We shall investigate that question in our treatment of lagged variable models in Chapter 6.
3. Is lobster catch a nonlinear function of temperature and fishing intensity? Can a nonlinear relationship be approximated by second-degree terms in those variables? We leave these questions as an exercise at the end of this chapter.
4. During the period 1942–1945 of World War II demand for seafoods increased, and lobster remained an unrationed commodity. Can wartime demand be expressed by a "dummy" variable

$$D_t = \begin{cases} 1 & t \text{ for } 1942\text{–}1945 \\ 0 & \text{all other years} \end{cases}$$

   incorporated into the regression model? We do this in the treatment of dummy variables in Chapter 7.
5. What more can we say about the two sets of residuals and predicted values in Table 4-5? This question is left as an exercise.

4-4 FITTING POLYNOMIALS WITH MULTIPLE OBSERVATIONS

Sometimes more than one observation of the time series variable is available at each time. For example, the series of mean scores on the GMAT test of matriculants at a graduate school of business could be replaced by the actual test scores for each incoming class. If trends in GMAT performance are to be represented by simple linear or polynomial models much more information would be provided by the individual scores. In a short time series of means the disturbance term variance would be estimated from the residuals of the polynomial fitted to the means, but with the complete scores an unbiased estimator could always be obtained from the within-class variation. In this section we develop the least squares estimators of the polynomial coefficients and give tests of hypotheses for the coefficients and the goodness of fit of the model.

Suppose that some variable Y is observed at M equally spaced times. At the jth time N_j independent observations y_{ij} have been collected. All observations are independent from time to time and have a common variance $\mathrm{Var}(y_{ij}) = \sigma^2$, so that the usual assumptions for least squares estimation hold. The polynomial model we use is

$$y_{ij} = \beta_0 + \beta_1\phi_1(j) + \cdots + \beta_p\phi_p(j) + e_{ij} \qquad i = 1, \ldots, N_j; \quad j = 1, \ldots, M \tag{1}$$

where $\phi_h(j)$ is the value of the hth orthogonal polynomial at time j. For hypothesis testing and confidence interval construction we shall assume that the disturbance terms e_{ij} are independently and normally distributed with mean zero and variance σ^2 for all combinations of i and j. Write the orthogonal polynomial values as the $M \times p$ matrix

$$\mathbf{T}_1 = \begin{bmatrix} \phi_1(1) & \cdots & \phi_p(1) \\ \cdot & & \cdot \\ \cdot & & \cdot \\ \cdot & & \cdot \\ \phi_1(M) & \cdots & \phi_p(M) \end{bmatrix} \tag{2}$$

which in turn can be partitioned by rows as

$$\mathbf{T}_1 = \begin{bmatrix} \boldsymbol{\phi}'(1) \\ \cdot \\ \cdot \\ \cdot \\ \boldsymbol{\phi}'(M) \end{bmatrix} \tag{3}$$

where $\boldsymbol{\phi}_1'(j) = [\phi_1(j), \ldots, \phi_p(j)]$. Denote the observations at the jth time by the $N_j \times 1$ vector $\mathbf{y}_j$. Then the polynomial model (1) for all $N = \sum N_j$ observations can be given in matrix form as

$$\begin{bmatrix} \mathbf{y}_1 \\ \cdot \\ \cdot \\ \cdot \\ \mathbf{y}_M \end{bmatrix} = \begin{bmatrix} \mathbf{j}_1 & \mathbf{j}_1\boldsymbol{\phi}'(1) \\ \cdot & \cdot \\ \cdot & \cdot \\ \cdot & \cdot \\ \mathbf{j}_M & \mathbf{j}_M\boldsymbol{\phi}'(M) \end{bmatrix} \begin{bmatrix} \beta_0 \\ \boldsymbol{\beta}_1 \end{bmatrix} + \mathbf{e} \tag{4}$$

where $\mathbf{j}_j$ is the $N_j \times 1$ vector of ones, and $\boldsymbol{\beta}_1' = [\beta_1, \ldots, \beta_p]$ is the vector of regression coefficients of the p orthogonal polynomials. The estimators of β_0 and $\boldsymbol{\beta}_1$ follow from the usual least squares formula. We first compute

$$\begin{aligned} \mathbf{X}'\mathbf{X} &= \begin{bmatrix} \mathbf{j}_1' & \cdots & \mathbf{j}_M' \\ \boldsymbol{\phi}(1)\mathbf{j}_1' & \cdots & \boldsymbol{\phi}(M)\mathbf{j}_M' \end{bmatrix} \begin{bmatrix} \mathbf{j}_1 & \mathbf{j}_1\boldsymbol{\phi}'(1) \\ \cdot & \cdot \\ \cdot & \cdot \\ \cdot & \cdot \\ \mathbf{j}_M & \mathbf{j}_M\boldsymbol{\phi}'(M) \end{bmatrix} \\ &= \begin{bmatrix} N & \sum N_j\boldsymbol{\phi}'(j) \\ \sum N_j\boldsymbol{\phi}(j) & \sum N_j\boldsymbol{\phi}(j)\boldsymbol{\phi}'(j) \end{bmatrix} \\ &= \begin{bmatrix} \mathbf{j}' \\ \mathbf{T}_1' \end{bmatrix} \mathbf{D}(N_j)[\mathbf{j} \quad \mathbf{T}_1] \end{aligned} \tag{5}$$

and

$$\mathbf{X}'\mathbf{y} = \begin{bmatrix} \mathbf{j}_1' & \cdots & \mathbf{j}_M' \\ \boldsymbol{\phi}(1)\mathbf{j}_1' & \cdots & \boldsymbol{\phi}(M)\mathbf{j}_M' \end{bmatrix} \begin{bmatrix} \mathbf{y}_1 \\ \cdot \\ \cdot \\ \cdot \\ \mathbf{y}_M \end{bmatrix}$$

$$= \begin{bmatrix} \sum N_j \bar{y}_j \\ \sum \phi(j) N_j \bar{y}_j \end{bmatrix}$$

$$= \begin{bmatrix} \mathbf{j}' \\ \mathbf{T}_1' \end{bmatrix} \mathbf{D}(N_j)\bar{\mathbf{y}} \tag{6}$$

$\mathbf{j}'$ is the $1 \times M$ vector of ones, while

$$\bar{\mathbf{y}} = \begin{bmatrix} \bar{y}_1 \\ \cdot \\ \cdot \\ \cdot \\ \bar{y}_M \end{bmatrix} \tag{7}$$

consists of the means

$$\bar{y}_j = \frac{1}{N_j} \sum_{i=1}^{N_j} y_{ij} \tag{8}$$

of the observations at the M times. The diagonal matrix

$$\mathbf{D}(N_j) = \begin{bmatrix} N_1 & \cdots & 0 \\ \cdot & & \cdot \\ \cdot & & \cdot \\ \cdot & & \cdot \\ 0 & \cdots & N_M \end{bmatrix} \tag{9}$$

contains the numbers of observations. The least squares estimators of the parameters of the polynomial model (1) are

$$\begin{bmatrix} \hat{\beta}_0 \\ \hat{\boldsymbol{\beta}}_1 \end{bmatrix} = \left(\begin{bmatrix} \mathbf{j}' \\ \mathbf{T}_1' \end{bmatrix} \mathbf{D}(N_j)[\mathbf{j} \quad \mathbf{T}_1] \right)^{-1} \begin{bmatrix} \mathbf{j}' \\ \mathbf{T}_1' \end{bmatrix} \mathbf{D}(N_j)\bar{\mathbf{y}}$$

$$= \begin{bmatrix} N & \sum N_j \boldsymbol{\phi}'(j) \\ \sum N_j \boldsymbol{\phi}(j) & \sum N_j \boldsymbol{\phi}(j)\boldsymbol{\phi}'(j) \end{bmatrix}^{-1} \begin{bmatrix} \sum N_j \bar{y}_j \\ \sum \boldsymbol{\phi}(j) N_j \bar{y}_j \end{bmatrix} \tag{10}$$

We should note immediately that the $p + 1$ estimators are independent of one another in the algebraic and statistical senses only if $N_1 = \cdots = N_M$. For general N_j the orthogonal polynomials do not lead directly to independent estimators of their coefficients, nor do they explain independent components of the total sum of squares of the y_{ij}. Why, then, did we frame the model (1) in terms of orthogonal polynomials rather than the first p powers of the time variable? Apart from the case of equal N_j, the orthogonalized model has a computational advantage: Unless the N_j are markedly different the matrix $\mathbf{X}'\mathbf{X}$ will have the *well-conditioned* property of diagonal elements whose magnitudes dominate the corresponding off-diagonal elements of their rows and columns.

Then the inverse of $\mathbf{X}'\mathbf{X}$ will be a well-defined matrix, and the estimators of the regression coefficients will tend to be nearly independent.

For the hypothesis tests on the parameters of the model it will be convenient to begin with the analysis of variance of Table 4-7. The first three lines

TABLE 4-7 Analysis of Variance for the Polynomial Model with Multiple Observations

Source	Sum of squares	d.f.	Mean square
Grand mean	$N\bar{y}^2$	1	$N\bar{y}^2$
pth-degree polynomial (adjusted for constant)	$SSR_a = \hat{\boldsymbol{\beta}}'\mathbf{X}'\mathbf{y} - N\bar{y}^2$	p	$\frac{SSR_a}{p}$
Constant and pth-degree polynomial	$SSR = \hat{\boldsymbol{\beta}}'\mathbf{X}'\mathbf{y}$	$p+1$	$\frac{SSR}{p+1}$
Lack of fit	$SSF = SSE - SSW$	$M-p-1$	$\frac{SSF}{M-p-1}$
Within times	$SSW = \sum_{j=1}^{M}\sum_{i=1}^{N_j}(y_{ij}-\bar{y}_j)^2$	$N-M$	$\frac{SSW}{N-M}$
Error	$SSE = \mathbf{y}'\mathbf{y} - \hat{\boldsymbol{\beta}}'\mathbf{X}'\mathbf{y}$	$N-p-1$	$\frac{SSE}{N-p-1}$
Total	$\mathbf{y}'\mathbf{y}$	N	—

of the table are the usual ones for testing an hypothesis on a subset of the regression parameters—in this case that of

$$H_0: \boldsymbol{\beta}_1 = \begin{bmatrix} \beta_1 \\ \cdot \\ \cdot \\ \cdot \\ \beta_p \end{bmatrix} = \begin{bmatrix} 0 \\ \cdot \\ \cdot \\ \cdot \\ 0 \end{bmatrix} \tag{11}$$

or no polynomial component in the Y observations. The adjusted sum of squares expression must be used rather than one in $\hat{\boldsymbol{\beta}}_1$, because with the unequal numbers of observations the last p columns of $\mathbf{X}$ no longer sum to zero. A conservative test for H_0 is provided by the statistic

$$F = \frac{N-M}{p}\frac{SSR_a}{SSW} \tag{12}$$

If F exceeds the upper 100α critical value of the F-distribution with p and $N-M$ degrees of freedom H_0 should be rejected at that level. The test is conservative in the sense that

$$E\left(\frac{SSW}{N-M}\right) = \sigma^2 \tag{13}$$

regardless of the validity of the pth-degree polynomial model. Of course, if

$N - M$ is small the power of the test will be poor. A more powerful test, albeit one dependent on the accuracy of the specification of the degree of the polynomial model, is given by computing

$$F = \frac{N - p - 1}{p} \frac{\text{SSR}_a}{\text{SSE}} \tag{14}$$

and referring it to the critical value $F_{\alpha; p, N-p-1}$ of the F-distribution. If

$$F > F_{\alpha; p, N-p-1} \tag{15}$$

we should reject H_0 in favor of the alternative that a real polynomial trend exists in the Y-series.

The tests of the hypotheses

$$H_0: \beta_j = \beta_{0j} \qquad j = 0, 1, \ldots, p \tag{16}$$

against appropriate alternatives can be made by the usual t-statistics

$$t_j = \frac{\hat{\beta}_j - \beta_{0j}}{\hat{\sigma}\sqrt{a_{j+1, j+1}}} \tag{17}$$

introduced in Section 2-3, where now $a_{j+1, j+1}$ is the $(j + 1)$th diagonal element of the inverse of the matrix $\mathbf{X}'\mathbf{X}$ of (5). If we are confident that a pth-degree polynomial model is the appropriate one for the time series, $\hat{\sigma}$ is given by

$$\hat{\sigma} = \sqrt{\frac{\text{SSE}}{N - p - 1}} \tag{18}$$

and each null hypothesis in (16) would be rejected in favor of the two-sided alternative

$$H_1: \beta_j \neq \beta_{0j} \tag{19}$$

if

$$|t_j| > t_{\alpha/2; N-p-1} \tag{20}$$

A more conservative test, and, accordingly, a less powerful one, is provided by taking

$$\hat{\sigma} = \sqrt{\frac{\text{SSW}}{N - M}} \tag{21}$$

in the computation of the t-statistics. Then we should reject the hypothesis (16) in favor of the alternative (19) if

$$|t| > t_{\alpha/2; N-M} \tag{22}$$

In either case the most common hypotheses for the p polynomial coefficients are

$$H_0: \beta_j = 0 \tag{23}$$

or those stating that the respective trend components are not present in the time series. We may also follow the example of Section 2-3 and use the general t-statistic (17) to construct $100(1 - \alpha)\%$ confidence intervals for the individual regression coefficients.

Now let us return to the analysis of variance of Table 4-7. The multiple observations have provided us with an additional test for the goodness of fit of the pth-degree polynomial model. The difference

$$\text{SSF} = \text{SSE} - \text{SSW} \tag{24}$$

of the conventional error sum of squares for the pth-degree polynomial and the sum of squared deviations

$$\text{SSW} = \sum_{j=1}^{M} \sum_{i=1}^{N_j} (y_{ij} - \bar{y}_j)^2 \tag{25}$$

of the observations at each time about their mean provides a measure of the validity of the pth-degree model. The mean square $\text{SSF}/(M - p - 1)$ has the expected value σ^2 only if the model holds true for the data. Otherwise, its expected mean square will exceed σ^2. To test the fit of the pth-degree model we compute

$$F = \frac{N - M}{M - p - 1} \frac{\text{SSF}}{\text{SSW}} \tag{26}$$

and reject the hypothesis of a pth-degree model at the α level if

$$F > F_{\alpha; M-p-1, N-M} \tag{27}$$

Guest (1954, 1956) has approached the problem of fitting orthogonal polynomials to multiple observations through grouping methods. He has considered unequally spaced data as well as the traditional one of a time series at equal intervals.

Example 4-2

The decline in scholastic aptitude test scores. Let us use the methods of this section to analyze a short time series formed by the mean Scholastic Aptitude Test—Verbal (SAT-V) scores of female high school seniors for the years 1965–1966 through 1975–1976. Although the decline is obvious and hardly needs a certification of "statistically significant," we would like to examine the components of the trend beyond the linear for possible changes in its rate and for the extent to which it might be represented by a polynomial of low degree. Such a multiple-observation time series appears to satisfy our assumptions for least squares estimation and normal-theory hypothesis testing: The individual scores are almost completely independent from year to year, and the variances scarcely change from time to time. However, the sample sizes are enormous, and we should be prepared for significant effects with little substantive meaning in the obscure parts of the model.

The mean scores, standard deviations, and numbers of examinees are shown in Table 4-8. From these we may compute the statistics

$$N = \sum_{j=1}^{11} N_j = 4{,}273{,}783$$

$$\bar{y} = \frac{1}{N} \sum_{j=1}^{11} N_j \bar{y}_j = 449.347766$$

$$\text{SSW} = \sum_{j=1}^{11} \sum_{i=1}^{N_j} (y_{ij} - \bar{y}_j)^2 = \sum_{j=1}^{11} (N_j - 1)(\text{SD}_j)^2 = 5.098503497 \times 10^{10}$$

From the plot of the means in Figure 4-2 it would appear that a polynomial model of low degree will not reproduce the means accurately. Let us then begin by fitting the

TABLE 4-8 Verbal SAT Statistics for Female High School Seniors[a]

Year	Mean SAT-V	Number of examinees	Standard deviation
1965–1966[b]	470	355,262	109
1966–1967	466	384,828	109
1967–1968	464	399,977	109
1968–1969	464	415,925	110
1969–1970	458	429,712	111
1970–1971	452	403,251	109
1971–1972	444	390,534	110
1972–1973	437	373,242	107
1973–1974	437	360,272	109
1974–1975	428	373,183	108
1975–1976	420	387,597	110

SOURCE: Data reproduced with the kind permission of the Educational Testing Service.

[a]The data are based on the scores for high school students who actually took the test as seniors; they do not include the scores for seniors who were tested before their senior year.

[b]Excluding the November 1965 test results.

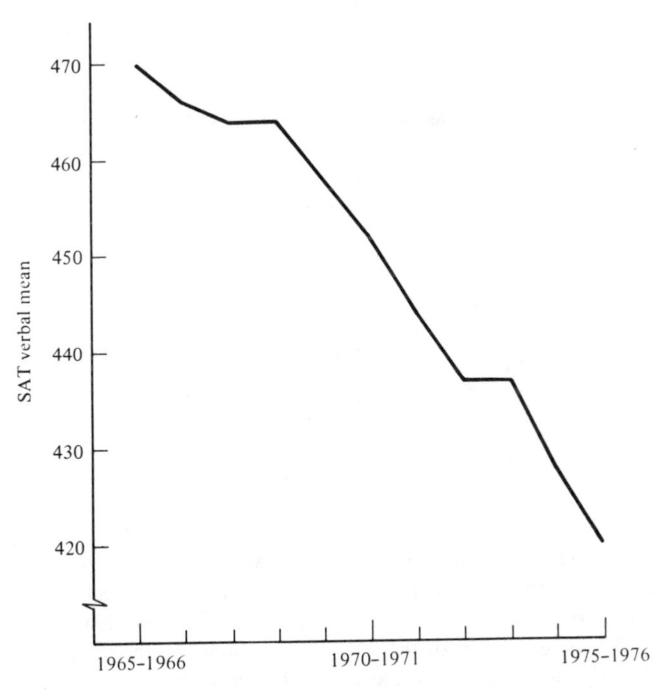

Figure 4-2 SAT verbal mean scores.

first six polynomials in terms of the orthogonal polynomial values for 11 points in Table 9 of Appendix A. The estimates (10) of the coefficients are given in the first column of Table 4-9. The estimates of the standard deviations of those coefficients were cal-

TABLE 4-9 Polynomial Regression Coefficients for the SAT-V Means

Polynomial	$\hat{\beta}_j$	$\sqrt{\widehat{\text{Var}}(\hat{\beta}_j)}$	t_j
Constant	449.09	0.0529	8487
Linear	−5.011	0.0170	−295
Quadratic	−0.269	0.0060	−45
Cubic	0.033	0.0027	12
Quartic	−0.035	0.0104	−3.3
Quintic	−0.379	0.0141	−27
Sextic	0.00838	0.00166	5.1

culated by the usual formula

$$\sqrt{\widehat{\text{Var}}(\hat{\beta}_j)} = \sqrt{\frac{\text{SSE}}{N-p-1} a_{j+1,\,j+1}} \tag{28}$$

and are shown in the second column of the table. The t-statistics (17) for testing the successive hypotheses of zero coefficients are given in the last column. We must conclude that even the highest-degree terms of the model have contributed to its prediction of the mean scores. The test that all six regression coefficients are simultaneously zero can be made with the adjusted sum of squares in the second line of Table 4-10 and

TABLE 4-10 Analysis of Variance for the SAT-V Scores

Source	Sum of squares	d.f.	Mean square
Grand mean	8.62934×10^{11}	1	—
First- through sixth-degree polynomials (adjusted)	1.090×10^{9}	6	1.816667×10^{8}
Intercept and polynomials	8.64024×10^{11}	7	—
Lack of fit	5.45007×10^{6}	4	1.36252×10^{6}
Within years	5.09850×10^{10}	4,273,772	11,930
Error	5.09905×10^{10}	4,273,776	11,931
Total	9.15014×10^{11}	4,273,783	—

either the within-year or error sum of squares, although its F-statistic in excess of 15,000 should exceed any imaginable critical value. In that connection we note that the two estimates of the disturbance term variance are virtually identical.

Finally, let us test the fit of the sixth-degree model to the eleven means by the F-statistic (26). The mean squares in the lack-of-fit and within-years lines of Table 4-10 give $F = 114$; since $F_{0.01;4,\infty} = 3.32$, we must conclude that much variation in the mean scores has been left unexplained by the polynomial model. The residuals of Table 4-11

TABLE 4-11 Estimated SAT-V Mean Scores and Residuals

Year	Estimated mean	Residual
1	470.17	−0.17
2	465.25	0.75
3	465.18	−1.18
4	463.33	0.67
5	458.26	−0.26
6	451.24	0.76
7	444.29	−0.29
8	438.75	−1.75
9	434.43	2.57
10	429.31	−1.31
11	419.76	0.24

indicate that the model was least successful in the years 1967–1968, 1972–1973, and 1974–1975.

Some further examination of Figure 4-2 suggests that a more appropriate model might consist of disjoint linear segments for the first three, the middle five, and the last three years. Such linear trends within those periods, or their shifts from one period to another, still await explanation in terms of the educational, psychological, and social forces impinging on test performance.

4-5 REFERENCES

ANDERSON, R. L., and E. E. HOUSEMAN (1942): *Tables of Orthogonal Polynomial Values Extended to $N = 104$*, Research Bulletin 297, Agricultural Experiment Station, Iowa State University, Ames, Iowa.

DOW, R. L. (1976): Yield trends of the American lobster resource with increased fishing effort, *Marine Technology Society Journal*, vol. 10, pp. 17–25.

FISHER, R. A. (1920): Studies in crop variation: I. An examination of the yield of dressed grain from Broadbalk, *Journal of Agricultural Science*, vol. 11, pp. 107–135.

FISHER, R. A. (1924): Influence of rainfall on the yield of wheat at Rothamstead, *Philosophical Transactions of the Royal Society of London, B*, vol. 213, pp. 89–142.

FISHER, R. A., and F. YATES (1953): *Statistical Tables*, Hafner Publishing Company, New York.

GREENWOOD, J. A., and H. O. HARTLEY (1962): *Guide to Tables in Mathematical Statistics*, Princeton University Press, Princeton, N.J.

GUEST, P. G. (1954): Grouping methods in the fitting of polynomials to equally spaced observations, *Biometrika*, vol. 41, pp. 62–76.

GUEST, P. G. (1956): Grouping methods in the fitting of polynomials to unequally spaced observations, *Biometrika*, vol. 43, pp. 149–160.

PEARSON, E. S., and H. O. HARTLEY (1966): *Biometrika Tables for Statisticians*, vol. 1, 3rd ed., Cambridge University Press, Cambridge, England.

PLACKETT, R. L. (1960): *Principles of Regression Analysis*, Oxford University Press, London.

ROBSON, D. S. (1959): A simple method for constructing orthogonal polynomials when the independent variable is unequally spaced, *Biometrics*, vol. 15, pp. 187–191.
WISHART, J., and T. METAKIDES (1953): Orthogonal polynomial fitting, *Biometrika*, vol. 40, pp. 361–369.

4-6 EXERCISES

1. These centered variables were added to the second model of Section 4-3 for Maine lobster yields:

$$X_4 = (X_2 - \bar{X}_2)^2 - \overline{(X_2 - \bar{X}_2)^2}$$

$$X_5 = (X_3 - \bar{X}_3)^2 - \overline{(X_3 - \bar{X}_3)^2}$$

Recall that X_2 is the number of fishermen, and X_3 is the mean annual sea temperature. These results were obtained from the regression analysis:

Variable	$\hat{\beta}_j$	t_j
Polynomial		
Linear	23.159	2.96
Quadratic	−7.898	−7.76
Cubic	−0.1199	−6.87
Fishermen	0.9095	5.88
Temperature	542.71	5.94
Quadratic fishermen	−0.0001043	−3.15
Quadratic temperature	−15.418	−0.24

$$\text{SSE} = 8{,}423{,}434 \qquad \hat{\sigma} = 483.72 \qquad R^2 = 0.9737$$

(a) Discuss the meaning of this model and its regression coefficients.
(b) Find the predicted annual catches and compare their residuals with those of the second model in Table 4-5.

2. A professional journal received these numbers of manuscripts during bimonthly periods from 1972 through 1975:

Year	Period JF	MA	MJ	JA	SO	ND	Total
1972	39	47	43	45	30	41	245
1973	55	52	46	45	38	49	285
1974	51	63	46	59	48	42	309
1975	46	37	51	64	47	53	298
Total	191	199	186	213	163	185	1137

Submissions in a given period may be regarded as a Poisson random variable, although one whose parameter changes seasonally and with trends of longer term. Two-month periods were chosen for these data in an attempt to smooth some of the fluctuations due to mail deliveries and working days in each month.

(a) Can the data be represented by an orthogonal polynomial model of fairly low degree?
(b) Will transformations of the observations yield values that can be modeled more precisely by orthogonal polynomials?
(c) Can the trends in the annual and bimonthly totals be represented by an orthogonal polynomial of low degree?

3. The professional journal in Exercise 2 had these percentages of articles rejected for publication by its editorial board during bimonthly periods of 1972–1975:

	Period					
Year	JF	MA	MJ	JA	SO	ND
1972	67	64	75	64	57	76
1973	62	69	67	62	79	86
1974	78	83	83	77	73	78
1975	80	68	69	72	70	73

The two-month periods were chosen to smooth the often-severe fluctuations in the rejection rate. Each manuscript was considered independently so that the assumption of independent disturbance terms in a regression model is reasonable.

(a) Fit orthogonal polynomials to the $N = 24$ percentages in the time series and determine whether statistically significant trends can be represented by the low-degree polynomials.
(b) The rejection rate appeared to shift upward abruptly in the last months of 1973, then generally decline. Can this behavior be described by separate regression models for the first 10 and last 14 data points?
(c) When the rejection percentages were calculated by quarters these values were obtained:

	Quarter			
Year	JFM	AMJ	JAS	OND
1972	67	70	59	74
1973	64	69	67	83
1974	80	83	74	78
1975	76	69	70	74

Can a trend in the rejection rates be represented by an orthogonal polynomial model?
(d) Compare the residuals and estimated variances from the models fitted in (a) and (c). What can you conclude about the effects of grouping by two- or three-month periods?

4. Fit the first few orthogonal polynomials to the time series of water temperatures given in Exercise 2, Section 1-9. Is the hypothesis of a linear trend in temperature tenable at the 0.05 level during the observation period?

5. Plot the residuals against the predicted lobster yields from the two models of Table 4-5. From the two plots examine (a) the assumption of a common variance for each

year, (b) trends in the residuals indicating inadequacies of the models, and (c) the comparative goodness of fit of the two models.

6. Statistics for the SAT-Mathematical Test* given to male senior students are shown below for the years 1965–1966 through 1975–1976:

Year	Number of students	Mean	Standard deviation
1965–1966[a]	453,489	510	114
1966–1967	475,148	511	113
1967–1968	490,737	509	117
1968–1969	498,462	510	116
1969–1970	504,588	506	118
1970–1971	460,715	503	117
1971–1972	430,208	498	118
1972–1973	400,108	498	117
1973–1974	369,627	495	120
1974–1975	379,095	489	118
1975–1976	386,896	490	124

[a]Excluding the November 1965 test results.

(a) Fit an orthogonal polynomial to the yearly mean scores.

(b) Use the within-year variation given by the standard deviations to test the successive hypotheses of linearity, a quadratic trend, . . . , in the means.

7. Use the SAT-Verbal statistics given in Section 4-4 to fit a linear model to the years 1968–1969 through 1972–1973. Test the adequacy of the fit of the linear model to the mean scores.

8. Three mathematical and statistical societies had these numbers of members for the years 1968–1974:

Year	A	B	C
1968	3,450	2,691	8,064
1969	3,669	2,857	10,055
1970	3,618	2,843	10,081
1971	3,434	2,812	10,277
1972	3,566	2,802	10,409
1973	3,703	2,965	10,785
1974	3,803	2,885	11,364

(a) Fit orthogonal polynomial models to the three time series.

(b) To what extent can one test hypotheses of equal coefficients for a given orthogonal polynomial in two or three of the series? Note that a formal analysis of variance test of such an hypothesis requires a common disturbance term variance in each series.

*See the footnote to Table 4-8.

chapter 5

Choosing a Set of Independent Variables

5-1 INTRODUCTION

In the previous chapters we always assumed that our regression analyses began with a fixed set of independent variables carefully chosen for their supposed relation to the dependent variable. Although an insignificant coefficient suggested implicitly that the variable should be dropped from the set, we did not give formal rules or algorithms for selecting variables from a larger group of candidates. In this chapter we describe four methods for composing predictor sets that are "good" in some sense. The first successively adds variables, the second removes them from the original full set, and the third, stepwise selection, is a combination of the forward and backward rules. The fourth method is based on the C_p measure for balancing subset size against the variation explained by the fitted model. The C_p criterion is a linear function of the adjusted coefficient of determination and appears to be a very effective way of finding good subsets.

5-2 BACKWARD ELIMINATION OF VARIABLES

Let us assume that we have chosen a moderate number of independent variables $X_1, \ldots, X_p$ for a regression analysis with some dependent variable Y. We selected the X_i carefully because we hypothesized that each might bear some relationship to the dependent variable. For some variables the relationship might be causal: If Y is a measure of the consistency of a cake, we would expect it to be a function of oven temperature, baking time, and perhaps idiosyncratic properties of the particular oven and its controls. If Y is a measure of cognitive intelligence given by a certain standardized test, we would expect it to be correlated with, but not necessarily driven by, the scores on other psychological tests with common substantive properties. The regression analysis will proceed with the set of p predictor variables. The wisdom of including each X_i will then be assayed from the individual regression coefficients, their t-statistics, and their degrees of statistical significance.

Now suppose that some of the t_j-statistics found in the regression on the p-predictors are close to zero, others are moderately large, and a third group of variables has $|t_j|$-statistics well in excess of the usual critical values. How should we use this information to drop variables from the model and thereby reduce our number of predictors with little loss of explained variation in Y? The procedure for finding a subset of important variables we shall describe in this section is called *backward elimination*, for it starts with the full multiple regression model and systematically eliminates variables. Backward elimination follows these steps:

1. For each variable X_j in the full regression model test the hypothesis

$$H_0: \beta_j = 0 \qquad j = 1, \ldots, p$$

   that its regression coefficient is zero or that it contributes nothing to the prediction of Y in the full model.

2. If the smallest $|t_j|$-statistic in step 1 is less than some critical value $t_{\alpha/2;\,N-p-1}$, drop its variable from the regression model and carry out the regression analysis for the remaining $p - 1$ variables. To avoid removing marginally significant variables which may contribute more in the later regression models, a large value of α (perhaps 0.10 or more) should be used.
3. Test each $H_0: \beta_j = 0$ in the second regression analysis, and drop the variable with smallest $|t_j|$ less than the critical value $t_{\alpha/2;\,N-p}$.
4. Continue in this fashion until all $|t_j|$-statistics are greater than the two-sided critical value with appropriate degrees of freedom. The remaining set of variables is that specified by backward elimination.

Example 5-1

We shall illustrate backward selection with the psychological test scores described in Example 3-3. As in that example Y will be the Draw a Person test score, and the $p = 11$ predictor variables will be the projective and cognitive test scores listed in Table 3-2 for the complete regression analysis. $N = 46$ subjects were tested on the 12 scales. We note immediately in Table 3-2 that the test statistics t_j for the predictors X_3, Homonyms, X_7, Sentence Completion, and X_{10}, Problem Situations, are exceptionally small. Rather than follow the backward elimination rule in its strict sense and drop X_7 we prefer to save steps and eliminate the three at once. The new coefficients and statistics are listed under regression 2 in Table 5-1. The coefficient of determination is scarcely less than the value $R^2 = 0.7159$ given in Section 3-3 for the 11 predictors. The elimi-

TABLE 5-1 Regression Analysis for Backward Elimination of Predictors

	Regression											
	2		3		4		5		6		7	
Variable	$\hat{\beta}_j$	t_j	$\hat{\beta}_j$	t_j	$\hat{\beta}_j$	t_j	$\hat{\beta}_j$	t_j	$\hat{\beta}_j$	t_j	$\hat{\beta}_j$	t_j
X_1	−0.14	−1.73	−0.15	−1.93	−0.16	−1.93	−0.15	−1.71	—	—	—	—
X_2	1.99	3.38	1.89	3.17	1.74	2.92	1.31	2.14	0.92	1.58	—	—
X_3	—	—	—	—	—	—	—	—	—	—	—	—
X_4	−1.27	−7.34	−1.27	−7.23	−1.20	−6.97	−1.05	−6.26	−0.87	−6.44	−0.75	−6.52
X_5	−0.35	−1.99	−0.37	−2.04	−0.32	−1.81	—	—	—	—	—	—
X_6	1.12	1.65	1.20	1.75	1.27	1.82	—	—	—	—	—	—
X_7	—	—	—	—	—	—	—	—	—	—	—	—
X_8	1.24	1.60	1.18	1.49	—	—	—	—	—	—	—	—
X_9	1.47	1.53	—	—	—	—	—	—	—	—	—	—
X_{10}	—	—	—	—	—	—	—	—	—	—	—	—
X_{11}	3.99	2.41	5.33	3.72	6.01	4.36	5.40	3.76	5.14	3.51	5.06	3.40
SSE	1210.89		1287.97		1363.62		1642.21		1759.93		1864.98	
SSR	3040.85		2963.77		2888.12		2609.53		2491.81		2386.76	
R^2	0.7152		0.6971		0.6793		0.6138		0.5861		0.5614	
$\hat{\sigma}^2$	32.73		33.89		34.96		40.05		41.90		43.37	

nation of the three variables has also hardly affected the regression coefficients or their t_j-statistics.

Since X_9 has the regression coefficient with the smallest $|t_j|$ in the second regression model it is eliminated for regression 3. In that regression X_8 has minimum $|t_j|$ and so is omitted for regression 4. We note that the coefficient of determination has only dropped less than four percentage points, yet the number of predictors has been reduced from 11 to 6. In regression 4 $|t_5|$ and $|t_6|$ are virtually equal, and so we omit both X_5 and X_6 for regression 5. At the same time R^2 will drop rather precipitously to 0.6138. Omission of X_1 leads to regression 6, wherein the t_j-statistic for X_2 is now much smaller than its counterpart in the preceding model. Elimination of X_2 gives the seventh regression function consisting of two predictors X_4 and X_{11}. Their highly significant coefficients imply that the elimination process should stop with that subset.

Conversely, we might have eliminated variables X_6, X_8, and X_9 together before fitting regression 3. That and the subsequent regressions are summarized in Table 5-2. The variables omitted in the second and third regressions are not listed. We are led to the same final subset of X_4 and X_{11}, although the use of a large ($\alpha > 0.05$) two-sided error rate would cause us to stop with the fourth regression subset of X_2, X_4, X_5, and X_{11}.

TABLE 5-2 Backward Elimination With X_6, X_8, and X_9 Omitted After Regression 2

	Regression							
	3		4		5		6	
Variable	$\hat{\beta}_j$	t_j	$\hat{\beta}_j$	t_j	$\hat{\beta}_j$	t_j	$\hat{\beta}_j$	t_j
X_1	−0.14	−1.65	—	—	—	—	—	—
X_2	1.46	2.47	1.11	1.96	—	—	—	—
X_4	−1.07	−6.65	−0.91	−6.95	−0.77	−6.81	−0.75	−6.52
X_5	−0.38	−2.10	−0.40	−2.16	−0.34	−1.82	—	—
X_{11}	5.45	3.94	5.21	3.71	5.11	3.52	5.06	3.40
R^2	0.6283		0.5934		0.5614		0.5614	

Regressions with all subsets of the predictors. Backward elimination or the other selection methods we describe in Sections 5-3 to 5-5 will not necessarily select the subset "best" in some sense of giving a large multiple correlation for a minimal number of predictors. Rigid application of selection rules makes no allowance for the range of sampling variation in the sums of squares and products of the variables, their correlations, and their various multiple and partial correlations. It is likely that a good small subset may be overlooked, or another small subset chosen as optimal when slightly larger sets with much higher R^2-values are present among the predictors. One obvious solution would consist of fitting all possible regression models with subsets of the independent variables. Efficient algorithms for computing the

$$2^p - 1 = \binom{p}{1} + \binom{p}{2} + \cdots + 1$$

regression analyses have been given by Garside (1965), Schatzoff et al. (1968), Furnival (1971), and Furnival and Wilson (1974). Frane (1977) has included the Furnival and Wilson method as Program P9R in the BMDP statistical computation package. That program finds the "best" subset in terms of R^2, adjusted R^2, or the Mallows C_p criterion of Section 5-5, and prints its regression analysis statistics. The program will print the regression analysis output for as many as 10 best predictor sets for successive values of the criterion measure.

Example 5-2

In Chapter 4 we investigated the effects of sea temperature and fishing intensity on annual catch of Maine lobsters. We shall consider now some data (Dow, 1976) from an earlier period in the Maine lobster fishery: the $N = 15$ years 1880, 1887–1889, 1892, and 1897–1906 (data for the missing years were not available). The dependent variable Y is annual lobster catch in metric tons, and the $p = 4$ independent variables are these:

$X_1 = t - \bar{t}$ (linear trend); $t = 1, 8, 9, 10, 13, 18, \ldots, 27$

$X_2 =$ number of lobster traps

$X_3 =$ number of fishermen

$X_4 =$ mean annual sea surface temperature (°C)

The matrix of sums of squares and products of the centered predictor variables is

$$\mathbf{X}_1'\mathbf{X}_1 = \begin{bmatrix} 842.933 & -4{,}579.37 & 16{,}777.07 & -9.91333 \\ & 67{,}522.50 & -76{,}157.30 & -108.3737 \\ & & 2{,}032{,}044.93 & -1469.787 \\ & & & 4.71333 \end{bmatrix}$$

and the vector of sums of products of the dependent and centered independent variables is

$$\mathbf{y}'\mathbf{X}_1' = [-132{,}797.73, \quad -1{,}703{,}130.8, \quad -7{,}473{,}998.3, \quad 10{,}972.653]$$

The correlations among the free variables formed this matrix:

$$\begin{bmatrix} 1 & -0.63 & -0.72 & -0.72 & 0.69 \\ & 1 & 0.87 & 0.41 & -0.16 \\ & & 1 & 0.72 & -0.21 \\ & & & 1 & -0.47 \\ & & & & 1 \end{bmatrix}$$

We note that in keeping with the time series analyses of Chapter 4 for lobster catch a variable measuring quadratic trend was also included; because its correlations with $Y, X_3,$ and X_4 were very small it was dropped before the present analyses. The quadratic trend variable also made no contribution to the multiple regression model fitted with all predictors.

Let us begin by listing the values of R^2 for the 15 possible regressions:

Predictors	R^2
X_1	0.3946
X_2	0.5200
X_3	0.5185
X_4	0.4814
X_1, X_2	0.5201
X_1, X_3	0.6538
X_1, X_4	0.7576
X_2, X_3	0.6038
X_2, X_4	0.8271
X_3, X_4	0.6784
X_1, X_2, X_3	0.6673
X_1, X_2, X_4	0.8278
X_1, X_3, X_4	0.8278
X_2, X_3, X_4	0.8279
X_1, X_2, X_3, X_4	0.8328

The predictors and their combinations can be ordered in an approximate fashion by their values of R^2: X_1 (linear trend) is the worst predictor, and of course the complete set of four variables has the maximum R^2 of 0.83. At the same time four other combinations (X_1, X_2, X_4), (X_1, X_3, X_4), (X_2, X_3, X_4), and (X_2, X_4) have coefficients of determination nearly identical with that value. We may conclude that the last pair of variables X_2, number of traps, and X_4, mean annual temperature, are the "best" set in the sense of maximal R^2 for the least number of predictors.

Now let us examine the regression analyses for the best subset and the complete set of predictors. In the latter case the following statistics were computed:

Predictor	$\hat{\beta}_j$	t_j
X_1: trend	−52.975	−0.54
X_2: traps	−6.851	−0.58
X_3: fishermen	−0.868	−0.55
X_4: temperature	1728.27	3.15

We note immediately that only X_4 has a regression coefficient statistically significant (at the 0.01 level for a two-sided test). The t_j-statistics for the other predictors are virtually equal. We would be justified to modify the backward elimination and drop X_1, X_2, and X_3. That would leave a single predictor explaining only 48.14% of the variation in lobster catch.

The importance of the other variables can also be demonstrated by a test of the hypothesis

$$H_0: \beta_1 = \beta_2 = \beta_3 = 0$$

by the analysis of variance of Table 3-3:

Source	Sum of squares	d.f.	Mean square	F
X_4 (unadjusted)	25,522,708	1	—	—
X_1, X_2, X_3 (adjusted)	18,634,111	3	6,211,270	7.01
X_1, X_2, X_3, X_4	44,156,819	4	11,036,705	12.45
Error	8,864,598	10	886,460	—
Total (about mean)	53,021,417	14	—	—

Since the F-statistic for H_0 exceeds $F_{0.01;\,3,\,10} = 6.55$ we should conclude that the first three variables (or at least some linear combination of them) will contribute significantly to a regression model for Y.

Finally, we consider the optimal subset regression analysis of Y with X_2 and X_4:

Variable	$\hat{\beta}_j$	t_j
X_2: traps	-13.405	-4.90
X_4: temperature	1900.47	4.62

$$\text{SSE}_{2,\,4} = 9{,}169{,}038 \qquad \text{SSR}_{2,\,4} = 43{,}852{,}379 \qquad \hat{\sigma}_{2,\,4} = 874.12$$

Both t-statistics exceed the two-sided critical value $t_{0.0005;\,12} = 4.318$, and we may conclude that the number of traps and mean annual sea temperature are important determinants of lobster yields. The negative sign of the traps regression coefficient seems to indicate the effect of overfishing. We also note that the estimate $\hat{\sigma}_{2,\,4}$ of the disturbance term standard deviation is less than the estimate $\hat{\sigma} = 941.52$ obtained from the error mean square in the four-predictor model; this difference is due to the greater degrees of freedom for error in the two-predictor model.

5-3 FORWARD SELECTION OF VARIABLES

Forward selection consists of building up the subset of predictors one at a time from a single variable. Although the matrices that are inverted will generally be smaller than those in the backward elimination method, the procedure seems to suffer more from the vagaries of the largest correlations of the dependent variable with the predictors. The successive subsets are also nested: Predictors are added only by forward selection and are retained in the set even though the t-statistics for their regression coefficients do not exceed a chosen critical value. A modification that permits such variables to be dropped from the set is known as *stepwise selection;* we discuss that method in Section 5-4.

Forward selection of the predictor variables follows these steps.

1. Select the first variable X_1^* of the subset as the one with the greatest correlation with Y.
2. Calculate the t-statistic for the test of $H_0: \beta_1^* = 0$ that Y is not related to X_1^* in a simple linear model. If H_0 cannot be rejected at some specified level we must conclude that the subset is empty and terminate the search. If $|t|$ exceeds some critical value $t_{\alpha/2; N-2}$ we proceed to the choice of a second variable.
3. To choose the second variable X_2^* we compute the partial correlations of Y and each remaining predictor, with X_1^* held constant. X_2^* is the variable with the largest absolute partial correlation. We calculate the t-statistic for testing $H_0: \beta_2^* = 0$ in the linear regression model with the two predictors X_1^* and X_2^*. If $|t| \leq t_{\alpha/2; N-3}$ for some error rate α, we must stop the selection process with the single-variable subset X_1^*. If $|t|$ exceeds that critical value, we move to the choice of a third variable.
4. We proceed in that fashion, recalculating successively higher-order partial correlations at each step and inserting the variable with highest absolute partial correlation when all of the preceding variables have been held constant. At each stage the coefficient of determination for Y with the selected variables and the F-statistic for the test of overall hypothesis of zero regression coefficients should be calculated. In that way the improvement in variation explained by the successive subsets can be tracked. The process should stop when the $|t|$-statistic for the last-selected variable does not exceed some designated critical value. The final subset consists of all variables before the last one with the insignificant $|t|$-statistics. We would expect that the F-statistic for the overall hypothesis of zero regression coefficients for those variables would exceed its critical value for some suitable significance level.

Example 5-3

Let us illustrate the forward selection procedure with the blood lead data described in Exercises 1 through 3, Chapter 3. We shall use this set of eight initial predictors:

X_1 = sex (1 = female, 0 = male)

X_2 = race (1 = black, 0 = white)

$X_3 = X_1 X_2$

X_4 = age of child (months)

X_5 = rank of child in family

X_6 = weight of child

X_7 = months in dwelling

X_8 = surface lead score for Type I rooms

We begin with the correlations of blood lead level (Y) and the predictors:

Predictor	X_1	X_2	X_3	X_4	X_5	X_6	X_7	X_8
Correlation	−0.01	0.25	0.21	−0.03	0.16	−0.18	−0.16	0.25

The correlations have been rounded to two places to save space; more accurate values were used to compute the subsequent partial correlations. We note that the correlations of X_2, race and X_8, surface lead level scores are equal. We shall alter the forward selection rule slightly and begin with a two-variable linear model containing both X_2 and X_8. These regression coefficients and related statistics were computed:

Predictor	$\hat{\beta}_j$	t_j
X_2: race	3.851	5.32
X_8: surface lead	0.0749	5.32

$$SSE_1 = 17{,}653.56 \qquad SSR_1 = 2489.14 \qquad R_1^2 = 0.1236 \qquad F_1 = 28.48$$

The degrees of freedom for SSE_1 are $407 - 2 - 1 = 404$. Both t_j and the F-statistic for the test of simultaneous zero coefficients exceed any reasonable critical values: X_2 and X_8 should remain as predictors.

Next we calculate the partial correlations of Y and the variables $X_1, X_3, \ldots, X_7$ with X_2 and X_8 held constant:

Predictor	X_1	X_3	X_4	X_5	X_6	X_7
Correlation	−0.03	0.09	−0.02	0.11	−0.15	−0.11

Variable X_6, weight, has the greatest absolute partial correlation with blood lead level, so it is selected for inclusion in the subset. The statistics for the three-predictor set are as follows:

Variable	$\hat{\beta}_j$	t_j
X_2: race	3.441	4.72
X_8: surface lead	0.075	5.39
X_6: weight	−0.067	−3.06

$$SSE_2 = 17{,}252.05 \qquad SSR_2 = 2890.65 \qquad R_2^2 = 0.1435 \qquad F_2 = 22.51$$

The various statistics again are highly significant.

To choose the next variable for the set we compute the partial correlations with X_2, X_6, and X_8 fixed:

Predictor	X_1	X_3	X_4	X_5	X_7
Correlation	−0.01	0.09	0.01	0.11	−0.07

X_5, rank of child, should be entered as the fourth variable in the subset. The regression analysis gave these results:

Variable	$\hat{\beta}_j$	t_j
X_2: race	3.496	4.79
X_8: surface lead	0.078	5.53
X_6: weight	−0.068	−3.09
X_5: rank	−0.233	−1.26

$$\text{SSE}_3 = 17{,}184.34 \qquad \text{SSR}_3 = 2958.37 \qquad R_3^2 = 0.1469 \qquad F = 17.30$$

This time the $|t|$-statistic for testing whether X_5 has a zero regression coefficient does not exceed even $t_{0.10;\,\infty} = 1.282$. X_5 should be dropped from the subset. The forward selection method has led us to the predictor subset (X_2, X_6, X_8) with a coefficient of determination $R^2 = 0.1435$.

However, we should not accept this result uncritically. In the exercise regressions of Chapter 3, X_5, rank, entered significantly, as did X_3, race × sex. We note that the last partial correlations of Y with X_3 and X_5 are scarcely different. Let us then enter both X_3 and X_5 into the subset. That five-predictor regression analysis gave these results:

Variable	$\hat{\beta}_j$	t_j
X_2: race	2.188	2.34
X_8: surface lead	0.0719	5.14
X_6: weight	−0.0664	−3.06
X_5: rank	0.418	2.27
X_3: race × sex	2.366	1.97

$$\text{SSE}_4 = 16{,}881.91 \qquad \text{SSR}_4 = 3260.80 \qquad R_4^2 = 0.1619 \qquad F = 15.49$$

In the same sense we might have entered X_7, months in dwelling, because its partial correlation was so close to those of X_3 and X_5. These coefficients and statistics would have been obtained:

Variable	$\hat{\beta}_j$	t_j
X_2: race	2.047	2.19
X_8: surface lead	0.0686	4.88
X_6: weight	−0.0541	−2.40
X_5: rank	0.472	2.54
X_3: race × sex	2.414	2.01
X_7: months in dwelling	−0.030	−1.89

$$\text{SSE}_5 = 16{,}733.26 \qquad \text{SSR}_5 = 3409.44 \quad R_5^2 = 0.1693 \qquad F = 13.58$$

X_7 is not quite statistically significant and probably should be omitted.

In a strict sense these last two regression analyses have confirmed the original forward selection optimal set consisting of X_2, X_6, and X_8. If we drop X_7 in the final six-predictor set we would revert to the five-predictor set in which t_3 only barely exceeds the two-sided 0.05 critical value. Removing X_3 leads to the four-predictor set in which X_5 does not have a significant regression coefficient, and thence to the three-predictor case. By choosing that set we have traded three marginally important predictors for a loss of 2.58% in variation explained by the regression model. Our final conclusion might be that blood lead level appears to be positively related to race, surface lead level in the dwelling, and some measure of the age of the child. We should also qualify this finding as more "suggestive" of a relationship of blood lead to environment than of usefulness for predicting blood lead level from demographic and environmental variables.

5-4 STEPWISE REGRESSION SELECTION

Stepwise regression analysis for the selection of a "good" subset of predictors is similar to forward selection, except that for every successive subset the hypotheses $H_0: \beta_j = 0$ are tested for *all* predictors, and those with $|t_j|$-statistics less than a specified critical value are dropped from the subset. The next variable is added to the subset by the same maximum partial correlation criterion of the forward selection method. Stepwise selection continues until a subset is attained with no variables whose $|t_j|$-statistics are less than some appropriate t critical value, and no variables remain to be entered into the subset. A program for stepwise selection by various criteria is contained in the BMDP Biomedical Computer Programs, P series (1977, Program P2R, Chap. 13.2, pp. 399–417).

Example 5-4

We shall illustrate stepwise regression with the same blood lead data used for forward selection in Example 5-3. Steps 1, 2, and the initial part of step 3 are identical with those of the forward selection process. However, in the subset X_2, X_5, X_6, X_8 determined by step 3 the t-statistic for X_5, rank in family, was only equal to -1.26: X_5 will be dropped from the subsequent sets of variables. We shall include X_5 among the variables held constant in the partial correlations of blood lead level with the other predictors.

We begin the fourth selection step by computing the partial correlations of Y and the remaining variables with X_2, X_5, X_6, and X_8 held constant:

Variable	X_1 (sex)	X_3 (race $\times$ sex)	X_4 (age)	X_7 (months in dwelling)
Correlation	-0.009	0.098	0.092	-0.091

Strictly interpreted, the stepwise algorithm would direct us to enter variable X_3, although its partial correlation with blood lead level is scarcely distinguishable from the magnitudes of those of X_4 and X_7. These results would be obtained:

Variable	$\hat{\beta}_j$	t_j
X_2: race	2.324	2.48
X_8: surface lead	0.0769	5.53
X_6: weight	−0.0673	−3.09
X_3: race × sex	2.288	1.89

$$\text{SSR}_4 = 3042.93 \qquad \text{SSE}_4 = 17{,}099.77 \qquad R_4^2 = 0.1511 \qquad F_4 = 17.88$$

If the 0.05 level is used for omitting variables X_3 would be dropped, and the stepwise selection would terminate with the variables X_2, X_6, and X_8.

If X_3, X_4, and X_7 had been added as a block of predictors in the previous step, these results would have been obtained:

Variable	$\hat{\beta}_j$	t_j
X_2: race	2.009	2.14
X_3: race × sex	2.436	2.03
X_4: age	0.0525	2.36
X_6: weight	−0.0756	−3.18
X_7: months in dwelling	−0.0540	−2.66
X_8: surface lead	0.0716	5.14

$$\text{SSR}_5 = 3373.00 \qquad \text{SSE}_5 = 16{,}769.70 \qquad R_5^2 = 0.1675 \qquad F_5 = 13.41$$

We have gained a slight increase in the proportion of variation explained, at the expense of including twice as many variables in the "best" subset. At the same time the regression coefficients in the six-variable model are all statistically significant in the two-sided sense at the 0.05 level, and we should not ignore the implications of their possible associations or relationships with blood lead level.

Finally, how do these stepwise selection subsets compare with those obtained by backward elimination? The full regression model with the eight predictors would have $R^2 = 0.1872$ and $t_1 = -1.65$, $t_2 = 1.18$ for X_1, sex, and X_2, race, respectively. The other six t_j-statistics would exceed 2.5 in absolute value. Elimination of X_2 as the variable with t_j of smallest magnitude would yield the statistics of regression 1 in Table 5-3. Each t_j exceeds the approximate 5% critical value of 1.96, and a strict interpretation of backward elimination would stop with the seven variables X_1, X_3–X_8. Alternatively, if we had eliminated both X_1 and X_2 after the initial model had been fitted we would have the six variables shown in regression 2. The coefficients and statistics of the common predictors are very similar. In either case backward elimination stopped short of the three-variable subset found by the strict interpretation of the forward selection rule. In the regression 2 variant backward selection produced a six-variable subset slightly different from the six-variable set X_2–X_4, X_6–X_8 given by the modified forward selection rule. We probably should conclude that the forward selection subset X_2, X_6, X_8 is misleading in its simplicity and that the other variables or combinations of them contribute significantly to the prediction of blood lead level.

TABLE 5-3 Two Variants of the Second-Stage Regressions in Backward Elimination

	Regression			
	1		2	
Variable	$\hat{\beta}_j$	t_j	$\hat{\beta}_j$	t_j
X_1	−1.5940	−2.27	—	—
X_2	—	—	—	—
X_3	4.9577	4.88	4.007	4.31
X_4	0.0582	2.66	0.0583	2.65
X_5	0.5044	2.75	0.5122	2.78
X_6	−0.0798	−3.42	−0.0790	−3.36
X_7	−0.0666	−3.31	−0.0662	−3.27
X_8	0.0645	4.60	0.06576	4.66
R^2	0.1843		0.1738	

5-5 THE C_p CRITERION

Another aid for choosing an "optimal" subset of predictors is the measure C_p proposed by Mallows (1964, 1973). C_p attempts to balance subset size against the effect of omitting important predictors. As we shall see, C_p is also related closely to the adjusted coefficient of determination defined in Section 2-2.

To motivate our definition of C_p we begin with the usual linear regression model

$$\mathbf{y} = \mathbf{X}\boldsymbol{\beta} + \mathbf{e} \tag{1}$$

in which $\mathbf{X}$ is an $N \times (k + 1)$ matrix of predictor values, and the N elements of $\mathbf{e}$ obey the ordinary least squares assumptions of zero expectations, independence, a common variance σ^2, and normality. We shall consider this model the "true" one for the dependent variable Y. Now partition $\mathbf{X}$ and $\boldsymbol{\beta}$ as

$$\mathbf{X} = [\mathbf{X}_P \quad \mathbf{X}_Q] \qquad \boldsymbol{\beta} = \begin{bmatrix} \boldsymbol{\beta}_P \\ \boldsymbol{\beta}_Q \end{bmatrix} \tag{2}$$

where $\mathbf{X}_p$ has dimensions $N \times p$, $\mathbf{X}_Q$ is $N \times (k + 1 - p)$, and $\boldsymbol{\beta}_P$, $\boldsymbol{\beta}_Q$ have conformable dimensions $p \times 1$ and $(k + 1 - p) \times 1$, respectively. The elements of $\mathbf{X}$ may or may not be centered as deviations about the predictor means. One column, usually in $\mathbf{X}_P$, will consist of ones to introduce the intercept parameter. We shall denote the p predictors in $\mathbf{X}_P$ by the set $P = \{i_1, \ldots, i_p\}$ of their subscripts. In the selection process we shall consider many such sets, but for the moment we concentrate on a particular subset P. Then the least squares estimator of $\boldsymbol{\beta}_P$ alone in the subset model

$$\mathbf{y} = \mathbf{X}_P\boldsymbol{\beta}_P + \mathbf{d} \tag{3}$$

is

$$\hat{\boldsymbol{\beta}}_P = (\mathbf{X}_P'\mathbf{X}_P)^{-1}\mathbf{X}_P'\mathbf{y} \tag{4}$$

and the error sum of squares for its model is

$$\text{SSE}_P = (\mathbf{y} - \mathbf{X}_P\hat{\boldsymbol{\beta}}_P)'(\mathbf{y} - \mathbf{X}_P\hat{\boldsymbol{\beta}}_P) \tag{5}$$

As a measure of the fit of the model using only the P-set predictors to the full model (1), Mallows (1973) proposed the *scaled sum of squared errors*

$$\begin{aligned} J_P &= \frac{(\mathbf{X}_P\hat{\boldsymbol{\beta}}_P - \mathbf{X}\boldsymbol{\beta})'(\mathbf{X}_P\hat{\boldsymbol{\beta}}_P - \mathbf{X}\boldsymbol{\beta})}{\sigma^2} \\ &= \frac{(\hat{\boldsymbol{\gamma}}_P - \boldsymbol{\beta})'\mathbf{X}'\mathbf{X}(\hat{\boldsymbol{\gamma}}_P - \boldsymbol{\beta})}{\sigma^2} \end{aligned} \tag{6}$$

where the $(k + 1) \times 1$ vector

$$\hat{\boldsymbol{\gamma}}_P = \begin{bmatrix} \hat{\boldsymbol{\beta}}_P \\ \mathbf{0} \end{bmatrix}$$

consists of $\hat{\boldsymbol{\beta}}_P$ followed by $k + 1 - p$ zeros corresponding to the omitted variables in the complement of P. Then

$$E(J_P) = \frac{p + B_P}{\sigma^2} \tag{7}$$

where

$$B_P = \boldsymbol{\beta}_Q'\mathbf{X}_P'[1 - \mathbf{X}_P(\mathbf{X}_P'\mathbf{X}_P)^{-1}\mathbf{X}_P']\mathbf{X}_P\boldsymbol{\beta}_Q \tag{8}$$

measures the bias due to fitting the subset model (3) rather than the complete model with $k + 1$ predictors. We may calculate

$$E(\text{SSE}_P) = (N - p)\sigma^2 + B_P \tag{9}$$

and the nature of B_P as the bias in the error sum of squares of the subset model is apparent.

Now we are ready to define C_p as

$$C_p = \frac{\text{SSE}_P}{\hat{\sigma}^2} - N + 2p \tag{10}$$

where

$$\begin{aligned} \hat{\sigma}^2 &= \frac{(\mathbf{y} - \mathbf{X}\hat{\boldsymbol{\beta}})'(\mathbf{y} - \mathbf{X}\hat{\boldsymbol{\beta}})}{N - k - 1} \\ &= \frac{\text{SSE}}{N - k - 1} \end{aligned} \tag{11}$$

or the estimate of the variance of the disturbance terms in the full model (1) with estimator

$$\hat{\boldsymbol{\beta}} = (\mathbf{X}'\mathbf{X})^{-1}\mathbf{X}'\mathbf{y}$$

C_p is a reasonable estimator for $E(J_P)$:

$$E(C_p) \approx \frac{p + B_P}{\sigma^2} \tag{12}$$

where the approximate equality is due to replacing σ^2 by $\hat{\sigma}^2$. The average value of C_p increases with the number of predictor variables in the subset, but the bias components decrease as the subsets approach the full model with

$k + 1$ independent variables. Those properties of C_p are intended to provide a means of balancing subset size and explained variation: Essentially, one should look for the subset with smallest value of C_p.

Example 5-5

C_p-statistics for all subset regressions in the lobster catch data of Example 5-2 have been plotted against their numbers of variables in Figure 5-1. The point labeled 0 is

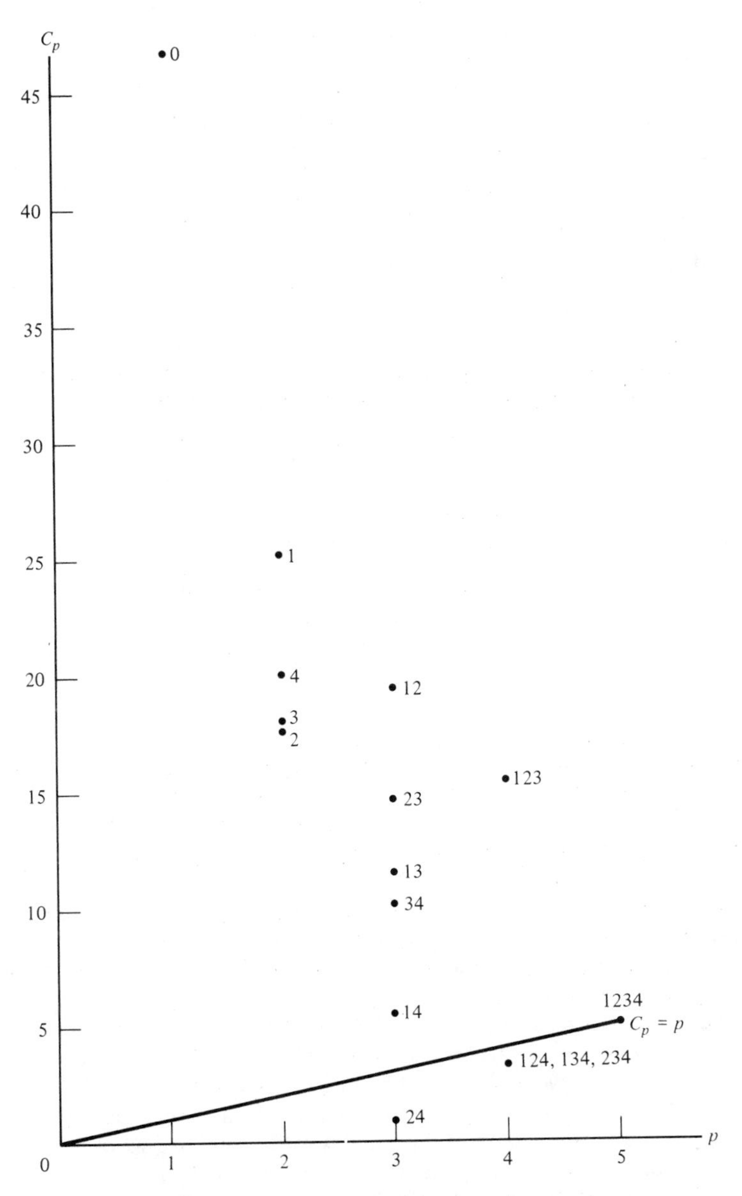

Figure 5-1 C_p values for the lobster catch regressions.

the value C_1 for the model with the intercept alone. All other C_p-values were computed with the intercept term in the models. The smallest value of C_p occurs for the set $P = \{0, 2, 4\}$, and that criterion would lead to the choice of X_2, number of traps and X_4, mean sea temperature as the optimal set of predictors. The line $C_p = p$ shows the approximate expectation of C_p when the bias term B_P is zero; the single-predictor sets and many of the other sets have large bias components.

Example 5-6

The C_p criterion proved less useful for subset selection in the blood lead level data. Because of the large number of predictors only a sampling of C_p-values was made, beginning with the full set of eight predictors and ending with the three-predictor subset case. The smaller values of C_p for selected subsets are shown in Table 5-4.

TABLE 5-4 Statistics for Selected Blood Lead Level Regressions

Predictor variables	p	R^2	C_p
1–8	9	0.187	9
1, 3–8	8	0.184	8.4
2–8	8	0.182	9.7
1–3, 5–8	8	0.174	13.3
1–4, 6–8	8	0.173	14.0
1, 2, 4–8	8	0.172	14.2
1–5, 7, 8	8	0.167	16.7
1–7	8	0.144	28.1
3–8	7	0.174	11.6
2, 4–8	7	0.172	12.3
2–3, 5–8	7	0.169	13.8
2–4, 6–8	7	0.168	14.7
2–3, 5–6, 8	6	0.162	15.4
3, 5–8	6	0.159	16.6
3–4, 6–8	6	0.158	17.4
2, 3, 6, 8	5	0.151	18.7
3, 5, 6, 8	5	0.150	19.0
3, 6, 7, 8	5	0.144	20.2
3, 6, 8	4	0.138	23.0
2, 3, 8	4	0.126	29.1
5, 6, 8	4	0.109	37.2

The smallest C_p of those computed would be attained with the variable set $P = \{0, 1, 3, 4, \ldots, 8\}$, although the set $P = \{0, 2, 3, \ldots, 8\}$ is a close competitor. The first set is that given by backward elimination in the latter part of Example 5-4 (Table 5-3). The minimum C_p regression for the $p = 7$ cases would have $P = \{0, 3, 4, 5, 6, 7, 8\}$ and would be the second subset found by the variant of backward elimination in Table 5-3. The C_p-statistics tend to increase rapidly for other subsets and smaller numbers of predictors, and little would be gained by their complete enumeration. The conclusions we may draw are similar to those from backward elimination: Perhaps variables X_1, sex, or X_2, race, can be omitted, but subsets with fewer than six of the remaining predictors are likely to explain significantly less variation in blood lead level.

Mallows (1973) has discussed C_p and its limitations as a means of determining optimal predictor sets. He has given hypothesis tests based on Scheffé simultaneous confidence intervals for the subset regression parameters. Kennard (1971) has shown that C_p is the linear function

$$C_p = \frac{(N-p)[1 - R_a^2(P)]}{1 - R_a^2(k+1)} - N + 2p \tag{13}$$

of the adjusted coefficient of determination

$$R_a^2(P) = 1 - \frac{N-1}{N-p}(1 - R_{Y \cdot P}^2) \tag{14}$$

for Y and the variables in the set P. $R_a^2(P)$ was originally given by (35), Section 2-2, for the whole model. The adjusted coefficient $R_a^2(k+1)$ for the complete predictor set is of course the same for all values of C_p.

Gorman and Toman (1966) have found the recursion formula

$$C_{p^*} = C_p - 2 + \frac{t_j^2(\mathrm{SSE}_P)}{\hat{\sigma}^2(N-p)} \tag{15}$$

for computing C_p with the variable X_j omitted. t_j is the t-statistic for the regression coefficient of X_j, and SSE_P is the error sum of squares for the variable set P. Hocking and Leslie (1967) have considered efficient computing algorithms for choosing subsets with C_p as the measure of optimality.

5-6 REFERENCES

Further descriptions and examples of variable selection methods can be found in the texts by Draper and Smith (1981) and Daniel and Wood (1980). Recent comparisons and surveys of the several procedures have been given by Hocking (1976) and Berk (1978).

Berk, K. N. (1978): Comparing subset regression procedures, *Technometrics*, vol. 20, pp. 1–6.

BMDP Biomedical Computer Programs—P-Series (1977), Health Sciences Computing Facility, University of California, University of California Press, Los Angeles.

Daniel, C., and F. S. Wood (1980): *Fitting Equations to Data: Computer Analysis of Multifactor Data*, 2nd ed., John Wiley & Sons, Inc., New York.

Dow, R. L. (1976): Yield trends of the American lobster resource with increased fishing effort, *Marine Technology Society Journal*, vol. 10, pp. 17–25.

Draper, N. R., and H. Smith (1981): *Applied Regression Analysis*, 2nd ed., John Wiley & Sons, Inc., New York.

Frane, J. (1977): All possible subsets regression, in W. J. Dixon and M. B. Brown (eds.), *BMDP Biomedical Computer Programs—P-Series*, pp. 418–436, Health Sciences Computing Facility, University of California, University of California Press, Los Angeles.

FURNIVAL, G. M. (1971): All possible regressions with less computation, *Technometrics*, vol. 13, pp. 403–408.

FURNIVAL, G. M., and R. W. WILSON (1974): Regression by leaps and bounds, *Technometrics*, vol. 16, pp. 499–511.

GARSIDE, M. J. (1965): The best subset in multiple regression analysis, *Applied Statistics*, vol. 14, pp. 196–200.

GORMAN, J. W., and R. J. TOMAN (1966): Selection of variables for fitting equations to data, *Technometrics*, vol. 8, pp. 27–51.

HOCKING, R. R. (1976): The analysis and selection of variables in linear regression, *Biometrics*, vol. 32, pp. 1–49.

HOCKING, R. R., and R. N. LESLIE (1967): Selection of the best subset in regression analysis, *Technometrics*, vol. 9, pp. 531–540.

KENNARD, R. W. (1971): A note on the C_p statistic, *Technometrics*, vol. 13, pp. 899–900.

MALLOWS, C. L. (1964): Choosing variables in a linear regression: a graphical aid. Presented at the Central Regional Meeting of the Institute of Mathematical Statistics, Manhattan, Kan., May 7–9.

MALLOWS, C. L. (1973): Some comments on C_p, *Technometrics*, vol. 15, pp. 661–675.

SCHATZOFF, M., S. FIENBERG, and R. TSAO (1968): Efficient calculations of all possible regressions, *Technometrics*, vol. 10, pp.769–779.

5-7 EXERCISES

1. Once again we shall analyze some course evaluations. The present set was obtained from $N = 32$ students in an M.B.A. statistics class. The original variables consisted of 11 instructor and course evaluation scales from 1 (outstanding) to 5 (poor), five dummy variables indicating the student's major, and three other dummy variables coded from written comments on the forms. The dependent variable Y was the sixth scale, overall evaluation of the instructor. An initial regression analysis with all 18 independent variables gave an unadjusted coefficient of determination equal to 0.9720 and regression coefficient $|t|$-statistics well below one for five evaluation scales and the dummy variables indicating majors in finance, marketing, or accounting. Those variables were eliminated. The following 10 independent variables were retained:

   1. Ability to present material clearly
   2. Stimulation of interest
   3. Knowledge of subject
   4. Apparent interest in students
   5. Overall evaluation of the course
   6. Management major (1 = yes, 0 = no)
   7. Decision science major (1 = yes, 0 = no)
   8. Use of humor and ancedotal material (1 = favorable comment, 0 = no comment)

9. Comments on mathematical level
   (1 = excessive, 0 = no comment)
10. Comments on text (−1 = negative, 0 = none, 1 = positive)

The data are given in Table 5-5. The results of the regression analysis with all 10 independent variables are given in Table 5-6. In that model

$$\text{SSR} = 19.042 \qquad \text{SSE} = 0.6765 \qquad F = 59.11 \qquad R^2 = 0.9657$$

Find subsets of the 10 independent variables that explain much of the variation in the overall evaluation ratings, but contain fewer variables.

TABLE 5-5 Course Evaluation Data

	Variable										
Student	Y	1	2	3	4	5	6	7	8	9	10
1	1	1	2	1	1	2	1	0	1	0	0
2	1	2	2	1	1	1	0	0	1	1	0
3	1	1	1	1	1	2	1	0	0	0	0
4	1	1	2	1	1	2	0	1	1	0	0
5	2	1	3	2	2	2	0	0	0	0	0
6	2	2	4	1	1	2	0	0	0	0	0
7	2	3	3	1	1	2	0	0	0	0	−1
8	2	3	4	1	2	3	0	1	0	0	0
9	2	2	3	1	3	3	1	0	0	0	0
10	2	2	2	2	2	2	0	0	0	0	0
11	2	2	3	2	1	2	0	0	1	0	1
12	2	2	2	3	3	2	0	0	0	1	0
13	2	2	2	1	1	2	0	0	0	0	0
14	2	2	4	2	2	2	1	0	0	0	0
15	2	3	3	1	1	3	1	0	0	0	0
16	2	3	4	1	1	2	0	0	0	1	0
17	2	3	2	1	1	2	0	0	0	0	0
18	3	4	4	3	2	2	0	0	0	0	−1
19	3	4	3	1	1	4	1	0	0	1	0
20	3	4	3	1	2	3	0	0	0	0	0
21	3	4	3	2	2	3	0	0	0	1	−1
22	3	3	4	2	3	3	0	0	0	0	0
23	3	3	4	2	3	3	0	0	0	0	0
24	3	4	3	1	1	2	0	0	0	0	1
25	3	4	5	1	1	3	0	0	1	1	0
26	3	3	5	1	2	3	0	0	0	0	0
27	3	4	4	1	2	3	0	0	0	1	0
28	3	4	4	1	1	3	0	0	0	1	0
29	3	3	3	2	1	3	0	0	0	0	0
30	3	3	5	1	1	2	0	0	0	1	1
31	4	5	5	2	3	4	0	0	0	0	0
32	4	4	5	2	3	4	0	0	0	1	1

TABLE 5-6 First Regression Analysis with p = 10 Independent Variables

Variable	Mean	Standard deviation	$\hat{\beta}_j$	t_j
Y	2.41	0.7976	—	—
1	2.84	1.0809	0.28	4.98
2	3.31	1.0906	0.15	3.54
3	1.44	0.6189	0.23	3.27
4	1.66	0.7874	−0.04	−0.66
5	2.53	0.7177	0.42	5.19
6	0.19	0.3966	−0.40	−3.76
7	0.06	0.2459	−0.57	−3.69
8	0.16	0.3689	−0.23	−2.17
9	0.31	0.4709	−0.09	−1.18
10	0.03	0.4741	0.24	3.29

2. A trout fisherman kept a record of his catch, weather, and water conditions for a five-year period.* From those data $N = 30$ trips were selected at random to determine if the number Y of fish caught was related to the following independent variables:

X_1: cloud proportion (0 = clear, 1 = completely overcast)
X_2: water temperature expressed as the absolute deviation from 56°F, the temperature at which trout have maximum metabolic rate, and hence greatest hunger
X_3: water level (1 = normal, 0 = otherwise)
X_4: water level (1 = high, 0 = otherwise)
X_5: water clarity (1 = normal clarity, 0 = discolored)
X_6: time (1 = morning, 0 = otherwise)
X_7: time (1 = afternoon, 0 = otherwise)

These statistics were computed for the eight variables:

Variable	Mean	Standard deviation
Y: catch	2.33	2.04
X_1: sky	0.25	0.38
X_2: temperature	4.53	3.58
X_3: water level	0.50	0.51
X_4: water level	0.37	0.49
X_5: water clarity	0.80	0.41
X_6: morning	0.20	0.41
X_7: afternoon	0.33	0.48

The correlation matrix of the variables had these values (rounded to two places and with subdiagonal entries omitted):

*I am indebted to Robert G. Pali for this example, and for kindly permitting the use of his analyses.

$$\begin{bmatrix} 1 & 0.17 & -0.09 & 0.73 & -0.44 & 0.42 & -0.08 & 0.34 \\ & 1 & -0.41 & 0.18 & 0.00 & 0.34 & -0.34 & 0.24 \\ & & 1 & -0.34 & -0.02 & -0.14 & 0.16 & -0.17 \\ & & & 1 & -0.76 & 0.50 & 0.00 & 0.14 \\ & & & & 1 & -0.48 & -0.03 & 0.05 \\ & & & & & 1 & -0.17 & 0.18 \\ & & & & & & 1 & -0.35 \\ & & & & & & & 1 \end{bmatrix}$$

The linear regression model with all seven independent variables gave these coefficients and analysis of variance:

Variable	$\hat{\beta}_j$	t_j
X_1: sky	0.189	0.25
X_2: temperature	0.212	2.52
X_3: water level	4.585	5.09
X_4: water level	1.921	2.17
X_5: water clarity	0.339	0.46
X_6: morning	−0.193	−0.29
X_7: afternoon	0.787	1.42

Source	Sum of squares	d.f.	Mean square	F	R^2
Regression	84.74	7	12.11	7.41	0.702
Error	35.92	22	1.63	—	—
Total	120.66	29	—	—	—

Can you find a smaller subset of useful independent variables?

chapter 6

Some Methods for Time Series

6-1 INTRODUCTION

In Chapter 4 we used polynomial models to describe the trends in time series data. To estimate the polynomial coefficients by least squares it was necessary to assume that the random disturbances in the model were uncorrelated from time to time. That condition usually is not satisfied in a time series, for the forces impinging on the variable usually persist in varying degrees beyond a single time. In this chapter we describe some ways of analyzing time series with correlated observations. We begin in Section 6-2 with the generalization of least squares estimation to that case. In Section 6-3 we describe a special case of autocorrelation and show its effects on the estimators of the intercept and slope of a linear trend. The Durbin–Watson test for autocorrelation and some applications of the preceding methods are given in Section 6-4. In Section 6-5 we consider some models for time series with lagged relations among the variables.

6-2 GENERALIZED LEAST SQUARES ESTIMATORS

Suppose that N observations on the dependent variable Y and p predictors can be related by the usual linear model

$$\mathbf{y} = \mathbf{X}\boldsymbol{\beta} + \mathbf{e} \tag{1}$$

in which the first column of the $N \times (p + 1)$ matrix $\mathbf{X}$ consists of ones for the inclusion of an intercept parameter. For the moment we shall not specify the remaining columns of $\mathbf{X}$ as centered as deviations from their means. Unlike the original linear model of Chapter 2 we shall now assume that the disturbances are correlated, or

$$\text{Cov}(\mathbf{e}, \mathbf{e}') = \mathbf{\Sigma} \tag{2}$$

Because the covariance matrix $\mathbf{\Sigma}$ will be required to be positive definite, there exists an $N \times N$ matrix $\mathbf{T}$ with the property

$$\mathbf{T\Sigma T}' = \lambda\mathbf{I} \tag{3}$$

described by equation (15), of Section B-4. Then the transformed disturbance vector

$$\mathbf{d} = \mathbf{Te} \tag{4}$$

has the covariance matrix

$$\begin{aligned} \text{Cov}(\mathbf{d}, \mathbf{d}') &= \text{Cov}(\mathbf{Te}, \mathbf{e}'\mathbf{T}') \\ &= \mathbf{T}\,\text{Cov}(\mathbf{e}, \mathbf{e}')\mathbf{T}' \\ &= \mathbf{T\Sigma T}' \\ &= \lambda\mathbf{I} \end{aligned} \tag{5}$$

or the usual one of the least squares model introduced in Chapter 2. We can estimate $\boldsymbol{\beta}$ by multiplying the linear model (1) by $\mathbf{T}$ to obtain

$$\mathbf{Ty} = (\mathbf{TX})\boldsymbol{\beta} + \mathbf{Te} \tag{6}$$

The transformed vector and matrix are used to compute the estimator:

$$\begin{aligned} \hat{\boldsymbol{\beta}} &= (\mathbf{X}'\mathbf{T}'\mathbf{TX})^{-1}\mathbf{X}'\mathbf{T}'\mathbf{Ty} \\ &= (\mathbf{X}'\mathbf{\Sigma}^{-1}\mathbf{X})^{-1}\mathbf{X}'\mathbf{\Sigma}^{-1}\mathbf{y} \end{aligned} \tag{7}$$

The common variance λ of the transformed disturbance variates canceled in the inverses of the matrices in (7).

The estimator (7) is the *generalized least squares*, or *Aitken, estimator*, after its discoverer (Aitken, 1934). In this chapter we refer to those estimators by the acronym GLSE, and we employ the notation OLSE for the *ordinary least squares estimator* used throughout the preceding five chapters. Aitken showed that his estimator has minimum variance among all unbiased estimators when $\mathbf{\Sigma}$ is the true covariance matrix of the disturbances. If $\mathbf{e}$ has the N-dimensional multinormal distribution, the GLSE (7) could also be obtained by the maximum likelihood principle.

The covariance matrix of the GLSE $\hat{\boldsymbol{\beta}}$ is

$$\text{Cov}(\hat{\boldsymbol{\beta}}, \hat{\boldsymbol{\beta}}') = (\mathbf{X}'\mathbf{\Sigma}^{-1}\mathbf{X})^{-1} \tag{8}$$

If $\mathbf{e}$ is a multinormal random variable, $\hat{\boldsymbol{\beta}}$ is multinormally distributed with mean $\boldsymbol{\beta}$ and covariance matrix (8); that result follows from $\hat{\boldsymbol{\beta}}$ being a linear function of other multinormal variates.

An important special case of the linear model with correlated disturbances occurs when the covariance matrix is known, apart from a positive constant. For simplicity we take the constant as the common variance σ^2 and the remaining term as the correlation matrix $\mathbf{P}$:

$$\mathbf{\Sigma} = \sigma^2 \mathbf{P} \tag{9}$$

By a slight extension of (15), Section B-4, we can write $\mathbf{P}$ as the product

$$\mathbf{P} = \mathbf{KK}' \tag{10}$$

of some nonsingular matrix and its transpose. Then

$$\mathbf{K}^{-1}\mathbf{P}(\mathbf{K}')^{-1} = \mathbf{I} \tag{11}$$

so that $\mathbf{K}^{-1}$ is akin to $\mathbf{T}$ in (5). The unbiased estimator of σ^2 can be shown to be

$$\begin{aligned}
\hat{\sigma}^2 &= \frac{1}{N-p-1}(\mathbf{K}^{-1}\mathbf{y} - \mathbf{K}^{-1}\mathbf{X}\hat{\boldsymbol{\beta}})'(\mathbf{K}^{-1}\mathbf{y} - \mathbf{K}^{-1}\mathbf{X}\hat{\boldsymbol{\beta}}) \\
&= \frac{1}{N-p-1}(\mathbf{y} - \mathbf{X}\hat{\boldsymbol{\beta}})'\mathbf{P}^{-1}(\mathbf{y} - \mathbf{X}\hat{\boldsymbol{\beta}}) \\
&= \frac{1}{N-p-1}\mathbf{y}'\mathbf{P}^{-1}[\mathbf{P} - \mathbf{X}(\mathbf{X}'\mathbf{P}^{-1}\mathbf{X})^{-1}X']\mathbf{P}^{-1}\mathbf{y}
\end{aligned} \tag{12}$$

Weighted least squares. Suppose that the N disturbance terms in the linear model (1) are uncorrelated but have different variances $\sigma^2\lambda_1, \ldots, \sigma^2\lambda_N$. The scale factors λ_i are known, but σ^2 is not. The generalized least squares estimator of $\boldsymbol{\beta}$ is

$$\hat{\boldsymbol{\beta}} = \left(\mathbf{X}'\mathbf{D}\left(\frac{1}{\lambda_i}\right)\mathbf{X}\right)^{-1}\mathbf{X}'\mathbf{D}\left(\frac{1}{\lambda_i}\right)\mathbf{y} \tag{13}$$

where $\mathbf{D}(1/\lambda_i)$ denotes the $N \times N$ diagonal matrix with ith element $1/\lambda_i$. This is equivalent to carrying out a least squares fit with the new variables

$$\mathbf{D}\left(\frac{1}{\sqrt{\lambda_i}}\right)\mathbf{X} \qquad \mathbf{D}\left(\frac{1}{\sqrt{\lambda_i}}\right)\mathbf{y} \tag{14}$$

in which the ith row of $\mathbf{X}$ and element of $\mathbf{y}$ have been divided by $\sqrt{\lambda_i}$. If the predictor variable values in $\mathbf{X}$ are centered or orthogonal, those properties no longer necessarily hold after this transformation.

Johnston (1972) has given examples of weighted least squares estimators with particular patterns of the λ_i and has compared their sampling variances with those of the ordinary least squares estimators.

Example 6-1

A special pattern of correlated disturbances. Let us see what effect the covariance structure

$$\mathbf{\Sigma} = \sigma^2 \begin{bmatrix} 1 & \cdots & \rho \\ \cdot & & \cdot \\ \cdot & & \cdot \\ \cdot & & \cdot \\ \rho & \cdots & 1 \end{bmatrix} \tag{15}$$

of equal variances and correlations has on the least squares estimation of the linear model parameters. The inverse of $\mathbf{\Sigma}$ has the same pattern of a common diagonal element and equal off-diagonal elements:

$$\mathbf{\Sigma}^{-1} = \left(\frac{d}{\sigma^2}\right)(\mathbf{I} - \rho g \mathbf{J}) \tag{16}$$

where

$$d = \frac{1}{1-\rho} \qquad g = \frac{1}{1+(N-1)\rho}$$

(Morrison, 1976, p. 70). Now assume that the model (1) has been expressed in partitioned and centered form as

$$\mathbf{y} = \alpha\mathbf{j} + \mathbf{X}_1\boldsymbol{\beta}_1 + \mathbf{e} \tag{17}$$

Then

$$\begin{aligned}\mathbf{X}'\boldsymbol{\Sigma}^{-1}\mathbf{X} &= \begin{bmatrix}\mathbf{j}'\boldsymbol{\Sigma}^{-1}\mathbf{j} & \mathbf{j}'\boldsymbol{\Sigma}^{-1}\mathbf{X}_1 \\ \mathbf{X}_1'\boldsymbol{\Sigma}^{-1}\mathbf{j} & \mathbf{X}_1'\boldsymbol{\Sigma}^{-1}\mathbf{X}_1\end{bmatrix} \\ &= \frac{1}{\sigma^2}\begin{bmatrix} gN & \mathbf{0}' \\ \mathbf{0} & d\mathbf{X}_1'\mathbf{X}_1\end{bmatrix}\end{aligned} \tag{18}$$

and

$$\mathbf{X}\boldsymbol{\Sigma}^{-1}\mathbf{y} = \begin{bmatrix}\mathbf{j}'\boldsymbol{\Sigma}^{-1}\mathbf{y} \\ \mathbf{X}_1'\boldsymbol{\Sigma}^{-1}\mathbf{y}\end{bmatrix} = \frac{1}{\sigma^2}\begin{bmatrix} gN\bar{y} \\ d\mathbf{X}_1'\mathbf{y}\end{bmatrix} \tag{19}$$

The simplification follows from the zero column sums, or $\mathbf{j}'\mathbf{X}_1 = \mathbf{0}'$, of the centered predictors. Then the generalized least squares estimators are merely

$$\hat{\alpha} = \bar{y} \qquad \hat{\boldsymbol{\beta}}_1 = (\mathbf{X}_1'\mathbf{X}_1)^{-1}\mathbf{X}_1'\mathbf{y} \tag{20}$$

or the usual ordinary least squares estimators. The equal-variance, equal-correlation structure of the disturbances has not affected the estimators.

Relative efficiency of ordinary and generalized least squares estimators. The ordinary least squares estimator $\hat{\boldsymbol{\beta}}_0$ of regression coefficients is still unbiased when the disturbance variates are correlated, but its variances and covariances are no longer the elements of $\sigma^2(\mathbf{X}'\mathbf{X})^{-1}$. If the disturbance vector has the positive definite covariance matrix $\boldsymbol{\Sigma}$, the covariance matrix of the OLS estimator $\hat{\boldsymbol{\beta}}_0$ is

$$\begin{aligned}\text{Cov}\,(\hat{\boldsymbol{\beta}}_0, \hat{\boldsymbol{\beta}}_0') &= \text{Cov}\,[(\mathbf{X}'\mathbf{X})^{-1}\mathbf{X}'\mathbf{y}, \mathbf{y}'\mathbf{X}(\mathbf{X}'\mathbf{X})^{-1}] \\ &= (\mathbf{X}'\mathbf{X})^{-1}\mathbf{X}'\,\text{Cov}\,(\mathbf{y}, \mathbf{y}')\mathbf{X}(\mathbf{X}'\mathbf{X})^{-1} \\ &= (\mathbf{X}'\mathbf{X})^{-1}\mathbf{X}'\boldsymbol{\Sigma}\mathbf{X}(\mathbf{X}'\mathbf{X})^{-1}\end{aligned} \tag{21}$$

Then the relative efficiency of the GLS and OLS estimators of the jth element of $\boldsymbol{\beta}$ is found by dividing the jth diagonal term of the GLSE covariance matrix (8) by the corresponding elements in (21). Since the GLS estimators are unbiased and of minimum variance, the OLSE variances will always be greater and the OLS estimators less efficient. We consider some examples of these variances and relative efficiency ratios in Section 6-3.

6-3 AUTOREGRESSIVE DISTURBANCES

In most analyses of short time series it is very unlikely that we would know the population covariance matrix of the model disturbance terms. Unless many independent replications of the series under identical conditions were available we would not have sufficient information to estimate a general covariance matrix.* The covariance matrix of a time series with N observations contains $N(N+1)/2$ parameters. We might begin by assuming a common variance σ^2 at each point and the same covariance for successive adjacent observations:

$$\text{Cov}\,(e_t, e_{t+1}) = \sigma^2\rho_1 \tag{1}$$

ρ_1 is called the first serial correlation or first-order autocorrelation of the e_t. Similarly, we might define the sth autocorrelation as

$$\text{Cov}\,(e_t, e_{t+s}) = \sigma^2\rho_s \qquad t = 1, \ldots, N - s \tag{2}$$

Then the successive diagonals of the matrix contain the same terms $\sigma^2\rho_1, \ldots, \sigma^2\rho_{N-1}$ as we move away from the main diagonal. By this simplification we have reduced the number of parameters to be estimated to N. We might estimate σ^2 by some mean square of the residuals from the fitted model and $\sigma^2\rho_1$ by the sample covariance of each term with its preceding neighbor. The higher-order serial covariances could be estimated by similar covariances of the observations lagged by two, three, . . . , positions in the series. Nevertheless, as the lag increases the number of product terms in the estimator decreases, and the high-order estimates become unstable and meaningless. It seems clear that we need a further specialization in our covariance structure.

One useful pattern follows by letting the correlation between the disturbance variates decay exponentially with their separation:

$$\text{Cov}\,(e_t, e_{t+s}) = \sigma^2\rho^{|s|} \tag{3}$$

Then $\rho_1 = \rho$, $\rho_2 = \rho^2$, and so on, where ρ is the single correlation parameter. The $N \times N$ autocorrelation matrix

$$\mathbf{P} = \begin{bmatrix} 1 & \rho & \rho^2 & \cdots & \rho^{N-1} \\ \rho & 1 & \rho & \cdots & \rho^{N-2} \\ \cdots & \cdots & \cdots & \cdots & \cdots \\ \rho^{N-1} & \rho^{N-2} & \rho^{N-3} & \cdots & 1 \end{bmatrix} \tag{4}$$

is said to have the *Markov pattern*. Through that correlation model we have reduced the number of parameters in the covariance structure to merely two.

Before examining some useful properties of the Markov matrix let us introduce a linear model for the disturbances that leads to the correlation structure. The autoregressive model

$$e_t = \rho e_{t-1} + u_t \tag{5}$$

says that the tth disturbance e_t is proportional to the $(t-1)$th, except for a random error u_t. We shall always assume that $|\rho| < 1$. The random errors are independently and normally distributed with a zero mean and a common variance σ_u^2. u_t is also independent of the preceding disturbance terms e_{t-1}, $e_{t-2}, \ldots$. Then we can write e_t recursively as

*Such a combination of time and "cross-sectional" data occurs in the analysis of growth curves for several organisms; methods for fitting growth curves have been described by Morrison (1976).

$$e_t = \rho(\rho e_{t-2} + u_{t-1}) + u_t$$
$$= u_t + \rho u_{t-1} + \rho^2 u_{t-2} + \rho^3 u_{t-3} + \cdots \quad (6)$$

or a linear function of all the independent errors beginning with some very early time. We note that the influence of the errors on e_t declines geometrically with the distance of the error from time t. By the formula for an infinitely long geometric series,

$$\text{Var}(e_t) = \text{Var}(u_t + \rho u_{t-1} + \rho^2 u_{t-2} + \cdots)$$
$$= \sigma_u^2(1 + \rho^2 + \rho^4 + \rho^6 + \cdots)$$
$$= \frac{\sigma_u^2}{1 - \rho^2} \quad (7)$$

As long as we can neglect the effect of the initial terms in the model, or assume that the difference with a finite geometric series sum is negligible, the variance (7) will hold for all values of the index t. The correlations in the matrix (4) follow from the recursive relationship:

$$\text{Cov}(e_t, e_{t+1}) = \text{Cov}(e_t, \rho e_t + u_{t+1})$$
$$= \rho\sigma^2$$
$$\text{Cov}(e_t, e_{t+2}) = \text{Cov}(e_t, \rho^2 e_t + \rho u_{t+1} + u_{t+2})$$
$$= \rho^2\sigma^2 \quad (8)$$
$$\vdots$$
$$\text{Cov}(e_t, e_{t+s}) = \text{Cov}(e_t, \rho^s e_t + \rho^{s-1} u_{t+1} + \cdots + u_{t+s})$$
$$= \rho^s\sigma^2$$

We see in general that disturbance terms e_t and $e_{t\pm s}$ separated in either direction by s time intervals have covariance $\rho^{|s|}\sigma^2$ or correlation $\rho^{|s|}$. The autoregressive parameter ρ in the original model (5) is the product moment correlation of adjacent disturbances. That correlation declines geometrically with the distance between the disturbance variates.

But why is the correlation structure called *Markov*, after the Russian probabilist A. A. Markov? Recall the definition of the simplest Markov chain in elementary probability: A random experiment can have outcome S (success) or F (failure) at each trial. If the conditional probability of an outcome depends only on the previous result, the process is called *Markovian*. Markov chains have constant conditional probabilities at each trial. Those transition probabilities and the initial probabilities of the states S and F determine the properties of the chain. The Markov process generalizes to random variables as well. Let the variates $e_1, \ldots, e_N$ describe a time series in which the subscripts show the usual ordering of the variates in time. The e_t constitute a Markov process if the conditional distribution of e_{t+1} for fixed $e_1, \ldots, e_t$ depends only on the immediately preceding variate e_t. That is, the entire history of the time series only affects future values through the most recent value: Its "memory" extends back but a single step. If $e_1, \ldots, e_N$ have a joint multivariate normal distribution with a common mean, the same variance σ^2, and the correlation matrix (4), they form a stationary Markov process.

Such random variables have interesting and useful properties. Since they are multinormal their dependence structure and conditional distribution can be described in terms of their correlations—simple, partial, and multiple. For example, the variates e_h and e_j are independently distributed if the intermediate random variable e_i, or one for which $h < i < j$, is held constant. Since the three variates have the trivariate normal distribution that independence property will hold if the partial correlation $\rho_{hj.i}$ is zero. The partial correlation can be computed from the simple correlations as

$$\rho_{hj.i} = \frac{\rho_{hj} - \rho_{hi}\rho_{ji}}{\sqrt{(1 - \rho_{hi}^2)(1 - \rho_{ji}^2)}} \tag{9}$$

In the Markov case

$$\begin{aligned} \rho_{hj.i} &= \frac{\rho^{j-h} - \rho^{i-h}\rho^{j-i}}{\sqrt{(1 - \rho^{2(i-h)})(1 - \rho^{2(j-i)})}} \\ &= 0 \end{aligned} \tag{10}$$

That property holds for any three distinct variates e_h, e_i, e_j with the time ordering $h < i < j$. It generalizes to the independence of sets of variates with one or more intermediate variates held fixed. The vanishing partial correlations demonstrate that the correlation among Markov variates passes from variable to variable in the time series.

A second property of Markov random variables involves the multiple regression functions of their joint distribution. Under the correlation structure (4) the multiple regression function for predicting e_i from the remaining $N - 1$ variates only involves e_{i-1} and e_{i+1}. All of the other variates have regression coefficients of zero. Once again we see how only the immediately adjacent variates in a Markov time series carry information for prediction. The derivation of the property follows from the connection of the multiple regression coefficients and correlations with the elements of the inverse of $\mathbf{P}$ (Morrison, 1976, p. 125). The details will be left as an exercise.

Let us find the generalized least squares estimators of the parameters of the linear trend line for equally spaced observations whose disturbance terms have the Markov correlation structure (4). The inverse of the matrix is

$$\mathbf{P}^{-1} = \frac{1}{1 - \rho^2}\begin{bmatrix} 1 & -\rho & 0 & \cdots & 0 & 0 \\ -\rho & 1 + \rho^2 & -\rho & \cdots & 0 & 0 \\ 0 & -\rho & 1 + \rho^2 & \cdots & 0 & 0 \\ \cdots & \cdots & \cdots & \cdots & \cdots & \cdots \\ 0 & 0 & 0 & \cdots & 1 + \rho^2 & -\rho \\ 0 & 0 & 0 & \cdots & -\rho & 1 \end{bmatrix} \tag{11}$$

For convenience we shall assume that the number of terms in the series is odd, or $N = 2m + 1$, so that the linear orthogonal polynomial values are given by

$$\boldsymbol{\phi}' = [-m, -(m-1), \ldots, -1, 0, 1, \ldots, m-1, m] \tag{12}$$

Then the matrix

$$\mathbf{X}'\mathbf{P}^{-1}\mathbf{X} = \begin{bmatrix} \mathbf{j}' \\ \boldsymbol{\phi}' \end{bmatrix} \mathbf{P}^{-1}[\mathbf{j} \quad \boldsymbol{\phi}] = \begin{bmatrix} a_{11} & a_{12} \\ a_{12} & a_{22} \end{bmatrix} \tag{13}$$

has the elements

$$\begin{aligned} a_{11} &= \frac{N(1-\rho) + 2\rho}{1+\rho} \\ a_{22} &= \frac{1-\rho}{1+\rho}\boldsymbol{\phi}'\boldsymbol{\phi} + \frac{\rho(N-1)^2}{2(1+\rho)} + \frac{\rho(N-1)}{1-\rho^2} \\ a_{12} &= 0 \end{aligned} \tag{14}$$

The intercept and linear trend estimators are uncorrelated. Unfortunately, that property does not carry over to the even-degree polynomial coefficients. The sum of squares of the linear polynomial values is

$$\boldsymbol{\phi}'\boldsymbol{\phi} = \frac{m(m+1)(2m+1)}{3} = \frac{N(N^2-1)}{12} \tag{15}$$

For comparison with the ordinary least squares estimators and their variances we shall prefer to leave that term as $\boldsymbol{\phi}'\boldsymbol{\phi}$. Next we need the vector

$$\mathbf{X}'\mathbf{P}^{-1}\mathbf{y} = \frac{1}{1+\rho}\begin{bmatrix} (1-\rho)\sum_{t=1}^{N} y_t - \rho(y_1 + y_N) \\ (1-\rho)\boldsymbol{\phi}'\mathbf{y} - [\rho(N-1)/2](y_N - y_1) \end{bmatrix} \tag{16}$$

Then the generalized least squares estimators of the parameters are

$$\begin{aligned} \hat{\alpha} &= \frac{\bar{y} + \dfrac{\rho}{(1-\rho)N}(y_1 + y_N)}{1 + \dfrac{2\rho}{(1-\rho)N}} \\ \hat{\beta} &= \frac{\dfrac{1}{\boldsymbol{\phi}'\boldsymbol{\phi}}\sum_{t=1}^{N} \phi_t y_t + \dfrac{\rho(N-1)}{2(1-\rho)\boldsymbol{\phi}'\boldsymbol{\phi}}(y_N - y_1)}{1 + \dfrac{\rho(N-1)}{\boldsymbol{\phi}'\boldsymbol{\phi}(1-\rho)}\left(\dfrac{N-1}{2} + \dfrac{1}{1+\rho}\right)} \end{aligned} \tag{17}$$

The estimators have an important property for large N: The initial terms in their respective numerators are the *ordinary* least squares estimators; as the series increases in length the two kinds of estimators are equivalent. With long time series and a Markov correlation structure the usual ordinary least squares estimators will suffice for fitting a linear trend. That result is a special case of a more general theorem for regression analysis with correlated disturbances; the property has been described by Anderson (1971, Chapter 10).

The variances of $\hat{\alpha}$ and $\hat{\beta}$ follow from (8), Section 6-2. Since the required matrix (13) is diagonal the variances are merely proportional to the reciprocals of its diagonal elements:

$$\begin{aligned}\operatorname{Var}(\hat{\alpha}) &= \frac{\sigma^2}{a_{11}} \\ &= \frac{\sigma^2(1+\rho)}{N(1-\rho)+2\rho}\end{aligned} \tag{18}$$

$$\begin{aligned}\operatorname{Var}(\hat{\beta}) &= \frac{\sigma^2}{a_{22}} \\ &= \sigma^2\Big/\left[\frac{(1-\rho)N(N^2-1)}{12(1+\rho)} + \frac{\rho(N-1)^2}{2(1+\rho)} + \frac{(N-1)\rho}{1-\rho^2}\right]\end{aligned} \tag{19}$$

It will be informative to compare these variances to those of the ordinary least squares estimators for different assumptions on the disturbance terms in the model. As N becomes large (18) and (19) tend to

$$\frac{1+\rho}{1-\rho}\frac{\sigma^2}{N} \quad \text{and} \quad \frac{1+\rho}{1-\rho}\frac{12\sigma^2}{N(N^2-1)} \tag{20}$$

or $(1+\rho)/(1-\rho)$ times the variances of the ordinary least squares estimators with uncorrelated disturbances. The positive autocorrelation ρ has increased the variances of our estimators. Effectively, we have fewer observations than a time series with uncorrelated random components.

The comparison of the large sample values of (18) and (19) with their OLS counterparts has only shown how autocorrelation reduces precision in the estimators. We should also compare the variances of the OLSEs and GLSEs under correlated disturbances. For the straight line with equally spaced points the OLSE covariance matrix (21) of Section 6-2 becomes

$$\begin{aligned}\operatorname{Cov}(\hat{\boldsymbol{\beta}}_0, \hat{\boldsymbol{\beta}}_0') &= \begin{bmatrix} 1/N & 0 \\ 0 & 1/\boldsymbol{\phi}'\boldsymbol{\phi} \end{bmatrix} \begin{bmatrix} \mathbf{j}' \\ \boldsymbol{\phi}' \end{bmatrix} \boldsymbol{\Sigma}[\mathbf{j} \quad \boldsymbol{\phi}] \begin{bmatrix} 1/N & 0 \\ 0 & 1/\boldsymbol{\phi}'\boldsymbol{\phi} \end{bmatrix} \\ &= \sigma^2 \begin{bmatrix} \mathbf{j}'\mathbf{P}\mathbf{j}/N^2 & 0 \\ 0 & \boldsymbol{\phi}'P\boldsymbol{\phi}/(\boldsymbol{\phi}'\boldsymbol{\phi})^2 \end{bmatrix}\end{aligned} \tag{21}$$

P is the Markov correlation matrix (4). We see that the OLS intercept and slope estimators are uncorrelated, but their variances are

$$\begin{aligned}\operatorname{Var}(\hat{\alpha}_0) &= \frac{\sigma^2\mathbf{j}'\mathbf{P}\mathbf{j}}{N^2} \\ \operatorname{Var}(\hat{\beta}_0) &= \frac{\sigma^2\boldsymbol{\phi}'\mathbf{P}\boldsymbol{\phi}}{(\boldsymbol{\phi}'\boldsymbol{\phi})^2}\end{aligned} \tag{22}$$

Values of those variances, the corresponding GLSE variances (18) and (19), and their relative efficiencies are shown in Table 6-1 for some values of the autocorrelation ρ and the series length N. We note that the efficiency of the intercept estimator decreases with N at first, then increases. The variance of the slope appears to behave in the same way, although more slowly for larger ρ. The equivalence of the OLS and GLS estimators requires much longer time series if the autocorrelation is strong.

TABLE 6-1 Variances and Relative Efficiencies of Least Squares Estimators with Autoregressive Disturbances

		Intercept: Var ($\hat{\alpha}$)			Slope: Var ($\hat{\beta}$)		
N	ρ	OLSE	GLSE	Efficiency	OLSE	GLSE	Efficiency
3	0.5	0.6111	0.6000	0.982	0.375	0.375	1
	0.8	0.8311	0.8182	0.984	0.1800	0.1800	1
5	0.5	0.445	0.4286	0.963	0.1200	0.1154	0.962
	0.8	0.7243	0.6923	0.956	0.07747	0.07377	0.952
9	0.5	0.2840	0.2727	0.960	0.02992	0.0278	0.928
	0.8	0.5725	0.5294	0.925	0.02809	0.02586	0.921
21	0.5	0.1338	0.1304	0.975	0.003169	0.002970	0.937
	0.8	0.3389	0.3103	0.916	0.005476	0.004569	0.834

Transformations for eliminating autocorrelation. In the preceding section we found the GLSE of $\boldsymbol{\beta}$ by transforming the disturbances to uncorrelated variates with a common variance. To compute the GLSE we must invert an $N \times N$ covariance matrix. We could avoid that inversion by computing from the transformed model

$$\mathbf{Ty} = (\mathbf{TX})\boldsymbol{\beta} + \mathbf{Te} \tag{23}$$

the OLSE

$$\hat{\boldsymbol{\beta}}_T = [(\mathbf{TX})'(\mathbf{TX})]^{-1}(\mathbf{TX})'(\mathbf{Ty}) \tag{24}$$

Of course, we must know the $N \times N$ matrix $\mathbf{T}$ that diagonalizes $\boldsymbol{\Sigma}$ as in (5), Section 6-2. Fortunately, $\mathbf{T}$ has a very simple form when the covariance structure has the Markov pattern. Then

$$\mathbf{T} = \begin{bmatrix} \sqrt{1-\rho^2} & 0 & 0 & \cdots & 0 & 0 \\ -\rho & 1 & 0 & \cdots & 0 & 0 \\ 0 & -\rho & 1 & \cdots & 0 & 0 \\ \cdots & \cdots & \cdots & \cdots & \cdots & \cdots \\ 0 & 0 & 0 & \cdots & 1 & 0 \\ 0 & 0 & 0 & \cdots & -\rho & 1 \end{bmatrix} \tag{25}$$

The new dependent variable values are

$$\mathbf{w}' = \mathbf{y}'\mathbf{T}' = [\sqrt{1-\rho^2}\, y_1, y_2 - \rho y_1, \ldots, y_N - \rho y_{N-1}] \tag{26}$$

The elements in the first column of the transformed $\mathbf{X}$ matrix are

$$[\sqrt{1-\rho^2}, 1-\rho, \ldots, 1-\rho] \tag{27}$$

and those in the jth column are

$$[\sqrt{1-\rho^2}\, x_{1j}, x_{2j} - \rho x_{1j}, \ldots, x_{Nj} - \rho x_{N-1,j}] \tag{28}$$

The transformation has replaced the observations by their weighted differences. We may verify that

$$\mathbf{TPT}' = (1-\rho^2)\mathbf{I} \tag{29}$$

If a time series has residuals whose autocorrelations appear to resemble the Markov correlation pattern, we might find the GLSEs of the model parameters by making the differencing transformation with a reasonable value of ρ, fitting a second linear model with estimated coefficients (24), and repeating that process until the error sum of squares computed from the differenced data is a minimum. That iterative procedure is due to Cochrane and Orcutt (1949). It avoids the nonlinear least squares equations given by direct differentiation with respect to ρ.

If ρ is large, the transformation matrix $\mathbf{T}$ amounts to finding first differences of the dependent and independent variables and estimating $\boldsymbol{\beta}$ by applying least squares to those values. As another simplification we might drop the first row in $\mathbf{T}$ and work with only the $N - 1$ weighted differences. The test of any of these methods seems to be empirical: how well a particular transformation leads to a model that fits the observed series closely. In Section 6-4 we apply these methods to some time series after we have addressed the question of testing for autocorrelation.

6-4 TESTING AND ESTIMATING AUTOCORRELATION

The Markov correlation structure of the disturbances in Section 6-3 was assumed to hold throughout the time series with a known correlation parameter ρ. In this section we describe a common test of the hypothesis that ρ is zero and give a simple estimator of ρ.

The test we use for

$$H_0: \rho = 0 \tag{1}$$

as opposed to the alternative of a positive autocorrelation coefficient is due to Durbin and Watson (1950, 1951, 1971). Let us suppose that we have fitted to the N equally spaced points of the time series a model with an intercept and p independent variables. Then the $N \times 1$ vector of residuals from the values $\hat{\mathbf{y}}$ predicted by the model is

$$\begin{aligned} \mathbf{z} &= \mathbf{y} - \hat{\mathbf{y}} \\ &= \mathbf{y} - \mathbf{X}\hat{\boldsymbol{\beta}} \\ &= (\mathbf{I} - \mathbf{X}(\mathbf{X}'\mathbf{X})^{-1}\mathbf{X}')\mathbf{y} \end{aligned} \tag{2}$$

$\hat{\boldsymbol{\beta}}$ is the OLSE of the parameter vector. If the tth element of $\mathbf{z}$ is denoted by z_t, the Durbin–Watson statistic is

$$d = \frac{\sum_{t=1}^{N-1} (z_t - z_{t+1})^2}{\sum_{t=1}^{N} z_t^2} \tag{3}$$

The statistic compares the variation in the residuals with that in their successive differences. If the residuals are highly correlated from time to time, we would expect their first differences to be about the same size, or have little variability.

A measure of serial correlation is d: We may write its formula as

$$d = \frac{\sum_{t=1}^{N-1} z_t^2 + \sum_{t=2}^{N} z_t^2 - 2\sum_{t=1}^{N-1} z_t z_{t+1}}{\sum_{t=1}^{N} z_t^2}$$

$$= 2\left(1 - \frac{\sum_{t=1}^{N-1} z_t z_{t+1}}{\sum_{t=1}^{N} z_t^2}\right) - \frac{z_1^2 + z_N^2}{\sum_{t=1}^{N} z_t^2} \tag{4}$$

Because $\sum_1^N z_t = 0$ from the definition of the residual terms, the second term in the parentheses is one approximate estimator

$$\hat{\rho} = \frac{\sum_{t=1}^{N-1} z_t z_{t+1}}{\sum_{t=1}^{N} z_t^2} \tag{5}$$

of the autoregressive model or Markov correlation matrix parameter. Apart from effects due to the first and last terms,

$$\hat{\rho} \approx 1 - \frac{d}{2} \tag{6}$$

When $d = 0$ the sample correlation is one, while $d = 2$ gives a zero autocorrelation. As N becomes large the upper bound on d tends to its limiting value of four.

The distribution of the statistic d is complicated and involves the matrix $\mathbf{X}$ of independent variable values as a parameter. However, Durbin and Watson (1950, 1951) found lower and upper bounds, d_L, d_U, for d whose distributions did not depend on $\mathbf{X}$, and tabulated the upper percentage points of the bounding variates. Those 5% and 1% critical values are given in Table 8, Appendix A. The test is carried out by this decision rule:

$$\begin{aligned} &\text{Reject } H_0: \rho = 0 \text{ if } d < d_L \\ &H_0: \rho = 0 \text{ is tenable if } d > d_U \\ &\text{The test is inconclusive if } d_L \leq d \leq d_U \end{aligned} \tag{7}$$

Durbin and Watson (1971) have calculated the exact percentage points of d for certain $\mathbf{X}$ matrices and have compared them with approximations proposed by various authors. The upper bound critical value d_U appears to be close to the exact critical value of d when the series is long and the number of independent variables small.

The Durbin–Watson critical values can also be used to test the hypothesis (1) against the alternative of negative autocorrelation. We refer $4 - d$ to Table 8 and use this decision rule:

$$\begin{aligned} &\text{Reject } H_0: \rho = 0 \text{ if } 4 - d < d_L \\ &H_0: \rho = 0 \text{ is tenable if } 4 - d > d_U \\ &\text{The test is inconclusive if } d_L \leq 4 - d \leq d_U \end{aligned} \tag{8}$$

The error due to using the large-sample upper limit of four is negligible for the values of N we should encounter in time series.

Example 6-2

We shall illustrate the Durbin–Watson test and some of the methods of the preceding section with the analyses of two time series of crime rates in the United States. The series are the numbers of homicides and rapes per 100,000 population.* The data are given in Table 6-2 and depicted in Figures 6-1 and 6-2. Let us suppose that we wish to

TABLE 6-2 Annual Rates of Homicide and Rape in the United States (per 100,000 population)

Year	Homicide	Rape
1950	5.1	5.7
1951	4.8	5.7
1952	4.8	6.0
1953	4.6	6.0
1954	4.5	6.3
1955	4.3	7.0
1956	4.5	8.1
1957	4.5	8.2
1958	4.5	8.4
1959	4.6	8.7
1960	4.6	8.8
1961	4.8	9.7
1962	4.8	9.9
1963	4.8	9.6
1964	5.4	11.6
1965	5.5	12.4
1966	6.0	14.2
1967	6.8	15.5
1968	7.8	17.9
1969	8.5	21.0
1970	9.3	22.4
1971	10.3	24.5
1972	10.7	27.3
1973	11.1	29.7
1974	11.4	31.5
1975	11.1	31.7
1976	9.9	30.8
1977	10.2	34.2
1978	10.1	36.0

SOURCE: Reprinted by permission of the publisher from *Forecasting Crime Data*, by James A. Fox (Boston: Lexington Books, D.C. Heath and Company; copyright 1978, D.C. Heath and Company).

*These data were kindly provided by James A. Fox.

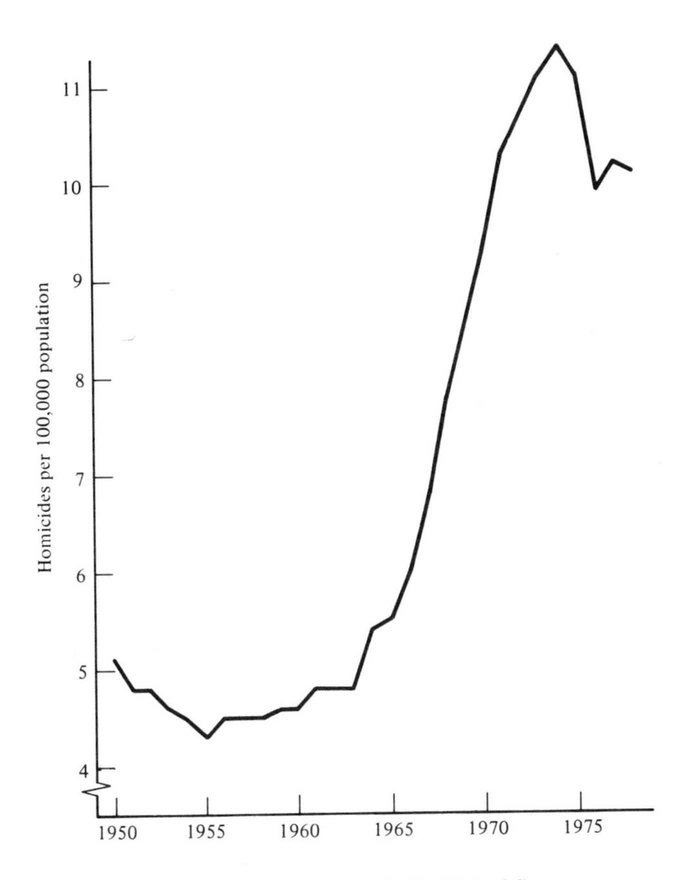

Figure 6-1 Homicide rates in the United States.

represent the series by low-degree polynomial models so that the residuals from the fitted polynomial trends can be correlated with other crime rates and social indicators for the same period. We shall not attempt to predict future crime rates; a more suitable approach for that purpose has been made via simultaneous equation models (Fox, 1976). The figures suggest that a third-or fourth-degree polynomial might be appropriate for the homicide series, whereas a quadratic or cubic might suffice for the rape series. We shall begin by fitting the first four orthogonal polynomials by ordinary least squares. The regression coefficients, their sums of squares, and other statistics are shown in Table 6-3. The estimates of the disturbance variances under the different polynomial models are given at the bottom of the table. We have omitted t- or F-statistics for testing the significance of the coefficients: Visual inspection and some mental division indicate that all (save perhaps for the cubic component in the rape series) would greatly exceed any of the usual critical values. We should test that the assumption of independent disturbances holds before making formal inferences.

The Durbin–Watson statistics for the autocorrelation tests based on the residuals of the quadratic, cubic, and quartic models are shown in Table 6-4. The residuals from the three models appear to be autocorrelated in each series, so that the classical tests and confidence intervals based on the OLS estimators would be invalid. Their signifi-

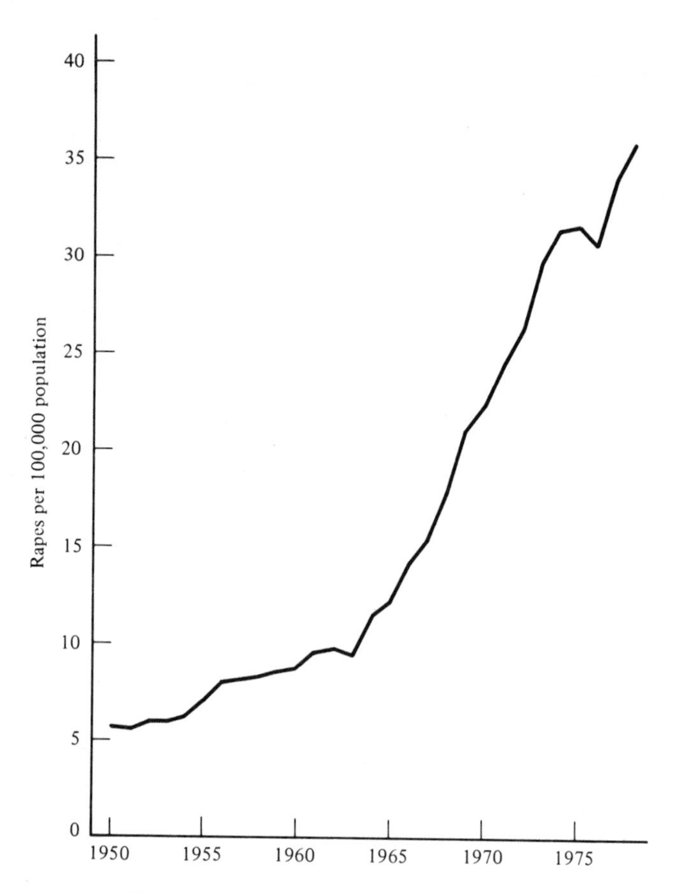

Figure 6-2 Rape rates in the United States.

cance would be greatly overstated or the intervals improperly short. We must take the autocorrelation into account in our estimation of the polynomial coefficients. But how can that be done with such short time series? We must begin by making some heroic assumptions about the covariance structure of the residuals. We let the variances be the same for the 29 years, although that may not be realistic for series that are changing slowly at some times and rising steeply at others. Next we assume that the correlations have the Markov pattern of the matrix (4), Section 6-3, so that the entire matrix is generated by the single parameter ρ. The value of ρ is estimated initially from the autocorrelation of the OLSE residuals. We might also try other values of ρ to make the generalized least squares residuals quadratic form

$$\begin{aligned} Q(\rho) &= (\mathbf{y} - \hat{\mathbf{y}})'\hat{P}^{-1}(\rho)(\mathbf{y} - \hat{\mathbf{y}}) \\ &= (N - p - 1)\hat{\sigma}^2 \end{aligned}$$

given in (12), Section 6-2, as small as possible.

We proceeded in that manner to reestimate the cubic and quartic polynomial model coefficients for the two series. We began with $\rho = 0.9$, as suggested by the

TABLE 6-3 Fourth-Degree Orthogonal Polynomials Fitted by Ordinary Least Squares

	Homicide			Rape		
Component	$\hat{\beta}_j$	Sum of squares	R^2 (cumulative)	$\hat{\beta}_j$	Sum of squares	R^2 (cumulative)
Linear	0.2776	156.42	0.7950	1.1378	2627.94	0.8998
Quadratic	0.01144	14.82	0.8703	0.04596	239.22	0.9817
Cubic	−0.001788	13.45	0.9387	−0.001037	4.52	0.9833
Quartic	−0.0002815	8.56	0.9822	−0.005125	28.37	0.9930
Error	—	3.51	—	—	20.47	—
Total (about mean)	—	196.76	—	—	2920.52	—
Mean	6.8724	—	—	16.1655	—	—

Error Mean Squares

	Homicide		Rape	
Model	$\hat{\sigma}^2$	d.f.	$\hat{\sigma}^2$	d.f.
Quadratic	0.9815	26	2.0523	26
Cubic	0.4828	25	1.9536	25
Quartic	0.1462	24	0.8529	24

TABLE 6-4 Durbin–Watson d-Statistics for the OLSE Residuals of the Three Polynomial Models

		Series	
Model		Homicide	Rape
Quadratic:	d	0.238**	0.460**
	$\hat{\rho}$	0.881	0.770
Cubic:	d	0.340**	0.490**
	$\hat{\rho}$	0.830	0.755
Quartic:	d	0.988**	1.145[a]
	$\hat{\rho}$	0.506	0.427

**Significant at the 0.01 level.

[a]Inconclusive; nearly significant at the 0.05 level.

quadratic model serial correlations in Table 6-4, generated the inverse matrix $\mathbf{P}^{-1}$ of (11), Section 6-3, and computed the cubic model GLS estimates, their standard deviations, and their t-ratios. Then the process was repeated for $\rho = 0.8, 0.7, 0.6$, and 0.5. Some of the results for two representative values of ρ are given in Table 6-5. Note that the coefficient estimates hardly change with ρ, but their estimated standard deviations decrease as ρ moves toward zero, or the OLS case. The t-ratios for both homicide rate models exceed their usual critical values even when ρ is equal to 0.8. The significance of the polynomial coefficients is *not* an artifact of the autocorrelated residuals. The same

TABLE 6-5 Some Examples of GLS Estimates for the Crime Rate Series

a. Homicide

1. Cubic model:

ρ	$\hat{\sigma}^2$		Constant	Linear	Quadratic	Cubic
0.8	0.4082	$\hat{\beta}$	6.82	0.2782	0.00756	−0.00173
		SD($\hat{\beta}$)	0.33	0.032	0.0032	0.00043
		t	20.9	9.6	2.4	−4.0
0.5	0.2778	$\hat{\beta}$	6.84	0.2782	0.00985	−0.00175
		SD($\hat{\beta}$)	0.165	0.019	0.0023	0.00035
		t	41	15	4.2	−5.0

2. Quartic model:

ρ	$\hat{\sigma}^2$		Constant	Linear	Quadratic	Cubic	Quartic
0.8	0.3064	$\hat{\beta}$	6.88	0.2782	0.0118	−0.00171	−0.00019
		SD($\hat{\beta}$)	0.28	0.028	0.0031	0.00037	0.000062
		t	24	10	3.8	−4.6	−3.1
0.5	0.1421	$\hat{\beta}$	6.88	0.2782	0.0120	−0.00175	−0.00023
		SD($\hat{\beta}$)	0.118	0.0134	0.0017	0.00025	0.000047
		t	58	21	7	−7	−5

b. Rape

1. Cubic model:

ρ	$\hat{\sigma}^2$		Constant	Linear	Quadratic	Cubic
0.8	2.3186	$\hat{\beta}$	16.09	1.138	0.040	−0.00097
		SD($\hat{\beta}$)	0.78	0.077	0.0076	0.0010
		t	21	15	5.2	−0.9
0.5	1.3064	$\hat{\beta}$	16.11	1.138	0.043	−0.0010
		SD($\hat{\beta}$)	0.36	0.041	0.0050	0.00059
		t	45	28	9	−1.7

2. Quartic model:

ρ	$\hat{\sigma}^2$		Constant	Linear	Quadratic	Cubic	Quartic
0.8	2.1071	$\hat{\beta}$	16.18	1.138	0.0468	−0.000969	−0.000307
		SD($\hat{\beta}$)	0.74	0.074	0.0081	0.00098	0.00016
		t	22	15	5.8	−1.0	−1.9
0.5	0.9194	$\hat{\beta}$	16.19	1.138	0.0472	−0.00101	0.000406
		SD($\hat{\beta}$)	0.30	0.034	0.0043	0.00064	0.00012
		t	54	33	11	−1.6	−3.4

is true for the constant, linear, and quadratic terms in both models for rape rates. However, the cubic term appears to lose significance when the autocorrelation is taken into account, as does the quartic term when ρ is high. A quadratic model might suffice if its disturbance variates are related by an autoregressive process.

Let us summarize our findings to this point. We have established that the homicide rate series would require at least a fourth-degree polynomial even when generalized least squares and the Markov pattern covariance matrix is used. We should not impute exact probability levels to the t-ratios of the GLS estimates, although it seems safe to conclude that their population values are different from zero. Under no circumstance should the polynomial model be used for extrapolation of homicide rates: That would be naive. However, these conclusions seem to follow from an intensive analysis of the data:

1. The relative smoothness of the homicide series appears to lead to runs of positive and negative residuals and consequent serial correlation. Ordinary least squares is inappropriate with such series, for the standard deviations of its estimates are too small.
2. Computation of serial correlations for several lags in the residuals of some of the OLSE models for the homicide series indicated that the Markov correlation pattern might be appropriate, at least for lags one, two, and three. The autocorrelations began to oscillate for higher lags, although one should not place much trust in such correlations computed from short time series.
3. The residuals of the homicide OLSE and GLSE quartic polynomials tended to be similar: It seems doubtful that one could select a "best" model from their inspection. Model differences seemed more apparent in the lower-degree (for example, quadratic) polynomials.
4. The downturn in homicide rates, and the abrupt dip in 1976, should be explained. One might model the time series by two or three separate linear or quadratic functions for descriptive or predictive purposes.
5. The rape rates are more amenable to modeling by a low-order polynomial, although the 1976 dip should be explained by someone familiar with contemporary demographic or social series. For predictive purposes we might concentrate on the nearly linear segment from 1963 to 1978; we shall consider that series in Example 6-3.

Example 6-3

The rape rates of Table 6-2 behave nearly linearly from 1963 to 1978, and a naive straight-line model might be used to project the prevalence ahead for a few years. Nevertheless, it seems likely that the disturbance terms in that model would be autocorrelated so that OLS estimation should not be used uncritically. Let us fit a straight line, then examine an alternative method for dealing with correlated residuals. The OLSE equation is

$$\hat{y}_t = 23.14 + 1.8264(t - 8.5) \qquad t = 1, \ldots, 16$$

where $t = 1$ corresponds to 1963. These statistics associated with the fitted line were computed:

$$\text{SSR} = 1134.24 \qquad \text{SSE} = 19.76 \qquad R^2 = 0.9829$$

$$\hat{\sigma}^2 = 1.4116 \qquad \hat{\sigma}(\hat{\beta}) = 0.06443 \qquad t(\hat{\beta}) = 28.4$$

$$\text{Durbin–Watson } d = 0.8379 \qquad \hat{\rho} = 0.58$$

The hypothesis of zero autocorrelation should be rejected just at the 0.01 level. We should take that correlation into account in the estimation of the straight-line parameters and the inferences about a linear trend.

Let us use a simplified version of the differencing transformation given by expressions (25)–(28) of Section 6-3, in which the initial term has been omitted. We shall let $\rho = 0.6$, as suggested by the approximate autocorrelation estimate $\hat{\rho} = 0.58$. The new dependent variable has values

$$w_t = y_t - 0.6y_{t-1} \qquad t = 2, \ldots, 16$$

The ones in the first column of the $\mathbf{X}$ matrix are replaced by

$$0.4 = 1 - (0.6 \times 1)$$

and the time indices $t - 8.5$ are replaced by

$$u_t = 0.4t - 2.8 = (t - 8.5) - 0.6(t - 1 - 8.5) \qquad t = 2, \ldots, 16$$

The new variable values are shown in Table 6-6. The regression line fitted to the transformed variables is

$$\hat{w}_t = 23.085 + 1.8009u_t$$

TABLE 6-6 Transformed Variables

t	w_t		u_t
2	5.84		−2.0
3	5.44		−1.6
4	6.76		−1.2
5	6.98		−0.8
6	8.60		−0.4
7	10.26		0
8	9.80		0.4
9	11.06		0.8
10	12.60		1.2
11	13.32		1.6
12	13.68		2.0
13	12.80		2.4
14	11.78		2.8
15	15.72		3.2
16	15.48		3.6
Mean	10.67		0.8
Sum of squares about mean	159.16		44.8
Sum of centered products		80.68	

The intercept and slope are nearly equal to those of the first line fitted by OLSE. If we compute the residuals $w_t - \hat{w}_t$ for the 15 pairs of transformed variables we may verify that the Durbin–Watson statistic is equal to 2.20. The estimated autocorrelation for those residuals is −0.10; the transformation has indeed led to uncorrelated residuals.

The similarity of the two fitted lines begins to disappear when we estimate the standard deviation of the slope $\hat{\beta}_{wu}$ computed from the transformed data. The error sum of squares is

$$\text{SSE}_w = 159.16 - \frac{(80.68)^2}{44.8} = 13.8640$$

Then

$$\hat{\sigma}_w^2 = \frac{\text{SSE}_w}{15 - 2} = 1.0665$$

and the estimated standard deviation of $\hat{\beta}_{wu}$ is

$$\hat{\sigma}(\hat{\beta}_{wu}) = \sqrt{\frac{1.0665}{44.8}} = 0.1543$$

The standard deviation is nearly two and one-half times that of the OLSE slope. Tests and confidence intervals using the estimates from the transformed variables would give a more accurate picture of the significance of the fitted model. Of course, if the necessary statistics are to have exact t- or F-distributions, the disturbances must have the Markov correlation pattern with a known parameter ρ.

6-5 LAGGED VARIABLES

The values in a time series are often related to their previous history and the earlier values of any independent variables in their model. Models that relate y_t to y_{t-1} and perhaps the independent variables x_t and x_{t-1} are called *lagged.* For example, the lobster populations producing the catches described in Section 4-3 are the results of several years of maturation. Since the growth of the lobsters depends on sea temperature, it is conceivable that the catch in year t will involve the temperature not only in that year but also in years $t - 1, t - 2, t - 3$, and $t - 4$ if maturation requires as many years. Similarly, sales of a product may be generated by a long period of institutional or low-keyed advertising whose intensity might fluctuate from quarter to quarter according to the budgets or whims of management.

Let us begin with the simplest model in which only a single independent variable has been lagged:

$$y_t = \alpha + \beta_1 x_t + \beta_2 x_{t-1} + \beta_3 x_{t-2} + \cdots + \beta_p x_{t-p+1} + e_t \qquad t = p, p + 1, \ldots, N \tag{1}$$

The lagged x_t-values are treated as fixed, and the disturbances e_t obey the usual assumptions of a common variance and zero correlation for ordinary least squares estimation. We might center each of the lagged variables about the mean of its respective observations for computational ease. Apart from the loss of the first $p - 1$ observations in the series our model and estimation methods are no different from those in the earlier chapters.

Example 6-4

Before pursuing extensions of the model (1) let us see if lagged values of the independent variables in the lobster catch series of Section 4-3 will contribute to the prediction of the annual catch. We begin with a simple model postulating catch as a linear function of the number of fishermen during that year and the preceding one, and the average sea surface temperatures in the present and last three years. The results of fitting that model are shown in the left-hand side of Table 6-7. Curiously, the number of fishermen in the catch year has a negative, although not statistically significant, coefficient. The coefficient of the number of fishermen in the preceding year is significantly different

TABLE 6-7 Lobster Catch Models With Lagged Independent Variables

	Model 1		Model 2	
Variable	$\hat{\beta}_j$	t_j	$\hat{\beta}_j$	t_j
Trend				
Linear	—	—	94.89	5.90
Quadratic	—	—	−12.59	−11.31
Fishermen				
Year t	−0.4210	−0.90	0.2416	1.13
Year $t-1$	1.2164	2.34	0.0237	0.10
Temperature				
Year t	696.21	1.71	514.97	3.00
Year $t-1$	−92.85	−0.21	−320.08	−1.70
Year $t-2$	210.27	0.47	92.37	0.50
Year $t-3$	536.64	1.35	137.23	0.82
Mean catch	7444.41	—	7444.41	—
R^2	0.6068		0.9366	

from zero at the 0.05 level. The temperature regression coefficient for the catch year is barely significant in the one-sided sense, whereas those for the preceding three years are not. Previous fishing intensity and temperature values do not appear to contribute to the amount of lobster catch.

Next we remove the linear and quadratic trends in lobster catch and fit the previous model to the residuals. The results are shown in the right-hand half of Table 6-7. As in the earlier unlagged models the polynomial trends have explained large parts of the variation in catch. The number of fishermen is no longer a significant predictor. Sea temperature for the year of the catch now has a significant regression coefficient, but the coefficients for the three preceding years are of negative or negligible value. The contribution of lagged predictor values seems at best inconclusive in these first two models.

An examination of the correlations of the lagged predictors suggested another approach. The product moment correlation of the current and previous numbers of fishermen is 0.93; we might replace those variables by their unweighted average. Similarly, the first-, second-, and third-order serial correlations of the sea temperatures are approximately 0.72, 0.66, and 0.47, respectively. We might substitute the average of the four temperatures or represent sea temperature as a four-year moving average of its values. We also noted that the two predictors have nearly equal correlations with catch for the concurrent and lagged years: That symmetry supports replacement by averages. These results were obtained with those averaged predictors:

Variable	β_j	t_j
Linear trend	92.23	5.47
Quadratic trend	−12.36	−11.23
Moving average of fishermen	0.3027	2.12
Moving average of temperature	390.81	2.29

$$R^2 = 0.9205$$

The proportion of variation explained by the four variables is nearly that of the second model. The regression coefficients of the averaged predictors are significant at the 0.05 level, as opposed to the mixed results in Table 6-7. We might conclude that whereas the lagged variables have not had a clear effect on catch, their moving averages appear to be significantly related to catch.

Models with lagged dependent variables. In many applications the time series variable may depend on its earlier values: Prices of a commodity or stock may be influenced by past prices, and the prices may be driven higher in anticipation of returns from the appreciated value. Lagged values of the dependent variable also occur if we make certain assumptions on the lagged independent variable in the model (1). In order to reduce the number of regression coefficients Koyck (1954) proposed his method of *distributed lags*: The coefficients of the lagged terms are assumed to decline in size geometrically, so that the jth parameter can be represented as

$$\beta_j = (1 - \lambda)\gamma\lambda^{j-1} \qquad j = 1, \ldots, p \tag{2}$$

Then the model is

$$y_t = \alpha + (1 - \lambda)\gamma[x_t + \lambda x_{t-1} + \lambda^2 x_{t-2} + \cdots + \lambda^{p-1}x_{t-p+1}] + e_t \tag{3}$$

Next we introduce the lag operator D. That operator replaces a variable value at time t with the one at time $t - 1$, or

$$Dx_t = x_{t-1} \qquad D^2x_t = Dx_{t-1} = x_{t-2} \qquad Dy_t = y_{t-1} \tag{4}$$

Unlike the earlier expectation, variance, and covariance operators, we have omitted parentheses around the variables, for we shall shortly treat the operator algebraically as a variable. Now the model becomes

$$\begin{aligned} y_t &= \alpha + (1 - \lambda)\gamma[x_t + \lambda Dx_t + \lambda^2 D^2 x_t + \cdots + \lambda^{p-1}D^{p-1}x_t] + e_t \\ &= \alpha + (1 - \lambda)\gamma[1 + \lambda D + (\lambda D)^2 + \cdots + (\lambda D)^{p-1}]x_t + e_t \end{aligned} \tag{5}$$

As p becomes large the geometric series tends to $1/(1 - \lambda D)$, and the model can be represented as

$$y_t = \alpha + \frac{(1 - \lambda)\gamma}{1 - \lambda D}x_t + e_t \tag{6}$$

or

$$(1 - \lambda D)y_t = (1 - \lambda D)\alpha + (1 - \lambda)\gamma x_t + (1 - \lambda D)e_t \tag{7}$$

Now apply the lag operator to each variable:

$$Dy_t = y_{t-1} \qquad D\alpha = \alpha \qquad De_t = e_{t-1} \tag{8}$$

Then the model is

$$\begin{aligned} y_t &= \lambda y_{t-1} + (1 - \lambda)\alpha + (1 - \lambda)\gamma x_t + e_t - \lambda e_{t-1} \\ &= \mu + \lambda y_{t-1} + \beta x_t + e_t - \lambda e_{t-1} \end{aligned} \tag{9}$$

The Koyck assumption for the independent variables has led to a lagged dependent variable and a single contemporary independent variable. The disturbance term

$$d_t = e_t - \lambda e_{t-1} \tag{10}$$

is autocorrelated:

$$\text{Cov}(d_t, d_{t\pm1}) = -\lambda\sigma^2$$
$$\text{Var}(e_t - \lambda e_{t-1}) = \sigma^2(1 + \lambda^2)$$

Disturbances that are not adjacent are uncorrelated. The d_t constitute an example of a *moving average* process.

Because the disturbances in the final model (9) are autocorrelated we should not use ordinary least squares to estimate, μ, λ, and β. However, the estimation problem is even more serious than that of an inefficient OLSE. It can be shown (Johnston, 1972) that the lagged dependent variable and the correlated disturbances lead to inconsistent OLS estimators. The Durbin–Watson test applied to the residuals from that fitted model will be biased in favor of uncorrelated disturbances. Koyck (1954) and Nerlove (1958) have given ways of obtaining consistent estimators; one method has been described by Goldberger (1964, pp. 276–278). Other aspects of the estimation problem under different assumptions on the disturbance terms have been treated at length by Johnston (1972, pp. 304–320).

Example 6-5

If the disturbances (10) in the Koyck model (9) are considered uncorrelated with a common variance, we can use ordinary least squares to estimate μ, λ, and β, although the estimators will be biased for small values of N (Johnston, 1972, pp. 304–307). We have fitted several models containing lagged catch values to the lobster data by ordinary least squares. The estimates from two models are shown in Table 6-8. In each case all variables were centered about their averages to reduce the size of their sums of squares and products. In the first model the quadratic function of the number of fishermen was included to test for declining catch due to overfishing as the number of fishermen

TABLE 6-8 Lagged Models for Lobster Catch

	Model 1		Model 2	
Variable	$\hat{\beta}_j$	t_j	$\hat{\beta}_j$	t_j
Catch				
Year $t-1$	0.4217	2.82	0.4225	3.11
Year $t-2$	0.2101	1.82	—	—
Trend				
Linear	—	—	32.190	1.34
Quadratic	—	—	−4.847	−2.83
Fishermen				
Year t	0.4333	2.57	1.268	1.99
Year $t-1$	—	—	5×10^{-7}	0.03
Quadratic, year t	−0.0001235	−2.93	-7.82×10^{-5}	−1.68
Temperature				
Year t	683.35	4.45	633.56	4.35
Year $t-1$	−362.29	−2.09	−363.66	−2.13
R^2	0.9425		0.9535	

increased. The negative coefficient and the appreciable t-statistic for the quadratic variable indicate that catches may indeed decline with intensive fishing. The catches of the two previous years appear to be related strongly to the current catch. Strangely, the sea surface temperature of the preceding year has a negative coefficient: Again, the hypothesis that high temperatures in previous years contribute to larger catches does not appear to be tenable. Because the model involves lagged values of the dependent variable and the strong OLSE assumption of uncorrelated disturbances, the t_j should only be interpreted descriptively rather than as test statistics with the Student–Fisher distribution.

Model 2 contained linear and quadratic orthogonal polynomial components as well as the number of fishermen in the preceding year. The explained variation was increased scarcely more than 1%. The lagged number of fishermen had no effect. The contributions of the linear and quadratic trends were small relative to the earlier models in which they appeared without the lagged dependent variable. One might conclude that catch is related to that of the preceding year and to the number of fishermen and the sea surface temperature in the current year. Beyond that point the regression coefficients seem too unstable, small, or of uncertain sign to merit inclusion of their variables.

6-6 REFERENCES

AITKEN, A. C. (1934): On least squares and linear combination of observations, *Proceedings of the Royal Society of Edinburgh A*, vol. 55, pp. 42–47.

ANDERSON, T. W. (1971): *The Statistical Analysis of Time Series*, John Wiley & Sons, Inc., New York.

COCHRANE, D., and G. H. ORCUTT (1949): Application of least-squares regressions to relationships containing auto-correlated error terms, *Journal of the American Statistical Association*, vol. 44, pp. 32–61.

DURBIN, J., and G. S. WATSON (1950): Testing for serial correlation in least squares regression: I, *Biometrika*, vol. 37, pp. 409–428.

DURBIN, J., and G. S. WATSON (1951): Testing for serial correlation in least squares regression: II, *Biometrika*, vol. 38, pp. 159–178.

DURBIN, J., and G. S. WATSON (1971): Testing for serial correlation in least squares regression: III, *Biometrika*, vol. 58, pp. 1–19.

FOX, J. A. (1976): An Econometric Analysis of Crime Data, Unpublished Ph.D. dissertation, University of Pennsylvania, Philadelphia.

GOLDBERGER, A. S. (1964): *Econometric Theory*, John Wiley & Sons, Inc., New York.

JOHNSTON, J. (1972): *Econometric Methods*, 2nd ed., McGraw-Hill Book Company, New York.

KOYCK, L. M. (1954): *Distributed Lags and Investment Analysis*, North-Holland Publishing Company, Amsterdam.

MORRISON, D. F. (1976): *Multivariate Statistical Methods*, 2nd ed., McGraw-Hill Book Company, New York.

NERLOVE, M. (1958): *Distributed Lags and Demand Analysis for Agricultural and Other Commodities*, U.S. Department of Agriculture, Washington, D.C.

6-7 EXERCISES

1. The College Retirement Equities Fund (CREF) had these values of its annuity unit on December 31 in its annuity fiscal years since 1952:

Year	Unit value	Year	Unit value
1952	$10.00	1966	$30.43
1953	9.46	1967	31.92
1954	10.74	1968	29.90
1955	14.11	1969	32.50
1956	18.51	1970	28.91
1957	16.88	1971	30.64
1958	16.71	1972	35.74
1959	22.03	1973	31.58
1960	22.18	1974	26.21
1961	26.25	1975	21.84
1962	26.13	1976	26.24
1963	22.68	1977	24.80
1964	26.48	1978	23.28
1965	28.21	1979	27.28

SOURCE: Data reproduced with the kind permission of the Teachers Insurance and Annuity Association and College Retirement Equities Fund.

Fit linear, quadratic, and cubic polynomials by ordinary least squares and test their residuals for autocorrelation by the Durbin–Watson statistic. Does the use of OLSE appear to be valid?

2. Let the $N \times N$ covariance matrix

$$\boldsymbol{\Sigma} = \begin{bmatrix} \sigma_{11} & \boldsymbol{\sigma}'_{12} \\ \boldsymbol{\sigma}_{12} & \boldsymbol{\Sigma}_{22} \end{bmatrix} \begin{matrix} 1 \\ N-1 \end{matrix}$$
$$\quad 1 \quad N-1$$

be partitioned as shown, where the row and column dimensions are shown beside and below the submatrices. The inverse of $\boldsymbol{\Sigma}$ can be found in terms of the submatrices (Morrison, 1976, Section 2-11) as

$$\boldsymbol{\Sigma}^{-1} = \begin{bmatrix} 1/\sigma_{11.2} & -(1/\sigma_{11.2})\boldsymbol{\sigma}'_{12}\boldsymbol{\Sigma}_{22}^{-1} \\ -(1/\sigma_{11.2})\boldsymbol{\Sigma}_{22}^{-1}\boldsymbol{\sigma}_{12} & \boldsymbol{\Sigma}_{22}^{-1} + (1/\sigma_{11.2})\boldsymbol{\Sigma}_{22}^{-1}\boldsymbol{\sigma}_{12}\boldsymbol{\sigma}'_{12}\boldsymbol{\Sigma}_{22}^{-1} \end{bmatrix}$$

where $\sigma_{11.2} = \sigma_{11} - \boldsymbol{\sigma}'_{12}\boldsymbol{\Sigma}_{22}^{-1}\boldsymbol{\sigma}_{12}$.

(a) Show that the (1, 1) element of $\boldsymbol{\Sigma}^{-1}$ is $(1/\sigma_{11})(1 - R_1^2)$, where $R_{1.2\ldots N}^2 = R_1^2$ is the squared multiple correlation of the first variate with the remaining $N - 1$.

(b) Verify that the off-diagonal elements in the first row and column of $\boldsymbol{\Sigma}^{-1}$ are proportional to the multiple regression coefficients of the first variate with the other $N - 1$ variates.

(c) Show how the results in (a) and (b) could be used to obtain all coefficients of determination and regression coefficient vectors of each variate with the remaining ones.

3. Consider a short time series with $N = 5$ points and a disturbance term with an inverse covariance matrix

$$\mathbf{\Sigma}^{-1} = c\begin{bmatrix} 1 & -\frac{1}{2} & \frac{1}{4} & -\frac{1}{8} & \frac{1}{16} \\ -\frac{1}{2} & 1 & -\frac{1}{2} & \frac{1}{4} & -\frac{1}{8} \\ \frac{1}{4} & -\frac{1}{2} & 1 & -\frac{1}{2} & \frac{1}{4} \\ -\frac{1}{8} & \frac{1}{4} & -\frac{1}{2} & 1 & -\frac{1}{2} \\ \frac{1}{16} & -\frac{1}{8} & \frac{1}{4} & -\frac{1}{2} & 1 \end{bmatrix}$$

c is an appropriate positive constant. Let

$$\mathbf{X}' = \begin{bmatrix} 1 & 1 & 1 & 1 & 1 \\ -2 & -1 & 0 & 1 & 2 \end{bmatrix}$$

be the usual centered matrix for fitting a straight line. Hence

$$\mathbf{X}'\mathbf{\Sigma}^{-1}\mathbf{X} = c\begin{bmatrix} \frac{17}{8} & 0 \\ 0 & 6 \end{bmatrix}$$

and

$$\mathbf{X}'\mathbf{\Sigma}^{-1}\mathbf{Y} = c\begin{bmatrix} (11y_1 + 2y_2 + 8y_3 + 2y_4 + 11y_5)/16 \\ (\frac{3}{2})(y_5 - y_1) \end{bmatrix}$$

(a) Give the explicit generalized least squares estimators of the straight line parameters α and β.
(b) Is $\hat{\beta}$ an intuitively reasonable estimator?
(c) Find the covariance structure of the disturbances.

4. Now let the time series

$$y_t = \alpha + \beta t + e_t \qquad t = 1, \ldots, 5$$

have disturbances with the *independent increments* covariance matrix

$$\mathbf{\Sigma} = \sigma^2\begin{bmatrix} 1 & 1 & 1 & 1 & 1 \\ 1 & 2 & 2 & 2 & 2 \\ 1 & 2 & 3 & 3 & 3 \\ 1 & 2 & 3 & 4 & 4 \\ 1 & 2 & 3 & 4 & 5 \end{bmatrix}$$

The inverse is

$$\mathbf{\Sigma}^{-1} = \frac{1}{\sigma^2}\begin{bmatrix} 2 & -1 & 0 & 0 & 0 \\ -1 & 2 & -1 & 0 & 0 \\ 0 & -1 & 2 & -1 & 0 \\ 0 & 0 & -1 & 2 & -1 \\ 0 & 0 & 0 & -1 & 1 \end{bmatrix}$$

(a) Show that the GLSEs of α and β are

$$\hat{\alpha} = \frac{5y_1 - y_5}{4}$$

$$\hat{\beta} = \frac{y_5 - y_1}{4}$$

(b) Verify that the increments

$$e_t - e_{t-1} \qquad \text{and} \qquad e_s - e_{s-1}$$

are uncorrelated for all unequal indices t and s.

(c) Show that ordinary least squares estimation applied to y_1 and the successive differences

$$y_t - y_{t-1} = \beta + e_t - e_{t-1} \qquad t = 2, \ldots, 5$$

gives the GLSEs of (a).

(d) Extend the estimators to a linear model with N equally spaced points and the $N \times N$ generalization of $\mathbf{\Sigma}$. Find the estimators by the transformation of (c) and OLSE.

chapter 7

Some Further Topics in Regression Analysis

7-1 INTRODUCTION

In this chapter we treat a few additional methods for the ordinary least squares model. In Section 7-2 we show how qualitative or categorical predictors can be coded numerically. In Section 7-3 we see how dependent variables that assume those coded values can be used to develop rules for classifying individuals into one of two groups based on their predictors. We describe principal components analysis in Section 7-4 as a means of dissecting the covariance and correlation structure of independent variable values. In Section 7-5 we investigate the problem of highly correlated or linearly related predictors, their effects on least squares estimators, and ways for dealing with such collinear observations.

7-2 DUMMY VARIABLES

We have generally assumed that the independent variables in the preceding least squares models were continuous, or perhaps scores assigned to categories ordered from "best" to "worst." Frequently, we wish to incorporate categorical variables such as sex, ethnic status, or the membership in a demographic, political, or social group. We may represent those nominal characteristics by *dummy*, or *indicator*, variables. A dummy variable assumes one value for membership in one category and another for all individuals not in the category. Usually, the two values assigned to membership and nonmembership are 1 and 0, although those scores are an arbitrary choice. For example, the dummy variable X used to denote sex in a study relating salary to gender and other factors might take on the values

$$X = \begin{cases} 1 & \text{if the respondent is female} \\ 0 & \text{if male} \end{cases} \tag{1}$$

If the variable has k categories its nominal values can be described by $k - 1$ dummy variables:

$$X_1 = \begin{cases} 1 & \text{if the sampling unit is in category 1} \\ 0 & \text{otherwise} \end{cases}$$

$$\vdots$$

$$X_{k-1} = \begin{cases} 1 & \text{if the sampling unit is in category } k - 1 \\ 0 & \text{otherwise} \end{cases}$$

Membership in the kth category would be indicated by zero values for all $k - 1$ dummy variables. If we had used a kth dummy variable to indicate a member of the last category, we would have a data matrix of less than full rank k, for every unit value of X_k would occur with the $k - 1$ zeros of the other variables. As an example of several dummy variables representing categories we might recall Exercise 3 of Section 2-9: Employees' salaries were related to gender, length of service, and educational level. As an expedient we coded the last variable by the integers 1 through 5 as a crude measure of formal education. Alternatively, we might have used these four indicator variables:

$$X_1 = \begin{cases} 1 & \text{high school graduate} \\ 0 & \text{otherwise} \end{cases}$$

$$X_2 = \begin{cases} 1 & \text{some college} \\ 0 & \text{otherwise} \end{cases}$$

$$X_3 = \begin{cases} 1 & \text{baccalaureate degree} \\ 0 & \text{otherwise} \end{cases}$$

$$X_4 = \begin{cases} 1 & \text{Master's degree} \\ 0 & \text{otherwise} \end{cases}$$

From the insignificant regression coefficient of the original educational level measure it seems unlikely that the new dummy variables would have a greatly different effect on the proportion of explained salary variation.

Example 7-1

During World War II lobster was not rationed, although demand for seafoods increased. Let us represent that period by the dummy variable

$$D_t = \begin{cases} 1 & \text{values of } t \text{ corresponding to 1942–1945} \\ 0 & \text{other years} \end{cases}$$

and include the variable in the second regression model of Section 4-3. These regression coefficients and other statistics were obtained:

Variable	$\hat{\beta}_j$	t_j
Linear	38.5815	5.78
Quadratic	−10.3075	−16.55
Cubic	−0.1366	−6.92
Fishermen	0.5935	5.33
Sea temperature	534.5703	5.41
1942–1945 dummy	410.8615	1.24

$$\hat{\sigma}^2 = 286{,}974 \qquad R^2 = 0.9668$$

Although the coefficient for the dummy variable is positive, its t-statistic does not exceed any conventional critical value. The increase in the proportion of explained variation is only about 0.0014. In the interest of simplicity in our model we would not include the dummy variable, although we might mention in a discussion of the analysis that the variable had a positive, if not significant, coefficient.

7-3 CLASSIFICATION BY THE LINEAR DISCRIMINANT FUNCTION

In Section 7-3 we provided for categorical independent variables by coding their categories with dummy variables. In many applications of statistical analysis we also wish to allow for categorical dependent variables, in the sense of predicting their membership in one or more classes from the values of several independent variables. For example, a botanist might wish to use petal and sepal dimensions of flowers to decide in some optimal fashion the species of the plant. We assume, of course, that data from identified species have been collected on the dimensions. One or more indices might be computed from the measurements of an unclassified plant and the plant assigned to a species according a rule based on the values of the indices. Similarly, a financial officer of a government agency or contracting firm might wish to classify potential vendors or subcontractors as high or low bankruptcy risks based on accounting information applied by the firms in their credit applications or contract proposals. The classification would be made in terms of a weighted sum of financial ratios or cash flow measures, where the weights would be determined by some kind of least squares criterion.

One classification index for making such decisions is the *linear discriminant function*. The measure was proposed by Fisher (1936) in answer to the request of a botanist for a numerical method for assigning iris flowers to one of three species. Fisher defined the coefficients of the linear function for two species as those that maximized the distances between the averages of the index in the groups, relative to the within-group variance of the index. Because our orientation is that of least squares and linear regression analysis we shall take a different motivation. We shall use the ordinary least squares model, in which the independent variable values are the characteristics of the unit being classified, and the dependent variable is a dummy variable indicating membership in the first or second group. We shall only consider the case of two distinct groups, although discrimination and classification has been extended by Fisher and others to the case of several populations.

Let us begin by assuming that we have drawn N_1 independent plants, persons, or other sampling units from the first population and have recorded the values of p dimensions, test scores, or measurements for each unit. In the regression context those are the independent variable observations, and we shall denote their values by the $N_1 \times p$ matrix $\mathbf{U}_1$. Similarly, the N_2 observation vectors from the second population will be written as the $N_2 \times p$ matrix $\mathbf{U}_2$. We shall denote the $p \times 1$ mean vectors from the two groups by $\bar{\mathbf{u}}_1$ and $\bar{\mathbf{u}}_2$ and the grand mean vector by

$$\bar{\mathbf{u}} = \frac{N_1\bar{\mathbf{u}}_1 + N_2\bar{\mathbf{u}}_2}{N_1 + N_2} \tag{1}$$

We shall indicate group membership by this dummy dependent variable:

$$y_i = \begin{cases} \dfrac{N_2}{N_1 + N_2} & i = 1, \ldots, N_1 \text{ (group 1)} \\ \dfrac{-N_1}{N_1 + N_2} & i = N_1 + 1, \ldots, N_1 + N_2 \text{ (group 2)} \end{cases} \tag{2}$$

These dummy scores were chosen over the ones and zeros of Section 7-2 for their mean of zero.

The matrix $\mathbf{X}_1'\mathbf{X}_1$ of sums of squares and products of the independent variables centered about their grand means is

$$\begin{aligned} \mathbf{X}_1'\mathbf{X}_1 &= \mathbf{U}_1'\mathbf{U}_1 + \mathbf{U}_2'\mathbf{U}_2 - (N_1 + N_2)(\bar{\mathbf{u}}\bar{\mathbf{u}}') \\ &= \mathbf{A}_1 + \mathbf{A}_2 + \frac{N_1N_2}{N_1 + N_2}(\bar{\mathbf{u}}_1 - \bar{\mathbf{u}}_2)(\bar{\mathbf{u}}_1 - \bar{\mathbf{u}}_2)' \end{aligned} \tag{3}$$

where

$$\mathbf{A}_1 = \mathbf{U}_1'\mathbf{U}_1 - N_1\bar{\mathbf{u}}_1\bar{\mathbf{u}}_1' \qquad \mathbf{A}_2 = \mathbf{U}_2'\mathbf{U}_2 - N_2\bar{\mathbf{u}}_2\bar{\mathbf{u}}_2' \tag{4}$$

are the sums of squares and products matrices computed within each group. If we let $\mathbf{A} = \mathbf{A}_1 + \mathbf{A}_2$ and apply Bartlett's formula (9), Section B-3, for the inverse of the resulting matrix,

$$(\mathbf{X}_1'\mathbf{X}_1)^{-1} = \mathbf{A}^{-1} - \frac{\dfrac{N_1N_2}{N_1 + N_2}\mathbf{A}^{-1}(\bar{\mathbf{u}}_1 - \bar{\mathbf{u}}_2)(\bar{\mathbf{u}}_1 - \bar{\mathbf{u}}_2)'\mathbf{A}^{-1}}{1 + \dfrac{N_1N_2}{N_1 + N_2}(\bar{\mathbf{u}}_1 - \bar{\mathbf{u}}_2)'\mathbf{A}^{-1}(\bar{\mathbf{u}}_1 - \bar{\mathbf{u}}_2)} \tag{5}$$

Also,

$$\mathbf{X}_1'\mathbf{y} = \frac{N_1N_2}{N_1 + N_2}(\bar{\mathbf{u}}_1 - \bar{\mathbf{u}}_2) \tag{6}$$

Now we are ready to compute the estimated regression coefficients for the prediction of the dummy variable values from the independent variable values. The vector of coefficients is

$$\begin{aligned} \hat{\boldsymbol{\beta}}_1 &= (\mathbf{X}_1'\mathbf{X}_1)^{-1}\mathbf{X}_1\mathbf{y} \\ &= \mathbf{A}^{-1}(\bar{\mathbf{u}}_1 - \bar{\mathbf{u}}_2)\frac{N_1N_2/(N_1 + N_2)}{1 + [N_1N_2/(N_1 + N_2)](\bar{\mathbf{u}}_1 - \bar{\mathbf{u}}_2)'\mathbf{A}^{-1}(\bar{\mathbf{u}}_1 - \bar{\mathbf{u}}_2)} \end{aligned} \tag{7}$$

We shall presently see that the constant factor can be dropped.

The least squares prediction of the dummy variable value for the ith individual's observations $\mathbf{x}_i$ is

$$y_i = \hat{\boldsymbol{\beta}}_1'(\mathbf{x}_i - \bar{\mathbf{u}}) \tag{8}$$

The means of the predicted values for the first and second groups are

$$\hat{\boldsymbol{\beta}}_1'(\bar{\mathbf{u}}_1 - \bar{\mathbf{u}}) \qquad \hat{\boldsymbol{\beta}}_1'(\bar{\mathbf{u}}_2 - \bar{\mathbf{u}}) \tag{9}$$

Sampling units should be classified as coming from the population whose mean is closer to the predicted value (8), or according to this rule:

Assign $\mathbf{x}_i$ to population 1 if

$$\hat{\boldsymbol{\beta}}_1'(\mathbf{x}_i - \bar{\mathbf{u}}) > \hat{\boldsymbol{\beta}}_1'[\tfrac{1}{2}(\bar{\mathbf{u}}_1 + \bar{\mathbf{u}}_2) - \bar{\mathbf{u}}] \tag{10}$$

and to population 2 otherwise

We note that the inequality is equivalent to

$$\hat{\boldsymbol{\beta}}_1'\mathbf{x}_i > \frac{\hat{\boldsymbol{\beta}}_1'(\bar{\mathbf{u}}_1 + \bar{\mathbf{u}}_2)}{2} \tag{11}$$

so that the regression coefficients can be scaled by any positive constant without affecting the assignment rule. We adopt a form of the coefficients common to multivariate analysis. The constant factor in (7) will be omitted and the matrix $\mathbf{A}$ replaced by the within-groups sample covariance matrix

$$\mathbf{S} = \frac{1}{N_1 + N_2 - 2}\mathbf{A} \tag{12}$$

Then the elements of the resulting vector

$$\mathbf{a} = \mathbf{S}^{-1}(\bar{\mathbf{u}}_1 - \bar{\mathbf{u}}_2) \tag{13}$$

will be referred to as the *linear discriminant function coefficients*, and the index

$$w = \mathbf{a}'\mathbf{x} = (\bar{\mathbf{u}}_1 - \bar{\mathbf{u}}_2)'\mathbf{S}^{-1}\mathbf{x} \tag{14}$$

as the *linear discriminant function*. The midpoint between the averages of w for the two samples of observations is

$$\frac{(\bar{\mathbf{u}}_1 - \bar{\mathbf{u}}_2)'\mathbf{S}^{-1}(\bar{\mathbf{u}}_1 + \bar{\mathbf{u}}_2)}{2} \tag{15}$$

The rule (10) becomes

Assign $\mathbf{x}_i$ to population 1 if

$$\mathbf{a}'\mathbf{x}_i > \frac{(\bar{\mathbf{u}}_1 - \bar{\mathbf{u}}_2)'\mathbf{S}^{-1}(\bar{\mathbf{u}}_1 + \bar{\mathbf{u}}_2)}{2} \tag{16}$$

and otherwise to population 2

Example 7-2

Let us find the linear discriminant function and a classification rule for two groups and a pair of predictor variables. The groups are samples of male and female management-level employees of a business firm. The predictors are annual salary in thousands of dollars and years of service with the firm. The data were adapted to preserve confidentiality; the actual observations will be omitted to conserve space. The statistics for the two groups are shown in Table 7-1. The linear discriminant function is

$$w = 0.03x_1 + 0.47x_2$$

TABLE 7-1 Salary and Service Characteristics

	Women ($N_1 = 16$)		Men ($N_2 = 36$)	
	Salary (thousands of dollars)	Service (years)	Salary (thousands of dollars)	Service (years)
Mean	18.3125	2.375	19.6944	3.806
Sums of squares and products matrix	$\begin{bmatrix} 33.4375 & 1.125 \\ 1.125 & 31.75 \end{bmatrix}$		$\begin{bmatrix} 109.6389 & 119.8611 \\ 119.8611 & 359.6389 \end{bmatrix}$	

Within-groups sample covariance matrix:

$$\mathbf{S} = \begin{bmatrix} 2.8615 & 2.4197 \\ 2.4197 & 7.8278 \end{bmatrix}$$

The average values for the discriminant index for the female and male observations are

$$\bar{u}_1 = 1.673 \qquad \bar{u}_2 = 2.393$$

and we would adopt the decision rule

Classify an employee as female if

$$w \leq 2.033$$

and as male if $w > 2.033$

Computation of the discriminant scores for each employee gave these results for the classification rule:

Classification rule	Actual Employee Groups: Female	Male	Total
Female	13	22	35
Male	3	14	17
Total	16	36	52

The linear discriminant function performed rather badly. Twenty-five individuals were misclassified, and in fact the sample sizes were nearly reversed. The fault lies in the years of service variable: Men with three or less years tend to be misclassified as women. In fact, if we dropped that variable and merely used salary as a criterion these results would obtain:

Discriminant rule	Actual: Female	Male	Total
Female	11	16	27
Male	5	20	25
Total	16	36	52

The total number of misclassifications has diminished, although more women are misclassified as men.

It is important to note that the error rates found by classifying the original observations have a severe downward bias. The estimation of misclassification probabilities has many aspects and has been investigated intensively. A survey of the problem has been given by Lachenbruch (1975, pp. 29–36).

Statistical significance of the linear discriminant function. We developed the linear discriminant coefficients by an application of the ordinary least squares model to a dependent variable assuming dummy values. The estimators of the regression coefficients are valid as long as the disturbance terms are uncorrelated random variables with mean zero and a constant variance. With the possible exception of the common variance condition, these assumptions were not unreasonable. However, their use for constructing hypothesis tests on the regression coefficients seems rather doubtful, at least in small samples. Nevertheless, we shall find the usual regression and error sums of squares, and from them the F-statistics for testing

$$H_0: \boldsymbol{\beta}_1 = \mathbf{0} \tag{17}$$

Remarkably, the statistic has the F-distribution when H_0 is true in spite of the discrete nature of the dependent variable.

The usual regression sum of squares for the null hypothesis can be computed from expressions (5) and (6). It is

$$\begin{aligned}\text{SSR} &= \mathbf{y}'\mathbf{X}_1(\mathbf{X}_1'\mathbf{X}_1)^{-1}\mathbf{X}_1'\mathbf{y} \\ &= \frac{\left(\dfrac{N_1 N_2}{N_1 + N_2}\right)^2 (\bar{\mathbf{u}}_1 - \bar{\mathbf{u}}_2)'\mathbf{A}^{-1}(\bar{\mathbf{u}}_1 - \bar{\mathbf{u}}_2)}{1 + \left(\dfrac{N_1 N_2}{N_1 + N_2}\right)(\bar{\mathbf{u}}_1 - \bar{\mathbf{u}}_2)'\mathbf{A}^{-1}(\bar{\mathbf{u}}_1 - \bar{\mathbf{u}}_2)}\end{aligned} \tag{18}$$

The total sum of squares of the dependent variable is

$$\begin{aligned}\sum_{i=1}^{N_1+N_2} y_i^2 &= N_1\left(\frac{N_2}{N_1 + N_2}\right)^2 + N_2\left(\frac{N_1}{N_1 + N_2}\right)^2 \\ &= \frac{N_1 N_2}{N_1 + N_2}\end{aligned} \tag{19}$$

We recall that $\bar{y} = 0$ from the choice of the dummy scores (2). Then by subtraction,

$$\begin{aligned}\text{SSE} &= \frac{N_1 N_2}{N_1 + N_2} - \text{SSR} \\ &= \frac{N_1 N_2}{N_1 + N_2} \frac{1}{1 + [N_1 N_2/(N_1 + N_2)](\bar{\mathbf{u}}_1 - \bar{\mathbf{u}}_2)'\mathbf{A}^{-1}(\bar{\mathbf{u}}_1 - \bar{\mathbf{u}}_2)}\end{aligned} \tag{20}$$

The test statistic is the mean square ratio

$$\begin{aligned}F &= \frac{\text{SSR}}{p} \bigg/ \frac{\text{SSE}}{N - p - 1} \\ &= \frac{(N_1 + N_2 - p - 1)N_1 N_2}{p(N_1 + N_2)}(\bar{\mathbf{u}}_1 - \bar{\mathbf{u}}_2)'\mathbf{A}^{-1}(\bar{\mathbf{u}}_1 - \bar{\mathbf{u}}_2)\end{aligned} \tag{21}$$

If the hypothesis of zero regression coefficients is true, F has the F-distribution with p and $N_1 + N_2 - p - 1$ degrees of freedom. For a test at the α level we should reject the null hypothesis if

$$F > F_{\alpha;\, p, N_1+N_2-p-1} \tag{22}$$

The distribution does not follow from the ordinary least squares developments in Chapters 2 and 3, although a certain geometrical duality of the two models leads to the same distribution under the null hypothesis. That duality has been described by Kendall (1957, pp. 161–162).

Let us recast the hypothesis (17) of no relation of the dependent variables to the dummy scores in a form more directly interpretable through the parameters of two p-dimensional multinormal distributions with respective mean vectors $\boldsymbol{\mu}_1$ and $\boldsymbol{\mu}_2$ and a common covariance matrix $\boldsymbol{\Sigma}$. The hypothesis is equivalent to

$$H_0 : \boldsymbol{\mu}_1 = \boldsymbol{\mu}_2 \tag{23}$$

of equal mean vectors. The test statistic is

$$T^2 = \frac{N_1 N_2}{N_1 + N_2}(\bar{\mathbf{u}}_1 - \bar{\mathbf{u}}_2)'\mathbf{S}^{-1}(\bar{\mathbf{u}}_1 - \bar{\mathbf{u}}_2) \tag{24}$$

where $\bar{\mathbf{u}}_1$ and $\bar{\mathbf{u}}_2$ are the sample mean vectors defined before, and $\mathbf{S}$ is the sample covariance matrix (12). Then

$$F = \frac{N_1 + N_2 - p - 1}{p(N_1 + N_2 - 2)} T^2 \tag{25}$$

is identical to (21) and has the F-distribution with the same degrees of freedom $p, N_1 + N_2 - p - 1$ when (23) is true. If instead the alternative of unequal mean vectors holds, the distribution is noncentral F with noncentrality parameter

$$\delta^2 = \frac{N_1 N_2}{N_1 + N_2}(\boldsymbol{\mu}_1 - \boldsymbol{\mu}_2)'\boldsymbol{\Sigma}^{-1}(\boldsymbol{\mu}_1 - \boldsymbol{\mu}_2) \tag{26}$$

Derivations of the T^2-distribution have been given by Anderson (1958) and Rao (1965); numerous applications to multivariate data have been described by Morrison (1976).

The linear discriminant function (14) has another interesting interpretation under the multinormal model. It is the linear compound that has the greatest two-sample t-statistic for the two groups. The discriminant function gives the greatest separation of the two samples relative to the within-sample variation. This was the original criterion Fisher (1936) used to derive linear discrimination.

Example 7-3

The T^2-statistic for the salary and years of service data in Example 7-2 is equal to 7.98, and its associated F-statistic is 3.91. Because F exceeds the 0.05 critical value with 2 and 49 degrees of freedom we should reject the hypothesis of common salary and service means for men and women and conclude that the samples are sufficiently separated to justify the use of the discriminant rule for classification. The significant T^2 only indicates that classification is feasible: The large number of misclassified persons in the 2×2 table of Example 7-2 indicates that the samples still overlap so much that the linear rule has little practical value.

A large literature exists on discrimination, classification, and inferences about multivariate mean vectors; Lachenbruch (1975) has given a comprehensive overview of the subject. A theoretical development by likelihood ratios that includes prior probabilities for the populations and costs of misclassification has been given by Anderson (1958). Classification into several groups and its connection with the multivariate analysis of variance has been described by Morrison (1976).

7-4 PRINCIPAL COMPONENTS ANALYSIS

Our development of least squares models has largely assumed that the independent variables were carefully chosen as a fixed set of predictors for the dependent variable. Exceptions were the treatment of variable selection methods in Chapter 5 and the emphasis in Chapters 2 and 3 on dropping variables whose regression coefficients were not significantly different from zero. We also used partial correlation in Section 2-6 to dissect the interrelations of several variables that might form a predictor set. In this section we describe a method for analyzing the correlation structure of a set of variables without assigning dependent or fixed roles to any variables.

Let us assume that we have collected N observations on p random variables with some kind of multidimensional distribution. The distribution need not be multinormal, but we shall require its mean vector and covariance matrix to be defined, that is, to have finite elements. The N observation vectors can be represented as a scatterplot in p-dimensional space. The shape of the plot should be generally ellipsoidal, and indeed it will be if the population is multinormal. Such a plot is shown in Figure 7-1. We observe that the points have a long axis denoted by W_1 and two shorter axes W_2 and W_3. In a sense those axes are the "natural" coordinate system for the sample of points, just as we would prefer to translate and rotate an ellipsoid into a standard location and orientation. To make that rotation we would first find the positions of the principal axes, or equivalently, their angles with the original axes. We proceed similarly with a data set, except that distances along a dimension are defined in terms of variance, so that the longest axis would be in the direction of greatest variation, the second longest would be in the direction of greatest variance perpendicular to the first new axis, and so on, for the complete rotation.

The principal axis rotation can also be used with data sets of less than full rank p. If the rank is r, or the data matrix contains only r linearly independent columns, the variation in the points can be explained by r new principal axes. In that case the rotation provides a rational rule for choosing a new coordinate system with as many dimensions as the number of original independent variables. But what of the case in which the last few dimensions account for very small

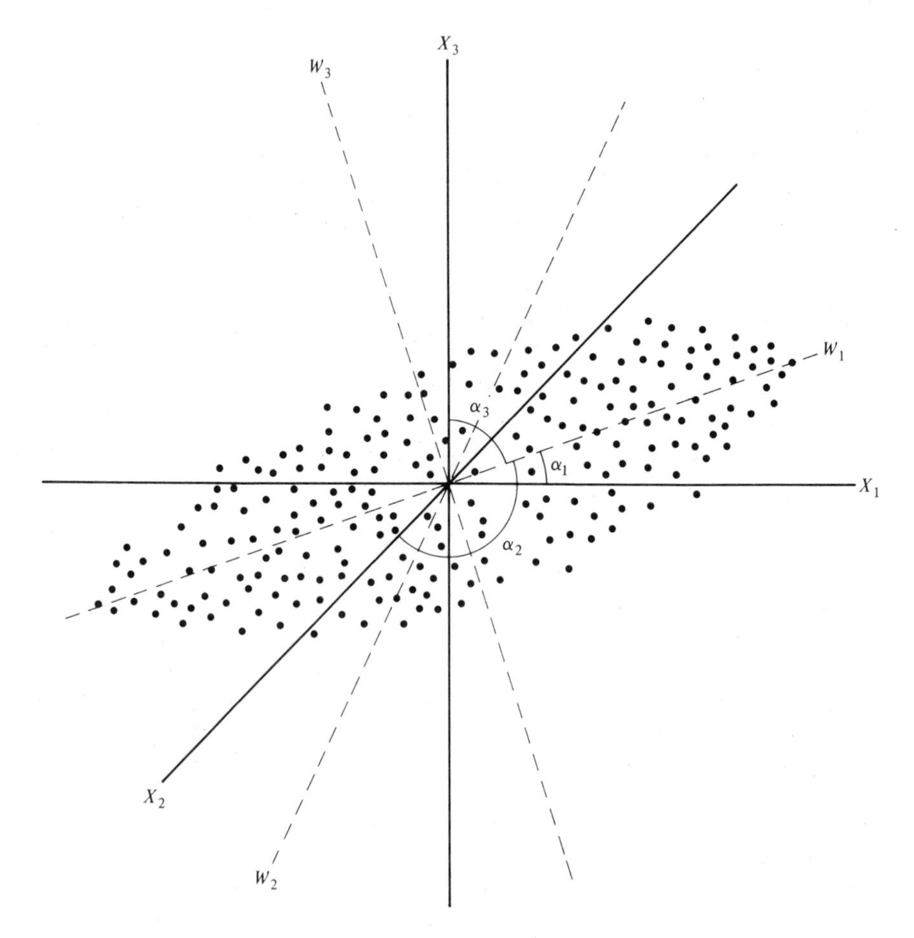

Figure 7-1 Observations on three variates and their principal axes.

proportions of variation? Should those axes be ignored? By dropping derived variables that have relatively little variance we may reduce the data to a simpler set of scores. We have found not only the principal axes of the complex but have also achieved parsimony in the number of variables.

Now let us find the first principal axis of the data points. The axis is defined by its angles $\alpha_1, \alpha_2, \ldots, \alpha_p$ with the original coordinate axes, or equivalently by the direction cosines

$$a_{11} = \cos \alpha_{11}, \ldots, a_{p1} = \cos \alpha_{p1} \tag{1}$$

of the angles. Direction cosines must satisfy the constraint

$$\sum_{j=1}^{p} a_{j1}^2 = 1 \tag{2}$$

If we take the origin of the coordinates at the means $\bar{\mathbf{x}}' = [\bar{x}_1, \ldots, \bar{x}_p]$ of the observations, the value of the ith observation vector on the first principal axis W_1 is

$$w_{i1} = a_{11}(x_{i1} - \bar{x}_1) + \cdots + a_{p1}(x_{ip} - \bar{x}_p) \tag{3}$$

We note immediately that the component values have a zero mean:

$$\bar{w}_1 = \frac{1}{N}\sum_{i=1}^{N} w_{i1} = a_{11}\sum_{i=1}^{N}(x_{i1} - \bar{x}_1) + \cdots + a_{p1}\sum_{i=1}^{N}(x_{ip} - \bar{x}_p) = 0 \tag{4}$$

To find the direction cosine vector $\mathbf{a}_1' = [a_{11}, \ldots, a_{p1}]$ that maximizes the variance of the w_{i1} it will be convenient to use matrix notation. The $p \times p$ sample covariance matrix of the original observations was introduced in expression (26) of Section 2-5; it is

$$\mathbf{S} = \frac{1}{N-1}\sum_{i=1}^{N}(\mathbf{x}_i - \bar{\mathbf{x}})(\mathbf{x}_i - \bar{\mathbf{x}})' \tag{5}$$

and from it the sample variance of the w_{i1} can be expressed as

$$\begin{aligned} s_{w_1}^2 &= \frac{1}{N-1}\sum_{i=1}^{N} w_{i1}^2 \\ &= \frac{1}{N-1}\sum_{i=1}^{N}[\mathbf{a}_1'(\mathbf{x}_i - \bar{\mathbf{x}})]^2 \\ &= \mathbf{a}_1'\mathbf{S}\mathbf{a}_1 \end{aligned} \tag{6}$$

For the maximization of the variance we introduce the constraint (2) by the Lagrange multiplier λ and form the Lagrangian function

$$f(\mathbf{a}_1, \lambda) = \mathbf{a}_1'\mathbf{S}\mathbf{a}_1 + \lambda(1 - \mathbf{a}_1'\mathbf{a}_1) \tag{7}$$

Then the cosine vector $\mathbf{a}_1$ that maximizes the function must satisfy

$$\frac{\partial f(\mathbf{a}_1, \lambda)}{\partial \mathbf{a}_1} = 2\mathbf{S}\mathbf{a}_1 - 2\lambda\mathbf{a}_1 = \mathbf{0} \tag{8}$$

or

$$[\mathbf{S} - \lambda\mathbf{I}]\mathbf{a}_1 = \mathbf{0} \tag{9}$$

For these homogeneous equations to have a nontrivial solution it is necessary that λ satisfy the determinantal equation

$$|\mathbf{S} - \lambda\mathbf{I}| = 0 \tag{10}$$

or that λ be a characteristic root of $\mathbf{S}$. Premultiplication of (9) by $\mathbf{a}_1'$ shows that

$$\begin{aligned} \lambda &= \frac{\mathbf{a}_1'\mathbf{S}\mathbf{a}_1}{\mathbf{a}_1'\mathbf{a}_1} \\ &= \mathbf{a}_1'\mathbf{S}\mathbf{a}_1 \\ &= s_{w_1}^2 \end{aligned} \tag{11}$$

or the sample variance of the first principal variable. λ must be the *largest* characteristic root of $\mathbf{S}$, and the direction cosines are the elements of its normalized characteristic vector. Similarly, the successive principal axes are defined by the characteristic vectors corresponding to the second, third, . . . , characteristic roots of $\mathbf{S}$. Determination of the principal axes of a sample of multivariate observations, or the transformation to principal components, is merely equivalent to finding the characteristic root and vector structure of the covariance matrix.

Sample principal components have a number of useful properties. The variances $\lambda_1, \ldots, \lambda_p$ of the principal variables sum to the total of the variances of the original ones, or

$$\lambda_1 + \cdots + \lambda_p = \text{tr}\,\mathbf{S} \tag{12}$$

The proportion of total variance explained by the jth component is

$$\frac{\lambda_j}{\text{tr}\,\mathbf{S}} \tag{13}$$

If the first four or five principal variables account for a large proportion of the total variance (for example, 80 or 90%), we might choose to work with those derived measures as independent variables rather than a larger set of the original observations.

If the variances of two principal variables are different their axes must be perpendicular. Multiply the equations

$$\begin{aligned} [\mathbf{S} - \lambda_i \mathbf{I}]\mathbf{a}_i &= \mathbf{0} \\ [\mathbf{S} - \lambda_j \mathbf{I}]\mathbf{a}_j &= \mathbf{0} \end{aligned} \tag{14}$$

defining the ith and jth component coefficients by $\mathbf{a}'_j$ and $\mathbf{a}'_i$, respectively. Then

$$\begin{aligned} \lambda_i \mathbf{a}'_j \mathbf{a}_i &= \mathbf{a}'_j \mathbf{S} \mathbf{a}_i \\ \lambda_j \mathbf{a}'_i \mathbf{a}_j &= \mathbf{a}'_i \mathbf{S} \mathbf{a}_j \end{aligned} \tag{15}$$

Since the right-hand sides are the same the two equations can hold for distinct λ_i, λ_j only if

$$\mathbf{a}'_i \mathbf{a}_j = 0 \tag{16}$$

or if the components are orthogonal. At the same time, because the sample covariance of $w_i = \mathbf{a}'_i \mathbf{X}$ and $w_j = \mathbf{a}'_j \mathbf{X}$ is $\mathbf{a}'_i \mathbf{S} \mathbf{a}_j$, the ith and jth component scores are uncorrelated. If some of the characteristic roots are equal, we can still arbitrarily choose their component axes to be orthogonal, so that their derived scores are uncorrelated. Principal component analysis has enabled us to transform the observed variables into uncorrelated linear combinations that explain progressively smaller amounts of the total variance in the data set.

To apply principal component analysis one must be able to give meanings and names to the individual component variables. As the direction cosine of a given component with one of the original axes approaches one the component and observed variable tend to be coincident: The nature of the component is very close to that of the variable. At the same time the normalization constraint (2) assures that the other variables will be nearly orthogonal, or unrelated, to the particular component. To name our derived components we should look for large elements in their coefficient vectors. We might also measure the association in terms of the correlation coefficients of the given component and the original variables. The sample covariances of the jth component w_j and the p original variables can be written as the vector

$$\frac{1}{N-1} \sum_{i=1}^{N} (\mathbf{x}_i - \bar{\mathbf{x}}) w_{ij} = \frac{1}{N-1} \sum_{i=1}^{N} (\mathbf{x}_i - \bar{\mathbf{x}})(\mathbf{x}_i - \bar{\mathbf{x}})' a_j = \mathbf{S}\mathbf{a}_j = \lambda_j \mathbf{a}_j \tag{17}$$

To find the correlations we divide the successive elements by the standard deviation $\sqrt{\lambda_j}$ of the jth component scores and the standard deviation of s_h of the appropriate original variable. The vector of correlations is

$$\begin{bmatrix} \sqrt{\lambda_j}\, a_{1j}/s_1 \\ \cdot \\ \cdot \\ \cdot \\ \sqrt{\lambda_j}\, a_{pj}/s_p \end{bmatrix} \tag{18}$$

If a correlation is large in absolute value we would associate the jth component with that original variable. Such correlations are called the *loadings* of the original variables on the jth component.

Example 7-4

We shall extract the principal components from variables x_1, x_2, x_3, x_4 of the projective psychological test scores in Table 2-5. The variances and covariances of those observations are given by the last four rows and columns of the matrix **S** in the display (33), Section 2-5. The same display contains the correlation matrix of the variables: The large absolute values of most correlations indicate that the scatter swarm of the data points will have an elongated shape. The differences among the correlations assure us that the configuration of points will have some well-defined principal axes. The direction cosines, characteristic roots, and loading correlations of the four components are shown in Table 7-2. The first component accounts for more than 80% of the total variance of the scores, the second 16.5%, and the third and fourth only miniscule amounts. The first component has an orientation closest to the aspiration level and total performance ratings axes. The component is highly correlated with the first variable and has an equally large negative correlation with the total ratings. If we ignore the second and third variables the first component appears to be a comparison of aspiration level and total ratings. Conversely, the second component is largely a weighted average of those variables, with a slight decrement due to the third homonym statements variable. The third and fourth components have axes virtually coincident with those of variables 3 and 2, respectively. However, because of their small variances the last two components are of little practical import.

TABLE 7-2 Principal Components of the Test Scores

	Component							
	Coefficient				Correlation			
	1	2	3	4	1	2	3	4
1. Aspiration level	0.70	0.70	0.12	−0.04	0.91	0.41	0.03	0.00
2. Proverbs rating	−0.07	0.14	−0.03	0.99	0.60	0.54	−0.05	0.59
3. Homonyms statements	0.28	−0.41	0.86	0.10	0.76	−0.51	0.40	0.02
4. Total performance ratings	−0.66	0.56	0.49	−0.11	−0.93	0.36	0.12	−0.01
Characteristic roots	452.94	92.54	13.14	2.16				
Percentage of variance	80.77	16.50	2.34	0.39				
Cumulative variance percentage	80.77	97.27	99.61	100.00				

We might find the regression function of the original dependent variable and the first two principal components. The sample covariances of Y, Draw a Person score and the first two component scores are

$$s_{Y,w_1} = 0.70s_{Y,X_1} - 0.07s_{Y,X_2} + 0.28s_{Y,X_3} - 0.66s_{Y,X_4} = 122.91$$
$$s_{Y,w_2} = 0.70s_{Y,X_1} + 0.14s_{Y,X_2} - 0.41s_{Y,X_3} + 0.56s_{Y,X_4} = -28.27$$

The sample covariance matrix of Y and the components is

$$\begin{bmatrix} 94.48 & 122.91 & -28.27 \\ 122.91 & 452.94 & 0 \\ -28.27 & 0 & 92.54 \end{bmatrix}$$

The regression coefficient vector is

$$\hat{\boldsymbol{\beta}}'_1 = [0.2714, \quad -0.3055]$$

and the coefficient of determination is $R^2 = 0.4444$. The explained variance has declined by about 9.5% when the first two component scores are used as predictors.

A formal treatment of regression analysis based on the principal components of the independent variables has been given by Massy (1965).

Because principal components were chosen to explain the variance in a set of multivariate observations their coefficients depend strongly on the units and variances of the original variables. In the extreme case of uncorrelated variates the components are the variates themselves in the order of descending variances. The scatter ellipses of such variates are shown in Figure 7-2. The first

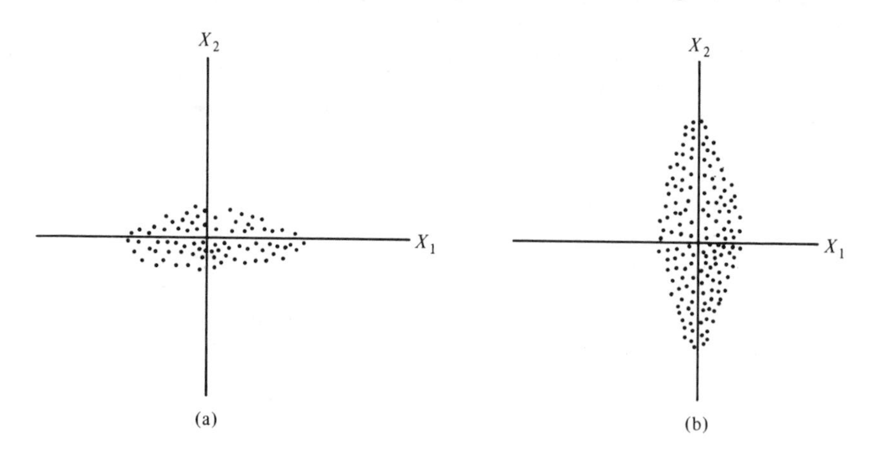

Figure 7-2 Principal axes of uncorrelated observations.

principal axes are clearly X_1 and X_2 in (a) and (b), respectively, but that knowledge is not especially helpful in reducing the number of variates in the systems. The effect of different variances carries over to correlated variables: Their principal components will be drawn in the directions with greatest variances. If our principal interest is in the way the variates are related through their components we might make transformations to equal or nearly equal variances before extracting the component vectors. The simplest transformation is that of standard scores:

$$z_{ij} = \frac{x_{ij} - \bar{x}_j}{s_j} \tag{19}$$

Since the covariance matrix of the scores is the sample correlation matrix of the original observations, principal component analysis is equivalent to extracting the characteristic roots and vectors from the correlation matrix. Alternatively, one might make the variances of highly skewed variates more equal by transforming to logarithms and taking components from the covariance matrix of the new variables.

Inferences about components. We have developed principal component analysis purely as a descriptive method for determining if a smaller number of linear functions of the variables will explain most of the system's variance. We have not addressed principal components of the population covariance or correlation matrix, or hypothesis tests and confidence intervals for them. The large-sample distributional properties of principal components have been found by Anderson (1963) and others. In particular, it is possible to test the hypothesis that the variance explained by the jth component has some specified value, that some component vector has a given set of elements, or that the variances of a succession of components are equal. Those tests and examples of their application to life and behavioral sciences data have been described by Morrison (1976). Unfortunately, other inferences we might wish to make about component parameters may lead to statistics with complicated and untabulated distributions or may involve other population components and variances. That is especially so for tests involving the components of correlation matrices.

Factor analysis. We have seen that principal component analysis consists of a rotation of the coordinate system of our variables into new axes through the directions of greatest variation. Our motivation was the explanation of total variance, although for the analysis of a dependence structure it would seem that we should try to reproduce the covariances or correlations instead. That is the purpose of the *factor analysis* of a set of variables. We postulate that the $p \times 1$ observable random vector $\mathbf{X}$ can be expressed as the linear function

$$\mathbf{X} = \boldsymbol{\mu} + \boldsymbol{\Lambda}\mathbf{W} + \mathbf{e} \tag{20}$$

of its mean vector $\boldsymbol{\mu}$, the unobservable $m \times 1$ *common factor variate* vector $\mathbf{W}$, and the vector of *specific factor variates* $\mathbf{e}$. $\boldsymbol{\Lambda}$ is a $p \times m$ matrix of coefficients called *factor loading* parameters. The common factors are uncorrelated, or

$$\text{Cov}(\mathbf{W}, \mathbf{W}') = \mathbf{I} \tag{21}$$

and the specific variates are uncorrelated, but with variances given by the diagonal elements of the matrix

$$\text{Cov}(\mathbf{e}, \mathbf{e}') = \boldsymbol{\Psi} \tag{22}$$

The differences of the diagonal elements of $\boldsymbol{\Sigma}$ and $\boldsymbol{\Psi}$, or the sums of squares of the loadings for each observed variate, are called the *communalities*. $\mathbf{W}$ and $\mathbf{e}$ are uncorrelated. If we put these results together, we have this model for the covariance matrix of the observable variates:

$$\text{Cov}(\mathbf{X}, \mathbf{X}') = \mathbf{\Sigma} = \mathbf{\Lambda\Lambda}' + \mathbf{\Psi} \tag{23}$$

This is the *factor model* for the covariance structure. Factor analysis consists of the estimation of the elements of the loading matrix $\mathbf{\Lambda}$ and, to a lesser extent, the specific variances in $\mathbf{\Psi}$. We may do so by maximum likelihood, as proposed orginally by D. N. Lawley and implemented for the computer by Jöreskog (1975), or by various approximate methods. One of the latter consists of estimating $\mathbf{\Psi}$ in some fashion and then extracting the first m principal components from the reduced matrix

$$\mathbf{S} - \hat{\mathbf{\Psi}} \tag{24}$$

where $\mathbf{S}$ is the usual sample covariance matrix. More directly, we might begin by replacing the unit diagonal elements in the sample correlation matrix $\mathbf{R}$ by estimates of the communalities and finding the m components of the reduced matrix. Common estimators of the communalities are the coefficients of determination of each variate with the $p - 1$ remaining ones, or the simple correlations with largest absolute value in the rows and columns for the successive variates. However, such approximations may lead to negative characteristic roots if the number of common factors m is not small.

For the approximate solutions the estimator of $\mathbf{\Lambda}$ is found by expressing the characteristic vectors in the correlation or loading form (18). Then we are attempting to reproduce $\mathbf{S}$ by the matrix

$$\hat{\mathbf{\Sigma}} = \hat{\mathbf{\Lambda}}\hat{\mathbf{\Lambda}}' + \hat{\mathbf{\Psi}} \tag{25}$$

The off-diagonal elements of $\mathbf{S} - \hat{\mathbf{\Sigma}}$ tell us how well we have succeeded in generating the observed covariances or correlations by the factor model and, in fact, are the basis for a goodness-of-fit test for the validity of the m-factor model.

A large literature exists on the computation and interpretation of factor analyses. A comprehensive survey is contained in Harman's text (1967). Some examples of maximum likelihood factor solutions have been given by Morrison (1976). Many of the current statistical computer packages contain programs for maximum likelihood factor analysis and several approximate variants of it.

7-5 COLLINEAR VARIABLES

Let us begin with an example. For an investigation* of the relationship of the price of gold in dollars to exchange rates for other currencies, values of the ratios

$$Y = \text{price of gold (dollars/ounce)}$$
$$X_1 = \text{dollars/Deutsche mark}$$
$$X_2 = \text{dollars/Swiss franc}$$

*I am indebted to Charles A. Miller for this example and for permission to use his results here.

were collected for $N = 27$ quarters from 1973 to July 1979. The variables had these correlations:

$$\mathbf{R} = \begin{bmatrix} 1 & 0.8740 & 0.8418 \\ & 1 & 0.9538 \\ & & 1 \end{bmatrix} \tag{1}$$

The price of gold appears to depend strongly on both predictors. These matrices of centered sums of squares and products were computed from the data:

$$\mathbf{X}_1'\mathbf{X}_1 = \begin{bmatrix} 0.09126 & 0.15342 \\ 0.15342 & 0.28366 \end{bmatrix} \qquad \mathbf{X}_1'\mathbf{Y} = \begin{bmatrix} 63.884 \\ 108.465 \end{bmatrix}$$

$$\sum (y_i - \bar{y})^2 = 58{,}554.61 \tag{2}$$

These regression coefficients and their t-statistics were computed:

Variable	$\hat{\beta}_j$	t_j
X_1	630.19	2.39
X_2	41.54	0.28

The coefficient of determination showed that the two variables explained 76.45% of the variation in the price of gold, whereas the statistic $F = 38.95$ for the overall regression test exceeded any conventional critical value. Nevertheless, the t-statistic for the second exchange rate is insignificant by any criterion, and t_1 is less than the 0.05 Scheffé simultaneous test critical value $\sqrt{2F_{0.05;\,2,24}} = 2.61$. The individual significance of the predictors seems in doubt, even though their joint significance is clear. The reason for this paradox is the high correlation between X_1 and X_2: Those variables are said to be nearly *collinear*, for the values of one are almost linear transformations of the other. Of course, if the variables were exactly collinear, their matrix $\mathbf{X}_1'\mathbf{X}_1$ would not have an inverse, and the least squares estimators would be undefined. If the collinearity were exact except for round-off error, meaningless estimates might be computed.

Let us see why the overall hypothesis of no regression was resoundingly rejected, yet the significance of the individual coefficients was marginal or nil. The answer is contained in the 95% joint confidence region for β_1 and β_2 shown by the ellipse in Figure 7-3. The overall hypothesis

$$H_0\colon \beta_1 = \beta_2 = 0$$

would not be tenable at the 0.05 level, for the confidence region is well removed from the origin of the β_1, β_2 space. However, if we project the ellipse onto the β_2 axis we note that $\beta_2 = 0$ is near the center of the projected values, so that $H_0\colon \beta_2 = 0$ cannot be rejected. The other projection onto the β_1 axis would just barely include $\beta_1 = 0$, so that the hypothesis $H_0\colon \beta_1 = 0$ would be tenable in the Scheffé simultaneous testing sense of Section 3-5. The narrow shape and downward orientation of the confidence ellipse is due to the large negative correlation -0.9535 of $\hat{\beta}_1$ and $\hat{\beta}_2$, which in turn was caused by the high correlation of the original predictors.

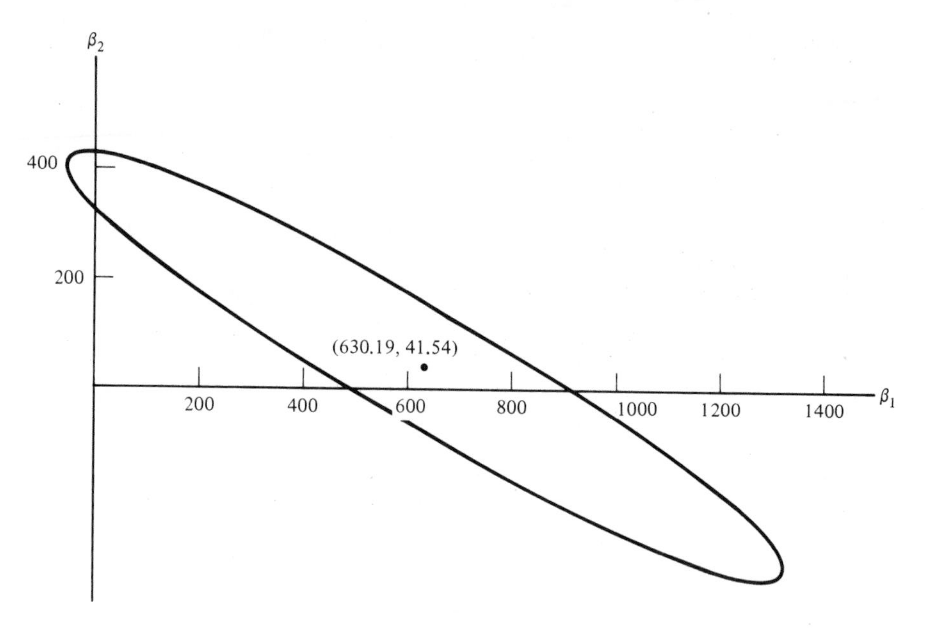

Figure 7-3 Joint confidence region for β_1 and β_2.

One direct way of dealing with the near-collinearity of two predictors is to omit one variable from the multiple regression model. In the present case these simple regression equations might be used to predict the price of gold in dollars:

1. X_1 = dollars/Deutsche mark

$\hat{\beta}$	t	$\hat{\sigma}$	R^2
700.02	8.99	23.52	0.7637

2. X_2 = dollars/Swiss franc

$\hat{\beta}$	t	$\hat{\sigma}$	R^2
382.38	7.79	26.14	0.7083

Both equations predict virtually the same proportion of variation in the dollar price of gold; the proportions are only slightly less than that in the original multiple regression model. Both slopes have t-statistics with highly significant values. Although each variable might be useful for the separate prediction of the price of gold, the high correlation only causes confusion when both are used together.

When groups of highly correlated variables are present the collinearity problem can be avoided by omitting all but one of the redundant variables in each set. Multicollinear variables present a more subtle detection problem: Their ordinary correlations may be of only moderate size, yet some of their multiple correlations are close to unity. One might begin by computing the determinant $|\mathbf{X}_1'\mathbf{X}_1|$ of the sums of squares and products matrix of the entire set of independent variables; if it is small relative to the product of the diagonal elements of $\mathbf{X}_1'\mathbf{X}_1$ multicollinearity is likely to be present somewhere in the set. For a more formal test we might compute the p coefficients of determination of each variable regressed on the remaining $p - 1$ predictors. Variables with high coefficients of determination would be presumed to be collinear with at least one or two other variables. From those multiple correlations and perhaps determinants of submatrices of $\mathbf{X}_1'\mathbf{X}_1$ we might locate groups of highly collinear predictors. One or more members of the group would be eliminated before carrying out the regression analysis with the original dependent variable.

An alternative way of dealing with collinearity is provided by the *ridge regression* method proposed by Hoerl (1962). A positive constant k is added to every diagonal element of $\mathbf{X}_1'\mathbf{X}_1$ to reduce the degree of ill-conditioning. The ridge estimator is then

$$\hat{\boldsymbol{\beta}}^* = (\mathbf{X}_1'\mathbf{X}_1 + k\mathbf{I})^{-1}\mathbf{X}_1'\mathbf{y} \tag{3}$$

Ridge estimators are necessarily biased, but their variances are considerably smaller than those of the ordinary least squares estimators. In practice one plots the ridge estimates for each predictor as a function of k and chooses the value of k as that at which the estimate curves have begun to flatten, or stabilize. The standard deviations of the values predicted from the ridge regression model are also plotted as a function of k to determine which constant minimizes the prediction standard deviation. Further motivation for ridge regression and several examples of its application in agriculture and chemical engineering have been described by Marquardt and Snee (1975).

7-6 REFERENCES

Anderson, T. W. (1958): *An Introduction to Multivariate Statistical Analysis*, John Wiley & Sons, Inc., New York.

Anderson, T. W. (1963): Asymptotic theory for principal component analvsis, *Annals of Mathematical Statistics*, vol. 34, pp. 122–148.

Fisher, R. A. (1936): The use of multiple measurements in taxonomic problems, *Annals of Eugenics*, vol. 7, pp. 179–188.

Harman, H. H. (1967): *Modern Factor Analysis*, 2nd ed., University of Chicago Press, Chicago.

Hoerl, A. E. (1962): Application of ridge analysis to regression problems, *Chemical Engineering Progress*, vol. 58, pp. 54–59.

Jolicoeur, P. (1963): The degree of generality of robustness in *Martes americana*, *Growth*, vol. 27, pp. 1–27.

JÖRESKOG, K. G. (1975): Factor analysis by least squares and maximum likelihood, in K. Enslein, A. Ralston, and H. Wilf (eds.), *Statistical Methods for Digital Computers*, John Wiley & Sons, Inc., New York.

KENDALL, M. G. (1957): *A Course in Multivariate Analysis*, Charles Griffin & Company Ltd., London.

LACHENBRUCH, P. A. (1975): *Discriminant Analysis*, Hafner Press, New York.

MARQUARDT, D. W., and R. D. SNEE (1975): Ridge regression in practice, *The American Statistician*, vol. 29, pp. 3–20.

MASSY, W. F. (1965): Principal components regression in exploratory statistical research, *Journal of the American Statistical Association*, vol. 60, pp. 234–256.

MORRISON, D. F. (1976): *Multivariate Statistical Methods*, 2nd ed., McGraw-Hill Book Company, New York.

RAO, C. R. (1965): *Linear Statistical Inference and Its Applications*, John Wiley & Sons, Inc., New York.

7-7 EXERCISES

1. Data were collected on $N = 55$ persons in a firm to investigate the relation of annual salary (Y) in thousands of dollars to these variables:

$$X_1 = \begin{cases} 1 & \text{if male} \\ 0 & \text{if female} \end{cases}$$

$$X_2 = \text{years of service}$$

$$X_3 = \begin{cases} 1 & \text{Bachelor's or other degree} \\ 0 & \text{no college degree} \end{cases}$$

$$X_4 = \begin{cases} 1 & \text{Bachelor's degree only} \\ 0 & \text{other (no degree or graduate degree)} \end{cases}$$

These statistics were computed:

Variable	Y	X_1	X_2	X_3	X_4
Mean	20.63	0.673	5.42	0.812	0.309

$$\mathbf{X}_1'\mathbf{X}_1 = \begin{bmatrix} 12.109 & 19.527 & -2.273 & -4.436 \\ & 463.382 & -35.818 & -10.109 \\ & & 8.182 & 3.091 \\ & & & 11.745 \end{bmatrix}$$

$$\mathbf{Y}'\mathbf{X}_1 = [22.788, \quad 220.881, \quad -17.603, \quad -16.409]$$

$$\sum (y_i - \bar{y})^2 = 277.371$$

Carry out a regression analysis for predicting annual salary and interpret the importance of the dummy variables for sex and degree status.

2. A study was conducted of $N = 117$ pulmonary X-rays of patients with lung cancer.* The sample was dichotomized into a first group of $N_1 = 62$ patients whose lesions were missed by the radiologists and another group of the remaining $N_2 = 55$ patients whose lesions were detected. The lesions were then described by $p = 8$ variables:

Variable	Definition
X_1	Size of the cancerous nodule (mm)
X_2	Completeness: percentage-circular
X_3	Completeness: steepness of first edge
X_4	Completeness: steepness of second edge
X_5	Surrounds: percentage of overlying shadows
X_6	Contrast: density of surroundings
X_7	Complexity: mean density change tangential to contour
X_8	Conspicuity: contrast

These statistics were computed for the two groups:

	Mean		Within-group standard	
Variable	Detected	Not detected	deviation	t
X_1	22.36	15.73	12.01	2.98
X_2	0.76	0.65	0.2209	2.62
X_3	1.89	3.02	1.2170	−4.99
X_4	2.56	2.94	1.2416	−1.62
X_5	0.58	0.64	0.2417	−1.18
X_6	15.65	11.05	9.2516	2.69
X_7	11.45	14.15	6.6668	−2.18
X_8	1.50	0.88	0.7716	4.34

Because the two-sample t-statistics for means of X_4 and X_5 did not exceed any conventional critical value those variables will be dropped. The within-groups covariance matrix of the remaining six variables is

$$\mathbf{S} = \begin{bmatrix} 144.2 & -0.7001 & -2.4463 & 38.323 & 14.283 & 1.8985 \\ & 0.0488 & -0.0565 & -0.5247 & -0.0625 & -0.0521 \\ & & 1.4811 & -1.3229 & -0.9080 & -0.1101 \\ & & & 85.592 & 27.532 & 4.1823 \\ & & & & 44.446 & -1.479 \\ & & & & & 0.5954 \end{bmatrix}$$

*This example is used with the kind permission of Dr. George Revesz, Diagnostic Radiology Research Laboratory, Temple University. The data were collected under NIH Grant No. CA24625. The example was suggested by Parvaneh Bonakdapour, who made an initial discriminant analysis.

(a) Verify that the linear discriminant coefficient vector is

$$\mathbf{a}' = [0.0406, \quad 3.0704, \quad -0.5567, \quad 0.0605, \quad -0.1050, \quad 0.3935]$$

(b) Test the hypothesis of equal mean vectors in the two populations.

(c) When the discriminant scores were evaluated for the original observations and the patients classified according to the scores these results were obtained:

	Original group	
Classification	Detected	Not detected
Detected	45	16
Not detected	10	46
Total	55	62

To what extent do the six measures classify a patient as detected or not detected to have lesions?

3. From the covariance matrix

$$\mathbf{S} = \begin{bmatrix} 10 & 2 & 5 & 1 \\ 2 & 10 & 6 & 3 \\ 5 & 6 & 10 & 8 \\ 1 & 3 & 8 & 10 \end{bmatrix}$$

the following principal component coefficients and characteristic roots were extracted:

	Component			
Variable	1	2	3	4
1	0.35	0.87	0.21	−0.29
2	0.45	−0.06	−0.85	−0.25
3	0.64	−0.03	0.12	0.76
4	0.51	−0.50	0.46	−0.53
Characteristic root	23.35	9.11	7.03	0.51

Verify either by a program for computing characteristic roots and vectors or by the simultaneous equations defining the coefficients that the components are those of the matrix **S**.

4. For an investigation of the dimensions of the limb bones of the North American marten, Jolicoeur (1963) carried out a principal component analysis of the logarithms of the humerus lengths, humerus widths, femur lengths, and femur widths

of $N = 92$ males.* The covariance matrix (multiplied by 10^4) of the measurements on the four variables was

$$\mathbf{S} = \begin{bmatrix} 1.1544 & 0.9109 & 1.0330 & 0.7993 \\ 0.9109 & 2.0381 & 0.7056 & 1.4083 \\ 1.0330 & 0.7056 & 1.2100 & 0.7953 \\ 0.7993 & 1.4083 & 0.7953 & 2.0277 \end{bmatrix}$$

These components were extracted from the matrix:

Dimension	Component 1	2	3	4
1. Humerus length	0.41	0.52	0.15	0.73
2. Humerus width	0.58	−0.40	0.68	−0.18
3. Femur length	0.39	0.64	−0.13	−0.65
4. Femur width	0.58	−0.40	−0.71	0.10
Characteristic root	4.548	1.116	0.645	0.121

(a) What proportions of the total variance of the four dimensions do the components explain?

(b) Do you agree with the interpretations of the first component as a measure of limb size and the second as an index of bone shape? Why or why not?

(c) Do you think that the fourth component and perhaps the third can be omitted from a description of the four bone dimensions?

5. In Section 7-4 we stated that the factor analysis of a covariance matrix began with the representation of the matrix as

$$\boldsymbol{\Sigma} = \boldsymbol{\Psi} + \boldsymbol{\Lambda}\boldsymbol{\Lambda}'$$

and consisted of the estimation of the diagonal matrix $\boldsymbol{\Psi}$ of specific variances and the factor loading matrix $\boldsymbol{\Lambda}$ by some means. $\boldsymbol{\Sigma}$ would of course be estimated by the sample covariance matrix. Factor analysis can be applied to correlation matrices by the same estimation process. The correlation matrix

$$\mathbf{R} = \begin{bmatrix} 1 & 0.49 & 0.39 & 0.28 & 0.17 \\ & 1 & 0.68 & 0.50 & 0.47 \\ & & 1 & 0.74 & 0.68 \\ & & & 1 & 0.73 \\ & & & & 1 \end{bmatrix}$$

of the reaction times of subjects under five conditions had the following estimated specific variances and loadings for single- and two-factor models:

*I am indebted to Pierre Jolicoeur and the managing editor of *Growth* for kindly permitting the use of these data.

	Model				
	Single-factor		Two-factor		
Condition variable	$\hat{\psi}_i$	$\hat{\lambda}_{i1}$	$\hat{\psi}_i$	$\hat{\lambda}_{i1}$	$\hat{\lambda}_{i2}$
1	0.84	0.40	0.64	0.42	0.44
2	0.51	0.70	0.29	0.73	0.42
3	0.18	0.91	0.20	0.89	0.06
4	0.32	0.83	0.24	0.84	−0.23
5	0.41	0.77	0.29	0.79	−0.29

(a) Use equation (25) of Section 7-4 to reproduce the variable correlations by each factor model. To what extent do one and two common factors explain the correlations?

(b) Unlike principal components, the vectors of factor loadings need not be orthogonal. Verify that such is the case for the two-factor solution.

6. Show that classification by the linear discriminant function with a single variable reduces to the rule

$$\text{Assign } X \text{ to one population if } x \geq \frac{\bar{u}_1 + \bar{u}_2}{2}$$

$$\text{and to the other population if } x < \frac{\bar{u}_1 + \bar{u}_2}{2}$$

That is, we assign x to the population with the closer sample mean.

7. In a physiological experiment measurements were made of blood pressure and four predictor variables describing heart performance in dogs. This correlation matrix was computed for the five variables from a moderately large sample of observations from several dogs and different times:

$$\begin{bmatrix} 1 & 0.4886 & 0.7342 & 0.2404 & -0.3148 \\ & 1 & 0.2149 & 0.3461 & -0.7000 \\ & & 1 & -0.4574 & -0.5578 \\ & & & 1 & 0.4059 \\ & & & & 1 \end{bmatrix}$$

The first row and column contain the correlations of blood pressure with the predictors. The first regression analysis led to peculiar coefficients and suggested that some of the predictors might be nearly collinear. By computing appropriate determinants or multiple correlations can you locate the culprit variables?

8. The course evaluation data of Exercise 1, Section 5-7, were taken from a larger set in which these six variables described the instructor's performance:

1. Ability to present course material
2. Ability to stimulate student interest
3. Knowledge of the subject matter
4. Interest in students
5. Objectivity and fairness
6. Overall evaluation

This correlation matrix was computed from the $N = 32$ students who returned evaluation forms:

$$\begin{bmatrix} 1 & 0.6174 & 0.0573 & 0.2002 & 0.3660 & 0.8618 \\ & 1 & 0.0777 & 0.3170 & 0.2641 & 0.7394 \\ & & 1 & 0.5833 & 0.4234 & 0.2818 \\ & & & 1 & 0.6838 & 0.4350 \\ & & & & 1 & 0.4798 \\ & & & & & 1 \end{bmatrix}$$

(a) Use a statistical program package or a characteristic root and vector routine to extract the principal components from the correlation matrix. Ignore the original evaluation ratings consisting of the integers 1 through 5 rather than values of continuous variables.

(b) Assign meanings and names to the component variables.

(c) Do you think the variation in the ratings can be described effectively in less than six dimensions?

9. Applicants for admission to a certain graduate program had these Graduate Record Examination quantitative and verbal aptitude test scores (divided by 10 for simplicity):

Admitted		Not admitted	
Q	V	Q	V
76	55	60	35
71	63	72	32
80	50	53	43
65	68	65	57
67	52	52	66
78	71	80	25
		67	48
		77	30

(a) Find the linear discriminant function for assigning an applicant to one of the two groups solely on the basis of aptitude test scores.

(b) Do you feel that the condition of a common population covariance matrix is supported by the observation pairs in the two groups? Do not attempt to carry out a formal test of that hypothesis.

10. To the observation pairs in Exercise 9 add the dummy variable

$$X_i = \begin{cases} 1 & \text{if the } i\text{th applicant was admitted} \\ 0 & \text{if the } i\text{th applicant was not admitted} \end{cases}$$

We wish to find the simple regression equation of quantitative test scores based on the dummy variable. The matrices for the uncentered linear model are

$$\mathbf{X}'\mathbf{X} = \begin{bmatrix} 14 & 6 \\ 6 & 6 \end{bmatrix} \qquad \mathbf{X}'\mathbf{y} = \begin{bmatrix} 963 \\ 437 \end{bmatrix}$$

(a) Find the estimates of the intercept and slope parameters of the line.

(b) Test the hypothesis of a zero slope parameter and show that the test is equivalent to the two-sample t-test for equal quantitative means.

(c) Repeat that process with the same dummy variable, but verbal scores as the dependent variable.

(d) Demonstrate algebraically that the zero-slope and equal-means test statistics are equivalent.

chapter 8

The Analysis of Variance for the One-Way Layout

8-1 INTRODUCTION

With this chapter we turn from linear regression analysis to the use of the linear model to detect differences among several conditions in an experiment. The method of testing hypotheses about the experimental conditions' effects on some response variable is known as the *analysis of variance*. Just as some hypothesis tests in the early chapters could be given in terms of an analysis of variance table, the test for equal condition effects consists of breaking sums of squares into independent components measuring the variation among condition means and the variation of the observations within each condition. Chapters 9 and 10 discuss such analyses of variance for experimental situations of greater complexity or linear models with different assumptions on their components.

We begin in Section 8-2 by motivating the test for equality of several means by expressing the two-sample t-statistics in terms of a ratio of variances. In Section 8-3 the linear model for observations obtained under k separate experimental conditions is developed, and the one-way analysis of variance test is given. We call the experimental design and analysis "one-way" because the k conditions or treatments are of a single kind: They are not necessarily cross-classified or nested in a hierarchy according to other criteria. Such higher-way designs are the subject of Chapter 9.

Some special aspects of estimation in the one-way linear model are discussed in Section 8-4, and the notion of a *contrast*, or comparison of treatment effects, is introduced. Section 8-5 deals with orthogonal contrasts, or independent comparisons. Various methods for making several tests or confidence statements with a fixed overall error probability are considered in Section 8-6. The use of the power function of the analysis of variance for determining sample sizes in the design of experiments is illustrated in Section 8-7.

In Section 8-8 we develop the one-way model in which the treatment effects are random variables rather than parameters. That is the simplest example of the variance components model. Under certain assumptions on the model the test for no treatment differences will be identical with the analysis of variance for the fixed-effects case. Nevertheless, the power function, and thereby the determination of sample size, will be different. The higher-way extensions of such models will also lead to interesting complications.

The new analyses of variance follow directly if we begin with linear regression models and frame our hypotheses of equal effects in terms of appropriate regression coefficients. In that sense the analysis of variance is a special case of regression analysis and could be developed by suitably parametrizing the model and applying the results of Chapters 2 and 3. However, we do not take that approach in our treatment. The linear models commonly used for experimental designs do not lead to a matrix $\mathbf{X}$ of independent variable values that is of full rank, so that $\mathbf{X}'\mathbf{X}$ has a unique inverse. To develop estimators and the analysis of variance we would need generalized inverses and other concepts of a mathematical level higher than that we have chosen for this book. Reparametrization of the model to one of full column rank would permit the use of our earlier results in regression analysis, but that device seems unnecessarily complex and a little asymmetric: Let us merely accept that the estimators and test statistics we give could have been derived in that fashion.

8-2 COMPARING TWO MEANS BY VARIANCES

Let us begin our development of the analysis of variance with the simplest case of two populations. As an example suppose that a small farm operator plants identical varieties of tomato plants on two plots within the farm boundaries. The soil characteristics, fertilizer applications, mulching, amounts of water, and weather exposures of the plots are the same. However, one plot is located within 100 feet of a moderately traveled road, whereas the other plot is 3000 feet distant from the road and any buildings of the farm. It is conjectured that the tomatoes grown near the highway may have elevated levels of lead from engine exhaust pollutions. At the same time in the season samples of tomatoes will be drawn at random from each plot, and each vegetable will be assayed for its lead level per unit volume. An effort will be made to make the samples comparable with respect to size and ripeness of the tomatoes without formal stratification or matching of the vegetables. From the sample measurements of lead level we wish to test the hypothesis that the lead level means of the two populations of tomatoes are the same, as opposed to the alternative that tomatoes grown near the highway have, on the average, a higher concentration of lead. If we can make certain assumptions about the statistical properties of those populations it will be possible to test these hypotheses.

As in the other methods of this text we assume that the two populations of lead levels can be described by normal distributions with a common unknown variance. We have available from each population independent random samples of observations, and we wish to use those data to test the hypothesis that the population means are equal. Basic statistical methodology would direct us to test that hypothesis by computing the two-sample t-statistic and referring it to the one-sided critical value of the t-distribution. Instead, we adopt a different approach to the test that will lead to the analysis of variance for testing the equality of several normal means.

To begin we need some notation for our population parameters, sample data, and statistics. These are summarized in Table 8-1. Essentially, the response variable of interest to us can be represented by a random variable described by a normal distribution. We have available a random sample of N_1 observations on that variable obtained under one condition: We call the source of these observations population 1. That population is normal with mean μ_1 and variance σ^2.

TABLE 8-1 Models and Statistics for Comparing Two Means

	Population	
	1	2
Distribution	$N(\mu_1, \sigma^2)$	$N(\mu_2, \sigma^2)$
Observations	$x_{11}, \ldots, x_{N_1 1}$	$x_{12}, \ldots, x_{N_2 2}$
Sample size	N_1	N_2
Mean	$\bar{x}_1$	$\bar{x}_2$
Sum of squares	$\sum_{i=1}^{N_1} (x_{i1} - \bar{x}_1)^2$	$\sum_{i=1}^{N_2} (x_{i2} - \bar{x}_2)^2$

Similarly, we have a second random sample of N_2 observations drawn independently of the first from another normal population with mean μ_2 and the same variance σ^2. In the context of the test for different lead levels the first population might consist of tomatoes grown near the highway, whereas the second population would be those grown on the distant plot. We wish to test the hypothesis

$$H_0: \mu_1 = \mu_2 \tag{1}$$

of equal means, as opposed to the two-sided alternative $H_1: \mu_1 \neq \mu_2$. Now the estimators of the means are

$$\begin{aligned} \hat{\mu}_1 &= \bar{x}_1 = \frac{1}{N_1} \sum_{i=1}^{N_1} x_{i1} \\ \hat{\mu}_2 &= \bar{x}_2 = \frac{1}{N_2} \sum_{i=1}^{N_2} x_{i2} \end{aligned} \tag{2}$$

and the estimators of the common variance σ^2 within the separate samples are the sample variances

$$s_1^2 = \frac{1}{N_1 - 1} \sum_{i=1}^{N_1} (x_{i1} - \bar{x}_1)^2$$
$$s_2^2 = \frac{1}{N_2 - 1} \sum_{i=1}^{N_2} (x_{i2} - \bar{x}_2)^2 \tag{3}$$

Because we have assumed a common variance we should combine s_1^2 and s_2^2 into a single unbiased estimator of σ^2:

$$\begin{aligned}\hat{\sigma}^2 &= \frac{1}{N_1 + N_2 - 2}\left[\sum_{i=1}^{N_1} (x_{i1} - \bar{x}_1)^2 + \sum_{i=1}^{N_2} (x_{i2} - \bar{x}_2)^2\right] \\ &= \frac{(N_1 - 1)s_1^2 + (N_2 - 1)s_2^2}{N_1 + N_2 - 2}\end{aligned} \tag{4}$$

This is the *within-samples mean square*; its numerator is called the *within-sample sum of squares*. Because those quantities are computed from sums of squared deviations from the individual sample means they are not affected by differences in the population means μ_1 and μ_2: $\hat{\sigma}^2$ is a valid estimator of σ^2 regardless of the truth of the hypothesis (1) that we are testing.

The properties of independent chi-squared variates imply that

$$\frac{\hat{\sigma}^2(N_1 + N_2 - 2)}{\sigma^2} \tag{5}$$

has the chi-squared distribution with $N_1 + N_2 - 2$ degrees of freedom. By computing the sums of squares as deviations from the individual sample means rather than deviations from the grand mean of the combined samples we have reduced the degrees of freedom parameter by one. The information in the differences of the sample means would appear to be associated with that lost degree of freedom, and indeed it is. To see this connection let us estimate σ^2 from the sample means alone. We denote the grand mean of the observations by

$$\begin{aligned}\bar{x} &= \frac{1}{N_1 + N_2}(N_1\bar{x}_1 + N_2\bar{x}_2) \\ &= \frac{1}{N_1 + N_2}\left(\sum_{i=1}^{N_1} x_{i1} + \sum_{i=1}^{N_2} x_{i2}\right)\end{aligned} \tag{6}$$

and the sum of squared deviations of the sample means about $\bar{x}$ by

$$\begin{aligned}\text{SST} &= N_1(\bar{x}_1 - \bar{x})^2 + N_2(\bar{x}_2 - \bar{x})^2 \\ &= \frac{N_1 N_2}{N_1 + N_2}(\bar{x}_1 - \bar{x}_2)^2\end{aligned} \tag{7}$$

It is necessary to weight the squared deviations by the sample sizes to reflect the different variances

$$\text{Var}(\bar{x}_1) = \frac{\sigma^2}{N_1} \qquad \text{Var}(\bar{x}_2) = \frac{\sigma^2}{N_2} \tag{8}$$

of the group means. SST is called the *between-groups* sum of squares. We have adopted the symbol SST to denote "between-treatments," as in the context of the experimental design layouts of the later sections. We note that $\bar{x}_1 - \bar{x}_2$ is normally distributed with mean $\mu_1 - \mu_2$ and variance

$$\operatorname{Var}(\bar{x}_1 - \bar{x}_2) = \sigma^2\left(\frac{1}{N_1} + \frac{1}{N_2}\right) \tag{9}$$

Hence

$$\begin{aligned} E(\text{SST}) &= \frac{N_1 N_2}{N_1 + N_2} E[(\bar{x}_1 - \bar{x}_2)^2] \\ &= \frac{N_1 N_2}{N_1 + N_2} \{\operatorname{Var}(\bar{x}_1 - \bar{x}_2) + [E(\bar{x}_1 - \bar{x}_2)]^2\} \\ &= \sigma^2 + \frac{N_1 N_2(\mu_1 - \mu_2)^2}{N_1 + N_2} \end{aligned} \tag{10}$$

We see that the between-groups sum of squares is an unbiased estimator of σ^2 if and only if the hypothesis of equal means is true. Otherwise, the expected value of SST will always be larger than σ^2. As the sample sizes N_1 and N_2 increase so will the upward bias in SST.

If the hypothesis (1) is true, SST/σ^2 is a chi-squared random variable with one degree of freedom. We can show by the theorems on quadratic forms in Section 3-2 that SST and $\hat{\sigma}^2$ are independently distributed. Then the ratio

$$F = \frac{\text{SST}}{\hat{\sigma}^2} \tag{11}$$

has the F-distribution with one and $N_1 + N_2 - 2$ degrees of freedom when the population means are equal, and we may test that hypothesis at the α level by the decision rule

Reject $H_0\colon \mu_1 = \mu_2$ in favor of $H_1\colon \mu_1 \neq \mu_2$ if $F > F_{\alpha;\,1,\,N_1+N_2-2}$

Our test criterion is this: We have compared the variations in the sample means as measured by SST with the internal variation of each sample as expressed by $\hat{\sigma}^2$. If the between-samples variance is much greater than that found within, we should conclude that it is probably due to differing population means. The exact F-distribution of the ratio of the two variances permits us to assign a significance level to the qualification "probably" for protection against Type I errors.

Let us establish that the F-test statistic is equivalent to the two-sample t-statistic

$$\begin{aligned} t &= \frac{\bar{x}_1 - \bar{x}_2}{\sqrt{\dfrac{\sum (x_{i1} - \bar{x}_1)^2 + \sum (x_{i2} - \bar{x}_2)^2}{N_1 + N_2 - 2}\left(\dfrac{1}{N_1} + \dfrac{1}{N_2}\right)}} \\ &= \frac{(\bar{x}_1 - \bar{x}_2)\sqrt{N_1 N_2/(N_1 + N_2)}}{\hat{\sigma}} \end{aligned} \tag{12}$$

We would reject $H_0\colon \mu_1 = \mu_2$ in favor of $H_1\colon \mu_1 \neq \mu_2$ if $|t| > t_{\alpha/2;N_1+N_2-2}$. But, it is easily seen from (4) and (7) that

$$t^2 = \frac{\text{SST}}{\hat{\sigma}^2} = F \tag{13}$$

so that the two statistics are functionally related. The equivalence of an F-variate with one and n degrees of freedom to the square of a Student–Fisher t

with n degrees of freedom also implies that

$$F_{\alpha;1,n} = t^2_{\alpha/2;n} \tag{14}$$

We note that the probability content $\alpha/2$ in each tail of the t-distribution is transferred into the right-hand tail of the F-distribution.

The F-test for the equal-means hypothesis is inherently two-sided, although our motivating example of the lead levels in the tomatoes had a one-sided alternative hypothesis of more lead in the vegetables grown near the road. For one-sided tests on means by the F-statistic we would have to observe that the sign of $\bar{x}_1 - \bar{x}_2$ agreed with the alternative hypothesis, and then refer F to the critical value $F_{2\alpha;N1+N2-2}$ for a one-sided test at level α.

Example 8-1

In an introductory statistics course the papers of students who completed a midterm examination before the end of the two-hour period were identified by a mark on the back and then graded along with the full-term papers. These scores (ordered in magnitude) were obtained for the two groups of students:

Early completion		Full term
56	95	68
69	95	70
70	95	72
77	100	75
83	100	84
84	100	85
90	100	92

The medians of the two groups are different and suggest that the students in each group may have come from different populations. Let us make the rather strong assumptions that those populations are normal with a common unknown variance, then test the hypothesis of equal means by the analysis of variance F-ratio. We shall need these statistics:

	Early	Full
Sample size	14	7
Mean	86.71	78
$\sum (x_{ij} - \bar{x}_j)^2$	2575	490
Variance	198.08	81.67

Then

$$\hat{\sigma}^2 = (\tfrac{1}{19})(2575 + 490) = 161.3158$$

$$\text{SST} = \frac{(14)(7)}{21}(8.71)^2 = 354.03$$

and $F = 2.19$. The 10% critical value of the F-distribution with 1 and 19 degrees of freedom is approximately 2.99. Because the sample F does not exceed that value we cannot reject the null hypothesis of a common population mean.

In the interests of thorough data analysis we should not leave the question of a possible significant difference between the early and full-term examinees at that point. Let us begin with the F-test and its alternative hypothesis of unequal means. A more appropriate alternative would be that of a *higher* mean score in the population of early finishers. We assume that completion of the test in less than the allotted time indicates greater competence rather than capitulation before the questions. For a one-sided test we must either refer F to the critical value $F_{2\alpha;1,N_1+N_2-2}$ or $t = \pm\sqrt{F}$ to the appropriate critical value $\pm t_{\alpha;N_1+N_2-2}$ of the t-distribution. For the first method $F_{0.25;1,19} = 1.41$, and our computed F would indicate that the alternative of a higher mean in the early population should be accepted at some level less than 0.125. For the second approach $t = 1.48$ exceeds $t_{0.10;19} = 1.328$, and the null hypothesis should be rejected in favor of the alternative at the 0.10 level.

However, some other more fundamental questions are suggested by the examination data. The scores of the early completion students do not appear to follow a normal distribution: half are in the range 95 through 100. With the possible exception of the lowest score the remainder appear to be similar to the full-term scores. The distributions may differ not only in their means or other location parameters, but in the first population's greater proportion of superior performers. We also note that the variance of the early group's scores is larger than that in the full-term group. Although the variance ratio $F = 198.08/81.67 = 2.43$ does not exceed the critical value $F_{0.10;13,6} = 2.89$, the assumption of a common variance underlying the F- and t-tests for equal means cannot be held tenaciously. One might pursue the hypothesis of equal distributions of scores for the two kinds of students by various nonparametric tests on the scores or transformations of them.

8-3 COMPARING SEVERAL MEANS BY THE ANALYSIS OF VARIANCE

We have seen how the test for the equality of the means of two normal populations can be expressed as a comparison of the variation between and within samples. We can extend that test to k normal populations by slight generalizations of the between- and within-samples sums of squares. However, the case of several groups contains some subtleties not encountered with two populations, and it will be essential to begin with a statement of a mathematical model for the data and their analysis.

As in the case of two groups we start with an interest in some phenomenon that can be described quantitatively, and whose levels, values, or measurements can be represented by a continuous random variable X. We shall refer to that variable and its observed values as the *response* variate, in the sense of the response of the sampling units to conditions imposed by an experimenter or the environment of the units. The response variate will be normally distributed. The experimenter or other investigator is interested in k conditions under which the response variate will be observed: For example, an agronomist might choose $k = 4$ variants of commercial chemical fertilizers for application to a standard strain of wheat. The response observations would be measures of the amounts of wheat harvested from the individual plots of ground at the end of the experiment. Similarly, a consumer research organization might select six models of moderate-size cars of different manufacturers for the determination of differences in their gas mileages. The weights of the cars are approximately equal, although the engine designs may be different. Lists of owners of the six models in a metropolitan area are available, and random samples of the owners of each model are invited to participate in a gas mileage trial for their car. The owner drives his or her car over a specified route within the city, and gas mileage is recorded. Here the experimental conditions are the six types of cars, the response variable is miles per gallon of gasoline, and the experimental units are the individual car and driver combinations.

Let us adopt this mathematical model for data collected under several experimental conditions: The observation x_{ij} from the ith sampling or experimental unit under the jth condition or treatment will be represented as

$$x_{ij} = \mu + \tau_j + e_{ij} \qquad \begin{array}{l} i = 1, \ldots, N_j \\ j = 1, \ldots, k \end{array} \tag{1}$$

where μ = parameter common to all sampling units
τ_j = effect of the jth experimental condition
e_{ij} = random variable expressing sampling variation and measurement error

μ is a kind of basal effect in the absence of any experimental treatment. The τ_j are parameters measuring the contributions of the treatments or experimental conditions to the observed responses. We shall assume that the e_{ij} are independently and normally distributed random variables with mean $E(e_{ij}) = 0$ and a common variance $\text{Var}(e_{ij}) = \sigma^2$. Of course

$$\text{Cov}(e_{ij}, e_{hl}) = 0 \qquad i \neq h \tag{2}$$

by the independence and normality assumptions. Then the x_{ij} are also independent normal random variables with mean

$$E(x_{ij}) = \mu + \tau_j \tag{3}$$

a common variance $\text{Var}(x_{ij}) = \sigma^2$ for all combinations of i and j, and

$$\text{Cov}(x_{ij}, x_{hl}) = 0 \qquad i \neq h \tag{4}$$

for observations obtained from different sampling units. The model is clearly a linear function of its parameters and the random disturbance term. Such linear models with additive error variates are the basis for the analysis of variance.

We chose to parametrize the model (1) in terms of the general effect μ and the k treatment effects $\tau_1, \ldots, \tau_k$ because the larger experimental layouts in the sequel will require that generality. A simpler model would merely be

$$x_{ij} = \mu_j + e_{ij} \qquad \begin{array}{l} i = 1, \ldots, N_j \\ j = 1, \ldots, k \end{array} \tag{5}$$

where $\mu_j = \mu + \tau_j$, and the e_{ij} obey the same distributional assumptions imposed in the model (1). Then the k treatment parameters would be uniquely estimated by the treatment means:

$$\hat{\mu}_j = \frac{1}{N_j} \sum_{i=1}^{N_j} x_{ij} = \bar{x}_j \tag{6}$$

Estimation of the $k + 1$ parameters in the original model (1) cannot be accomplished so easily; that matter will be deferred until we have considered tests of hypotheses for the treatment effects.

The hypothesis of equal treatment effects is

$$H_0: \quad \tau_1 = \cdots = \tau_k \tag{7}$$

or in terms of the alternative model (3),

$$H_0: \quad \mu_1 = \cdots = \mu_k \tag{8}$$

For the present we shall think of the alternative hypothesis as merely that of different τ_j or μ_j for at least some of the groups. We shall see subsequently in our treatments of estimable functions and power that the true alternative is more general. Let us develop a test of the equal-treatments hypothesis by extending the two-group test of Section 8-2 to k groups. We shall begin by estimating the common variance σ^2 of the k normal distributions. Define the sum of squares about the treament sample means as

$$\begin{aligned} \text{SSE} &= \sum_{i=1}^{N_1} (x_{i1} - \bar{x}_1)^2 + \cdots + \sum_{i=1}^{N_k} (x_{ik} - \bar{x}_k)^2 \\ &= \sum_{j=1}^{k} \sum_{i=1}^{N_j} (x_{ij} - \bar{x}_j)^2 \end{aligned} \tag{9}$$

where the symbol SSE denotes "sum of squares due to error" as in regression analysis. Then the unbiased estimator of σ^2 is the error mean square

$$\hat{\sigma}^2 = \text{MSE} = \frac{\text{SSE}}{N - k} \tag{10}$$

As in the two-group case MSE is a measure of the sampling variation of the random variable X that is unaffected by differences among the means of the treatment groups. We note that SSE/σ^2 is distributed as a chi-squared random variable with $N - k$ degrees of freedom if the assumptions of independence and normality of the x_{ij} hold. The measure of the variability in the treatment means is the analogue of (7), Section 8-2, for k groups:

$$\begin{aligned} \text{SST} &= \sum_{j=1}^{k} N_j(\bar{x}_j - \bar{x})^2 \\ &= \sum_{j=1}^{k} \frac{T_j^2}{N_j} - \frac{G^2}{N} \end{aligned} \tag{11}$$

where, in keeping with the usual analysis of variance conventions, T_j is the sum of observations obtained under treatment j:

$$T_j = \sum_{i=1}^{N_j} x_{ij} \tag{12}$$

G is the grand total of all observations:

$$
\begin{aligned}
G &= \sum_{j=1}^{k} T_j \\
&= \sum_{j=1}^{k} \sum_{i=1}^{N_j} x_{ij} \\
\bar{x} &= \frac{G}{N}
\end{aligned}
\tag{13}
$$

and $N = N_1 + \cdots + N_k$. The second line of (11) is more convenient for use with small calculators. If we can show that SST/σ^2 has a chi-squared distribution only when the equal treatment effects hypothesis is true, and is independently distributed of SSE, we shall have the basis for an F-test statistic for the hypothesis.

We may accomplish these tasks with the aid of Theorems 3-1 and 3-3. To establish the chi-squared distribution of SST/σ^2 by Theorem 3-1 we note that the k treatment sample means $\bar{x}_1, \ldots, \bar{x}_k$ are independently normally distributed with mean vector and covariance matrix

$$
\boldsymbol{\mu} = \begin{bmatrix} \mu_1 \\ \cdot \\ \cdot \\ \cdot \\ \mu_k \end{bmatrix} \qquad \boldsymbol{\Sigma} = \sigma^2 \begin{bmatrix} 1/N_1 & \cdots & 0 \\ \cdot & & \cdot \\ \cdot & & \cdot \\ \cdot & & \cdot \\ 0 & \cdots & 1/N_k \end{bmatrix} \tag{14}
$$

The treatment sum of squares scaled by σ^2 can be written as the quadratic form

$$
\frac{\text{SST}}{\sigma^2} = \bar{\mathbf{x}}'\mathbf{A}\bar{\mathbf{x}} \tag{15}
$$

with matrix

$$
\mathbf{A} = \frac{1}{N\sigma^2} \begin{bmatrix} N_1(N - N_1) & -N_1N_2 & \cdots & -N_1N_k \\ -N_1N_2 & N_2(N - N_2) & \cdots & -N_2N_k \\ \cdot & \cdot & & \cdot \\ \cdot & \cdot & & \cdot \\ \cdot & \cdot & & \cdot \\ -N_1N_k & -N_2N_k & \cdots & N_k(N - N_k) \end{bmatrix} \tag{16}
$$

The matrix

$$
\mathbf{A\Sigma} = \frac{1}{N} \begin{bmatrix} N - N_1 & -N_1 & -N_1 & \cdots & -N_1 \\ -N_2 & N - N_2 & -N_2 & \cdots & -N_2 \\ -N_3 & -N_3 & N - N_3 & \cdots & -N_3 \\ \cdot & \cdot & \cdot & & \cdot \\ \cdot & \cdot & \cdot & & \cdot \\ \cdot & \cdot & \cdot & & \cdot \\ -N_k & -N_k & -N_k & \cdots & N - N_k \end{bmatrix} \tag{17}
$$

can be shown to have the idempotency property $(\mathbf{A\Sigma})^2 = \mathbf{A\Sigma}$, so that by Theorem 3-1 SST/σ^2 has the noncentral chi-squared distribution with degrees of freedom given by the rank of $\mathbf{A}$, or equivalently, the rank of $\mathbf{A\Sigma}$. Because $\mathbf{A\Sigma}$ is idempotent its rank will be equal to its trace, or

$$\begin{aligned} \operatorname{tr}(\mathbf{A\Sigma}) &= \frac{(N - N_1) + (N - N_2) + \cdots + (N - N_k)}{N} \\ &= \frac{Nk - N}{N} = k - 1 \end{aligned} \tag{18}$$

If the null hypothesis $H_0: \tau_1 = \cdots = \tau_k$ of equal treatment effects is true then $\mu_1 = \cdots = \mu_k$, and

$$\boldsymbol{\mu}' = \mu[1, \ldots, 1] = \mu\mathbf{j}' \tag{19}$$

The noncentrality parameter of the distribution of SST$/\sigma^2$ will be

$$\lambda = \boldsymbol{\mu}'\mathbf{A}\boldsymbol{\mu} = \mu^2\mathbf{j}'\mathbf{A}\mathbf{j} = 0 \tag{20}$$

since each row of $\mathbf{A}$ (and each column) sums to zero. When H_0 is true SST$/\sigma^2$ has an ordinary chi-squared distribution and is distributed independently of the within-groups sum of squares.

Finally, we must show that SST and SSE are independently distributed. That can be accomplished by Theorem 3-3 and its corollary of Section 3-2: Write SSE as the sum of K within-groups quadratic forms

$$\begin{aligned} \text{SSE} &= \sum_{i=1}^{N_1} (x_{i1} - \bar{x}_1)^2 + \cdots + \sum_{i=1}^{N_k} (x_{ik} - \bar{x}_k)^2 \\ &= \mathbf{x}_1'\mathbf{A}_1\mathbf{x}_1 + \cdots + \mathbf{x}_k'\mathbf{A}_k\mathbf{x}_k \end{aligned} \tag{21}$$

where $\mathbf{x}_j$ is the $N_j \times 1$ vector of the jth treatment observations, and

$$\mathbf{A}_j = \frac{1}{N_j}\begin{bmatrix} N_j - 1 & -1 & \cdots & -1 \\ -1 & N_j - 1 & \cdots & -1 \\ \cdot & \cdot & & \cdot \\ \cdot & \cdot & & \cdot \\ \cdot & \cdot & & \cdot \\ -1 & -1 & \cdots & N_j - 1 \end{bmatrix} \tag{22}$$

Then SST can be written in terms of the group means

$$\bar{x}_j = \frac{1}{N_j}\sum_{i=1}^{N_j} x_{ij} = \mathbf{b}_j'\mathbf{x}_j \tag{23}$$

as the first line of expression (11). The grand mean is of course

$$\bar{x} = \frac{1}{N}\sum_{j=1}^{k} N_j\mathbf{b}_j'\mathbf{x}_j \tag{24}$$

Hence, SST will be distributed independently of SSE if each group mean $\bar{x}_j$ is distributed independently of its within-group sum of squares $\mathbf{x}_j'\mathbf{A}_j\mathbf{x}_j$. By the corollary of Theorem 3-3 we must have

$$\mathbf{A}_j\mathbf{b}_j = \mathbf{0} \qquad j = 1, \ldots, k \tag{25}$$

Because each row of $\mathbf{A}_j$ sums to zero that property holds, and SSE and SST are independent. Hence when $H_0: \tau_1 = \cdots = \tau_k$ is true

$$F = \frac{\text{SST}/\sigma^2(k-1)}{\text{SSE}/\sigma^2(N-k)}$$

$$= \frac{N-k}{k-1}\frac{\text{SST}}{\text{SSE}}$$

$$= \frac{\text{SST}/(k-1)}{\text{MSE}} \tag{26}$$

has the F-distribution with $k-1$ and $N-k$ degrees of freedom. For a test of equal means at the α level we reject the null hypothesis if

$$F > F_{\alpha; k-1, N-k} \tag{27}$$

The analysis of variance. Our F-test was constructed by finding two independent estimators of σ^2: One, based on intragroup variation, is always unbiased, while the other estimator is only unbiased if the treatment effects are equal. Such a partitioning of sums of squares into within- and between-group components is an example of the *analysis of variance*. A total sum of squares is decomposed, or analyzed, into independent components measuring sampling variation or differences among experimental conditions. Let us make such a decomposition of the total sum of squares about the grand mean $\bar{x}$ for all N observations. We write

$$\sum_{j=1}^{k}\sum_{i=1}^{N_j}(x_{ij}-\bar{x})^2 = \sum_{j=1}^{k}\sum_{i=1}^{N_j}[(x_{ij}-\bar{x}_j)+\bar{x}_j-\bar{x}]^2 \tag{28}$$

by subtracting and adding $\bar{x}_j$. Then, if we expand the term in square brackets and note that the crossproduct term sums to zero, we have

$$\begin{aligned}\sum_{j=1}^{k}\sum_{i=1}^{N_j}(x_{ij}-\bar{x})^2 &= \sum_{j=1}^{k}\sum_{i=1}^{N_j}(x_{ij}-\bar{x}_j)^2 + 2\sum_{j=1}^{k}\sum_{i=1}^{N_j}(x_{ij}-\bar{x}_j)(\bar{x}_j-\bar{x}) \\ &\quad + \sum_{j=1}^{k}\sum_{i=1}^{N_j}(\bar{x}_j-\bar{x})^2 \\ &= \sum_{j=1}^{k}\sum_{i=1}^{N_j}(x_{ij}-\bar{x}_j)^2 + \sum_{j=1}^{k}N_j(\bar{x}_j-\bar{x})^2 \\ &= \text{SSE} + \text{SST}\end{aligned} \tag{29}$$

The total sum of squares about the grand mean is equal to the sum of the within-groups and between-groups sums of squares. Alternatively, we might decompose the uncorrected total sum of squares as

$$\begin{aligned}\sum_{j=1}^{k}\sum_{i=1}^{N_j}x_{ij}^2 &= \sum_{j=1}^{k}\sum_{i=1}^{N_j}[(x_{ij}-\bar{x}_j)+(\bar{x}_j-\bar{x})+\bar{x}]^2 \\ &= \text{SSE} + \text{SST} + \frac{G^2}{N}\end{aligned} \tag{30}$$

We have shown by the corollary of Theorem 3-3 that the sums of squares SSE and SST are independent, and a similar invocation of that corollary will show that they are independently distributed of the grand mean $\bar{x}$ and its sum of squares G^2/N. Because we rarely wish to test the hypothesis of a zero population grand mean we shall summarize the first decomposition in the analysis of variance, or ANOVA, of Table 8-2.

TABLE 8-2 One-Way Analysis of Variance

Source	Sum of squares	d.f.	Mean square	Expected mean square
Treatments	$\text{SST} = \sum_{j=1}^{k} N_j(\bar{x}_j - \bar{x})^2$	$k-1$	$\frac{\text{SST}}{k-1}$	$\sigma^2 + \frac{\sum N_j(\tau_j - \bar{\tau})^2}{k-1}$
Within treatments	$\text{SSE} = \sum_{j=1}^{k} \sum_{i=1}^{N_j} (x_{ij} - \bar{x}_j)^2$	$N-k$	$\frac{\text{SSE}}{N-k}$	σ^2
Total	$\sum_{j=1}^{k} \sum_{i=1}^{N_j} x_{ij}^2 - \frac{G^2}{N}$	$N-1$	—	—

As in the hypothesis tests of regression models the analysis of variance table provides a systematic means of computing the statistics required for the F-test. In practice one usually computes SST and the total sum of squares and then obtains SSE by subtraction. The additivity of the degrees of freedom parameters follows from the independence of SST and SSE, as Cochran's theorem (Theorem 3-4) indicates. The last column contains the expected values of the mean squares. In $E(\text{MST})$,

$$\bar{\tau} = \frac{1}{N} \sum_{j=1}^{k} N_j \tau_j \tag{31}$$

or the weighted average of the treatment effects. We see that unequal treatment effects can only increase the expected treatment mean square.

We may compute the expected value of the treatments mean square by means of the results on quadratic forms and the noncentral chi-squared distribution given in Section 3-2. We have shown that $\text{SST}/\sigma^2 = \bar{\mathbf{x}}'\mathbf{A}\bar{\mathbf{x}}$ has the noncentral chi-squared distribution with $k-1$ degrees of freedom. When the hypothesis of equal treatment effects is not true that distribution has noncentrality parameter

$$\lambda = \boldsymbol{\mu}'\mathbf{A}\boldsymbol{\mu}$$

where $\mathbf{A}$ is the matrix (16) and

$$\boldsymbol{\mu}' = [\mu + \tau_1, \ldots, \mu + \tau_k] \tag{32}$$

Let us evaluate λ. We know from (20) that the common effect μ will be eliminated in the computations. Hence

$$\lambda = \frac{1}{N\sigma^2}[\tau_1, \ldots, \tau_k] \begin{bmatrix} N_1(N - N_1) & \cdots & -N_1 N_k \\ \vdots & & \vdots \\ -N_1 N_k & \cdots & N_k(N - N_k) \end{bmatrix} \begin{bmatrix} \tau_1 \\ \vdots \\ \tau_k \end{bmatrix}$$

$$= \frac{1}{\sigma^2} \sum_{j=1}^{k} N_j(\tau_j - \bar{\tau})^2 \tag{33}$$

By the results in Appendix C for the expected value of a noncentral chi-squared variate,

$$E(\text{SST}) = \sigma^2(k-1) + \sum_{j=1}^{k} N_j(\tau_j - \bar{\tau})^2 \tag{34}$$

and of course $E[\text{SST}/(k-1)]$ follows by dividing both sides by $k-1$.

Example 8-2

As an illustration of the analysis of variance we shall use the dependent variable scores for the second course in Example 3-9. The "treatment" groups will be the years in which the course was given. The dependent variable was the rating scale Overall Evaluation of the Instructor and had the scores 1 (outstanding 10%), . . . , 5 (poor). The data are shown in Table 8-3 as the frequencies of the scores and certain summary statistics.

TABLE 8-3 Rating Score Frequencies

	Year				
Score	0	1	2	3	Total
1	1	6	4	4	15
2	6	5	4	4	19
3	1	0	0	0	1
4	0	0	0	0	0
5	1	0	0	0	1
N_j	9	11	8	8	36
Mean	2.33	1.45	1.50	1.50	—
$\sum (x_{ij} - \bar{x}_j)^2$	10	2.7273	2	2	16.7273
Variance	1.25	0.2727	0.2857	0.2857	—

We may compute

$$G = \sum\sum x_{ij} = 61 \qquad \sum\sum x_{ij}^2 = 125$$

and obtain this analysis of variance table:

Source	Sum of squares	d.f.	Mean square	F
Years	4.9116	3	1.6372	3.13
Within years	16.7273	32	0.5227	—
Total	21.6389	35	—	—

Because the nearest tabulated F critical value is $F_{0.05;3,30} = 2.92$, the hypothesis of equal means in the hypothetical populations of each year is not tenable at the 0.05 level. A reasonable alternative hypothesis would be that of a higher mean in year 0 than in the later three years. To say more at this point would anticipate such matters as contrasts and multiple comparisons of the year effects; we shall return to these data when we discuss those topics.

As in the earlier example we should not take the conclusions of the analysis of variance F-test uncritically. The means for years 1, 2, 3 are virtually equal, while much of the increase in the mean of year 0 is due to the single score 5 in that group. We might treat that score as an unrealistic or aberrant observation, omit it, and see what effect its omission has upon the analysis of variance for the test of equal year means. Then for year 0

$$\bar{x}_1 = 2 \qquad \sum\sum (x_{i1} - \bar{x}_1)^2 = 2$$

and the four within-years variances are very nearly equal. The new F-statistic has value 1.98 with 3 and 31 degrees of freedom and does not exceed any conventional critical value. Although the new mean of year 0 is still higher than those of the other years, we cannot conclude that its value is beyond that attributable to random variation. Apparently, the rating 5 accounts for much of the significance in the first analysis of variance F-test.

8-4 ESTIMATION IN THE LINEAR MODEL

In our treatment of the one-way layout we proceeded directly from the linear model to the analysis of variance for testing the equality of the treatment effects. We did not consider ways of estimating those parameters. Unlike the least squares estimators in linear regression, the linear model estimators may not be attainable directly without imposing certain constraints or restricting the kinds of parametric functions to be estimated. Such properly defined functions are called *estimable*. In this section we define the concept of estimability and consider an important class of estimable functions known as *contrasts*, or measures for comparing treatment effects.

Let us begin by seeing why estimation in the linear model of the preceding section has an elusive quality. Assume that the $N = N_1 + \cdots + N_k$ observations x_{ij} obtained under the respective treatments can be written in a matrix form of the linear model as

$$\begin{bmatrix} x_{11} \\ \cdot \\ \cdot \\ \cdot \\ x_{N_1 1} \\ x_{12} \\ \cdot \\ \cdot \\ \cdot \\ x_{N_2 2} \\ \cdot \\ \cdot \\ \cdot \\ x_{1k} \\ \cdot \\ \cdot \\ \cdot \\ x_{N_k k} \end{bmatrix} = \begin{bmatrix} 1 & 1 & 0 & \cdots & 0 \\ \cdot & \cdot & \cdot & & \cdot \\ \cdot & \cdot & \cdot & & \cdot \\ \cdot & \cdot & \cdot & & \cdot \\ 1 & 1 & 0 & \cdots & 0 \\ 1 & 0 & 1 & \cdots & 0 \\ \cdot & \cdot & \cdot & & \cdot \\ \cdot & \cdot & \cdot & & \cdot \\ \cdot & \cdot & \cdot & & \cdot \\ 1 & 0 & 1 & \cdots & 0 \\ \cdot & \cdot & \cdot & & \cdot \\ \cdot & \cdot & \cdot & & \cdot \\ \cdot & \cdot & \cdot & & \cdot \\ 1 & 0 & 0 & \cdots & 1 \\ \cdot & \cdot & \cdot & & \cdot \\ \cdot & \cdot & \cdot & & \cdot \\ \cdot & \cdot & \cdot & & \cdot \\ 1 & 0 & 0 & \cdots & 1 \end{bmatrix} \begin{bmatrix} \mu \\ \tau_1 \\ \cdot \\ \cdot \\ \cdot \\ \tau_k \end{bmatrix} + \begin{bmatrix} e_{11} \\ \cdot \\ \cdot \\ \cdot \\ e_{N_1 1} \\ e_{12} \\ \cdot \\ \cdot \\ \cdot \\ e_{N_2 2} \\ \cdot \\ \cdot \\ \cdot \\ e_{1k} \\ \cdot \\ \cdot \\ \cdot \\ e_{N_k k} \end{bmatrix} \tag{1}$$

The matrix on the right consisting of ones and zeros is called the *design matrix*, for it describes the plan of the experimental design from which the observations were obtained. Each observation contains μ, whereas only the N_1 from the first treatment group contain τ_1, and so on, for the k treatments. The design matrix is equivalent to $\mathbf{X}$ in an uncentered multiple regression model based on dummy variables. However, unlike a proper dummy variable scheme we have as many variables as treatments, and our design matrix has rank less than its $k + 1$ columns. The second through $(k + 1)$th columns sum to equal the first column, and the rank is only k. The matrix $\mathbf{X}'\mathbf{X}$ would not have an inverse in the conventional sense, so that the usual regression estimators of Chapter 2 would not exist.

Estimation and testing can be developed in these *less than full-rank* models by means of the generalized inverse of $\mathbf{X}'\mathbf{X}$. A thorough treatment of that approach to the analysis of variance has been given by Searle (1971). Because those topics are at a higher mathematical level than that chosen for this text we shall not pursue them. However, the notion of estimability is a fundamental one in experimental design models, and we introduce it now for a general linear model. Let the $N \times 1$ observation vector $\mathbf{x}$ be related to the $p \times 1$ parameter vector $\boldsymbol{\beta}$ by the linear model

$$\mathbf{x} = \mathbf{A}\boldsymbol{\beta} + \mathbf{e} \tag{2}$$

where the $N \times p$ matrix $\mathbf{A}$ has rank r, and the vector $\mathbf{e}$ consists of uncorrelated random disturbances with zero expectations and a common variance σ^2. The linear function

$$\psi = \mathbf{c}'\boldsymbol{\beta} \tag{3}$$

is said to be estimable if there exists some $N \times 1$ vector of constants $\mathbf{t}$ such that

$$E(\mathbf{t}'\mathbf{x}) = \psi \tag{4}$$

Then

$$\begin{aligned} \hat{\psi} &= \widehat{\mathbf{c}'\boldsymbol{\beta}} \\ &= \mathbf{t}'\mathbf{x} \end{aligned} \tag{5}$$

is an estimator of $\mathbf{c}'\boldsymbol{\beta}$.

Let us examine some examples of estimable functions in the one-way layout. The parameter

$$\mu_j = \mu + \tau_j \tag{6}$$

is estimable because

$$\begin{aligned} E(\bar{x}_j) &= \frac{1}{N_j} \sum_{i=1}^{N_j} E(x_{ij}) \\ &= \frac{1}{N_j} \sum_{i=1}^{N_j} (\mu + \tau_j) \\ &= \mu + \tau_j \end{aligned} \tag{7}$$

The mean $\bar{x}_j$ of the observations from the jth treatment group is the estimator of $\mu + \tau_j$. Similarly, the treatment difference

$$\tau_j - \tau_h \qquad j \neq h \tag{8}$$

is estimable because

$$E(\bar{x}_j - \bar{x}_h) = \tau_j - \tau_h \tag{9}$$

However, μ and the τ_j alone are *not* estimable functions: It is impossible to find linear compounds of the observations whose expected values are those parameters. We should remark, though, that they can be made estimable by imposing appropriate constraints on the parameters, for example, the condition $\tau_1 + \cdots + \tau_k = 0$ that the treatment effects sum to zero.

Graybill (1961, Section 11.2; 1976, Section 13.2) and Searle (1971, Sections 5.4, 6.2) have discussed estimable functions for general linear models and, in addition to defining large classes of estimable parametric functions, have given tests for estimability. Of those results one will be especially useful for our purposes in the later sections. Define a *contrast* as the parametric function

$$\psi = \sum_{j=1}^{k} c_j \tau_j \tag{10}$$

with the property

$$\sum_{j=1}^{k} c_j = 0$$

Then *every contrast of the treatment effects is estimable*. The estimator is merely the same contrast of the treatment means, or

$$\hat{\psi} = \sum_{j=1}^{k} c_j \bar{x}_j \tag{11}$$

The estimability of ψ is evident:

$$\begin{aligned} E(\hat{\psi}) &= \sum_{j=1}^{k} c_j E(\bar{x}_j) \\ &= \sum_{j=1}^{k} c_j(\mu + \tau_j) \\ &= \sum_{j=1}^{k} c_j \tau_j \end{aligned} \tag{12}$$

Under the usual normal distribution model $\hat{\psi}$ is a normal random variable with mean ψ and variance

$$\text{Var}(\hat{\psi}) = \sigma^2 \sum_{j=1}^{k} \frac{c_j^2}{N_j} \tag{13}$$

When σ^2 is estimated by the error mean square (10) of Section 8-3,

$$t = (\hat{\psi} - \psi)/\hat{\sigma}\sqrt{\sum_{j=1}^{k} c_j^2/N_j} \tag{14}$$

has the Student–Fisher t-distribution with $N - k$ degrees of freedom. With that statistic we may test the hypothesis

$$H_0\colon \quad \psi = \psi_0 \tag{15}$$

at the α level against some appropriate alternative. For the usual two-sided alternatives these decision rules would be used:

Alternative	Decision rule
$H_1^{(1)}: \psi \neq \psi_0$	Reject H_0 if $\lvert t \rvert > t_{\alpha/2;N-k}$
$H_1^{(2)}: \psi > \psi_0$	Reject H_0 if $t > t_{\alpha;N-k}$
$H_1^{(3)}: \psi < \psi_0$	Reject H_0 if $t < -t_{\alpha;N-k}$

Alternatively, we might wish to find the $100(1-\alpha)\%$ confidence interval for ψ:

$$\sum_{j=1}^{k} c_j\bar{x}_j - t_{\alpha/2;N-k}\,\hat{\sigma}\sqrt{\sum_{j=1}^{k}\frac{c_j^2}{N_j}} \leq \sum_{j=1}^{k} c_j\tau_j$$
$$\leq \sum_{j=1}^{k} c_j\bar{x}_j + t_{\alpha/2;N-k}\hat{\sigma}\sqrt{\sum_{j=1}^{k}\frac{c_j^2}{N_j}} \tag{16}$$

Then the hypothesis (15) would be accepted if the confidence interval contained the value ψ_0 and rejected in favor of the two-sided alternative $H_1^{(1)}$ if ψ_0 is not included.

It is important to remember that the test level α and the confidence coefficient $1-\alpha$ refer to the Type I error rate of the test for a *single* contrast chosen before an examination of the treatment averages. Multiple tests and hypotheses generated a posteriori from the data require simultaneous inference methods: We discuss those in Section 8-6.

Example 8-3

In the course evaluation scores of Example 8-2 the means for years 1, 2, 3 were equal or nearly so, whereas that for year 0 was larger. The significance of the F-statistic would appear to be due to the difference of year 0 with the other years. That difference may be expressed by the contrast

$$\psi = \tau_1 - (\tfrac{1}{3})(\tau_2 + \tau_3 + \tau_4)$$

where the usual subscript convention has been used rather than the 0, 1, 2, 3 designations of the years. The estimate of $\hat{\psi}$ is

$$\hat{\psi} = 2.33 - (\tfrac{1}{3})(1.45 + 1.50 + 1.50) = 0.85$$

and its estimated variance is

$$\begin{aligned}\widehat{\text{Var}(\hat{\psi})} &= \hat{\sigma}^2[\tfrac{1}{9} + \tfrac{1}{9}(\tfrac{1}{11} + \tfrac{1}{8} + \tfrac{1}{8})] \\ &= (0.5227)(0.1490) \\ &= 0.07788\end{aligned}$$

Let us test the hypothesis that $\psi = 0$, or that no difference exists between year 0 and the other three means taken as a single group. The test statistic (14) with $\psi_0 = 0$ is

$$t = \frac{0.85}{\sqrt{0.07788}} = 3.05$$

Since we have no reason to believe that the ratings of year 0 should be higher or lower than those of the other years we shall adopt the two-sided alternative with the 1% critical value $t_{0.005;32} = 2.74$. Because the computed t exceeds that critical value we should reject the zero-contrast hypothesis. Apparently, the year 0 effect stands out from the other years.

As a second contrast we might estimate $\psi = \tau_1 - \tau_2$, and find its 95% confidence interval. The estimate of the contrast is $\hat{\psi} = 0.88$, and its estimated variance is

$$\widehat{\text{Var}}(\hat{\psi}) = \hat{\sigma}^2(\tfrac{1}{9} + \tfrac{1}{11}) = 0.1056$$

The confidence interval is

$$0.88 - (2.038)(0.3250) \leq \psi \leq 0.88 + (2.038)(0.3250)$$
$$0.22 \leq \psi \leq 1.54$$

Because the interval does not contain zero we would reject the hypothesis $H_0: \tau_1 = \tau_2$ at the 5% level.

We might continue to make tests and confidence statements in this fashion, but we must keep in mind that their α levels refer only to the individual inferences. If the contrasts are similar, so that their estimates are highly correlated, the conclusions derived from them will tend to be related. When the treatment group sizes are equal we may protect against this dependence with *orthogonal contrasts*; those are treated in the next section. If the contrasts cannot be made to be orthogonal, we should avoid the hazards of multiple tests on the same set of means with the simultaneous inference methods of Section 8-6.

8-5 ORTHOGONAL CONTRASTS

The nature of the treatments in the one-way layout for experimental or observational data may suggest important contrasts. For example, if the first treatment is a control condition that should not affect the response variable, we might wish to make inferences about the $k - 1$ contrasts comparing the actual treatments with the control. Another case is that in which four treatments can be cross-classified according to the presence or absence of two basic conditions, or *factors*. Such experimental plans are really examples of the 2×2 factorial designs we shall encounter in Chapter 9, but we consider them now in the context of contrasts. If the numbers of observations under the four treatments are equal, the estimates of the contrasts measuring the basic factor effects will be uncorrelated and independently distributed under the usual normality assumptions. Such contrasts are said to be *orthogonal*. Similarly, if the treatments in the one-way design are spaced at equal intervals on a time, temperature, or other scale, and if their sample sizes are equal, we might decompose the treatment sum of squares into linear, quadratic, or higher polynomials by the method of orthogonal polynomials given in Section 4-4. The independence of the contrast estimates enables one to concentrate on individual hypotheses and in that way meet at least one of the common objections to multiple testing. In this section we discuss some applications of orthogonal contrasts.

Let us begin by defining orthogonal contrasts in a general one-way layout. The contrasts

$$\psi_h = \sum_{j=1}^{k} c_{jh}\tau_j \qquad \psi_i = \sum_{j=1}^{k} c_{ji}\tau_j \tag{1}$$

are said to be orthogonal if

$$\sum_{j=1}^{k} \frac{c_{jh}c_{ji}}{N_j} = 0 \qquad h \neq i \tag{2}$$

The estimators of the contrasts are

$$\hat{\psi}_h = \sum_{j=1}^{k} c_{jh}\bar{x}_j \qquad \hat{\psi}_i = \sum_{j=1}^{k} c_{ji}\bar{x}_j \tag{3}$$

and the covariance of the estimators is

$$\begin{aligned}\operatorname{Cov}(\hat{\psi}_h, \hat{\psi}_i) &= \operatorname{Cov}\left(\sum_{j=1}^{k} c_{jh}\bar{x}_j, \sum_{j=1}^{k} c_{ji}\bar{x}_j\right)\\ &= \sigma^2 \sum_{j=1}^{k} \frac{c_{jh}c_{ji}}{N_j}\\ &= 0 \qquad h \neq i\end{aligned} \tag{4}$$

If $N_1 = \cdots = N_k$, or a balanced layout, orthogonal contrasts have orthogonal coefficient vectors:

$$\sum_{j=1}^{k} c_{jh}c_{ji} = 0 \qquad h \neq i \tag{5}$$

The contrast coefficients are said to be normalized if

$$\sum_{j=1}^{k} \frac{c_{jh}^2}{N_j} = 1 \tag{6}$$

By Cochran's theorem given in Section 3-2 it is possible to show that the treatment sum of squares can be decomposed into the sum of $k - 1$ squared normalized orthogonal contrasts:

$$\text{SST} = \left(\sum_{j=1}^{k} c_{j1}\bar{x}_j\right)^2 + \cdots + \left(\sum_{j=1}^{k} c_{j,k-1}\bar{x}_j\right)^2 \tag{7}$$

The contrasts are not unique for three or more treatments, but their number is exactly $k - 1$, the degrees of freedom for the treatments sum of squares.

Let us apply these results for orthogonal contrasts to a one-way balanced layout with N observations in each of its $k = 4$ treatment groups. We have chosen a balanced design for simplicity. In our illustrative example we shall extend these results to unequal sample sizes. The four treatments are the combinations of two basic factors A and B:

Treatment	Factor A	Factor B
1	Absent	Absent
2	Absent	Present
3	Present	Absent
4	Present	Present

For example, in an experiment conducted by a psychologist to measure creativity in problem-solving dyads factor A might be the quality of high self-esteem in the member of the pair designated as "leader" although factor B would be high self-esteem in the other member. The response variable would be the number of acceptable solution proposals per dyad in each of the four combinations. Similarly, in an agricultural field trial of a new variety of potato plots of ground of the same size would be treated with two levels of nitrogen (factor A) and two of phosphate (factor B); the response would be the yield in kilograms of potatoes in each of the $4N$ plots. If we denote the treatment effects by $\tau_1, \ldots, \tau_4$, the contrasts of Table 8-4 would be of particular interest.

TABLE 8-4 Contrasts for a Two-Factor Layout

Contrast	Coefficient	Interpretation
1	$-\frac{1}{2}\ -\frac{1}{2}\ \ \frac{1}{2}\ \frac{1}{2}$	Factor A effect
2	$-\frac{1}{2}\ \ \frac{1}{2}\ -\frac{1}{2}\ \frac{1}{2}$	Factor B effect
3	$\frac{1}{2}\ -\frac{1}{2}\ -\frac{1}{2}\ \frac{1}{2}$	Comparison of A effect with and without B
4	$-1\ \ 1\ \ 0\ 0$	Factor B with A absent
5	$0\ \ 0\ -1\ 1$	Factor B with A present

The first contrast is the average of the treatment effects with factor A present less the average of those two without factor A: It is a direct measure of the effect of the two levels of factor A. The second contrast is a similar measure of the effect of factor B. Contrast 3 is a bit different: It compares the effect of factor B with factor A absent and present. It is called the *interaction effect*, for it measures the degree to which the treatment means are additive functions of the two factors. Large values of contrast 3 would imply that the factors do not act together in an additive fashion. We shall have more to say about the concept of interaction in Chapter 9.

The first three contrasts are mutually orthogonal. If we write their estimators as the contrasts of the treatment means, or

$$
\begin{aligned}
\hat{\psi}_1 &= \tfrac{1}{2}(-\bar{x}_1 - \bar{x}_2 + \bar{x}_3 + \bar{x}_4) \\
\hat{\psi}_2 &= \tfrac{1}{2}(-\bar{x}_1 + \bar{x}_2 - \bar{x}_3 + \bar{x}_4) \\
\hat{\psi}_3 &= \tfrac{1}{2}(\bar{x}_1 - \bar{x}_2 - \bar{x}_3 + \bar{x}_4)
\end{aligned} \tag{8}
$$

then

$$\text{SST} = N(\hat{\psi}_1^2 + \hat{\psi}_2^2 + \hat{\psi}_3^2) \tag{9}$$

The three degrees of freedom for the treatment sum of squares can be ascribed to the independent factor A, factor B, and A $\times$ B interaction effects.

Contrasts 4 and 5 are orthogonal. They measure the factor B effect with A absent and present, respectively. As such, they are also orthogonal to contrast 1. We note that the difference of contrasts 4 and 5 is twice the interaction contrast 3. We might continue to generate similar contrasts within the context of the 2×2 factorial layout, but only three could be linearly independent or orthogonal.

Example 8-4

In the Pittsburgh study of blood lead levels referred to in other examples (Urban, 1976), children were cross-classified by race and sex for the comparison of the mean blood lead levels in those four groups. The sample sizes, means, sums of squares, and variances are shown for the groups in Table 8-5. From those statistics we can compute the grand

TABLE 8-5 Blood Lead Levels (100 μg/ml) of Children in the Pittsburgh Survey

Ethnic group	Sex	
	Male	Female
White		
Sample size	$N_{11} = 151$	$N_{12} = 139$
Mean	$\bar{x}_{11} = 21.40$	$\bar{x}_{12} = 20.39$
Variance	$s_{11}^2 = 48.6550$	$s_{12}^2 = 35.50$
Nonwhite		
Sample size	$N_{21} = 60$	$N_{22} = 57$
Mean	$\bar{x}_{21} = 23.92$	$\bar{x}_{22} = 25.70$
Variance	$s_{21}^2 = 49.9085$	$s_{22}^2 = 63.9989$

mean $\bar{x} = 22.0288$, the between-cells sum of squares, and the within-cells sum of squares for this analysis of variance table:

Source	Sum of squares	d.f.	Mean square	F
Cells	1,415.83	3	471.94	10.16
Within cells	18,725.79	403	46.46	—
Total	20,141.62	406	—	

The F ratio exceeds any reasonable critical value with three and 120 degrees of freedom, and we should be confident that the treatment, or cell, effects are almost certainly different.

Let us denote the treatment effects for the cells (1, 1), (1, 2), (2, 1), (2, 2) by τ_1, τ_2, τ_3, τ_4, respectively. The first contrast we examine is a measure of the difference in average blood lead level between male and female children. The treatment effects have been weighted by the cell sample sizes, so that the sex contrast is

$$\psi_1 = \frac{1}{N_{11} + N_{21}}(N_{11}\tau_1 + N_{21}\tau_3) - \frac{1}{N_{12} + N_{22}}(N_{12}\tau_2 + N_{22}\tau_4) \tag{10}$$

and its estimator is

$$\hat{\psi}_1 = \frac{1}{N_{11} + N_{21}}(N_{11}\bar{x}_{11} + N_{21}\bar{x}_{21}) - \frac{1}{N_{12} + N_{22}}(N_{12}\bar{x}_{12} + N_{22}\bar{x}_{22}) \tag{11}$$

The justification for the weighted contrast rather than the unweighted vector $\mathbf{c}_1' = [1, \ -1, \ 1, \ -1]$ will follow from our treatment of unbalanced two-way layouts in Chapter 9. For the present we merely note that $\hat{\psi}_1$ has smaller variance than the estimator based on $\mathbf{c}_1'$. Then

$$\hat{\psi}_1 = 0.1824 \qquad \hat{\sigma}(\hat{\psi}_1) = 0.6762$$

where $\hat{\sigma}(\hat{\psi}_1) = \sqrt{\text{Var}\,(\hat{\psi}_1)}$. The statistic for testing $H_0: \psi_1 = 0$ is $t = 0.22$, and we should conclude that the average blood lead levels of boys and girls are not different.

The contrast measuring a racial difference is

$$\psi_2 = \frac{1}{N_{21} + N_{22}}(N_{21}\tau_3 + N_{22}\tau_4) - \frac{1}{N_{11} + N_{12}}(N_{11}\tau_1 + N_{12}\tau_2) \tag{12}$$

Its estimate and associated standard deviation are

$$\hat{\psi}_2 = 3.8713 \qquad \hat{\sigma}(\hat{\psi}_2) = 0.7465$$

The 99% confidence interval for ψ_2 is

$$1.95 \leq \psi_2 \leq 5.79$$

We should conclude that some kind of difference obtains between blood lead levels of white and black children in the sample.

Now we pursue the ethnic difference separately for each sex. That contrast for male children, its estimate, and the estimated standard deviation are, respectively,

$$\psi_3 = \tau_3 - \tau_1 \qquad \hat{\psi}_3 = 2.52 \qquad \hat{\sigma}(\hat{\psi}_3) = 1.0402$$

The statistic for testing $H_0\colon \tau_3 = \tau_1$ is $t = 2.42$. We should conclude that the white and nonwhite means are different at the 0.05 significance level. Similarly, the racial contrast for females is

$$\psi_4 = \tau_4 - \tau_2 \qquad \hat{\psi}_4 = 5.31 \qquad \hat{\sigma}(\hat{\psi}_4) = 1.0721$$

and its 99% confidence interval is

$$2.55 \leq \psi_4 \leq 8.07$$

The ethnic difference for females is larger than that for males. We may estimate that difference by the sample contrast

$$\begin{aligned} \hat{\psi}_5 &= \bar{x}_{11} - \bar{x}_{12} - \bar{x}_{21} + \bar{x}_{22} \\ &= 2.79 \end{aligned} \tag{13}$$

Then $\hat{\sigma}(\hat{\psi}_5) = 1.4939$, and the t-statistic for testing the hypothesis

$$H_0\colon \tau_1 - \tau_2 = \tau_3 - \tau_4 \tag{14}$$

of the same racial mean difference for each sex has the value 1.87. The statistic does not exceed the critical value 1.96 for the two-sided alternative, but would exceed the one-sided critical value $t_{0.05;\infty} = 1.645$ if we had decided a priori that the alternative hypothesis should be a greater racial difference for females.

We emphasize again that the significance level or confidence coefficient of each of the contrast inferences applies only to that contrast. Even though some of the later contrasts were suggested by the sample means, we employed individual error rates rather than the forthcoming multiple comparison procedures of Section 8-6. We also note that the contrasts ψ_1 and ψ_2 are not orthogonal because of the unequal sample sizes. Nevertheless, their correlation is small. Contrast ψ_1 is orthogonal with ψ_3, ψ_4, and ψ_5. Similarly, the interaction contrast ψ_5 is orthogonal with ψ_2, and ψ_3 and ψ_4 are orthogonal. We shall have more to say about these properties of unbalanced 2×2 layouts in Chapter 9.

Orthogonal polynomial contrasts. In Section 4-4 we fitted polynomial models to the means of observations taken at different times. Frequently, the treatments in the one-way layout are values of times, temperatures, amounts of a chemical reagent, or other points on a scale. For example, in an agricultural field trial the $k = 3$ treatments might be 0, 40 or 80 pounds of superphosphate applied to each acre under cultivation. In an experiment measuring gasoline consumption of passenger cars the $k = 4$ treatments might be the engine volumes (in round numbers) of 300, 350, 400, and 450 cubic inches. When the treatments have equally spaced scale values and the same number of independent observations at each value we may use the orthogonal polynomial weights to decompose the treatment sum of squares into linearity and curvature of the plot of the means against the treatment scale values.

Let us assume that the observations x_{ij} from the one-way layout have been arranged in the usual $N \times k$ table with the treatments as columns. In keeping with the notation of Chapter 4 we denote the orthogonal polynomial values by

$$\begin{matrix} \phi_1(1), \ldots, \phi_1(k) \\ \vdots \quad\quad \vdots \\ \phi_{k-1}(1), \ldots, \phi_{k-1}(k) \end{matrix} \tag{15}$$

where $\phi_h(t)$ denotes the value of the hth-degree polynomial at the tth treatment. Then the regression coefficients of the first, . . . , $(k - 1)$th orthogonal polynomial are the estimated contrasts

$$\begin{aligned} \hat{\psi}_1 &= \left[\sum_{t=1}^{k} \phi_1(t)\bar{x}_t\right] \Big/ \left[\sum_{t=1}^{k} \phi_1^2(t)\right] \\ &\vdots \\ \hat{\psi}_{k-1} &= \left[\sum_{t=1}^{k} \phi_{k-1}(t)\bar{x}_t\right] \Big/ \left[\sum_{t=1}^{k} \phi_{k-1}^2(t)\right] \end{aligned} \tag{16}$$

The sum of squares due to the k treatments can be decomposed as

$$\begin{aligned} \text{SST} &= N \sum_{j=1}^{k} (\bar{x}_j - \bar{x})^2 \\ &= N \sum_{h=1}^{k-1} \hat{\psi}_h^2 \left[\sum_{t=1}^{k} \phi_k^2(t)\right] \\ &= \frac{N \sum_{h=1}^{k-1} \left[\sum_{t=1}^{k} \phi_h(t)\bar{x}_t\right]^2}{\sum_{t=1}^{k} \phi_h^2(t)} \end{aligned} \tag{17}$$

Because the contrasts are orthogonal the scaled sums of squares

$$S_h/\sigma^2 = \frac{\hat{\psi}_h^2 \left[\sum_{t=1}^{k} \phi_h^2(t)\right]}{\sigma^2} \tag{18}$$

are distributed as independent chi-squared variates with one degree of freedom. We may test the hypothesis of no linear component in the treatment mean plot by referring

$$F = \frac{S_1}{\hat{\sigma}^2} \tag{19}$$

to the critical values of the F-distribution with one and $N - k$ degrees of freedom or, equivalently, by referring

$$t = \hat{\psi}_1 \sqrt{\frac{\sum \phi_h^2(t)}{\hat{\sigma}^2}} \tag{20}$$

to the t-distribution with $N - k$ degrees of freedom. Similarly, we test the hypothesis of no quadratic, cubic, . . . , $(k - 1)$th-degree trends in the means by computing the respective F-statistics

$$F = \frac{S_2}{\hat{\sigma}^2} \cdots F = \frac{S_{k-1}}{\hat{\sigma}^2} \tag{21}$$

or the t-statistics corresponding to (20).

It is essential to note that the data must obey the one-way fixed-effects model with Nk distinct sampling units and independent observations. If instead the data have been obtained by successive measurements on N individuals, we must employ more complex procedures for the analysis of growth data or other examples of repeated measurements with scalable treatments. Those methods may involve generalized least squares or multivariate statistical models beyond the scope of our treatment; one approach to them has been described by Morrison (1976).

Example 8-5

We shall illustrate orthogonal polynomial contrasts with some fictitious, though realistic, data. Suppose that a consumer research group wishes to study the gasoline consumption of large eight-cylinder passenger cars of a given model year. The cars of interest have been classified by their engine sizes of approximately 300, 350, 400, and 450 cubic inches. From each engine size "treatment" population four cars were drawn at random, and the owner of each car drove over a standard urban route three times. These miles per gallon of fuel were recorded for the 16 cars:

	Engine size			
	300	350	400	450
	16.6	14.4	12.4	11.5
	16.9	14.9	12.7	12.8
	15.8	14.2	13.3	12.1
	15.5	14.1	13.6	12.0
Mean	16.2	14.4	13.0	12.1
$\sum (x_{ij} - \bar{x}_j)^2$	1.3	0.38	0.9	0.86

The analysis of variance for the four engine sizes gave these statistics:

Source	Sum of squares	d.f.	Mean square	F
Engine size	38.35	3	12.7833	44.59
Error	3.44	12	0.2867	—

The significant difference among the engine size means is evident. Let us now determine whether the decreasing gas mileage is a linear or more complex function of engine volume.

The linear, quadratic, and cubic orthogonal polynomials have the values given by the contrast vectors

$$\mathbf{c}_1' = [-3, \quad -1, \quad 1, \quad 3]$$
$$\mathbf{c}_2' = [1, \quad -1, \quad -1, \quad 1]$$
$$\mathbf{c}_3' = [-1, \quad 3, \quad -3, \quad 1]$$

The sums of squares (18) and F-statistics (19) and (21) for the three contrasts have these values:

Contrast	S_h	F_h
Linear	37.538	131
Quadratic	0.810	2.83
Cubic	0.002	0.01
Total	38.350	—

The linear trend contrast is overwhelmingly significant with 1 and 12 degrees of freedom. The quadratic contrast, or polynomial coefficient, is less than the 0.10 F critical value. We cannot conclude that the curvature in the plot of the means against volume is more than a sampling artifact. The differences among the means seem to be explainable by a linear trend alone.

8-6 MULTIPLE COMPARISONS AMONG TREATMENT MEANS

The analysis of variance for testing the hypothesis of equal treatment effects is akin to the test in Section 2-4 that all regression coefficients are zero. Large values of the respective F-statistics do not tell us *which* treatments or predictors led to rejection of the hypotheses, and we must follow the test with individual inferences about the treatment effects of the regression coefficients. In the present one-way layout we may make the comparisons with appropriate contrasts and t-statistics as in the preceding section, although with the limitation possessed by those methods: The *significance level or confidence coefficient refers only to the individual test or interval.* If we are generating the contrasts by searching for "interesting" differences among the treatment means, or if we are making a large number of inferences, we should protect ourselves against false rejections by adopting an error rate for the entire family of inferences. If the family error rate is α, and we make m tests or confidence statements $S_1, \ldots, S_m$, then we wish

$$P[(S_1 \text{ true}) \cap (S_2 \text{ true}) \cap \cdots \cap (S_m \text{ true})] = 1 - \alpha \tag{1}$$

or at least

$$P[(S_1 \text{ true}) \cap (S_2 \text{ true}) \cap \cdots \cap (S_m \text{ true})] \geq 1 - \alpha \tag{2}$$

when the null hypothesis of no treatment differences holds true. Such simultaneous inferences or tests will be called *multiple comparisons*, because in the analysis of variance we shall only be interested in comparisons, or contrasts, among the treatments.

In this section we deal with two especially important multiple comparison procedures. The first is due to Scheffé (1953) and permits inferences about all possible contrasts of the treatment effects. The second, due to Tukey (1953), enables one to make multiple comparisons among all pairs of treatments. We also make use of the Bonferroni probability inequality to obtain a family of tests for a specified number of comparisons.

Scheffé multiple comparisons for all contrasts. Suppose that

$$\psi = \sum_{j=1}^{k} c_j \tau_j \tag{3}$$

is a contrast of the k treatment effects in the one-way layout. Its estimator is of course the contrast

$$\hat{\psi} = \sum_{j=1}^{k} c_j \bar{x}_j \tag{4}$$

given by (11), Section 8-4. $\hat{\psi}$ has the estimated standard deviation

$$\hat{\sigma}(\hat{\psi}) = \hat{\sigma}\sqrt{\sum_{j=1}^{k} \frac{c_j^2}{N_j}}$$

where $\hat{\sigma}^2$ is the error mean square from the analysis of variance table. Scheffé (1953, 1959) has shown that the family of all possible contrasts can be enclosed by simultaneous confidence intervals with some family confidence coefficient $1 - \alpha$. The general member of that family of intervals is given by the expression

$$\hat{\psi} - \hat{\sigma}(\hat{\psi})\sqrt{(k-1)F_{\alpha;k-1,N-k}} \leq \psi \leq \hat{\psi} + \hat{\sigma}(\hat{\psi})\sqrt{(k-1)F_{\alpha;k-1,N-k}} \tag{5}$$

or its more explicit form

$$\sum_{j=1}^{k} c_j\bar{x}_j - \hat{\sigma}\sqrt{(k-1)F_{\alpha;k-1,N-k}\sum_{j=1}^{k}\frac{c_j^2}{N_j}} \leq \sum_{j=1}^{k} c_j\tau_j$$
$$\leq \sum_{j=1}^{k} c_j\bar{x}_j + \hat{\sigma}\sqrt{(k-1)F_{\alpha;k-1,N-k}\sum_{j=1}^{k}\frac{c_j^2}{N_j}} \tag{6}$$

The totality of such intervals covers all the contrast values with probability $1 - \alpha$. That is, we may form contrasts of the k treatment effects indefinitely with the assurance that our family coefficient is still $1 - \alpha$.

Alternatively, we may test the hypotheses

$$H_0\colon \sum_{j=1}^{k} c_j\tau_j = 0 \tag{7}$$

for all possible contrasts by referring

$$t = \frac{\hat{\psi}}{\hat{\sigma}\sqrt{\sum_{j=1}^{k} c_j^2/N_j}} \tag{8}$$

to the Scheffé multiple comparison critical value

$$S_{\alpha; k-1, N-k} = \sqrt{(k-1)F_{\alpha; k-1, N-k}} \tag{9}$$

If

$$|t| > S_{\alpha; k-1, N-k}$$

we should reject the particular hypothesis (7) of a zero value for the contrast in favor of the alternative of a nonzero value. In searching a set of treatment means for statistically significant contrasts it is usually simpler to compute the statistics (8) and refer them to the Scheffé critical value. For reporting purposes the simultaneous confidence intervals are more informative.

Example 8-6

Suppose that instructor ratings in the recitation sections of a large college course have been collected for comparison of the $k = 6$ different section instructors. The ratings of the overall competence of the instructors are on the usual five-point scale, with a rating of one indicating highest performance. These statistics were computed for the sections:

	Section					
	1	2	3	4	5	6
Mean	2.52	3.40	2.40	1.65	2.24	2.10
Variance	0.42	0.67	0.37	0.32	0.52	0.59
Students	20	25	30	24	28	21

From them we may compute this analysis of variance:

Source	Sum of squares	d.f.	Mean square	F
Sections	41.3819	5	8.2764	17.29
Within sections	67.99	142	0.4788	

The F-statistic exceeds any reasonable critical value of the F-distribution with 5 and 142 degrees of freedom, and we should clearly conclude that the differences among the section averages are much greater than we would expect from sampling variation.

Now, let us use the Scheffé multiple comparison method to determine which sections are different. We do that by computing the statistics (8)

$$|t_{ij}| = \frac{|\bar{x}_i - \bar{x}_j|}{\hat{\sigma}\sqrt{(1/N_i) + (1/N_j)}}$$

for the $k(k-1)/2 = 15$ distinct pairs of mean differences, and referring them to appropriate Scheffé critical values. The 5% and 1% values are approximately

$$\sqrt{(k-1)F_{0.05;5,120}} = 3.38 \qquad \sqrt{(k-1)F_{0.01;5,120}} = 3.98$$

The second degrees of freedom of 120 was chosen as the entry closest to 148; because it is smaller the critical values are conservative in the sense of leading to error probabilities slightly less than the stated rates 0.05 and 0.01. The values of the $|t_{ij}|$-statistics are shown below, with the recitation sections ordered by their average ratings:

	Section					
	4	6	5	3	1	2
4	—	2.18	3.07	3.96	4.15	8.85
6		—	0.70	1.52	1.94	6.35
5			—	0.88	1.38	6.09
3				—	0.60	5.34
1					—	4.24
2						—

We may represent the significance and nonsignificance of the section differences at the 0.05 level by this diagram:

Section					
4	6	5	3	1	2
1.65	2.10	2.24	2.40	2.52	3.40

The solid lines connecting the sections indicate that their means are not significantly different at the 0.05 level. Gaps in the lines show probable differences in the population means. We see that sections 4, 5, and 6 are not significantly different from one another. At the same time sections 1, 3, 5, and 6 form a set whose means are not different at the 0.05 level. Those ambiguous groupings mean that section 4 is probably different from sections 1 and 3, but the remaining sections 1, 3, 5, and 6 are indistinguishable at the present significance level. Such overlapping groups are a common occurrence in multiple comparisons of means. Finally, we note that section 2 stands alone; its mean is significantly higher than each of the other sections.

Scheffé multiple comparisons and the analysis of variance. The Scheffé simultaneous tests and confidence intervals for all contrasts are related to the analysis of variance for equal treatment effects in these ways:

1. If the F-statistic of the one-way analysis of variance exceeds $F_{\alpha; k-1, N-k}$, then some Scheffé statistic (8) must exceed the simultaneous critical value $S_{\alpha; k-1, N-k}$ defined by (9). Conversely, if the analysis of variance hypothesis cannot be rejected, one cannot find a contrast significant in the Scheffé sense at the same level α.
2. If a contrast has a statistic (8) greater than $S_{\alpha; k-1, N-k}$, the analysis of variance F must exceed $F_{\alpha; k-1, N-k}$. Similarly, if all possible contrasts in the family have statistics smaller than (9), F cannot exceed its critical value.

We see that the analysis of variance and the totality of all Scheffé tests and confidence intervals enjoy a one-to-one relationship. Finding a single significant contrast implies that the equal-means hypothesis will be rejected at the same level, whereas an insignificant F means that a Scheffé multiple comparison search for nonzero contrasts of the treatment effects will be futile. The proof of the relationship follows from algebraic or geometrical derivations of the Scheffé tests and intervals. The geometrical interpretation of the simultaneous confidence intervals as the family of all planes tangent to the $100(1-\alpha)\%$ joint confidence ellipsoid of the contrasts is particularly illuminating; it can be found in Scheffé's text (1959).

Occasionally, the analysis of variance F will exceed its 100α critical value, yet pairwise or other contrasts will not be significant in the Scheffé sense. In that anomalous case we may wish to find the contrast that has caused the rejection of the equal-means hypothesis. It is defined by the contrast vector $\mathbf{c}^*$ such that

$$t^2(\mathbf{c}^*) = \frac{\left(\sum_{j=1}^{k} c_j^* \bar{x}_j\right)^2}{\hat{\sigma}^2 \sum_{j=1}^{k} c_j^{*2}/N_j} \tag{10}$$

is a maximum. We find $\mathbf{c}^*$ by maximizing $t^2(\mathbf{c})$ with the aid of Lagrange multipliers expressing the contrast constraint and a constraint on the denominator term. The derivation will be left as an exercise. The maximizing vector can be found to be

$$\mathbf{c}^* = \begin{bmatrix} N_1(\bar{x}_1 - \bar{x}) \\ \cdot \\ \cdot \\ \cdot \\ N_k(\bar{x}_k - \bar{x}) \end{bmatrix} \tag{11}$$

We note immediately that

$$\begin{aligned} t^2(\mathbf{c}^*) &= \frac{\sum_{j=1}^{k} N_j(\bar{x}_j - \bar{x})^2}{\hat{\sigma}^2} \\ &= (k-1)F \end{aligned} \tag{12}$$

so that the greatest squared contrast statistic is proportional to the analysis of variance F-ratio. Of course, for interpretation and explanation to the investigator who produced the experiment and data we would hope that the contrast $\mathbf{c}^*$ has a meaning with respect to the nature of the k treatments.

Bonferroni multiple comparisons. In Section 3-5 we showed how the Bonferroni inequality for probabilities could be used to obtain multiple tests or simultaneous confidence intervals with a family error rate no greater than some nominal value α. For a family of K tests we carried out each at the α/K significance level. If K simultaneous confidence intervals are to be constructed, each would have confidence coefficient $1-\alpha/K$. If the tests are made on K distinct versions of the hypotheses (7), we would compute the t-statistic (8) and reject the jth hypothesis of a zero value for the contrast if

$$|t_j| > t_{\alpha/2K;N-k} \tag{13}$$

Here the alternative hypothesis is two-sided, as most multiple comparison tests should be if a clear direction for the difference in the treatment means cannot be specified beforehand. Similarly, the $100(1-\alpha)\%$ family of confidence intervals would have the general member

$$\sum_{j=1}^{k} c_j\bar{x}_j - t_{\alpha/2K;N-k}\hat{\sigma}\sqrt{\sum_{j=1}^{k}\frac{c_j^2}{N_j}} \leq \sum_{j=1}^{k} c_j\tau_j \leq \sum_{j=1}^{k} c_j\bar{x}_j + t_{\alpha/2K;N-k}\hat{\sigma}\sqrt{\sum_{j=1}^{k}\frac{c_j^2}{N_j}} \quad (14)$$

In particular, if we wished to make inferences only about the $K = \binom{k}{2} = k(k-1)/2$ contrasts of pairs of treatments, we would compute

$$t_{ij} = \frac{\bar{x}_i - \bar{x}_j}{\hat{\sigma}\sqrt{1/N_i + 1/N_j}} \quad (15)$$

and reject $H_0: \tau_i = \tau_j$ if

$$|t_{ij}| > t_{\alpha/k(k-1);N-k} \quad (16)$$

The corresponding general confidence interval for the ijth treatment effect difference would be

$$\bar{x}_i - \bar{x}_j - t_{\alpha/k(k-1);N-k}\hat{\sigma}\sqrt{1/N_i + 1/N_j} \leq \tau_i - \tau_j \leq \bar{x}_i - \bar{x}_j + t_{\alpha/k(k-1);N-k}\hat{\sigma}\sqrt{1/N_i + 1/N_j} \quad (17)$$

Example 8-7

Let us use the Bonferroni multiple comparison method to determine which section means in Example 8-6 are significantly different. The conservative 0.05 and 0.01 Bonferroni critical values of the t-distribution are approximately

$$t_{0.00167;120} = 2.99 \qquad t_{0.000333;120} = 3.50$$

The significance of the mean differences of a family of tests at the 0.05 level can be represented by this diagram:

Section					
4	6	5	3	1	2
———	———				———
	———	———	———	———	

As in the Scheffé comparisons sections 1, 3, 5, and 6 are not different in the multiple testing sense. Sections 4 and 6 do not differ, but again section 2 stands alone as having a significantly higher mean than any of the others.

Let us illustrate the Bonferroni simultaneous confidence intervals (17) with the five intervals for the successive differences of the ordered section means. We shall use a family confidence coefficient of 0.95. Because the means are ordered we still should consider the number of potential comparisons as $k(k-1)/2 = 15$. These are the confidence intervals:

$$-0.17 \leq \tau_6 - \tau_4 \leq 1.07$$
$$-0.46 \leq \tau_5 - \tau_6 \leq 0.74$$
$$-0.38 \leq \tau_3 - \tau_5 \leq 0.70$$
$$-0.48 \leq \tau_1 - \tau_3 \leq 0.72$$
$$0.26 \leq \tau_2 - \tau_1 \leq 1.50$$

Because the first interval contains zero the hypothesis of equal means for the section 4 and 6 populations is tenable. The zeros in the next three intervals agree with the conclusions from the Bonferroni multiple tests.

Tukey comparisons of means. In some one-way experimental layouts the investigator may only be interested in the $\binom{k}{2} = k(k-1)/2$ comparisons of pairs of treatment means. If the treatments have a common number n of independent observations, we may use a multiple comparison method due to Tukey (1953) based on the range of the treatment means. Now the distribution of the range will depend upon the variance σ^2 of the disturbance terms in the one-way model, as well as the treatment effects τ_j when the null hypothesis is not true. We can eliminate σ^2 by using the *studentized range*

$$q = \frac{\max(\bar{x}_i - \bar{x}_j)}{\hat{\sigma}\sqrt{n}} \tag{18}$$

where $\max(\bar{x}_i - \bar{x}_j)$ is the greatest difference among the k means, and the standard deviation of each mean is

$$\frac{\hat{\sigma}}{\sqrt{n}} = \sqrt{\frac{\text{MSE}}{n}} \tag{19}$$

Studentization refers to the method initiated by W. S. Gosset ("Student") of replacing the unknown standard deviation in a statistic by its sample estimate. When the population treatment means are equal the distribution of q involves only the number of means k and the degrees of freedom $\nu = N - k$ of the within-treatments mean square MSE. The upper 5% and 1% critical values $q_{\alpha;k,\nu}$ of the studentized range distribution have been calculated by Pearson and Hartley (1943) and have been reproduced as Table 7, Appendix A. Because the studentized range is the greatest standardized difference of the treatment means, the joint probability equality

$$P\left(\text{all } i \neq j \, \frac{|\bar{x}_i - \bar{x}_j|}{\hat{\sigma}/\sqrt{n}} \leq q_{\alpha;k,N-k}\right) = 1 - \alpha \tag{20}$$

holds when the null hypothesis is true. That statement enables us to test the $k(k-1)/2$ hypotheses

$$H_0(i,j)\colon \tau_i = \tau_j \qquad i \neq j \tag{21}$$

with a family error rate α: We reject $H_0(i,j)$ if

$$q_{ij} = \frac{|\bar{x}_i - \bar{x}_j|\sqrt{n}}{\hat{\sigma}} \tag{22}$$

exceeds $q_{\alpha:k,N-k}$. Similarly, when the alternative hypothesis of general treatment effects is true the joint probability statement

$$P\left(\text{all } i \neq j \, \frac{|\bar{x}_i - \bar{x}_j - (\tau_i - \tau_j)|}{\hat{\sigma}/\sqrt{n}} \leq q_{\alpha;k,N-k}\right) = 1 - \alpha \tag{23}$$

will hold, and we may invert the double inequality to give the family of simultaneous confidence intervals on all $k(k-1)/2$ treatment effect differences. The general member of the family is

$$\bar{x}_i - \bar{x}_j - \frac{q_{\alpha;k,N-k}\hat{\sigma}}{\sqrt{n}} \leq \tau_i - \tau_j \leq \bar{x}_i - \bar{x}_j + \frac{q_{\alpha;k,N-k}\hat{\sigma}}{\sqrt{n}} \tag{24}$$

By convention we assume that the treatments i and j have been ordered so that the mean differences $\bar{x}_i - \bar{x}_j$ are always positive. We may test the hypothesis (21) by the simultaneous confidence interval in the usual way: If the interval includes zero $H_0(i, j)$ is tenable at the α level. Otherwise, the null hypothesis should be rejected in favor of a difference in the ith and jth treatment effects.

Example 8-8

At a certain university eight graduate programs are grouped together for administrative purposes. Enrollments in the programs for three consecutive years are shown in this table:

	Program							
Year	A	B	C	D	E	F	G	H
1	1	4	7	2	9	2	3	3
2	3	6	12	1	13	6	6	3
3	5	9	7	3	14	4	3	4
Mean	3	6.33	8.67	2	12	4	4	3.33

The analysis of variance for program differences gave these results:

Source	Sum of squares	d.f.	Mean square	F
Program	241.83	7	34.55	8.13
Error	68.00	16	4.25	—
Total	309.83	23	—	—

The hypothesis of equal program means should be rejected at the 0.001 level. Now let us use the Tukey multiple comparison method to determine which programs are different from one another. The 0.01 upper critical value for the studentized range of $k = 8$ means with an estimate of σ^2 based on 16 degrees of freedom is $q_{0.01;8,16} = 6.08$. To test the 28 equal-mean hypotheses we should compare the mean differences to

$$\frac{q_{0.01;8,16}\hat{\sigma}}{\sqrt{n}} = 6.08\sqrt{\frac{4.25}{3}} = 7.24$$

For convenience we start by ordering the programs by their mean enrollments:

D	A	H	F	G	B	C	E
2	3	3.33	4	4	6.33	8.67	12

We see that programs D–C have mean differences less than 7.24. We cannot reject the hypotheses that their means are equal. As with the Scheffé method we shall indicate that grouping by the solid line under the seven means and a line under the program E mean. However, we note that programs B, C, and E have mean differences less than 7.24: They form a homogeneous group indicated by the third solid line. Once again the multiple comparison procedure has not led to disjoint groups of similar means.

Comparison of the simultaneous inference methods. Because the Scheffé family includes *all* contrasts its critical values will always exceed those for the Tukey, or studentized range, method for the same parameters and $k \geq 3$. In the preceding example the Scheffé critical value for comparing the means directly is

$$\sqrt{\hat{\sigma}^2 \frac{2}{n}(k-1)F_{0.01;k-1,k(n-1)}} = 8.94$$

Comparisons among the program means by that criteria would lead to the same disjoint groups (D, A, H, F, G, B, C) and E, but the overlapping group would now consist of programs H, F, G, B, C, and E. If the treatment sample sizes are equal and general treatment contrasts have little meaning the Tukey method is preferable for its more sensitive tests and confidence intervals of shorter average lengths. Whether the Bonferroni inequality method gives shorter confidence intervals for a particular set of treatments will depend on the number K of comparisons to be made.

Some further methods. The Tukey method has been extended to the family of all contrasts (Scheffé, 1959, Section 3-6) and to the case of unequal sample sizes by Spjøtvoll and Stoline (1973), Hochberg (1975, 1976), Genizi and Hochberg (1978), and Stoline and Ury (1979). Dunnett (1980a, 1980b) has investigated pairwise multiple comparisons with unequal sample sizes and different variances. With unequal sample sizes it is probably most convenient to resort to the Scheffé tests and intervals for all contrasts. The Tukey procedure is an extension of one due to Newman (1939) and Keuls (1952), in which preselected subsets of the means were chosen for comparison. The Newman–Keuls method is less conservative than the Tukey inferences for the whole family of comparisons.

Dunnett (1955, 1964) has given methods for comparing $k - 1$ treatments against a control. That situation often occurs in drug trials, in which several new drugs or therapeutic agents must be compared to a standard drug or placebo.

Many other variations on the multiple comparisons problem have been discussed in the large literature on the subject. The standard source is the book by Miller (1981), which contains his survey article (1977) on developments in the period 1966–1976. Other authors have written expository articles (Chew, 1976) or monographs (Chew, 1977).

8-7 POWER AND SAMPLE SIZE DETERMINATION

When the treatment effects are unequal the F-statistic in the one-way analysis of variance has the noncentral F-distribution described in Section 3-2 and Appendix C. The parameters of the distribution are $k - 1$ and $N - k$ degrees of freedom, and the noncentrality parameter

$$\lambda = \frac{1}{\sigma^2} \sum_{j=1}^{k} N_j(\tau_j - \bar{\tau})^2 \tag{1}$$

in which

$$\bar{\tau} = \frac{1}{N} \sum_{j=1}^{k} N_j \tau_j \tag{2}$$

for $N = N_1 + \cdots + N_k$. If each treatment has the same number of observations $n = N_1 = \cdots = N_k$ the noncentrality parameter becomes

$$\lambda = \frac{n}{\sigma^2} \sum_{j=1}^{k} \tau_j^2 \tag{3}$$

where for simplicity we have assumed that the treatment effects sum to zero. That can always be accomplished by properly defining the general effect μ in the model. The power of the one-way analysis of variance is the probability of rejecting the equal-treatment effects hypothesis at a given level α when the alternative of unequal effects is true:

$$\begin{aligned} \text{Power} &= P(\text{reject } H_0\colon \tau_1 = \cdots = \tau_k \mid \tau_j \text{ unequal}) \\ &= P(F' > F_{\alpha; k-1, N-k} \mid \lambda, k - 1, N - k) \end{aligned} \tag{4}$$

F' denotes the noncentral F-variate with parameters λ, $k - 1$, and $N - k$; in the present context it is the F-ratio for testing equality of the treatment effects. Power probabilities for the noncentral F-distribution can be obtained from Charts 1 through 10 of Appendix A (Pearson and Hartley, 1951, 1972) or the Tiku tables (1967). The 10 charts correspond to the values $\nu_1 = k - 1$ of the first degrees of freedom parameter. The two families of curves in each chart refer to the levels $\alpha = 0.05$ and $\alpha = 0.01$ of the hypothesis test. The individual curves pertain to values of the second degrees of freedom parameter $\nu_2 = N - k$. As in the multiple regression case of Chapter 3 we enter each chart with the modified noncentrality parameter

$$\begin{aligned} \phi &= \sqrt{\frac{\lambda}{\nu_1 + 1}} = \sqrt{\frac{\lambda}{k}} \\ &= \frac{\sqrt{\sum N_j(\tau_j - \bar{\tau})^2}}{\sigma\sqrt{k}} \end{aligned} \tag{5}$$

We read up to the curve for $N - k$, then across to the vertical axis for the power probability.

Let us illustrate the power calculations for a small one-way layout in which each of the $k = 4$ treatments had $n = 6$ observations. The analysis of variance for testing the hypothesis of equal treatment effects will be carried out at the

$\alpha = 0.01$ level. We shall use the third Pearson–Hartley chart, or that for $\nu_1 = 3$, and the curve in the right-hand family corresponding to $k(n-1) = 20$. Now, what values should we take for the treatment effects? First, we express the τ_j as deviations about their mean, so that $\bar{\tau} = 0$. The τ_j will also be expressed as units of the unknown population standard deviation σ. From the nature of the response variable and the treatments we must propose values for the standardized effects τ_j/σ. A representative selection is given in the left-hand column of Table 8-6. When the greatest difference in the standardized treatment effects is about $\frac{4}{3}$ or $\frac{3}{2}$ the power probability is in the vicinity of 0.30. The probability of detecting such small differences in treatment effects is only about 1 in 3. When the greatest difference increases to 2σ the probability of its detection lies in the vicinity of 0.50 to 0.73. If the greatest difference is of the order of 3σ, the power of the test may be quite high.

TABLE 8-6 Power Probabilities for a One-Way Layout With $k = 4$ Treatments ($\alpha = 0.01$)

Treatment effects, τ_j/σ						
1	2	3	4	λ	ϕ	Power
1	$-\frac{1}{3}$	$-\frac{1}{3}$	$-\frac{1}{3}$	8	1.41	0.30
1	$-\frac{1}{2}$	$-\frac{1}{2}$	0	9	1.5	0.35
1	-1	0	0	12	1.73	0.50
$\frac{4}{3}$	$-\frac{2}{3}$	$-\frac{2}{3}$	0	16	2	0.67
$\frac{3}{2}$	$-\frac{1}{2}$	$-\frac{1}{2}$	$-\frac{1}{2}$	18	2.12	0.73
$\frac{3}{2}$	$-\frac{3}{4}$	$-\frac{3}{4}$	0	20.25	2.25	0.80
$\frac{3}{2}$	$-\frac{3}{2}$	0	0	27	2.6	0.92
2	$-\frac{2}{3}$	$-\frac{2}{3}$	$-\frac{2}{3}$	32	2.83	0.96
2	-1	-1	0	36	3	0.98

We note from the charts that the power probabilities increase monotonically with the noncentrality parameters λ or ϕ. That is, as the common sample size n increases the probability of detecting treatment differences approaches certainty. Similarly, the probability goes to 1 as the treatment effects become more different.

For the determination of the sample size n for designing a single-way experiment the noncentrality parameter λ is not the most convenient way of describing the alternative hypothesis of different treatment effects. We are rarely able to specify all the effects τ_j or their values in units of σ. Instead, we would prefer to have a lower bound on λ when two of the treatment effects differ by δ or more. δ is the smallest absolute treatment difference we wish to be able to detect with a specified power probability. Because we are entering the charts with a lower bound on λ (and similarly on ϕ) the power probabilities will be *conservative*. More extreme differences among the treatment effects can only increase the power. The lower bound on λ as a function of

$$\delta = \min_{i \neq j} |\tau_i - \tau_j| \tag{6}$$

is

$$\frac{n\delta^2}{2\sigma^2} \leq \lambda \tag{7}$$

and

$$\frac{\delta}{\sigma}\sqrt{\frac{n}{2k}} \leq \phi \tag{8}$$

The bound follows from the physical and geometrical meaning of the second central moment; Scheffé (1959, pp. 63–64) has given an intuitively appealing derivation of it. We may describe the power of a particular design in terms of its least value for detecting a treatment difference as small as δ.

The sample size n for each treatment can also be obtained conveniently from the lower bound (8) and the Pearson–Hartley charts. We begin by specifying the smallest difference δ we wish to detect with a given power $1 - \beta$. We then must pick a trial value n_0 of the common treatment sample size, compute the second degrees of freedom parameter $k(n_0 - 1)$, and see if the power requirement is satisfied. If the power is less than $1 - \beta$ we must increase n_0 until the probability is at least $1 - \beta$. If our initial power exceeds $1 - \beta$ we may reduce n_0 until we just meet the requirement. Because n_0 and the curve corresponding to $k(n_0 - 1)$ are both changing it is necessary to proceed in this trial-and-error, or iterative, fashion. We illustrate the process in the next example.

Example 8-9

Suppose that we are faced with the following experimental design. A cardiologist wishes to study the differences in heart rate changes among $k = 5$ diagnostic groups after administration of a drug. The drug is known to have therapeutic benefits for patients who have undergone cardiac surgery, but it has not been determined whether the drug's effects will differ according to the diagnostic category of the patient. It is proposed that a one-way fixed-effects design be used to search for such differences. The response variable will be the post-drug–pre-drug difference in heart rate. Concomitant variables such as age, sex, and medical history will be controlled as carefully as possible. At the same time other variables in addition to heart rate will be measured to test other hypotheses about the diagnostic groups. Nevertheless, we shall only be interested in heart rate change for our determination of sample size.

From previous data let the standard deviation of heart rate change be $\sigma = 12$. The smallest group difference we wish to detect with power 0.95 will be 18. Then the lower bound (8) on the Pearson–Hartley chart parameter will be $0.4743\sqrt{n}$. These power probabilities can be obtained from the family of curves for $\alpha = 0.05$ by starting with the sample size $n = 10$:

n	Lower bound on ϕ	$5(n-1)$	Power
10	1.5	45	0.76
12	1.64	55	0.86
14	1.77	65	0.91
16	1.90	75	0.95

For a probability at least 0.90 of detecting a difference as small as 18 beats/minute we would need $n = 14$ patients in each diagnostic group. If the power is to be 0.95, we would need $n = 16$ patients in each group. Had we chosen $\alpha = 0.01$ for the level of the analysis of variance test we would need even larger sample sizes.

Because the preceding results were obtained from the lower bound on the noncentrality parameter their sample sizes may be unnecessarily large. Let us postulate more complete specifications of the treatment effects than the requirement that the smallest real difference is 18. Suppose that the groups had these parameters:

Diagnostic group	τ_j
1	10.8
2	10.8
3	−7.2
4	−7.2
5	−7.2
$\sum \tau^2$	388.8

$$\lambda = 2.7n \qquad \phi = 0.735\sqrt{n}$$

Let us determine the sample size as before for $\alpha = 0.05$ and $1 - \beta = 0.90$:

n	ϕ	$5(n-1)$	Power
5	1.64	20	0.75
6	1.80	25	0.85
7	1.94	30	0.92
8	2.08	35	0.95

$n = 7$ subjects per group would meet that power requirement—essentially half the sample size required by the lower bound approach. A more efficient experiment can be designed if we are able to hypothesize the individual treatment effects.

8-8 THE RANDOM-EFFECTS MODEL

In Sections 8-2 to 8-7 our treatment of the analysis of variance has been motivated by this model: Some variable of interest is observed under a fixed number k of experimental conditions to determine if the conditions differ in their effects on the variable. We are interested only in the k treatments, and we assume that their effects change the population means of the observed random variable. The observations collected under each treatment are independently distributed, and their sampling variation is reflected in the error, or disturbance, term e_{ij} in the model

$$x_{ij} = \mu + \tau_j + e_{ij}$$

for the ith observation under the jth condition. Our queries about treatment differences reduced to inferences about the *parameters* $\tau_1, \ldots, \tau_k$, or equivalently, about the means $\mu + \tau_j$ of the x_{ij}. Our conclusions referred only to the k conditions of the experimental design.

Now we modify our experimental design and model to one in which the treatment effects are not parameters, but realizations of random variables. Let us begin with an experimental situation which will lead to that model. Suppose that a large number of workers in a manufacturing plant operate machines, interact with a computer display, or carry out similar tasks that depend on fast reaction time, manual dexterity, or other skills. The operators for a particular task have been chosen carefully and qualified by experience, but we would like to determine to what extent they are different in their average reaction times to visual stimuli. The number of operators is too large to measure the reaction times of each, so we elect to draw a random sample of k operators out of the entire population and obtain several independent observations of reaction time to the stimulus for each. The time x_{ij} of the jth operator at the ith independent presentation of the stimulus will be represented by the model

$$x_{ij} = \mu + u_j + e_{ij} \qquad i = 1, \ldots, n; \quad j = 1, \ldots, k \tag{1}$$

In our initial development of the random-effects model we shall assume that the data are balanced with the same number of observations n for each treatment. Because unbalanced designs lead to various complications we shall only treat them briefly later in the section. In the model μ is a general location parameter for the reaction times. u_j is a measure of the jth operator's reaction time: Because that operator was selected at random from a large population of operators u_j is a random variable; its mean and variance are

$$E(u_j) = 0 \qquad \text{Var}\,(u_j) = \sigma_T^2 \tag{2}$$

We also assume that u_j is normally distributed. As in the fixed-effects model the individual disturbance terms are independently distributed as normal variates with mean zero and a common variance σ^2 for all i and j. The treatment effects u_j and the e_{ij} are also independently distributed. Because the random-effects model (1) is linear the x_{ij} are normally distributed with mean μ and variance

$$\text{Var}\,(x_{ij}) = \sigma_T^2 + \sigma^2 \tag{3}$$

We see that the variance has components due to operator variation (σ_T^2) and experimental error (σ^2). The random model is often referred to as the *variance components model*, for the usual hypothesis of equal treatment effects will now be defined in terms of a variance. It is also known as Model II of the analysis of variance, as opposed to the fixed-effects Model I.

Unlike the observations in the fixed-effects model the x_{ij} for the same treatment (operator) are correlated. Let us determine the covariance of the ith and hth observations for the jth operator:

$$\begin{aligned} \text{Cov}\,(x_{ij}, x_{hj}) &= \text{Cov}\,(\mu + u_j + e_{ij}, \mu + u_j + e_{hj}) \\ &= \text{Cov}\,(u_j, u_j) + \text{Cov}\,(e_{ij}, e_{hj}) \\ &= \begin{cases} \sigma_T^2 + \sigma^2 & i = h \\ \sigma_T^2 & i \neq h \end{cases} \end{aligned} \tag{4}$$

We note the covariances of distinct observations are all equal to σ_T^2, and of course the variance is the common value given by (3). The vector of observations $\mathbf{x}_j' = [x_{1j}, \ldots, x_{nj}]$ obtained under the jth treatment has the multinormal distribution with the equal-variance, equal-covariance matrix. We shall need the properties of that distribution in our test for equal treatment effects.

Because the treatment effects are random variables we state the hypothesis of no difference among treatments as

$$H_0: \sigma_T^2 = 0 \tag{5}$$

or that the treatments do not contribute to the variation in the observations; the alternative hypothesis is of course

$$H_1: \sigma_T^2 > 0 \tag{6}$$

or that the treatment effects do indeed vary. Remarkably, the test of H_0 is carried out by the same F-statistic as in the fixed-effects analysis of variance in Section 8-3. We compute the treatment and error sums of squares

$$\begin{aligned} \text{SST} &= n \sum_{j=1}^{k} (\bar{x}_j - \bar{x})^2 \\ \text{SSE} &= \sum_{j=1}^{k} \sum_{i=1}^{n} (x_{ij} - \bar{x}_j)^2 \end{aligned} \tag{7}$$

the mean squares

$$\text{MST} = \frac{\text{SST}}{k-1} \qquad \text{MSE} = \frac{\text{SSE}}{k(n-1)}$$

as before, and the test statistic

$$\begin{aligned} F &= \frac{k(n-1)}{k-1} \frac{\text{SST}}{\text{SSE}} \\ &= \frac{\text{MST}}{\text{MSE}} \end{aligned} \tag{8}$$

When $H_0: \sigma_T^2 = 0$ is true F has the central F-distribution with degrees of freedom $k - 1$ and $k(n - 1)$. We reject the hypothesis of no treatment variance component at the α level if

$$F > F_{\alpha; k-1, k(n-1)} \tag{9}$$

The distribution of the statistic (8) follows from the application of the results on quadratic forms given in Section 3-2. Let us begin by showing that SSE/σ^2 has the chi-squared distribution with $k(n - 1)$ degrees of freedom. Write

$$\text{SSE} = \text{SSE}_1 + \cdots + \text{SSE}_k \tag{10}$$

in terms of the sums of squares

$$\text{SSE}_j = \sum_{i=1}^{n} (x_{ij} - \bar{x}_j)^2 \tag{11}$$

within each treatment group. Now each quadratic form SSE_j has the $n \times n$ idempotent matrix

$$\mathbf{A} = \frac{1}{n} \begin{bmatrix} n-1 & -1 & \cdots & -1 \\ -1 & n-1 & \cdots & -1 \\ \cdots & \cdots & \cdots & \cdots \\ -1 & -1 & \cdots & n-1 \end{bmatrix} \tag{12}$$

and its variates have the $n \times n$ covariance matrix

$$\mathbf{\Sigma} = \begin{bmatrix} \sigma_T^2 + \sigma^2 & \sigma_T^2 & \cdots & \sigma_T^2 \\ \sigma_T^2 & \sigma_T^2 + \sigma^2 & \cdots & \sigma_T^2 \\ \cdots & \cdots & \cdots & \cdots \\ \sigma_T^2 & \sigma_T^2 & \cdots & \sigma_T^2 + \sigma^2 \end{bmatrix} \tag{13}$$

as specified by (4). Then the conditions of Theorem 3-1 for a chi-squared distribution are satisfied:

$$\left(\frac{\mathbf{A}}{\sigma^2}\right)\mathbf{\Sigma} = \mathbf{A} \tag{14}$$

is an idempotent matrix, and because the variates have the same expectation μ the noncentrality parameter is zero. SSE_j/σ^2 is a central chi-squared variate with $n - 1$ degrees of freedom, and the sum SSE/σ^2 is chi-squared with $k(n - 1)$ degrees of freedom. The within-treatment covariances have not affected the distribution of the error sum of squares.

The distribution of the treatment sum of squares follows more directly. The matrix of its quadratic form is

$$\mathbf{B} = \frac{n}{k}\begin{bmatrix} k-1 & -1 & \cdots & -1 \\ -1 & k-1 & \cdots & -1 \\ \cdots & \cdots & \cdots & \cdots \\ -1 & -1 & \cdots & k-1 \end{bmatrix} \tag{15}$$

The k treatment means are independently and normally distributed with a common mean μ and variance

$$\text{Var}(\bar{x}_j) = \sigma_T^2 + \left(\frac{1}{n}\right)\sigma^2 \tag{16}$$

Then

$$\frac{\text{SST}}{n\sigma_T^2 + \sigma^2} \tag{17}$$

has the associated matrix product required for Theorem 3-1 equal to

$$\frac{\mathbf{B}}{n\sigma_T^2 + \sigma^2}\left(\sigma_T^2 + \frac{1}{n}\sigma^2\right) = \frac{1}{k}\begin{bmatrix} k-1 & -1 & \cdots & -1 \\ -1 & k-1 & \cdots & -1 \\ \cdots & \cdots & \cdots & \cdots \\ -1 & -1 & \cdots & k-1 \end{bmatrix} \tag{18}$$

The matrix is idempotent, and the scaled treatment sum of squares (17) has the central chi-squared distribution with $k - 1$ degrees of freedom. Each mean $\bar{x}_j$ is independent of its within-treatment sum of squares SSE_j and of the sums of squares within the other groups. SST and SSE are independently distributed, and

$$W = \frac{\sigma^2}{n\sigma_T^2 + \sigma^2}\,\frac{\text{MST}}{\text{MSE}} \tag{19}$$

has the central F-distribution with $k - 1$ and $k(n - 1)$ degrees of freedom. When $\sigma_T^2 = 0$ the test statistic (8) follows. We shall presently return to the case of the alternative hypothesis and the calculation of power probabilities.

We may estimate the two variance components of the random-effects model by equating the mean squares to their expectations and solving for the component estimators. The expected error mean square is

$$E(\text{MSE}) = \sigma^2 \tag{20}$$

and clearly

$$\hat{\sigma}^2 = \text{MSE} \tag{21}$$

Similarly, the expectation of the treatment mean square is

$$E\left(\frac{\text{MST}}{k-1}\right) = \frac{(k-1)(n\sigma_T^2 + \sigma^2)}{k-1} = n\sigma_T^2 + \sigma^2 \tag{22}$$

and the estimator of the treatment effect variance is

$$\hat{\sigma}_T^2 = \frac{1}{n}(\text{MST} - \hat{\sigma}^2) \tag{23}$$

Unfortunately, the estimator can be negative—hardly a reasonable value for a variance. Negative estimates occur whenever $\text{MST} < \hat{\sigma}^2$ or if the F-statistic (8) is less than one. If the null hypothesis is true, such values of F are highly probable, and in the case of slight variation among treatments one should expect negative estimates of σ_T^2. The probability of a negative estimate can be calculated from (19) or from the results in the sequel on the power function. Perhaps the best conclusion from negative estimates of σ_T^2 is that the treatments have no real effect, or whatever real effects are present are slight relative to the within-group variance.

Example 8-10

Suppose that $k = 5$ drivers for a delivery service have been drawn at random from a large list of the company's drivers with similar ages and safety records. Among the many tests administered to the drivers was one measuring reaction time to a visual stimulus in hundredths of a second. $n = 8$ reaction times were recorded for each driver. These were in fact the medians of sets of three reaction times. The medians were chosen to eliminate extreme times. The first few trials were discarded to reduce the effect of the unfamiliarity of the experimental situation. These data were obtained from the experiment:

	Driver				
	1	2	3	4	5
	18	25	20	19	20
	22	24	21	15	17
	18	25	23	16	17
	20	20	24	17	18
	21	23	21	21	16
	19	24	19	18	15
	22	20	22	18	18
	20	23	18	20	15
Mean	20	23	21	18	17
$\sum (x_{ij} - \bar{x}_j)^2$	18	28	28	28	20

This analysis of variance was computed:

Source	Sum of squares	d.f.	Mean square	F
Drivers	182.4	4	45.6	13.08
Within drivers	122.0	35	3.4857	—
Total	304.4	39	—	—

The F-ratio exceeds the nearest critical value $F_{0.01;4,30} = 4.02$, and we should conclude that the hypothesis of no variation among drivers is untenable. Real differences in reaction time exist in the population of drivers. The estimates of the variance components are

$$\hat{\sigma}^2 = 3.49 \qquad \hat{\sigma}_T^2 = 5.26$$

We note that the within-driver variance is of the same order of magnitude as that among drivers.

Our inferences about σ_T^2 usually stop with the hypothesis test and estimate, although confidence intervals can be constructed for σ^2, σ_T^2, and certain functions of them (Searle, 1971, Section 9.9). Unlike the fixed model, we do not have the multiple comparisons problem of deciding which treatments are different. Our inferences are being made about a larger population of treatments and the degree of variation in that population.

Power and sample size determination. Recall from expression (19) that the scaled variance ratio

$$W = \frac{\sigma_T^2}{n\sigma_T^2 + \sigma^2} \frac{\text{MST}}{\text{MSE}} \tag{24}$$

has the F-distribution with $k - 1$ and $k(n - 1)$ degrees of freedom. Then the probability that the Model II test statistic $F = \text{MST}/\text{MSE}$ introduced in (8) exceeds its $100\alpha\%$ critical value is the power of the test

$$\begin{aligned} 1 - \beta &= P(F \geq F_{\alpha;k-1,k(n-1)}) \\ &= P\left(\frac{\sigma^2}{n\sigma_T^2 + \sigma^2} F \geq \frac{\sigma^2}{n\sigma_T^2 + \sigma^2} F_{\alpha;k-1,k(n-1)}\right) \\ &= P\left\{W \geq \frac{\sigma^2}{n\sigma_T^2 + \sigma^2} F_{\alpha;k-1,k(n-1)}\right\} \end{aligned} \tag{25}$$

The power function of the random-effects model test can be computed from extensive tables of the ordinary F-distribution or the incomplete beta function. Charts of the power have been given by Bowker and Lieberman (1972). Scheffé (1959, pp. 237–238) has considered the problem of the optimal allocation of nK sampling units to k treatments with n observations for each treatment.

As in the fixed-effects model the sample size n guaranteeing a test of specified power $1 - \beta$ must be determined iteratively because it appears both in the noncentrality factor and the degrees of freedom of the critical value. However, we can simplify the algebra of the equation a bit. For a given power $1 - \beta$ the F-variate W defined by (24) must satisfy

$$1 - \beta = P(W \geq F_{1-\beta;k-1,k(n-1)}) \tag{26}$$

and for the power requirement to hold we must have

$$F_{1-\beta;k-1,k(n-1)} = \frac{\sigma^2}{\sigma^2 + n\sigma_T^2} F_{\alpha;k-1,k(n-1)} \tag{27}$$

But $1 - \beta$ is usually a probability 0.9 or larger, and we should transform the critical value into the upper percentage point usually tabulated for the F-distribution. That is accomplished by reversing the degrees of freedom and taking the reciprocal:

$$F_{\beta;k(n-1),k-1} = \frac{1}{F_{1-\beta;k-1,k(n-1)}} \tag{28}$$

Then we may express the implicit equation (27) for n as

$$n = \frac{\sigma^2}{\sigma_T^2}(F_{\beta;k(n-1),k-1}F_{\alpha;k-1,k(n-1)} - 1) \tag{29}$$

To find n for α, β, and given ratio σ^2/σ_T^2 of the within and treatment variances we proceed in the following manner. We pick a trial value n_0, and use it to find the two critical values in the right-hand side of (29). If the resulting number exceeds n_0, we try large values of n until the formula yields an integer no larger than the preceding trial value. Let us illustrate the process in the following example.

Example 8-11

Mixes of concrete for highway and building construction are prepared in a batching plant and loaded in trucks for mixing in transit to the construction site. On a large highway project we shall measure the crushing strength of cubes of the concrete from different trucks to test the hypothesis that the quality of the concrete does not vary from truck to truck and to estimate the truck variance component. Because a large number of transit-mix trucks are used it will be necessary to draw a random sample of them and to employ the random-effects model to make inferences to the population of trucks. $k = 4$ trucks will be sampled. We wish to determine the number of 1-inch cubes of concrete that should be molded from the mix of each truck and cured 30 days for the crushing test. We shall specify that the hypothesis $H_0: \sigma_T^2 = 0$ of no truck-to-truck variation should be rejected by a test of level $\alpha = 0.01$ with power probability 0.95 when the alternative hypothesis

$$H_1: \sigma_T^2 = \lambda\sigma^2$$

is true for a given multiple λ.

The sample size n meeting these requirements is the smallest integer satisfying the implicit equation (29):

$$n = \frac{1}{\lambda}(F_{0.05;4(n-1),3}F_{0.01;3,4(n-1)} - 1)$$

Suppose that $\lambda = 1$, or the truck variance is equal to the variance within the mix. We shall begin with the trial value $n_0 = 16$. Then

$$n = (F_{0.05;60,3}F_{0.01;3,60} - 1) = 34.39$$

or a much larger value than the initial value; we must try a larger sample size. If $n_0 = 31$,

$$n = (F_{0.05:120,3}F_{0.01;3,120} - 1) = 33.77$$

so that the sample size should be a little larger. If we use values of 32 and 33 observations with crude linear extrapolation of the percentage points beyond those tabulated with 120 degrees of freedom we may verify that we need at least $n = 33$ concrete cubes from each truck to detect a truck variance component of σ^2 or larger.

Sample sizes required for the same power and level probabilities are shown below for different values $\lambda = \sigma_T^2/\sigma^2$:

λ	Minimum observations from each truck
0.5	65
1	33
2	17
3	12
4	10
8	6

Some further aspects of the random model. We shall encounter the random-effects model in Chapter 9 for two-way layouts, as well as in the *mixed model* case in which some treatments are fixed and others random. However, even the one-way random-effects model has a number of interesting variations and features. We shall describe some of these briefly.

Unbalanced data. We restricted our attention to a common number n of observations for each of the k treatments to simplify the estimation of σ_T^2 and the power function of the analysis of variance. If the sample sizes are $N_1, \ldots, N_k$, it is possible to show that the expectation of the treatments mean square is

$$E\left(\frac{\text{SST}}{k-1}\right) = \sigma^2 + n_0\sigma_T^2 \tag{30}$$

where, as in the fixed-effects model,

$$\text{SST} = \sum_{j=1}^{k} N_j(\bar{x}_j - \bar{x})^2 \tag{31}$$

and

$$n_0 = \frac{N^2 \sum_{j=1}^{k} N_j^2}{N(k-1)} \qquad N = \sum_{j=1}^{k} N_j \tag{32}$$

n_0 is found by calculating the expectation of SST under the random-effects model and collecting terms in σ^2 and σ_T^2; the derivation is given in various texts on the theory of linear models [e.g., Graybill (1961, Section 16.5)]. In practice we may estimate σ_T^2 by

$$\hat{\sigma}_T^2 = \frac{1}{n_0}\left(\frac{\text{SST}}{k-1} - \hat{\sigma}^2\right) \tag{33}$$

by the same argument that led to (23) in the balanced layout. As in that case $\hat{\sigma}_T^2$ may be negative. Other methods in estimating $\hat{\sigma}_T^2$ have been described *in extenso* by Searle (1971, Chapter 10).

Intraclass correlation. The correlation between any two observations under the same treatment is

$$\rho = \frac{\text{Cov}(x_{ij}, x_{hj})}{\sqrt{\text{Var}(x_{ij})\,\text{Var}(x_{hj})}}$$

$$= \frac{\sigma_T^2}{\sigma_T^2 + \sigma^2} \qquad i \neq h \tag{34}$$

ρ is called the *intraclass correlation coefficient*, for it is the common correlation of observations within any of the treatment classes. It is interpretable in the random-effects model as the proportion of observation variance due to the random treatment effect. Because of that variance component definition ρ cannot be negative, although the $n \times n$ covariance matrix will still be positive semidefinite if $-1/(n-1) \leq \rho \leq 1$. We may estimate the intraclass correlation coefficient by replacing the variance components in (34) by their estimators:

$$\hat{\rho} = \frac{\text{MST} - \text{MSE}}{\text{MST} + (n-1)\text{MSE}} \tag{35}$$

$\hat{\rho}$ is a biased estimator. We note that

$$\hat{\rho} = \frac{(\text{MST}/\text{MSE}) - 1}{(\text{MST}/\text{MSE}) + n - 1} = \frac{F-1}{F+n-1} \tag{36}$$

in which F is the usual observed statistic (8) for testing $H_0: \sigma_T^2 = 0$.

The test for zero intraclass correlation is identical with that for no treatment variation, as the definition (34) should clearly imply. Confidence intervals for ρ are useful adjuncts to its estimate, and we can obtain exact ones from the central F variate defined by expression (24). In that result

$$W = \frac{\sigma^2}{\sigma^2 + n\sigma_T^2}F$$

$$= \frac{(1-\rho)F}{1+\rho(n-1)} \tag{37}$$

had the F-distribution. By definition,

$$P\left\{F_{1-\alpha/2;k-1,k(n-1)} \leq \frac{(1-\rho)F}{1+\rho(n-1)} \leq F_{\alpha/2;k-1,k(n-1)}\right\} = 1-\alpha \tag{38}$$

where the left- and right-hand limits are the respective lower and upper $100(\alpha/2)\%$ critical values of the F-distribution with $k-1$ and $k(n-1)$ degrees of freedom. What we must do to find the $100(1-\alpha)\%$ confidence limits on ρ is to express the inequalities of (38) as ones on ρ. Successive manipulations lead to this confidence interval:

$$\frac{F/F_{\alpha/2;k-1,k(n-1)} - 1}{(F/F_{\alpha/2;k-1,k(n-1)} - 1) + n} \leq \rho \leq \frac{F/F_{1-\alpha/2;k-1,k(n-1)} - 1}{(F/F_{1-\alpha/2;k-1,k(n-1)} - 1) + n} \tag{39}$$

Recall that

$$F_{1-\alpha/2;k-1,k(n-1)} = 1/F_{\alpha/2;k(n-1),k-1}$$

The equal division of the error rate α between the two ends of the F-distribution is neither sacred nor necessarily optimum. For example, we might assign zero probability to one tail and obtain a one-sided confidence interval on ρ.

Example 8-12

Let us calculate the intraclass correlation and its 99% confidence interval for the reaction time data of Example 8-10:

$$\hat{\rho} = \frac{5.26}{5.26 + 3.49} = 0.6011$$

About 60% of the variance of the reaction time variate would be due to the individual driver's variance. For the confidence interval we need

$$F_{0.005;4,35} = 4.48 \qquad F_{0.995;4,35} = 1/F_{0.005;35,4} = 0.0504$$

The confidence interval is

$$0.1935 \leq \rho \leq 0.970$$

As we should expect from the large observed F-ratio the confidence interval is well removed from zero, and extends nearly to 1. It is not especially informative about the likely values of the population correlation.

8-9 REFERENCES

BOWKER, A. H., and G. J. LIEBERMAN (1972): *Engineering Statistics*, 2nd ed., Prentice-Hall, Inc., Englewood Cliffs, N.J.

CHEW, V. (1976): Comparing treatment means: a compendium, *HortScience*, vol. 11, pp. 348–357.

CHEW, V. (1977): *Comparisons among Treatment Means in an Analysis of Variance*, U.S. Department of Agriculture, Washington, D.C.

DUNNETT, C. W. (1955): A multiple comparison procedure for comparing several treatments with a control, *Journal of the American Statistical Association*, vol. 50, pp. 1096–1121.

DUNNETT, C. W. (1964): New tables for multiple comparisons with a control, *Biometrics*, vol. 20, pp. 482–491.

DUNNETT, C. W. (1980a): Pairwise multiple comparisons in the homogeneous variance, unequal sample size case, *Journal of the American Statistical Association*, vol. 75, pp. 789–795.

DUNNETT, C. W. (1980b): Pairwise multiple comparisons in the unequal variance case, *Journal of the American Statistical Association*, vol. 75, pp. 796–800.

GENIZI, A., and Y. HOCHBERG (1978): On improved extensions of the T method of multiple comparisons for unbalanced designs, *Journal of the American Statistical Association*, vol. 73, pp. 879–884.

GRAYBILL, F. A. (1961): *An Introduction to Linear Statistical Models*, vol. 1, McGraw-Hill Book Company, New York.

GRAYBILL, F. A. (1976): *Theory and Application of the Linear Model*, Duxbury Press, North Scituate, Mass.

HOCHBERG, Y. (1975): An extension of the *T*-method to general unbalanced models of fixed effects, *Journal of the Royal Statistical Society, Series B*, vol. 37, pp. 426–433.

HOCHBERG, Y. (1976): A modification of the *T*-method of multiple comparisons for a one-way layout with unequal variances, *Journal of the American Statistical Association*, vol. 71, pp. 200–203.

KEULS, M. (1952): The use of the "studentized range" in connection with an analysis of variance, *Euphytica*, vol. 1, pp. 112–122.

LANSDELL, H. (1968): Effect of temporal lobe ablations on two lateralized deficits, *Physiology and Behavior*, vol. 3, pp. 271–273.

MILLER, R. G., Jr. (1977): Developments in multiple comparisons, 1966–1976, *Journal of the American Statistical Association*, vol. 72, pp. 779–788.

MILLER, R. G., Jr. (1981): *Simultaneous Statistical Inference*, 2nd ed., Springer-Verlag, New York.

MORRISON, D. F. (1976): *Multivariate Statistical Methods*, 2nd ed., McGraw-Hill Book Company, New York.

NEWMAN, D. (1939): The distribution of the range in samples from a normal population, expressed in terms of an independent estimate of standard deviation, *Biometrika*, vol. 31, pp. 20–30.

PEARSON, E. S., and H. O. HARTLEY (1943): Tables of the probability integral of the studentized range, *Biometrika*, vol. 33, pp. 89–99.

PEARSON, E. S., and H. O. HARTLEY (1951): Charts of the power function of the analysis of variance tests, derived from the noncentral *F*-distribution, *Biometrika*, vol. 38, pp. 112–130.

PEARSON, E. S., and H. O. HARTLEY (1972): *Biometrika Tables for Statisticians*, vol. 2, Cambridge University Press, Cambridge, England.

SCHEFFÉ, H. (1953): A method for judging all contrasts in the analysis of variance, *Biometrika*, vol. 47, pp. 381–400.

SCHEFFÉ, H. (1959): *The Analysis of Variance*, John Wiley & Sons, Inc., New York.

SEARLE, S. R. (1971): *Linear Models*, John Wiley & Sons, Inc., New York.

SPJØTVOLL, E., and M. R. STOLINE (1973): An extension of the *T*-method of multiple comparison to include the cases with unequal sample sizes, *Journal of the American Statistical Association*, vol. 68, pp. 975–978.

STEPHENSON, L. W., L. H. EDMUNDS, Jr., R. RAPHAELY, D. F. MORRISON, W. S. HOFFMAN, and L. J. RUBIS (1979): Effects of nitroprusside and dopamine on pulmonary arterial vasculature in children after cardiac surgery, *Cardiovascular Surgery*, 1977, Supplement to *Circulation*, vol. 60, pp. I-104–I-110.

STOLINE, M. R., and H. K. URY (1979): Tables of the Studentized maximum modulus distribution and an application to multiple comparisons among means, *Technometrics*, vol. 21, pp. 87–93.

TIKU, M. L. (1967): Tables of the power of the *F* test, *Journal of the American Statistical Association*, vol. 62, pp. 525–539.

TUKEY, J. W. (1953): The Problem of Multiple Comparisons, Unpublished manuscript, Princeton, N.J.

URBAN, W. D. (1976): *Statistical Analysis of Blood Lead Levels of Children Surveyed in Pittsburgh, Pennsylvania: Analytical Methodology and Summary Results*, NBSIR 76-1024, Institute of Applied Technology, National Bureau of Standards, Washington, D.C.

WILSDON, B.H. (1934): Discrimination by specification statistically considered and illustrated by the standard specification for Portland cement, *Supplement to the Journal of the Royal Statistical Society, Series B*, vol. 1, pp. 152–206.

8-10 EXERCISES

1. Five teaching assistants for the recitation sections of a large basic statistics course were rated by their students with respect to overall ability. The ratings on the five-point scale had the following frequencies:

	Teaching assistant					
Scale value	A	B	C	D	E	Total
1 (highest)	20	12	14	10	16	72
2	10	17	18	24	30	99
3	4	6	9	8	14	41
4	0	1	2	4	4	11
5 (worst)	0	0	0	0	1	1
N_j	34	36	43	46	65	224
Mean	1.53	1.89	1.98	2.13	2.14	—
$\sum (x_{ij} - \bar{x}_j)^2$	16.4706	21.5556	30.9767	33.2174	53.7538	—

We shall assume that inferences are to be made only to the five instructors; the fixed-effects model should be used.

(a) Complete the analysis of variance for the hypothesis of equal teaching assistant means.

(b) Use the Scheffé and Bonferroni methods to determine which instructors are different.

(c) What are some other contrasts of the sample means that are "significant" at the 0.05 level?

(d) What is the contrast with the greatest Scheffé test statistic? Show that the square of that statistic is proportional to the F-ratio of the analysis of variance in (a).

2. Treat the data of Exercise 1 as if the teaching assistants came from a large population of instructors.

(a) Estimate the error and teaching assistant components of variance.

(b) Estimate the intraclass correlation from its definition in terms of the two variance components. Note that the common sample size n must be replaced by the value n_0.

3. In the blood lead data mentioned in Chapter 3 (Urban, 1976) these statistics for blood lead level (μg/ml) were computed for three degrees of father's education:

Education	Children	Mean	Within-group sum of squares
Elementary	29	25.66	1376
High school	101	21.47	3993
Some college	53	20.08	1360
Total	183	21.73	—

(a) Test the hypothesis of equal levels of blood lead in each of the educational level populations.

(b) If the hypothesis in (a) is rejected, use an appropriate multiple comparison procedure to determine which groups differ in average blood lead level.

4. Another statistics course had two teaching assistants for its recitation sections. Their ratings on the same scale as in Exercise 1 had the following values:

	Teaching assistant	
Scale	1	2
1	22	5
2	5	10
3	0	1
N_j	27	16
Mean	1.19	1.75

Carry out a random-effects analysis of variance and estimate the variance component of the teaching assistants.

5. Wilsdon (1934) reported the following statistics for crushing strengths in pounds per square inches of cubes of mortar made from dry- and wet-consistency cement:

Dry consistency			Wet consistency		
Lab	Mean	Sum of squares	Lab	Mean	Sum of squares
R	8,433	148,332	S	3,510	61,500
T	8,200	564,798	T	3,648	39,024
V	7,933	898,398	U	3,479	115,140
W	8,120	270,222	X	3,628	51,876
Y	7,971	530,880	Y	3,447	35,148
Z	8,263	299,526	Z	2,824	68,748

SOURCE: Data reproduced with the kind permission of the Royal Statistical Society.

"Consistency" refers to the amount of water in the mix. Each laboratory tested $n = 6$ cubes of wet and dry mix at the end of seven days. The sums of squares are those of the observations about the laboratory means. We have in effect two balanced one-way layouts, although the treatments are not identical in each.

(a) Test the hypotheses of no laboratory differences separately for the wet- and dry-consistency mixes.
(b) Now treat the laboratories as a sample from a larger population and estimate the variance components for laboratories separately for the wet and dry mixes.
(c) Estimate the intraclass correlation coefficients in the wet and dry cases and find the 90% confidence intervals for each.

6. Six treatments are to be set out in a one-way fixed-effects layout, and the common sample size for each treatment is to be determined so that the probability of detecting certain differences among the treatment effects is at least 0.95 when the analysis of variance is carried out at the 0.05 level. Let us assume that the vector of standardized treatment effects is

$$\frac{\boldsymbol{\tau}'}{\sigma} = [1, \quad -1, \quad 0, \quad 0, \quad 0, \quad 0]$$

(a) Find the sample size that will give a test with power 0.95 against the given alternative of treatment effects.
(b) Use the lower bound on the noncentrality parameter to determine the sample size and compare it with the size found in (a).
(c) Modify the elements of $\boldsymbol{\tau}'/\sigma$ slightly (subject to the constraint that they have sum zero) and find the sample sizes required for power 0.95. For example, you might begin with these vectors:

$$[1, \quad -\tfrac{1}{2}, \quad -\tfrac{1}{2}, \quad 0, \quad 0, \quad 0] \quad [\tfrac{3}{2}, \quad -\tfrac{3}{2}, \quad 0, \quad 0, \quad 0, \quad 0]$$

7. A new psychotropic drug is supposed to have a beneficial effect on persons suffering from a chronic anxiety syndrome. An experiment is to be conducted at a psychiatric outpatient clinic to determine whether the drug has a greater effect than an inert placebo: Two groups of male patients will be matched for similar ages and other characteristics and assigned the drug or placebo at random. The predrug–postdrug changes on a standardized mood-status questionnaire will be compared for the drug and placebo groups by a fixed-effects analysis of variance. Let us assume that the drug–placebo effect is only of the order

$$0.5 \leq \frac{\mu_1 - \mu_2}{\sigma} \leq 1$$

when measured in units of the standard deviation of the pre–post scale change. Construct a table of proposed sample sizes n for different combinations of significance levels 0.05 and 0.01 and probabilities of detecting the given small differences of 0.90 and 0.95.

8. A fast-food chain is concerned about its average waiting time for service in its outlets. For a study of waiting times it plans to draw k restaurants at random from a population with similar volumes and kinds of locations and collect n observations on waiting time at each. A large amount of additional data will also be collected, but we will not be concerned with it here.

(a) Let $k = 5$. What sample size n should be used for each outlet if we wish to reject H_0: $\sigma_0^2 = 0$ at the 0.01 level with probability 0.90 when σ_0^2/σ^2 is 2 or more?
(b) Repeat the sample size determination of (a) if $k = 3$ restaurants are to be studied.

9. Recall that the noncentrality parameter (3) of Section 8-7 for the fixed-effects model is

$$\lambda = \frac{n}{\sigma^2} \sum_{j=1}^{k} \tau_j^2$$

If we write

$$\sigma_T^2 = \frac{1}{k} \sum_{j=1}^{k} \tau_j^2$$

for the "variance" of the treatment parameters we may relate the fixed- and random-effects power parameters by

$$\lambda = \frac{nk\sigma_T^2}{\sigma^2}$$

Compare the powers given by the two models for tests at the 0.05 level and some representative combinations of n, k, and the ratio σ_T^2/σ^2.

10. The expected lengths of the Scheffé and Tukey simultaneous confidence intervals are, respectively,

$$2E\left(\sqrt{\frac{\hat{\sigma}^2}{n}}\right)\sqrt{2(k-1)F_{\alpha;k-1,k(n-1)}}$$

$$2E\left(\sqrt{\frac{\hat{\sigma}^2}{n}}\right)q_{\alpha;k,k(n-1)}$$

in the one-way balanced layout with k treatments and n observations for each.

(a) Find the ratios of the expected lengths for some common values of α, k, and n. Note that it is not necessary to calculate the expectation, for it appears in the same form in each expected length.

(b) Repeat the comparisons for the Bonferroni confidence intervals for the $k(k-1)/2$ pair contrasts based on the critical value $t_{\alpha/k(k-1);k(n-1)}$.

11. Lansdell (1968) has studied the effect of temporal lobe ablation in epileptic patients on their cognitive abilities and has reported these statistics for the postoperative performance of 52 persons on the Atwell and Wells Wide Range Vocabulary Test:

Group	N_j	Mean	$\sum (x_{ij} - \bar{x}_j)^2$
1. Left temporal, males	11	55.5	2624
2. Left temporal, females	8	47.6	1451
3. Right temporal, males	13	65.2	2958
4. Right temporal, females	20	64.1	6573

(a) Treat the data as a one-way layout with $k = 4$ treatments and make an analysis of variance for no treatment differences.

(b) The data actually constitute a 2×2 cross-classified layout, with sex and side of temporal lobe removal as the ways of classification. We cover such unbalanced layouts in Chapter 9. In the meantime, how much can you infer about male–female and left–right differences from appropriate contrasts applied to the four treatment parameters of the one-way model? Use the same error mean square found in (a) for tests on the contrasts.

(c) Do the same inferences for male–female differences hold for left temporal removals as for right removals? Similarly, are left–right differences probably the same for males and females? Express those questions as hypotheses on contrasts of the four means and test them by appropriate t-statistics.

(d) Which of the contrasts in (a) through (c) are orthogonal to one another?

12. Stephenson et al. (1979) studied the hemodynamic characteristics of patients in five diagnostic groups who were treated with nitroprusside and dopamine following cardiac surgery. These statistics were obtained for the nitroprusside-control changes in stroke volume index for the groups:

Group	N_j	$\bar{x}_j$	$\sum (x_{ij} - \bar{x}_j)^2$
A	7	−5.11	195.6486
T	6	2.83	50.3333
H	5	2.72	146.1080
V	3	−6.77	69.7867
P	5	−0.52	185.7680
Total	26	—	647.6446

(a) Do the average changes appear to be different for the five diagnostic groups?

(b) If the groups differ, use an appropriate multiple comparison method to determine which pairs of means are different.

13. For Exercises 3, 11, and 12 find the contrasts of the treatment means with the greatest Scheffé multiple test statistics.

14. Use the Lagrangian multiplier method to maximize the statistic (10) of Section 8-6.

15. Use expression (24) of Section 8-8 to find the probability that the estimator (23) of the Model II treatment variance component is negative.

chapter 9

The Analysis of Variance for Higher-Way Layouts

9-1 INTRODUCTION

In Chapter 8 we introduced the analysis of variance for determining whether the means of several normal populations were different. The populations corresponded to the treatments in an experiment or categories of individuals in a survey or study. In either case the groups were formed with respect to a single criterion, or way of classification, and we referred to the method as the *one-way* analysis of variance. The "way" of classification implied only a nominal scaling of the several categories; known numerical scores for the categories would lead to a natural treatment by regression analysis for grouped data. Now we are going to develop the analysis of variance for treatment groups in two or more ways of classification. Generally, we think of the ways as *crossed:* For two ways our groups will be all combinations of the treatments in the first way with those in the second. The layout for the data has the appearance of a two-way table, with all row and column combinations, or cells, capable of yielding observations on the response variable. Alternatively, the treatments of the second way of classification might be contained entirely within each treatment of the first way. Such *nested* layouts and their analysis are treated briefly. We devote most of our attention to two-dimensional layouts. They will be sufficient for most of our examples and applications and to illustrate the special qualities of an analysis of variance with more than one way of classification.

We begin by introducing the simplest two-way layout with one observation in each cell in Section 9-2. In Section 9-3 we extend that model to n independent observations in each cell, or the *balanced* two-way layout. That case allows not only for individual effects of the row and column treatments, but also for contributions due to their interaction. The analysis of variance provides independent measures of those three sources of differences among the cell means. In Section 9-4 we discuss contrasts for measuring interaction and other effects in the two-way model and for the fitting of polynomials to the row or column means when the treatments correspond to the equally spaced values of a variable. In Section 9-5 we extend the Bonferroni, Scheffé, and Tukey multiple comparison methods to the two-way model.

In the two-way analysis of variance with a single observation in each cell the interaction and error mean squares are identical, and a test for the presence of interaction effects in the linear model is not possible. However, a particular kind of nonadditivity in the data can be detected by an hypothesis test, and we treat that method in Section 9-6.

The linear model and its analysis of variance can be extended easily to balanced layouts in more than three dimensions. For simplicity we discuss only the general three-way layout, its analysis of variance, and its multiple comparison methods in Section 9-7. In Section 9-8 we develop a special case of the multidimensional layout in which each treatment occurs at two levels: for example, presence and absence of the treatment, or *factor*. Such two-level factorial designs have many special and convenient properties. For example, the main effects and interactions can be expressed as single contrasts and their hypothesis tests carried out by linear rather than variance ratio statistics.

The restriction to balanced data is a severe one, for nature and sampling variation often do not leave every cell of the cross-classified layout with the same number of independent observations. If a random sample of N subjects is drawn and then cross-classified with respect to several types of attributes, the cell numbers are almost certain to be unequal, unless the frequencies of the attributes obey certain relationships. In Section 9-9 we give one method of analyzing unbalanced cross-classified layouts with at least one observation in each cell.

Some experimental designs do not have cross-classified treatments but instead are arranged hierarchically. Such layouts are called nested, for the successive treatments are subsets of those at the next higher level in the hierarchy. For example, in comparing the performance of fourth-grade students at three elementary schools on a standardized achievement test we might use the individual fourth-grade classes at the schools as nested treatments and the schools as the primary treatments. A model for nested treatments and its analysis of variance is described in Section 9-10.

In Section 9-11 the two-way layouts are extended to the case of random treatment effects, or Model II, in a balanced plan. As in the case of Section 8-8, Model II is appropriate when the row and column treatments are random samples from larger populations of conditions. The fixed and random models are combined in Section 9-12 as one kind of mixed model for a balanced two-way experiment. The mixed model is especially useful when the column treatments are a fixed set of repeated measurements obtained from the same sampling units designated by the random row conditions.

9-2 THE SIMPLEST TWO-WAY LAYOUT

Investigations involving only a single kind of experimental condition are usually rare in practice. Experiments on an industrial or chemical process, a field trial of several fertilizers and plant varieties, or the behavior of pairs of problem solvers are almost always designed to measure the effects of several types of variables and their interactions. Such experimentation is usually costly and time consuming, and the investigator must extract as much information as possible about the determinants of the values of the response variable. In this section we introduce the case of experimental conditions that can be cross-classified, with a single observation on the response in each combination, or cell, of the two ways of classification.

For example, the two sets of observations on the crushing strength of mortar in Exercise 5, Section 8-10, form an example of a two-way design with 3 rows (the common laboratories T, Y, Z) and two columns (dry or wet consistency of the mortar mix). The crushing strengths in the 6 cells are actually the averages of the strengths of six separate mortar cubes, but in the present context we ignore the information given by the sums of squares within each cell and regard the averages as single data points. A two-way layout with r rows, c columns, and a single observation in each cell is displayed in Table 9-1. We assume that the observation in the ith row and jth column has been

TABLE 9-1 Data in a Two-Way Layout

	Column treatment			
Row treatment	1	$\cdots$	c	Row mean
1	x_{11}	$\cdots$	x_{1c}	$\bar{x}_{1.}$
$\vdots$	$\vdots$		$\vdots$	$\vdots$
r	x_{r1}	$\cdots$	x_{rc}	$\bar{x}_{r.}$
Column mean	$\bar{x}_{.1}$	$\cdots$	$\bar{x}_{.c}$	$\bar{x}_{..}$

generated by this model:

$$x_{ij} = \mu + \alpha_i + \beta_j + e_{ij} \qquad i = 1, \ldots, r; \quad j = 1, \ldots, c \tag{1}$$

μ is a general effect common to all observations. α_i is the effect of the treatment in the ith row, whereas β_j is that due to the jth column effect. In the mortar strength example α_i would be the contribution of the ith laboratory to the crushing strength, whereas β_1 and β_2 would be the respective effects due to dry or wet mix consistency. e_{ij} is the usual random disturbance. For estimation and inferential purposes we assume that the e_{ij} are independently and normally distributed random variables with mean zero and a common variance σ^2. As in all of our earlier models, the effects and disturbances are additive, although in the two-way layout additivity takes on a special meaning: The expected value of every x_{ij} is essentially the sum of row and column components. As we shall see, this model is often rather simplistic for some data sets.

As in the one-way layout μ, α_i, and β_j are not estimable parameters, although if we impose the constraints

$$\sum_{i=1}^{r} \alpha_i = \sum_{j=1}^{c} \beta_j = 0 \tag{2}$$

on the row and column effects, they become estimable. Then their least squares estimators are

$$\begin{aligned}
\hat{\mu} &= \frac{1}{rc}\sum_{i=1}^{r}\sum_{j=1}^{c} x_{ij} = \bar{x}_{..} \\
\hat{\alpha}_i &= \frac{1}{c}\sum_{j=1}^{c} x_{ij} - \hat{\mu} = \bar{x}_{i.} - \bar{x}_{..} \\
\hat{\beta}_j &= \frac{1}{r}\sum_{i=1}^{r} x_{ij} - \hat{\mu} = \bar{x}_{.j} - \bar{x}_{..}
\end{aligned} \tag{3}$$

$\bar{x}_{..}$ is the grand mean of all rc observations: Its double dots denote summation over rows and columns. $\bar{x}_{i.}$ is the mean of the observations in the ith row, and $\bar{x}_{.j}$ is that for the data in the jth column. The predicted value of the observation in the ith row and jth column would be

$$\hat{x}_{ij} = \bar{x}_{..} + (\bar{x}_{i.} - \bar{x}_{..}) + (\bar{x}_{.j} - \bar{x}_{..}) \tag{4}$$

We confine our attention largely to estimable functions that are contrasts of the row or column effects, or

$$\psi_R = \sum_{i=1}^{r} c_i\alpha_i \qquad \psi_C = \sum_{j=1}^{c} d_j\beta_j \tag{5}$$

where

$$\sum_{i=1}^{r} c_i = \sum_{j=1}^{c} d_j = 0 \tag{6}$$

by the contrast property. The estimates are merely

$$\hat{\psi}_R = \sum_{i=1}^{r} c_i\bar{x}_{i.} \qquad \hat{\psi}_C = \sum_{j=1}^{c} d_j\bar{x}_{.j} \tag{7}$$

These follow by substituting the estimators (3) into the contrasts; the $\bar{x}_{..}$ terms vanish by the contrast property (6) of the c_i and d_j.

Next we should like an estimator of the variance σ^2 of the disturbance terms. Those random variables can be estimated by the residuals

$$x_{ij} - \hat{x}_{ij} = x_{ij} - \bar{x}_{..} - (\bar{x}_{i.} - \bar{x}_{..}) - (\bar{x}_{.j} - \bar{x}_{..}) \tag{8}$$

from the predicted values, and an estimator of σ^2 would be given by the sum of squares

$$\begin{aligned}
\text{SSE} &= \sum_{i=1}^{r}\sum_{j=1}^{c} (x_{ij} - \hat{x}_{ij})^2 \\
&= \sum_{i=1}^{r}\sum_{j=1}^{c} [x_{ij} - \bar{x}_{..} - (\bar{x}_{i.} - \bar{x}_{..}) - (\bar{x}_{.j} - \bar{x}_{..})]^2 \\
&= \sum_{i=1}^{r}\sum_{j=1}^{c} (x_{ij} - \bar{x}_{i.} - \bar{x}_{.j} + \bar{x}_{..})^2
\end{aligned} \tag{9}$$

divided by some appropriate integer to give an unbiased estimator. That divisor is the degrees of freedom of SSE, or $(r-1)(c-1)$, and

$$\hat{\sigma}^2 = \text{MSE} = \frac{\text{SSE}}{(r-1)(c-1)} \tag{10}$$

$\hat{\sigma}^2$ is distributed independently of $\bar{x}_{..}$, $\bar{x}_{i.}$, and $\bar{x}_{.j}$ for all i and j. The quantity

$$\frac{\text{SSE}}{\sigma^2} = \frac{\text{MSE}(r-1)(c-1)}{\sigma^2} \tag{11}$$

has the chi-squared distribution with $(r-1)(c-1)$ degrees of freedom when the additive model (1) is true, and the rc e_{ij} are independently and normally distributed with mean zero and common variance σ^2. Then the sample contrasts of (7) are normally distributed with respective parameters

$$\begin{aligned} E(\hat{\psi}_R) &= \psi_R = \sum_{i=1}^{r} c_i\alpha_i \quad \text{Var}(\hat{\psi}_R) = \frac{\sigma^2}{r}\sum_{i=1}^{r} c_i^2 \\ E(\hat{\psi}_C) &= \psi_C = \sum_{j=1}^{c} d_j\alpha_j \quad \text{Var}(\hat{\psi}_C) = \frac{\sigma^2}{c}\sum_{j=1}^{c} d_j^2 \end{aligned} \tag{12}$$

and are independent of $\hat{\sigma}^2$. We shall see presently how those results may be used to find tests and confidence intervals for row and column effect contrasts. In the meantime let us consider the analysis of variance for the two-way layout.

If the hypothesis

$$H_0: \alpha_1 = \cdots = \alpha_r \tag{13}$$

of equal row effects is true, the standardized row sum of squares

$$\begin{aligned} \text{SSR}/\sigma^2 &= \frac{c\sum_{i=1}^{r}(\bar{x}_{i.} - \bar{x}_{..})^2}{\sigma^2} \\ &= \frac{c\sum_{i=1}^{r}\bar{x}_{i.}^2 - rc\bar{x}_{..}^2}{\sigma^2} \end{aligned} \tag{14}$$

is distributed as a chi-squared variate with $r-1$ degrees of freedom. SSR and the error sum of squares SSE defined by (9) are independently distributed regardless of the truth of H_0. When the row effects are unequal SSR/σ^2 has the noncentral chi-squared distribution with the same degrees of freedom and noncentrality parameter

$$\lambda_\alpha = \frac{c\sum_{i=1}^{r}(\alpha_i - \bar{\alpha})^2}{\sigma^2} \tag{15}$$

in which $\bar{\alpha} = (1/r)\sum \alpha_i$. We may test the equal row effects hypothesis by computing the ratio

$$F = \frac{\text{SSR}/(r-1)}{\text{MSE}} \tag{16}$$

of the error mean square (10) and the row mean square and referring it to the upper $100\alpha\%$ critical value $F_{\alpha; r-1, (r-1)(c-1)}$ of the F-distribution with

$r-1$ and $(r-1)(c-1)$ degrees of freedom. If

$$F > F_{\alpha; r-1, (r-1)(c-1)} \tag{17}$$

we should conclude that the row effects are different at the α level.

The hypothesis of equal column effects,

$$H_0: \beta_1 = \cdots = \beta_c \tag{18}$$

is tested in the same manner by comparing the variation among the column means to the error mean square. We compute

$$\begin{aligned} \text{SSC} &= r \sum_{j=1}^{c} (\bar{x}_{ij} - \bar{x}_{..})^2 \\ &= r \sum_{j=1}^{c} \bar{x}_{.j}^2 - rc\bar{x}_{..}^2 \end{aligned} \tag{19}$$

and its mean square SSC/$(c-1)$. SSC/σ^2 is distributed as a chi-squared variate with $c-1$ degrees of freedom when the equal-column effects hypothesis is true. When the column effects are different SSC/σ^2 has the noncentral chi-squared distribution with noncentrality parameter

$$\lambda_\beta = r \sum_{j=1}^{c} (\beta_j - \bar{\beta})^2 \tag{20}$$

in which $\bar{\beta} = (1/c) \sum \beta_j$. SSC is distributed independently of the error mean square and SSR: *In that sense the row and column effect tests are independent of one another.* We test the equal-column effect hypothesis (18) by computing

$$F = \frac{\text{SSC}/(c-1)}{\text{MSE}} \tag{21}$$

and referring it to the F critical value $F_{\alpha; c-1, (r-1)(c-1)}$. If F exceeds that point we should conclude that the column treatments are different at the α level. We can summarize the sums of squares and other components of the tests in the analysis of variance of Table 9-2. In practice one usually computes SSE by subtracting SSR and SSC from the total sum of squares

$$\sum_{i=1}^{r} \sum_{j=1}^{c} (x_{ij} - \bar{x}_{..})^2 = \sum_{i=1}^{r} \sum_{j=1}^{c} x_{ij} - \frac{G^2}{rc} \tag{22}$$

where $G = \sum\sum x_{ij}$ is the usual grand total of all rc observations in the two-way layout. However, it might also be efficient to compute SSE from its definition (9). On the way we can calculate and store the residuals of each cell observation.

The expected mean squares follow from the quadratic form results in Appendix C. We may see from them that departures from the respective hypotheses will only increase the average value of the F-variate and that each expected value involves only the effects being tested. We may use the noncentrality parameters associated with the expected values to determine power probabilities or sample sizes as in Section 8-7. We defer those calculations until Section 9-3 and the case of n observations in each cell of the two-way layout.

TABLE 9-2 Two-Way Analysis of Variance with One Observation in Each Cell

Source	Sum of squares	d.f.	Mean square	Expected mean square
Rows	$SSR = c \sum_{i=1}^{r} (\bar{x}_{i.} - \bar{x}_{..})^2$	$r - 1$	$\frac{SSR}{r-1}$	$\sigma^2 + \frac{c}{r-1} \sum_{i=1}^{r} (\alpha_i - \bar{\alpha})^2$
Columns	$SSC = r \sum_{j=1}^{c} (\bar{x}_{.j} - \bar{x}_{..})^2$	$c - 1$	$\frac{SSC}{c-1}$	$\sigma^2 + \frac{r}{c-1} \sum_{j=1}^{c} (\beta_j - \bar{\beta})^2$
Error	$SSE = \sum_{i=1}^{r} \sum_{j=1}^{c} (x_{ij} - \bar{x}_{i.} - \bar{x}_{.j} + \bar{x}_{..})^2$	$(r-1)(c-1)$	$\frac{SSE}{(r-1)(c-1)}$	σ^2
Total	$\sum_{i=1}^{r} \sum_{j=1}^{c} (x_{ij} - \bar{x}_{..})^2$	$rc - 1$	—	—

Example 9-1

Let us treat the graduate program enrollment data of Example 8-8 as a two-way layout with three rows (years) and eight columns (programs). We would like to determine whether (a) the differences among the program means are greater than one would expect from chance variation and (b) if the three annual means are different. We shall assume that a given enrollment number is the sum of an annual effect, a program effect, an amount common to all years and programs, and a random variable reflecting fluctuations not explained by the additive model. Although the data consist of integers from 1 to 14, we shall still make the assumption of normality and a constant variance for the random components.

The column means required for the two-way analysis of variance can be found in Example 8-8, and will not be repeated here. We shall also need the row means and some statistics for computing the total sum of squares:

Year	Mean
1	3.875
2	6.250
3	6.125

$$G = \sum_{i=1}^{3} \sum_{j=1}^{8} x_{ij} = 130 \qquad \sum_{i=1}^{3} \sum_{j=1}^{8} x_{ij}^2 = 1014$$

$$\bar{x}_{..} = \frac{G}{24} = 5.417$$

Then the years and programs sums of squares can be computed to be

$$\text{SSR} = 8 \sum_{i=1}^{3} (\bar{x}_{i.} - \bar{x}_{..})^2 = 28.58$$

$$\text{SSC} = 3 \sum_{j=1}^{8} (\bar{x}_{.j} - \bar{x}_{..})^2 = 241.83$$

This analysis of variance table was obtained:

Source	Sum of squares	d.f.	Mean square	F
Years	28.58	2	14.29	5.07
Programs	241.83	7	34.55	12.27
Error	39.42	14	2.816	—
Total	309.83	23	—	—

The F-ratio for the equal-program means test exceeds the critical value $F_{0.005;7,14} = 5.03$, and we should conclude that the programs are different at any reasonable level of significance. The F-ratio for years exceeds $F_{0.025;2,14} = 4.86$, and we should conclude that the annual enrollments are different at better than the traditional 0.05 level. Inspection of the annual means suggests that the difference is due to the low average for year 1, although we should confirm that impression with a formal Scheffé or Tukey test or confidence interval; that will be left until the next section. The reason for the highly significant variation among the program means is less clear, and will also await the application of simultaneous methods in Section 9-5.

We might estimate the parameters of the two-way linear model and from them determine how well the model represents the actual enrollments. We shall impose the constraints (2) that the row and column effects sum to zero. Then the general effect estimate is $\hat{\mu} = \bar{x}_{..} = 5.42$. The new parameter estimates are

$$\hat{\alpha}_1 = \bar{x}_{1.} - \bar{x}_{..} = -1.54$$
$$\hat{\alpha}_2 = \bar{x}_{2.} - x_{..} = 0.83$$
$$\hat{\alpha}_3 = \bar{x}_{3.} - x_{..} = 0.71$$

The program effect estimates are given by $\hat{\beta}_j = \bar{x}_{.j} - \bar{x}_{..}$:

Program	A	B	C	D	E	F	G	H
$\hat{\beta}_j$	−2.42	0.92	3.25	−3.42	6.58	−1.42	−1.42	−2.08

Then the predicted enrollment for the jth program in the ith year is

$$\hat{x}_{ij} = 5.42 + \hat{\alpha}_i + \hat{\beta}_j$$

The residuals $x_{ij} - \hat{x}_{ij}$ of the actual and predicted enrollments are shown in this table.

	Program							
Year	A	B	C	D	E	F	G	H
1	−0.46	−0.79	−0.12	1.54	−1.46	−0.46	0.54	1.21
2	−0.83	−1.17	2.50	−1.83	0.17	1.17	1.17	−1.17
3	1.29	1.96	−2.38	0.29	1.29	0.71	−1.71	−0.04

The sum of the squared residuals is within round-off error of the error sum of squares in the analysis of variance table. It is possible to show that

$$\text{Var}\,(x_{ij} - \bar{x}_{i.} - \bar{x}_{.j} + \bar{x}_{..}) = \frac{\sigma^2(r-1)(c-1)}{rc}$$

or 0.5833 σ^2 in the present example. The estimated standard deviation of each residual can be obtained by replacing σ^2 by the error mean square of 2.816; it is equal to 1.2817. We may use that value as a crude standard for assessing the importance of the extreme residuals; those for program C in years 2 and 3 are less than two standard deviations removed from the mean of zero. Such a deviation is probably acceptable, although we should *not* refer the residual divided by its estimated standard deviation to critical values of the t-distribution. The two quantities are not independently distributed, so that their quotient is not a t-variate. Similarly, the residuals themselves are not independent, and cannot be treated as a random sample from a normal population.

Certain aspects of this two-way analysis of variance should not be accepted uncritically. First, the enrollment figures are actually counts free of measurement error. They constitute a *census* rather than a random sample from a larger population. What is our justification for treating them as observations on random variables? The number of student matriculating each year can be viewed as the result of a random process, for example, a Poisson random variable whose parameter may fluctuate a bit from year to year. Year-to-year variation in the enrollment of a program reflects the popularity of its discipline, competition for students in the marketplace of programs, and the variety and availability of financial support. If the Poisson model is appropriate, the assumption of normally distributed disturbance terms does not seem unreasonable. But what of the independence requirement for those variates? Observations recorded over time are usually autocorrelated. Enrollment patterns and trends persist from year to year, and the number of admissions last year may influence the matriculations of the present term. Correlations among enrollments of different programs seem less likely, unless due to trends common to all programs. Although such failures of the classical linear model assumptions affect the conclusions of the analysis of variance it is rather difficult to test their validity in small two-way layouts. We shall accept the conclusions from the F-statistics in the present data as if the model holds.

With the exception of observational data that naturally fall into a two-way layout, examples of schemes with single observations in each cell are rather rare. If observations are expensive or difficult to obtain, such an experimental design may be necessary. A more common example of the layout is that of the *randomized complete block design:* rc experimental units are divided into r blocks of c comparable units. Each block has been formed or chosen so that its experimental units are as homogeneous as possible. For example, in an agricultural field trial the blocks might consist of strips of land with nearly constant degrees of fertility. The blocks are also chosen for maximum block-to-block variation or heterogeneity. Because of the way in which the blocks are chosen the test for equal block effects has little meaning, although its sum of squares is independent of that for treatment effects. Such block designs are called *complete* because each block receives the full set of c treatments and *randomized* because the treatments must be assigned to the c experimental units at random in each block. For the assignment we number the treatments from 1 to c and draw one of the $c!$ permutations of those integers from a table of random permutations (Cochran and Cox, 1957). Random permutations are drawn for each of the r blocks. As an example, we might obtain this layout for four treatments, A, B, C, D, set out in five blocks:

	Unit			
Block	1	2	3	4
1	A	D	C	B
2	A	B	D	C
3	B	C	D	A
4	C	A	D	B
5	B	D	A	C

The analysis of variance proceeds as in Table 9-2 when the data have been obtained from the 20 cells of the blocks-by-treatments layout.

9-3 THE BALANCED TWO-WAY LAYOUT

We have seen in Section 9-2 how kinds of treatments can be arranged in a two-dimensional layout to provide independent sums of squares for hypothesis tests on the treatments' effects. For simplicity in our introduction of the model and analysis of variance we considered the case with just one observation in each cell of the cross-classified layout. In most planned experiments we would prefer to have several independent observations in each cell; then the estimate of the sampling or measurement error variance can be obtained from the variation within the cells. Not only is that estimate free of treatment effects, but it can be used to test the validity of the additive model of row and column parameters. The multiple observations in each cell not only increase the precision of our estimates and the power of the tests but make possible other inferences about the treatments and model.

The data from an $r \times c$ layout are shown in Table 9-3. The model for the hth observation in the ith row and jth column is

$$x_{ijh} = \mu + \alpha_i + \beta_j + \gamma_{ij} + e_{ijh} \qquad \begin{aligned} i &= 1, \ldots, r \\ j &= 1, \ldots, c \\ h &= 1, \ldots, n \end{aligned} \tag{1}$$

TABLE 9-3 Two-Way Layout

Row treatment	Column treatment 1	$\cdots$	Column treatment c	Row mean
1	x_{111} . . . x_{11n}	$\cdots$	x_{1c1} . . . x_{1cn}	$\bar{x}_{1..}$
. . .	. . . .			. . .
r	x_{r11} . . . x_{r1n}	$\cdots$	x_{rc1} . . . x_{rcn}	$\bar{x}_{r..}$
Column mean	$\bar{x}_{.1.}$	$\cdots$	$\bar{x}_{.c.}$	$\bar{x}_{...}$

As in Section 9-2 μ, α_i, and β_j are the general, row, and column effects, respectively. But now we have a fourth parameter, γ_{ij}. That term measures the departure of the observations from the simple additive model. If the γ_{ij} are not zero, we say that some of the row and column treatments have *interacted* to produce observations different from mere sums of their row and column effects. In the earlier two-way model any interaction effects would be indistinguishable from the disturbance terms.

The disturbance term e_{ijh} is a normal random variable with mean zero and common variance σ^2 for all combinations of i, j, and h. We may estimate σ^2 in the individual cells via the within-cells sums of squares

$$\text{SSE}_{ij} = \sum_{h=1}^{n} (x_{ijh} - \bar{x}_{ij.})^2 \tag{2}$$

The estimator for the ijth cell is

$$\hat{\sigma}_{ij}^2 = \frac{\text{SSE}_{ij}}{n-1} \tag{3}$$

These separate estimators can be used later to test the assumption of a common variance. Under that assumption the estimator of σ^2 is

$$\begin{aligned}\hat{\sigma}^2 &= \frac{1}{rc(n-1)} \sum_{i=1}^{r} \sum_{j=1}^{c} \sum_{h=1}^{n} (x_{ijh} - \bar{x}_{ij.})^2 \\ &= \frac{1}{rc(n-1)} \sum_{i=1}^{r} \sum_{j=1}^{c} \text{SSE}_{ij} \\ &= \frac{1}{rc(n-1)} \text{SSE}\end{aligned} \tag{4}$$

We shall write

$$\begin{aligned}\text{SSE} &= \sum_{i=1}^{r} \sum_{j=1}^{c} \sum_{h=1}^{n} (x_{ijh} - \bar{x}_{ij.})^2 \\ &= \sum_{i=1}^{r} \sum_{j=1}^{c} \left(\sum_{j=1}^{n} x_{ijh}^2 - \frac{C_{ij}^2}{n} \right)\end{aligned} \tag{5}$$

for the within-cells sum of squares pooled for all rc cells. It will be convenient to denote the sum of the observations in the ijth cell by

$$C_{ij} = \sum_{h=1}^{n} x_{ijh} \tag{6}$$

so that

$$\bar{x}_{ij.} = \frac{C_{ij}}{n} \tag{7}$$

Now let us consider tests of the respective hypotheses

$$H_0: \alpha_1 = \cdots = \alpha_r \tag{8}$$

and

$$H_0: \beta_1 = \cdots = \beta_c \tag{9}$$

of no differences in the row and column effects. We carry out those tests by the sums of squares for row and column effects computed from the row and column means on the margins of Table 9-3. The row sum of squares is

$$\begin{aligned}\text{SSR} &= nc \sum_{i=1}^{r} (\bar{x}_{i..} - \bar{x}_{...})^2 \\ &= \frac{\sum_{i=1}^{r} \left(\sum_{j=1}^{c} \sum_{h=1}^{n} x_{ijh} \right)^2}{nc} - \frac{G^2}{rcn}\end{aligned} \tag{10}$$

where, as in our previous notation,

$$G = \sum_{i=1}^{r} \sum_{j=1}^{c} \sum_{h=1}^{n} x_{ijh} \tag{11}$$

Note that SSR measures the variation only among the row means: The individual observations and the cells of each row have been summed into a single average. The present row sum of squares differs from (14) in Section 9-2 for one observation per cell only in the sense that its row means were computed from two levels of classification. As in the case of $n = 1$ SSR is distributed independently of the error sum of squares (5). When the hypothesis (8) of equal row effects is true SSR/σ^2 has the chi-squared distribution with $r - 1$ degrees of freedom. We test the hypothesis by referring

$$F = \frac{\text{SSR}/(r-1)}{\text{SSE}/rc(n-1)} = \frac{\text{SSR}/(r-1)}{\hat{\sigma}^2} \tag{12}$$

to critical values of the F-distribution with $r - 1$ and $rc(n - 1)$ degrees of freedom. For a test at the α level we would reject the equal row effects hypothesis if

$$F > F_{\alpha; r-1, rc(n-1)} \tag{13}$$

When the alternative hypothesis of general row effects $\alpha_1, \ldots, \alpha_r$ is true the statistic (12) has the noncentral F-distribution with $r - 1$ and $rc(n - 1)$ degrees of freedom and noncentrality parameter

$$\lambda_R = \frac{nc \sum_{i=1}^{r} (\alpha_i - \bar{\alpha})^2}{\sigma^2} \tag{14}$$

wherein $\bar{\alpha} = (1/r) \sum \alpha_i$. We shall return to the use of the noncentral distribution when we consider power and sample size calculations.

For the test of the hypothesis (9) of equal column effects we compute the column sum of squares

$$\text{SSC} = nr \sum_{j=1}^{c} (\bar{x}_{.j.} - \bar{x}_{...})^2 = \frac{\sum_{j=1}^{c} \left(\sum_{i=1}^{r} \sum_{h=1}^{n} x_{ijh} \right)^2}{nr} - \frac{G^2}{rcn} \tag{15}$$

SSC is distributed independently of both SSR and the error sum of squares. If the hypothesis (9) is true SSC/σ^2 has the chi-squared distribution with $c - 1$ degrees of freedom, and

$$F = \frac{\text{SSC}/(c-1)}{\text{SSE}/rc(n-1)} = \frac{\text{SSC}/(c-1)}{\hat{\sigma}^2} \tag{16}$$

is a central F-variate with degrees of freedom $c - 1$ and $rc(n - 1)$. We should reject the hypothesis if

$$F > F_{\alpha; c-1, rc(n-1)} \tag{17}$$

for a test at the α level. When the alternative of unequal column effects is true F is a noncentral F-variate with the same degrees of freedom and noncentrality parameter

$$\lambda_C = \frac{nr \sum_{j=1}^{c} (\beta_j - \bar{\beta})^2}{\sigma^2} \tag{18}$$

in which $\bar{\beta} = (1/c) \sum \beta_j$.

Finally let us consider the test of the hypothesis

$$H_0: \gamma_{ij} = 0 \tag{19}$$

that the model for the expected value of x_{ijh} is additive:

$$H_0: E(x_{ijh}) = \mu + \alpha_i + \beta_j \tag{20}$$

Either representation of the hypothesis says that the row and column effects do not *interact* to produce an effect greater than their individual sum, and for that reason we refer to (19) as the hypothesis of *no interaction*. The γ_{ij} are called the *interaction parameters* in the two-way linear model.

Before proceeding to the sum of squares for testing the interaction hypothesis it will be useful to discuss some common examples of interactions and ways of measuring and representing interactions graphically. The first is mundane yet vivid to any parent of small children.* A child playing alone may generate a certain amount of noise which might be measured as the integral of its record in decibels over time. However, the sum of such noise measures of two separate children would probably be much less than the measure for the same children playing together: Their social interaction has greatly increased their production of noise. At a more technical level interaction effects are common in chemical reactions or in the effects of drugs upon an organism. For example, alcohol and some tranquilizing drugs may produce similar affective changes yet in combination their effect may be exceedingly dangerous. In a psychological experiment measuring the reaction times of normal volunteers to stimuli the difference between the reaction times for auditory and visual stimuli might differ significantly as other conditions in the experiment changed.

Interactions in two-way layouts can be illustrated and studied conveniently by plots of the cell means. The means of a given row are plotted against the column numbers and connected by line segments. The plot is repeated in each row. If the r corresponding line segments for successive columns are parallel, or at least parallel within sampling variation, no interaction is present. The plots, or *row profiles*, are displaced by equal amounts as in Figure 9-1. The choice of the plot by rows is arbitrary: We could have made it by columns equally as well. Because the plot consists of sample means the segments of the three profiles have been depicted as slightly less than parallel.

*This example was apparently due to Bessie Day Mauss.

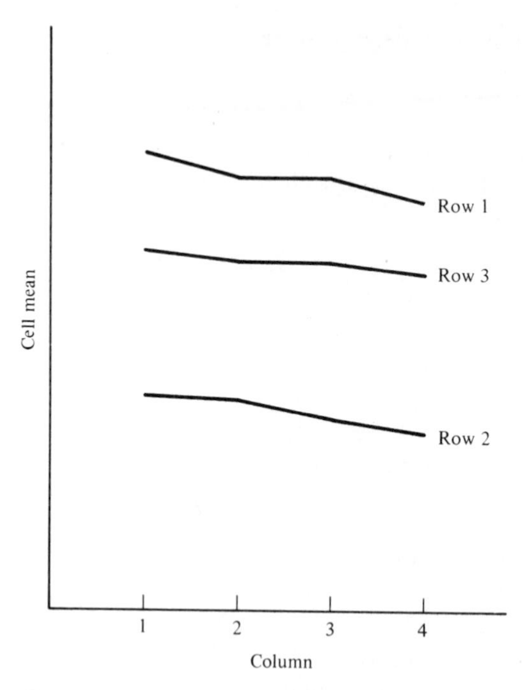

Figure 9-1 Plot of cell means with no interaction.

Exactly parallel mean profiles should engender as much interest (or circumspection) as statistical significance. For an example of an interaction in a 2×2 layout let us use some actual means reported by Lansdell (1968) in an investigation of the effect on cognitive ability of temporal lobe surgery for the treatment of epilepsy. Lansdell reported these Mooney Closure Faces scores (percent correct) for male and female patients classified by the side of the surgical ablation:

	Sex	
Temporal lobe	Male	Female
Left	79.1	89.6
Right	72.8	71.1

The means are plotted in Figure 9-2. The lack of parallel segments suggests different male–female effects for left- and right-temporal lobe patients: the sex difference is 10.5% for patients with ablation of the left lobe and -1.7% for those whose right lobes required surgery. The statistical significance of the difference of the slopes of the two segments still must be determined; one method is given in Section 9-9.

Now we need a sum of squares for a formal test of the significance of the interaction terms in the model. Let us begin with the special case of two rows and two columns, or the 2×2 layout. Then we can measure the interac-

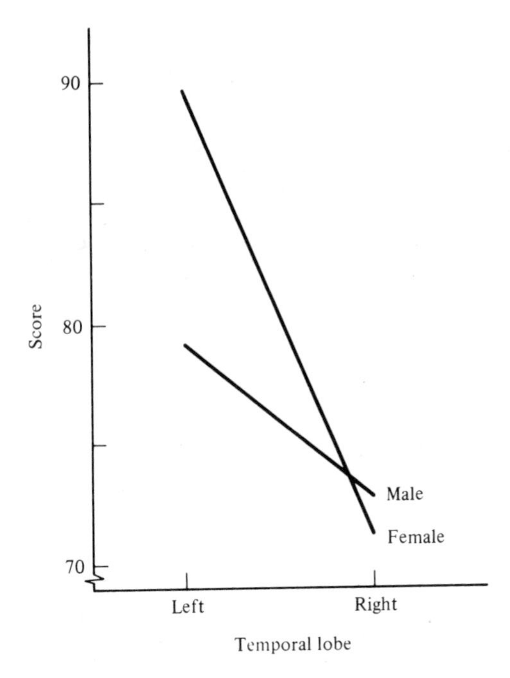

Figure 9-2 Closure scores for the four surgery groups.

tion effect by the difference in the slopes of the line segments of the column 1 and column 2 means:

$$\hat{\psi}_I = (\bar{x}_{11.} - \bar{x}_{21.}) - (\bar{x}_{12.} - \bar{x}_{22.}) \tag{21}$$

When the additive model of row and column effects holds, or if $\gamma_{11} = \gamma_{12} = \gamma_{21} = \gamma_{22} = 0$, $\hat{\psi}_I$ will have mean zero and variance

$$\text{Var}(\hat{\psi}_I) = \left(\frac{4}{n}\right)\sigma^2 \tag{22}$$

If the usual assumptions of independence and normality hold for the disturbance terms e_{ijh}, $\hat{\psi}_I$ will of course be normally distributed with those parameters. Then we may test the hypothesis of no interaction by referring the statistic

$$\begin{aligned} t &= \frac{\hat{\psi}_I}{\sqrt{4\hat{\sigma}^2/n}} \\ &= \frac{\hat{\psi}_I\sqrt{n}}{2\hat{\sigma}} \end{aligned} \tag{23}$$

to an appropriate critical value of the t-distribution with $4(n-1)$ degrees of freedom. $\hat{\sigma}^2$ is the same within-cell error mean square defined by (4). Usually, we adopt the two-sided rule

$$\text{Reject } H_0: \gamma_{ij} = 0 \text{ if } |t| \geq t_{\alpha/2;\,4(n-1)}$$

unless we have some reason to believe that one slope should be steeper than the other when the hypothesis is not true. We note that the test involves a *single* linear function of the four cell means, even though the hypothesis specifies four zero interaction parameters. We shall have more to say of such single-degrees-of-freedom contrasts in our treatment of factorial layouts.

With more than two rows or columns we must test for an interaction with a sum of squares and F-ratio. The interaction sum of squares should measure the departure of the cell means $\bar{x}_{ij.}$ from their values predicted by the additive model. To develop those predicted values let us start with some estimates of the row and column effects α_i and β_j obtained under the constraints $\sum \alpha_i = \sum \beta_j = 0$ of the sort introduced in (2) of Section 9-2; then

$$\begin{aligned} \hat{\alpha}_i &= \bar{x}_{i..} - \bar{x}_{...} \qquad i = 1, \ldots, r \\ \hat{\beta}_j &= \bar{x}_{.j.} - \bar{x}_{...} \qquad j = 1, \ldots, c \end{aligned} \tag{24}$$

and when the additive model and constraints both hold,

$$\hat{\mu} = \bar{x}_{...}$$

The predicted value of the ijth cell mean under additivity is

$$\begin{aligned} \hat{\mu}_{ij.} &= \bar{x}_{...} + (\bar{x}_{i..} - \bar{x}_{...}) + (\bar{x}_{.j.} - \bar{x}_{...}) \\ &= -\bar{x}_{...} + \bar{x}_{i..} + \bar{x}_{.j.} \end{aligned} \tag{25}$$

where $\mu_{ij} = \mu + \alpha_i + \beta_j = E(x_{ijh})$. We can measure the degree of non-additivity by the interaction sum of squares

$$\begin{aligned} \text{SSI} &= n \sum_{i=1}^{r} \sum_{j=1}^{c} (\bar{x}_{ij.} - \bar{x}_{i..} - \bar{x}_{.j.} + \bar{x}_{...})^2 \\ &= n \sum_{i=1}^{r} \sum_{j=1}^{c} (\bar{x}_{ij.} - \bar{x}_{...})^2 - \text{SSR} - \text{SSC} \end{aligned} \tag{26}$$

When

$$H_0 \colon \gamma_{ij} = 0 \qquad i = 1, \ldots, r; \quad j = 1, \ldots, c \tag{27}$$

is true SSI/σ^2 has the chi-squared distribution with $(r-1)(c-1)$ degrees of freedom. We test H_0 by referring

$$F = \frac{\text{SSI}/(r-1)(c-1)}{\hat{\sigma}^2} \tag{28}$$

to critical values of the F-distribution with $(r-1)(c-1)$ and $rc(n-1)$ degrees of freedom. If

$$F > F_{\alpha;\,(r-1)(c-1),\,rc(n-1)} \tag{29}$$

we should reject the additive model hypothesis, and conclude that the row and column treatments have interacted. Under that alternative the statistic (28) has the noncentral F-distribution with the same degrees of freedom, and noncentrality parameter

$$\lambda_I = \frac{n \sum_{i=1}^{r} \sum_{j=1}^{c} (\gamma_{ij} - \bar{\gamma}_{i.} - \bar{\gamma}_{.j} + \bar{\gamma}_{..})^2}{\sigma^2} \tag{30}$$

in which $\bar{\gamma}_{i.}$, $\bar{\gamma}_{.j}$, and $\bar{\gamma}_{..}$ are the row, column, and grand averages, respectively, of the γ_{ij}.

We may summarize the preceding tests in the analysis of variance for the balanced two-way layout in Table 9-4. Because the sums of squares for the

TABLE 9-4 Analysis of Variance for the Balanced Two-Way Layout

Source	Sum of squares	d.f.	Mean square	F	Expected mean square
Rows	$SSR = nc\sum_{i=1}^{r}(\bar{x}_{i..} - \bar{x}_{...})^2$	$r - 1$	$MSR = \frac{SSR}{r-1}$	$\frac{MSR}{\hat{\sigma}^2}$	$\sigma^2 + \frac{nc}{r-1}\sum_{i=1}^{r}(\alpha_i - \bar{\alpha})^2$
Columns	$SSC = nr\sum_{j=1}^{c}(\bar{x}_{.j.} - \bar{x}_{...})^2$	$c - 1$	$MSC = \frac{SSC}{(c-1)}$	$\frac{MSC}{\hat{\sigma}^2}$	$\sigma^2 + \frac{nr}{c-1}\sum_{j=1}^{c}(\beta_j - \bar{\beta})^2$
Interaction	$SSI = n\sum_{i=1}^{r}\sum_{j=1}^{c}(\bar{x}_{ij.} - \bar{x}_{i..} - \bar{x}_{.j.} + \bar{x}_{...})^2$	$(r-1)(c-1)$	$MSI = \frac{SSI}{(r-1)(c-1)}$	$\frac{MSI}{\hat{\sigma}^2}$	$\sigma^2 + \frac{n}{(r-1)(c-1)} \cdot \sum_{i=1}^{r}\sum_{j=1}^{c}(\gamma_{ij} - \bar{\gamma}_{i.} - \bar{\gamma}_{.j} + \bar{\gamma}_{..})^2$
Error	$SSE = \sum_{i=1}^{r}\sum_{j=1}^{c}\sum_{h=1}^{n}(x_{ijh} - \bar{x}_{ij.})^2$	$rc(n-1)$	$\hat{\sigma}^2 = \frac{SSE}{rc(n-1)}$	—	σ^2
Total	$\sum_{i=1}^{r}\sum_{j=1}^{c}\sum_{h=1}^{n}(x_{ijh} - \bar{x}_{...})^2$	$rcn - 1$	—	—	—

three hypotheses and the within-cells variation add to the total sum of squares Cochran's theorem assures that they are independently distributed. When the row, column, and interaction alternative hypotheses are true the expected mean squares can only increase from their value σ^2 under the null hypotheses of no effects. We note that when $n = 1$, or a single observation in each cell, the error sum of squares and its degrees of freedom vanish. Then the interaction mean square must be used for the row and column effects tests, as in Table 9-2. In the context of our present linear model we cannot distinguish error variation from a lack of additivity of row and column effects. In Section 9-6 we shall see how a special kind of nonadditivity can be tested with a single observation in each cell.

Whereas the sums of squares for row, column, and interaction effects are independently distributed, their F-statistics are dependent because of the common divisor $\hat{\sigma}^2$ in each ratio. However, as Kimball (1951) has shown, the dependence of the F-statistics has a benign effect on the probability that all three statistics are less than their respective critical values when the null hypotheses are true: That probability is always greater than the product of the probabilities of accepting the individual hypotheses. If we denote the row, column, and interaction F-statistics by F_R, F_C, and F_I, then

$$P[(F_R \leq F_{\alpha;r-1,rc(n-1)}) \cap (F_C \leq F_{\alpha;c-1,rc(n-1)}) \cap (F_I \leq F_{\alpha;(r-1)(c-1),rc(n-1)})]$$
$$> P(F_R \leq F_{\alpha;r-1,rc(n-1)})P(F_C \leq F_{\alpha;c-1,rc(n-1)})P(F_I \leq F_{\alpha;(r-1)(c-1),rc(n-1)}) \quad (31)$$

The common divisor has increased the chance that all three decisions about the hypotheses are correct. For example, if $\alpha = 0.01$ for each test the probability of simultaneously accepting each hypothesis when all are true is still greater than $(0.99)^3 = 0.9703$.

If the F-statistics for the hypothesis of no interaction exceeds its critical value, we cannot blithely turn our attention to the row and column tests and their significance. Let us see why that is so with the aid of a "worst case" example. Suppose that these sample means were obtained from the cells of a balanced 2×2 layout:

	Column treatment		
Row treatment	1	2	Row mean
1	60	40	50
2	38	58	48
Column mean	49	49	49

The means are plotted in Figure 9-3. The column means are equal, while the row means are virtually so. However, the column 1–column 2 contrast for row 1 is 20, whereas for row 2 it is −20. The column effects change direction according to the row condition, although an examination of the means on the margins of the table would indicate minimal or nonexistent main effects.

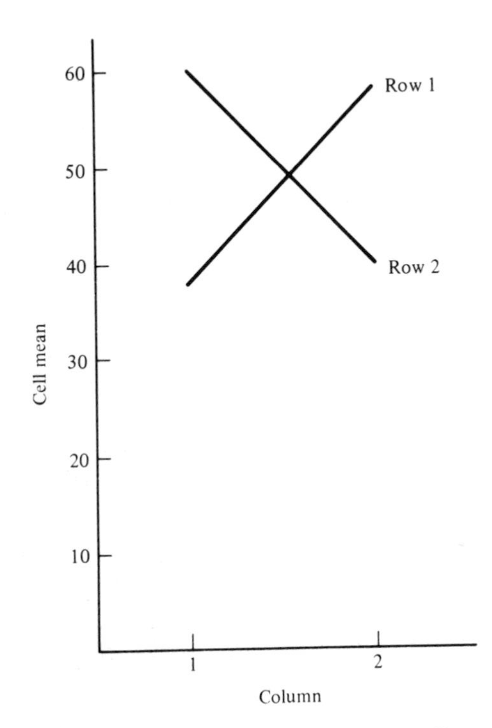

Figure 9-3 Row and column interaction.

Clearly, for such a significant interaction we should test for column effects separately for each row and for row effects separately in each column.

To test the hypothesis

$$H_0: E(x_{i1h}) = \cdots = E(x_{ich}) \qquad h = 1, \ldots, n \tag{32}$$

of equal cell expectations in the ith row we compute the treatment sum of squares

$$\text{SSC}_i = n \sum_{j=1}^{c} (\bar{x}_{ij.} - \bar{x}_{i..})^2 \qquad i = 1, \ldots, r \tag{33}$$

for that row and its mean square $\text{MSC}_i = \text{SSC}_i/(c - 1)$. We reject the hypothesis (32) at the α level if

$$F = \frac{\text{MSC}_i}{\hat{\sigma}^2} \tag{34}$$

exceeds the critical value $F_{\alpha; c-1, rc(n-1)}$ of the F-distribution. $\hat{\sigma}^2$ is the usual error mean square of expression (4) and Table 9-4 for the balanced two-way layout. We may show by Cochran's theorem that

$$\text{SSC}_1 + \cdots + \text{SSC}_r = \text{SSC} + \text{SSI} \tag{35}$$

We have decomposed the sum of squares due to columns and the rows by columns interaction into r independent components, each with $c - 1$ degrees of freedom.

Similar expressions follow for testing the hypothesis

$$H_0: E(x_{1jh}) = \cdots = E(x_{rjh}) \qquad h = 1, \ldots, n \tag{36}$$

of equal means in the cells of the jth column. We compute in succession

$$\text{SSR}_j = n \sum_{i=1}^{r} (\bar{x}_{ij.} - \bar{x}_{.j.})^2$$

$$\text{MSR}_j = \frac{\text{SSR}_j}{r-1} \tag{37}$$

$$F = \frac{\text{MSR}_j}{\hat{\sigma}^2}$$

and refer F to the critical value $F_{\alpha; r-1, rc(n-1)}$ for a test at the α level. As in the previous case of the column means hypothesis,

$$\text{SSR}_1 + \cdots + \text{SSR}_c = \text{SSR} + \text{SSI} \tag{38}$$

Whereas the c individual sums of squares are independent, the $r + c$ sums for all row and column tests required in the case of a significant interaction are *not* independently distributed. One runs a distinct risk of meeting the same significant effects twice in row and column guise.

Finally, we note that the preceding methods were based on the usual two-way layout with a common variance σ^2 of the disturbance terms in each cell. If the common variance assumption seems unrealistic, we might conduct r separate one-way analyses of variance for the rows and c analyses for the columns. However, the second degrees of freedom parameters are smaller, and the separate tests are less sensitive.

Power and sample size calculations. The power probabilities of the row, column, and interaction tests can be determined from the Pearson–Hartley charts in the same manner as the one-way powers were found in Section 8-7. For the test of equal row effects we must begin by specifying alternative values of the row effects,

$$\alpha_1 - \bar{\alpha}, \ldots, \alpha_r - \bar{\alpha}$$

as deviations from their average, that is, effects defined to have a zero sum. We must also arrive at a value of the disturbance variance σ^2 from previous studies, pilot experiments, or subjective guessing. Then we may evaluate the noncentrality parameter (14)

$$\lambda_R = nc \frac{\sum_{i=1}^{r} (\alpha_i - \bar{\alpha})^2}{\sigma^2}$$

and from it the parameter

$$\phi_R = \sqrt{\frac{\lambda_R}{r}}$$

for entering the Pearson–Hartley charts with $\nu_1 = r - 1$, $\nu_2 = rc(n - 1)$, and a choice of $\alpha = 0.05$ or $\alpha = 0.01$. If the layout has a single observation in each cell $\nu_2 = (r - 1)(c - 1)$. Similarly, we can compute the power of the equal column effects test for a given set of effects

$$\frac{\beta_1 - \bar{\beta}}{\sigma}, \ldots, \frac{\beta_c - \bar{\beta}}{\sigma}$$

expressed in units of the disturbance standard deviation. We evaluate the noncentrality parameter (18):

$$\lambda_c = nr \frac{\sum_{j=1}^{c} (\beta_j - \bar{\beta})^2}{\sigma^2}$$

and from it compute

$$\phi_c = \sqrt{\frac{\lambda_c}{c}}$$

for the Pearson–Hartley charts with $\nu_1 = c - 1$, $\nu_2 = rc(n - 1)$. The power of the interaction test can be computed in like manner from the noncentrality parameter λ_I defined by expression (30), although that probability usually is less important than the row and column powers.

As an illustration of the power computations and their use in sample size determination, let us consider a two-way layout with $r = 4$ row and $c = 3$ column treatments. The probabilities of rejecting the hypotheses of equal row or column effects when certain alternatives are true are shown in Table 9-5.

TABLE 9-5 Power Probabilities for Row and Column Tests in a 4 × 3 Layout ($\alpha = 0.05$)

a. Equal Row Effects

	Alternative								
	α_1/σ	α_2/σ	α_3/σ	α_4/σ	n	ν_2	λ	ϕ	Power
1.	−0.5	−0.5	0.5	0.5	2	12	6	1.22	0.38
					4	36	12	1.73	0.80
					6*	60	18	2.12	0.94
					8*	84	24	2.45	0.99
2.	0	0	−0.5	0.5	6	60	9	1.50	0.70
					8	84	12	1.73	0.82
					12*	132	18	2.12	0.95
3.	0	0	−1	1	2	12	12	1.73	0.69
					4*	36	24	2.45	0.98
4.	0	0	−1.5	1.5	1	6	13.5	1.84	0.60
					2*	12	27	2.60	0.97
5.	0	0	−2.45	2.45	1*	6	36	3	0.96

b. Equal Column Effects

	Alternative							
	β_1/σ	β_2/σ	β_3/σ	n	ν_2	λ	ϕ	Power
1.	0	−0.5	0.5	4	36	8	1.63	0.61
				9*	96	18	2.45	0.96
2.	0	−1	1	1	6	8	1.63	0.48
				3*	24	24	2.83	0.99
3.	0	−1.5	1.5	1	6	18	2.45	0.83
				2*	12	36	3.46	0.99+
4.	0	−2	2	1*	6	32	3.27	0.95+

*Power at least 0.94.

The cell sizes n giving powers at least equal to 0.94 are starred. To detect the unequal row effects

$$\frac{\alpha_1}{\sigma} = 0 \qquad \frac{\alpha_2}{\sigma} = 0 \qquad \frac{\alpha_3}{\sigma} = -0.5 \qquad \frac{\alpha_4}{\sigma} = 0.5$$

with a maximum difference of one standard deviation we would need 12 observations in each cell, or $rcn = 144$ sampling units in the layout. A similar difference among the column effects, or the first alternative hypothesis

$$\frac{\beta_1}{\sigma} = 0 \qquad \frac{\beta_2}{\sigma} = -0.5 \qquad \frac{\beta_3}{\sigma} = 0.5$$

would require only nine observations per cell. With a single observation in each cell we would need a range of nearly five standard deviations between the two extreme row effects; for the similar column alternative hypothesis (4) a range of four standard deviations would be required for rejection with power 0.95. Designs with one observation per cell are not sensitive to small row or column effects.

In practice we can determine appropriate cell sample sizes in much the same way as for the one-way layout. We specify the minimum differences in standard deviations among the centered row and column effects that we wish to detect with some probability, then search by trial and error for the smallest value of n satisfying that power.

Example 9-2

Let us pause now and illustrate the two-way methods by an example. We shall use the mortar crushing strength statistics compiled by Wilsdon (1934) and reported in Exercise 5 of Section 8-10. We shall use only the laboratories with observations on dry- and wet-consistency mixes, so that we may have a balanced 3×2 layout. For convenience we shall repeat the means and within-cell sums of squares SSE_{ij} here:

	Dry		Wet		
Lab	$\bar{x}_{i1.}$	SSE_{i1}	$\bar{x}_{i2.}$	SSE_{i2}	$\bar{x}_{i..}$
T	8,200	564,798	3,648	39,024	5,924
Y	7,971	530,880	3,447	35,148	5,709
Z	8,263	299,526	2,824	68,748	5,543.5
$\bar{x}_{.j.}$	8,144.67	—	3,306.33	—	5,725.5

From those statistics we may compute the following analysis of variance table:

Source	Sum of squares	d.f.	Mean square	F
Labs	873,582	2	436,791	8.52
Consistency	210,685,225	1	210,685,225	4,109
Interaction	1,624,778	2	812,389	15.85
Error	1,538,124	30	51,271	—

Because we have no need for the total sum of squares line in the table it has been omitted. All three F-statistics exceed any reasonable critical values. Because an interaction between laboratory and consistency appears to be present we must make further analyses of variance among the laboratories for dry and wet consistencies. Crushing strengths of dry- and wet-consistency mortar are greatly different, and need no statistical justification. The dry–wet mean differences for the three laboratories have these values:

Lab	T	Y	Z
Mean difference	4552	4524	5439

The differences are about the same for laboratories T and Y, but Z has a larger difference. The interaction seems to be due to the greater difference for laboratory Z. We may verify that by a formal significance test of the sample contrasts

$$\hat{\psi}_{ij} = (\bar{x}_{i1.} - \bar{x}_{i2.}) - (\bar{x}_{j1.} - \bar{x}_{j2.}) \qquad i < j$$

Since the six cell means are independently distributed,

$$\text{Var}\,(\hat{\psi}_{ij}) = \left(\frac{4}{n}\right)\sigma^2 = \left(\frac{2}{3}\right)\sigma^2$$

and its estimate is $(\frac{2}{3})\hat{\sigma}^2 = 34{,}181$. The contrasts and their t-ratios had these values:

Lab contrast	$\hat{\psi}_{ij}$	$t = \hat{\psi}_{ij}/\sqrt{2\hat{\sigma}^2/3}$
T–Y	28	0.15
T–Z	−887	−4.80
Y–Z	−915	−4.95

The t-statistics confirm laboratory Z unequivocally as the culprit; their magnitudes are sufficiently extreme that we need not be concerned with multiple comparison issues or critical values.

Finally, let us determine if the laboratories differ in crushing strength for the separate mixes. Should we use the error mean square from the complete layout analysis of variance, or should we assume different variances for the dry- and wet-consistency specimens? The estimates of those variances are

$$\hat{\sigma}^2_{\text{dry}} = \frac{564{,}798 + 530{,}880 + 299{,}526}{18 - 3} = 93{,}014$$

$$\hat{\sigma}^2_{\text{wet}} = \frac{39{,}024 + 35{,}148 + 68{,}748}{18 - 3} = 9528$$

We may test the hypothesis $H_0\colon \sigma^2_{\text{dry}} = \sigma^2_{\text{wet}}$ against the alternative of different variances by referring

$$F = \frac{\hat{\sigma}^2_{\text{dry}}}{\hat{\sigma}^2_{\text{wet}}} = 9.76$$

to critical values of the F-distribution with 15 and 15 degrees of freedom. We may verify that the ratio exceeds any of the tabulated values, and separate analyses of the two mixes would be more appropriate.

The dry consistency mixes gave this analysis of variance:

Source	Sum of squares	d.f.	Mean square	F
Labs	283,348	2	141,674	1.52
Error	1,395,204	15	93,014	—

Since $F_{0.10;2,15} = 2.70$, we cannot conclude that the laboratories differ in the crushing strengths of the dry-consistency mortar cubes. The analysis of variance for the wet mixes follows:

Source	Sum of squares	d.f.	Mean square	F
Labs	2,215,012	2	1,107,506	116.2
Error	142,920	15	9,528	—

The laboratories are clearly different. We may apply the usual one-way layout multiple comparison procedures of Section 8-6 to their means. These paired contrasts and statistics were obtained:

Contrast	Mean	t
T–Y	201	3.57
T–Z	824	14.62
Y–Z	623	11.05

The 5% studentized range critical value for the Tukey multiple comparison method is $q_{0.05;3,15} = 3.67$. We should conclude that laboratory Z has a lower mean strength than T and Y and that the means of T and Y are almost significantly different at the 0.05 level.

We have examined the model for the balanced two-way layout and have illustrated it with some actual data from an engineering application. Several other questions remain to be answered about cross-classified data. These include the following:

1. How are contrasts used in the two-way layout? How are the Scheffé and Tukey multiple comparison methods employed with them?
2. How can orthogonal polynomials be fitted to the means of two-way data?
3. How may the two-way analysis of variance be extended to higher-way plans?

We shall address these questions and others in the next few sections.

9-4 CONTRASTS AND ORTHOGONAL POLYNOMIALS

In Section 9-2 we encountered row and column effect contrasts briefly in the estimation of the parameters of the simplest two-way model. Now we treat contrasts more generally for the comparison of row or column treatment combinations, or interaction effects. In this section we think of the contrasts as chosen *before* our examination of the data; the problem of multiple inferences by the Bonferroni, Tukey, or Scheffé methods will be addressed in Section 9-5. Among the contrasts we shall consider here are those given by the orthogonal polynomial values for observations equally spaced on a time, space, or dosage scale. By those contrasts we shall be able to fit polynomial models to the row or column means when their treatments have the scale property.

We have already met some simple contrasts. In Example 9-2 we computed the differences of the average crushing strengths for laboratories T and Y, T and Z, and Y and Z to determine the source of the significant F-statistic in the analysis of variance. In the two-way model for the mortar strengths we also considered comparisons of the wet and dry consistency means for pairs of laboratories, or contrasts of the form

$$(T_{\text{dry}} - T_{\text{wet}}) - (Y_{\text{dry}} - Y_{\text{wet}}) \tag{1}$$

in which the effects or cell means have been replaced by the laboratory symbols. Now we give expressions for general row and column contrasts. For our discussion we think of the layout means as arranged in Table 9-6. For the row parameters that contrast is

$$\psi_R = \sum_{i=1}^{r} c_i \alpha_i \tag{2}$$

and its estimator is

$$\hat{\psi}_R = \sum_{i=1}^{r} c_i \bar{x}_{i..} \tag{3}$$

By definition $\sum c_i = 0$. The mean and variance of the estimator are

$$E(\hat{\psi}_R) = \psi_R \qquad \text{Var}\,(\hat{\psi}_R) = \left(\frac{\sigma^2}{nc}\right) \sum_{i=1}^{r} c_i^2 \tag{4}$$

TABLE 9-6 Means of the Two-Way Layout

	Column treatment			
Row treatment	1	...	c	Row mean
1	$\bar{x}_{11.}$	...	$\bar{x}_{1c.}$	$\bar{x}_{1..}$
⋮	....			⋮
r	$\bar{x}_{r1.}$	...	$\bar{x}_{rc.}$	$\bar{x}_{r..}$
Column mean	$\bar{x}_{.1.}$	...	$\bar{x}_{.c.}$	$\bar{x}_{...}$

and of course under the usual linear model and normal disturbance assumptions $\hat{\psi}_R$ is a normal random variable with the parameters (4). Because it is composed of the rc cell means it is distributed independently of the error sum of squares and mean square. Hence the ratio

$$t = \frac{\hat{\psi}_R - \psi_R}{\sqrt{(\sum c_i^2)(\hat{\sigma}^2/nc)}} \tag{5}$$

has the t-distribution with $rc(n-1)$ degrees of freedom. If the contrast were chosen a priori rather than by examination of the row means for likely differences, we might use (5) to test the hypothesis that ψ_R has a particular value, for example, zero, or to form a confidence interval for ψ_R.

Similar contrasts follow for the column means merely by interchanging the row and column positions in the two-way layout, or by some corresponding changes in our notation. Denote the general column contrast by

$$\psi_C = \sum_{j=1}^{c} d_j \beta_j \tag{6}$$

and its estimator by

$$\hat{\psi}_C = \sum_{j=1}^{c} d_j \bar{x}_{.j.} \tag{7}$$

where of course $\sum d_j = 0$. Then $\hat{\psi}_C$ is a normal random variable with respective mean and variance

$$E(\hat{\psi}_C) = \psi_C \qquad \mathrm{Var}\,(\hat{\psi}_C) = \frac{\sigma^2}{nr} \sum_{j=1}^{c} d_j^2 \tag{8}$$

and we may test hypotheses or make confidence statements about the β_j by means of the statistic

$$t = \frac{\hat{\psi}_C - \psi_C}{\sqrt{(\sum d_j^2)(\hat{\sigma}^2/nr)}} \tag{9}$$

Since t involves the same error mean square its degrees of freedom are $rc(n-1)$.

Before turning to contrasts for measuring the interaction of the row and column treatments let us show that the row and column contrast estimators (3) and (7) are independently distributed. Because $\hat{\psi}_R$ and $\hat{\psi}_C$ have the bivariate normal distribution they will be independent if their covariance is zero. Let us calculate the covariance:

$$\begin{aligned}
\mathrm{Cov}\,(\hat{\psi}_R, \hat{\psi}_C) &= \mathrm{Cov}\,(\textstyle\sum c_i \bar{x}_{i..}, \sum d_j \bar{x}_{.j.}) \\
&= \mathrm{Cov}\left[\frac{\sum c_i(\sum_j \bar{x}_{ij.})}{c}, \frac{\sum d_j(\sum_i \bar{x}_{ij.})}{r}\right] \\
&= \mathrm{Cov}\left(\frac{\sum\sum c_i x_{ij.}}{c}, \frac{\sum\sum d_j x_{ij.}}{r}\right) \\
&= \frac{\sum\sum c_i d_j \,\mathrm{Var}\,(\bar{x}_{ij.})}{rc} \\
&= \frac{\sigma^2}{rcn} \sum\sum c_i\, d_j \\
&= 0
\end{aligned} \tag{10}$$

In the second line we expressed the row and column means in terms of the cell means $\bar{x}_{ij.}$. Those means from different rows or columns are uncorrelated and have a common variance σ^2/rcn. The covariance of zero follows from the two contrast properties in the penultimate line.

The independence property is fundamental and useful. It tells us that any sample contrast on one dimension will be independent of all the contrasts on the other way of classification. We may proceed with our row inferences content in the assurance that they are not influenced or confounded with column effects. The row and column conditions are said to be *orthogonal*, just as individual contrasts can be constructed to have orthogonal vectors of coefficients. Balanced two-way designs are examples of *orthogonal* layouts, or ones in which the row and column hypotheses being tested have independent sums of squares capable of being broken down into independent, or orthogonal, contrasts.

Now let us find some contrasts that measure the interaction effects of the row and column treatments. Recall from the preceding section that no interaction in the 2×2 layout is equivalent to the same column effect in row 1 as in row 2, or parallelism of the profiles of the cell means in Figure 9-1. In the general $r \times c$ layout we might search for interactions by computing the sample contrasts

$$\hat{\psi} = \bar{x}_{ij.} - x_{ih.} - (\bar{x}_{rj.} - \bar{x}_{rh.}) \tag{11}$$

for $j \neq h$, $i \neq r$, and as many combinations of i, h, j, and r as seem necessary from a plot of the means. For example, consider the means depicted in Figure 9-4. The cell means within each row have been joined by line segments. The segments of rows 1 and 2 are essentially parallel, although we really would have to compare the slope differences to their standard deviations for a definitive answer. Row 3 is different: Its first and second segments are pitched differently

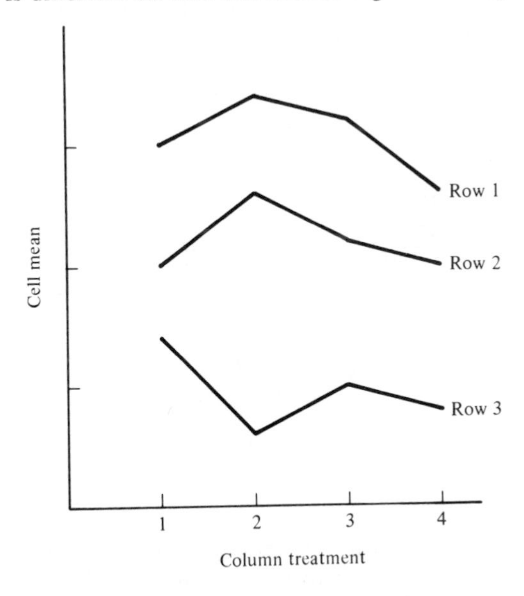

Figure 9-4 Cell means of a 3×4 layout.

from their counterparts in rows 1 and 2. The third segments of rows 2 and 3 are parallel, although the segment of row 1 has a steeper slope. We might compute these sample contrasts:

$$\begin{aligned}
\hat{\psi}_1 &= \bar{x}_{11.} - \bar{x}_{12.} - (\bar{x}_{31.} - \bar{x}_{32.}) \\
\hat{\psi}_2 &= \bar{x}_{21.} - \bar{x}_{22.} - (\bar{x}_{31.} - \bar{x}_{32.}) \\
\hat{\psi}_3 &= \bar{x}_{12.} - \bar{x}_{13.} - (\bar{x}_{32.} - \bar{x}_{33.}) \\
\hat{\psi}_4 &= \bar{x}_{22.} - \bar{x}_{23.} - (\bar{x}_{32.} - \bar{x}_{33.})
\end{aligned} \tag{12}$$

Because the cell means are independent each contrast has the same variance:

$$\text{Var}(\hat{\psi}_i) = \left(\frac{4}{n}\right)\sigma^2 \qquad i = 1, \ldots, 4$$

We may test the hypothesis of parallelism, or no interaction within given pairs of row and column treatments, by referring

$$t = \hat{\psi}_i \sqrt{\frac{n}{4\hat{\sigma}^2}} \tag{13}$$

to a critical value of the t-distribution with $rc(n - 1)$ degrees of freedom. The choice of one- or two-sided tests and their decision rules follow in the usual ways, and we shall not give space to them now. Once again we note that such tests based on the t critical values offer protection with a given Type I error rate only when the contrasts were chosen a priori, or if only one contrast is of interest. More appropriate critical values are given by the Scheffé or Bonferroni multiple comparison methods. We address the contrasts (12) again when we discuss simultaneous inference in Section 9-5.

Many other contrasts can be constructed for the interaction effects, but we only treat one at this time. Recall from expression (25) of Section 9-3 that the deviation of the ijth cell mean from its value predicted by the additive model can be expressed as

$$\begin{aligned}
\hat{\psi} &= \bar{x}_{ij.} - [\bar{x}_{...} + (\bar{x}_{i..} - \bar{x}_{...}) + (\bar{x}_{.j.} - \bar{x}_{...})] \\
&= \bar{x}_{ij.} - \bar{x}_{i..} - \bar{x}_{.j.} + \bar{x}_{...}
\end{aligned} \tag{14}$$

The variance of the contrast is

$$\text{Var}(\hat{\psi}) = \frac{\sigma^2(r - 1)(c - 1)}{nrc} \tag{15}$$

We might form the statistic

$$t = \hat{\psi} \sqrt{\frac{nrc}{\hat{\sigma}^2(r - 1)(c - 1)}} \tag{16}$$

and test for the additive model at a given cell mean selected a priori by referring (16) to a critical value of the t-distribution with $rc(n - 1)$ degrees of freedom.

Example 9-3

Let us calculate some contrasts in a simple two-way layout. A photographic light meter was used to record illumination intensity in units of "light values" under different levels of two sources of artificial light. The first source, corresponding to the columns of the layout, was a three-way table lamp. Its levels, or column treatments, were Off, 50 watts, 100 watts, and 150 watts. The row treatments were the levels Off, 20 watts, and 40 watts of a fluorescent desk lamp to the left and slightly to the rear of the observer. The readings were taken as uniformly as possible from the same position facing the table lamp over an hour's period. Two readings were obtained under each combination of illumination. The data are given in Table 9-7. The within-cell sums of squares can

TABLE 9-7 Light Intensity Values

	Table lamp				
Desk lamp	Off	50	100	150	Mean
Off	6.0, 5.9	7.8, 7.5	8.3, 8.0	9.2, 8.5	7.65
20	6.3, 5.9	7.8, 7.3	8.5, 8.0	9.0, 8.5	7.66
40	6.6, 6.3	7.5, 7.6	8.2, 8.3	9.2, 8.9	7.82
Mean	6.17	7.58	8.22	8.88	7.71

be calculated easily by the formula

$$\text{SSE}_{ij} = \tfrac{1}{2}(x_{ij1} - x_{ij2})^2$$

The values of the cell means are as follows:

	Off	50	100	150
Off	5.95	7.65	8.15	8.85
20	6.10	7.55	8.25	8.75
40	6.45	7.55	8.25	9.05

This analysis of variance was computed from the data:

Source	Sum of squares	d.f.	Mean square	F
Desk lamp	0.1525	2	0.0762	1.02
Table lamp	24.1879	3	8.0626	108.1
Interaction	0.2310	6	0.0385	0.52
Error	0.8950	12	0.0746	—
Total	25.4664	23	—	—

The column means for the four levels of the table lamp illumination are highly different. The row means and the interaction F-statistics do not even approach the conventional critical values: The desk lamp seems to have no significant effect on the light values.

Now let us examine some contrasts suggested by the data. We note from Figure 9-5 that the slopes of the segments joining the column means at 50, 100, and 150 watts are virtually the same, but the segment from the Off mean to the 50 watt mean has a

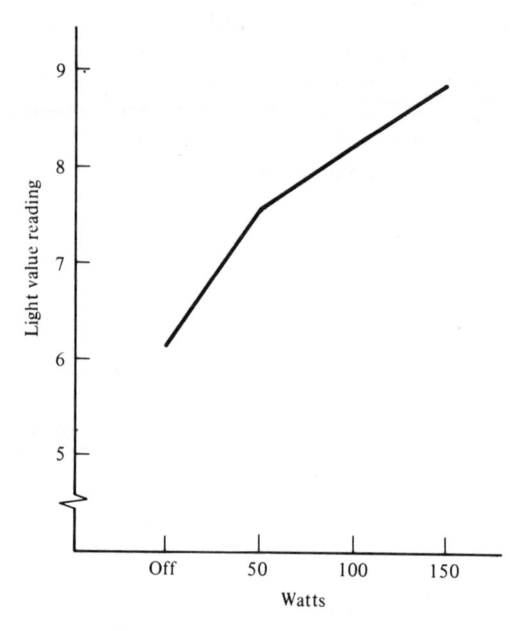

Figure 9-5 Mean light values.

steeper slope. Are the two slopes significantly different? Let us estimate the contrast

$$\psi_1 = \beta_2 - \beta_1 - \tfrac{1}{2}(\beta_4 - \beta_2)$$

by

$$\hat{\psi}_1 = \bar{x}_{.2.} - \bar{x}_{.1.} - \tfrac{1}{2}(\bar{x}_{.4.} - \bar{x}_{.2.}) = 0.76$$

The variance of the contrast is

$$\begin{aligned}\text{Var}(\hat{\psi}_1) &= \text{Var}\left(-\bar{x}_{.1.} + \frac{3}{2}\bar{x}_{.2.} - \frac{1}{2}\bar{x}_{.4.}\right)\\ &= \frac{\sigma^2}{rn}\left(\frac{9}{4} + 1 + \frac{1}{4}\right)\\ &= \frac{7}{12}\sigma^2\end{aligned}$$

and the estimated standard derivation is

$$\sqrt{\widehat{\text{Var}}(\hat{\psi}_1)} = \sqrt{(\tfrac{7}{12})(0.0746)} = 0.2086$$

Because

$$t = \frac{\hat{\psi}_1}{\sqrt{\widehat{\text{Var}}(\hat{\psi}_1)}} = 3.64$$

exceeds the 0.01 critical value of a t-statistic with 12 degrees of freedom the line segments probably have different slopes. The response of the light values does not appear to be a linear function of watts consumed.

Although we could not reject the hypothesis of equal row, or desk lamp illumination, effects, let us still examine a contrast of those parameters. We shall choose

$$\psi_2 = \alpha_3 - \tfrac{1}{2}(\alpha_1 + \alpha_2)$$

Its estimate is

$$\hat{\psi}_2 = \bar{x}_{3..} - \tfrac{1}{2}(\bar{x}_{1..} + \bar{x}_{2..}) = 0.165$$

The variance of the estimate is

$$\text{Var}(\hat{\psi}_2) = \frac{\sigma^2}{nc}\left(\frac{3}{2}\right) = \frac{3}{16}\sigma^2$$

and the estimated standard deviation is

$$\sqrt{\widehat{\text{Var}(\hat{\psi}_2)}} = \sqrt{(\tfrac{3}{16})(0.0746)} = 0.1183$$

The quantity

$$t = \frac{\hat{\psi}_2}{\sqrt{\widehat{\text{Var}(\hat{\psi}_2)}}} = 1.40$$

is not excessive. The hypothesis of a zero value for ψ_2 seems tenable. The illumination effect of a 40-watt desk lamp does not appear to be different from the average of the 0- and 20-watt effects.

Finally, we might study some contrasts measuring the row and column interactions. For example, let us compare the 50 watt–Off contrasts for 0- and 40-watt illumination from the desk lamp. The sample contrast is

$$\hat{\psi}_3 = \bar{x}_{11.} - \bar{x}_{12.} - (\bar{x}_{31.} - \bar{x}_{32.}) = 0.60$$

and its estimated standard deviation is

$$\sqrt{\widehat{\text{Var}(\hat{\psi}_3)}} = \sqrt{(4/2)\hat{\sigma}^2} = 0.3862$$

$\hat{\psi}_3$ is less than twice its standard deviation. As the insignificant interaction F-statistic would portend, we have no basis for contending that the initial amount of light with the desk lamp off is different from that with the desk lamp fully lighted. Similarly, we could compare the 50-watt–Off and 150-watt–100-watt differences in the third row of the layout:

$$\hat{\psi}_4 = \bar{x}_{34.} - \bar{x}_{33.} - (\bar{x}_{32.} - \bar{x}_{31.}) = -0.30$$

$$\sqrt{\widehat{\text{Var}(\hat{\psi}_4)}} = 0.3862$$

The difference is hardly significant.

Orthogonal polynomials. We may fit polynomial models to the row, column, and cell means of a two-way layout if the treatments are located on some scale. If they are equally spaced we may use orthogonal polynomials as in Chapter 4 and Section 8-5 for the one-way layout. The orthogonal polynomial regression coefficients are examples of orthogonal contrasts of the means that may be used to decompose the row or column sum of squares into independent sums of squares with a single degree of freedom apiece. As an illustration we might fit orthogonal polynomials to the four column means of the light-value data of Example 9-3 to see to what extent the profile of the means could be explained by a second-degree or third-degree function. We have already determined by a suitable contrast that a single straight line will not suffice, but it would be useful to express the nonlinearity in terms of quadratic and cubic components.

As in Chapter 4 let $\phi_h(t)$ denote the value of the hth-degree orthogonal polynomial at the point t on the underlying scale of the treatments. The row means can be represented exactly by the $r-1$ orthogonal polynomials:

$$\bar{x}_{i..} = \bar{x}_{...} + \hat{\beta}_1\phi_1(i) + \cdots + \hat{\beta}_{r-1}\phi_{r-1}(i) \tag{17}$$

The estimated regression coefficient of the ith polynomial is the contrast

$$\hat{\beta}_i = \frac{\sum_{t=1}^{r} \phi_i(t)\bar{x}_{t..}}{\sum_{t=1}^{r} \phi_i^2(t)} \qquad i = 1, \ldots, r-1 \tag{18}$$

Because the polynomials are orthogonal, that is,

$$\sum_{t=1}^{r} \phi_i(t)\phi_j(t) = 0 \qquad i \neq j$$

the contrasts are orthogonal and thus independent. We can decompose the row sum of squares into $r-1$ independent components:

$$\begin{aligned} \text{SSR} &= nc\hat{\beta}_1^2\left[\sum_{t=1}^{r} \phi_1^2(t)\right] + \cdots + nc\hat{\beta}_{r-1}^2\left[\sum_{t=1}^{r} \phi_{r-1}^2(t)\right] \\ &= \text{SSR}_1 + \cdots + \text{SSR}_{r-1} \end{aligned} \tag{19}$$

Each standardized contrast

$$t_i = \hat{\beta}_i \frac{\sqrt{cn \sum_{t=1}^{r} \phi_i^2(t)}}{\hat{\sigma}} \tag{20}$$

has the t-distribution with $rc(n-1)$ degrees of freedom. We may test the hypothesis of no ith-degree polynomial component among the row means or

$$H_0: \beta_i = 0$$

by referring (20) to a one- or two-sided critical value of the t-distribution with $rc(n-1)$ degrees of freedom.

In like manner we can fit polynomials in the column means. Their estimated coefficients are the sample contrasts

$$\hat{\beta}_j = \frac{\sum_{t=1}^{c} \phi_j(t)\bar{x}_{.t.}}{\sum_{t=1}^{c} \phi_j^2(t)} \qquad j = 1, \ldots, c-1 \tag{21}$$

The polynomial values $\phi_j(t)$ are not the same as those for rows, unless of course $r = c$. The column sum of squares can be decomposed as

$$\begin{aligned} \text{SSC} &= nr\hat{\beta}_1^2\left[\sum_{t=1}^{c} \phi_1^2(t)\right] + \cdots + nr\hat{\beta}_{c-1}^2\left[\sum_{t=1}^{c} \phi_{c-1}^2(t)\right] \\ &= \text{SSC}_1 + \cdots + \text{SSC}_{c-1} \end{aligned} \tag{22}$$

and we can test the hypothesis

$$H_j: \beta_j = 0$$

of no jth-degree polynomial component in the column means by referring

$$t_j = \hat{\beta}_j \frac{\sqrt{rn \sum_{t=1}^{c} \phi_j^2(t)}}{\hat{\sigma}} \tag{23}$$

to a critical value of the t-distribution with $rc(n-1)$ degrees of freedom.

We can also fit polynomial models to the cell means in any row or column of Table 9-6. However, the t-tests (20) and (23) and the decompositions (19) and (22) of the row and column sums of squares are valid only if every cell has $n \geq 2$ independent observations.

Example 9-4

Let us break up the column sum of squares for the data of Example 9-3 into components for linear, quadratic, and cubic trends in the means. The orthogonal polynomial values for $c = 4$ points can be found in Table 9, Appendix A; it will be convenient to write them below the column means in this table:

	Column			
	1	2	3	4
Mean	6.17	7.58	8.22	8.88
$\phi_1(t)$	−3	−1	1	3
$\phi_2(t)$	1	−1	−1	1
$\phi_3(t)$	−1	3	−3	1

We may summarize the polynomial regression coefficients, their contributions to the columns sum of squares, and their F- and t-statistics in this table:

j	Polynomial	$\hat{\beta}_j$	SSC_j	F_j	t_j
1	Linear	0.4392	23.1441	310.31	17.62
2	Quadratic	−0.1875	0.8437	11.31	−3.36
3	Cubic	0.04083	0.200	2.68	1.64
Sum		—	24.1879	—	—

The F-statistics are given by

$$F_j = \frac{SSC_j}{\hat{\sigma}^2}$$

and are equal to the squares of the t_j defined by (23). The linear trend component is overwhelmingly significant, whereas $|t_2|$ exceeds $t_{0.005;12} = 3.055$: The quadratic component is significant at the 0.01 level. A second-degree polynomial should suffice as a model for illumination as a function of the power consumed by the table lamp.

The insignificant F-ratio for the row means suggests that fitting a polynomial to those averages would not be worthwhile. However, we might consider some polynomial models for the cell means. If particular, we might test whether the trend in the four means of the first row is linear. That is, does the table lamp illumination increase in equal increments when the desk lamp is not lighted? We may apply formula (21) and the orthogonal polynomial values used for the column means to obtain these contrast coefficients and their test statistics:

j	Polynomial	$\hat{\beta}_j$	t_j
1	Linear	0.46	10.65
2	Quadratic	−0.25	−2.59
3	Cubic	0.07	1.62

Again at least the quadratic term is statistically significant: The response does not appear to be linear. We might confirm that finding by pooling the quadratic and cubic single-degree-of-freedom sums of squares and computing the F-statistic for nonlinearity:

$$F = \frac{(\text{SSC}_2 + \text{SSC}_3)/2}{\hat{\sigma}^2}$$

$$= \frac{(0.50 + 0.196)/2}{0.0746}$$

$$= 4.66$$

Since F exceeds the 0.05 critical value with 2 and 12 degrees of freedom we should conclude again that the response is not linear.

Example 9-5

Let us treat the manuscript submission data of Exercise 2 of Section 4-6 as a two-way layout and express the trends in its row and column means in terms of orthogonal polynomials. Let us begin with the analysis of variance table for the data:

Source	Sum of squares	d.f.	Mean square	F
Rows (years)	390.458	3	130.15	2.66
Columns (periods)	344.875	5	68.98	1.41
Error	734.292	15	48.95	—
Total	1469.625	23	—	—

The two-month periods do not have significantly different means, and the four years are only different at the 0.10 level. Nevertheless, let us decompose the years sum of squares into linear, quadratic, and cubic components. The annual means had these values:

Year	Mean submissions per period
1972	40.83
1973	47.50
1974	51.50
1975	49.67

Since we have $r = 4$ years we may use the same orthogonal polynomial values of Example 9-3. The contrasts, their sums of squares components, and their F-statistics had these values:

i	Polynomial	$\hat{\beta}_i$	SSC_i	F_i
1	Linear	1.525	279.075	5.70
2	Quadratic	−2.125	108.375	2.21
3	Cubic	−0.1533	3.008	0.06
Sum		—	390.458	—

Most of the rows sum of squares is due to a linear trend. The linear orthogonal polynomial coefficient is indeed significant at the 0.05 level, although the overall rows F-ratio with 3 and 15 degrees of freedom was not.

The conclusion of a linear trend in the annual means or totals is probably more correct and informative than that of unequal means suggested by the marginally significant F-ratio in the original analysis of variance. Because the independent polynomial contrasts were selected a priori the matter of multiple tests does not appear to be a serious one in this instance. In Section 9-5 we consider ways of handling that multiple comparison problem.

9-5 MULTIPLE COMPARISONS

Simultaneous tests and confidence intervals for comparing many treatment effects in the two-way layout follow directly from the methods in Section 8-6 with some changes in notation and parameters. We give their formulas for an $r \times c$ layout with n independent observations in each cell. The case of the layout with a single datum in each cell follows by setting $n = 1$ and replacing the within-cells error mean square $\hat{\sigma}^2$ by the row by column interaction mean square.

Let us begin with the Tukey multiple comparison method for the row and column effects. Recall from Section 8-6 that this procedure is based on the range of the treatment means when the unknown population variance must be estimated by the sample error mean square. The family of row comparisons consists of the $r(r - 1)/2$ hypotheses

$$H_0: \quad \alpha_i = \alpha_j \tag{1}$$

that are to be tested against the individual alternatives of unequal effects. We may do so with a family error rate not in excess of α by referring

$$q_{ij} = |\bar{x}_{i..} - \bar{x}_{j..}| \frac{\sqrt{nc}}{\hat{\sigma}} \tag{2}$$

to the studentized range upper 100α critical values $q_{\alpha; r, rc(n-1)}$ in Table 7, Appendix A. If $q_{ij} > q_{\alpha; r, rc(n-1)}$, we should reject H_0 and conclude that the ith and jth row effects are different. Equivalently, we may construct as many of the $r(r - 1)/2$ confidence intervals on the effect differences as we deem necessary or interesting and conclude those treatments are significantly different whose intervals do not enclose zero. The general interval in the family is

$$\bar{x}_{i..} - \bar{x}_{j..} - \frac{q_{\alpha;\, r,\, rc(n-1)}\hat{\sigma}}{\sqrt{nc}} \le \alpha_i - \alpha_j \le \bar{x}_{i..} - \bar{x}_{j..} + \frac{q_{\alpha;\, r,\, rc(n-1)}\hat{\sigma}}{\sqrt{nc}} \tag{3}$$

Similarly, for all $c(c - 1)/2$ tests of the column effects hypotheses

$$H_0: \quad \beta_i = \beta_j \tag{4}$$

we may refer

$$q_{ij} = \frac{|\bar{x}_{.i.} - \bar{x}_{.j.}|\sqrt{nr}}{\hat{\sigma}} \tag{5}$$

to the $100\alpha\%$ upper critical value $q_{\alpha;\, c,\, rc(n-1)}$ of the studentized range distribution with parameters c and $rc(n - 1)$. If q_{ij} execeds that critical value for a particular pair of columns, we should consider their effects to be different. The $100(1 - \alpha)\%$ simultaneous confidence interval family for the column effects has the general member

$$\bar{x}_{.i.} - \bar{x}_{.j.} - \frac{q_{\alpha;\, c,\, rc(n-1)}\hat{\sigma}}{\sqrt{rn}} \le \beta_i - \beta_j \le \bar{x}_{.i.} - \bar{x}_{.j.} + \frac{q_{\alpha;\, c,\, rc(n-1)}\hat{\sigma}}{\sqrt{rn}} \tag{6}$$

We should conclude that the ith and jth column treatments have different effects if that interval does not contain zero.

The Bonferroni simultaneous tests and intervals can be obtained from the preceding expressions by replacing the critical values $q_{\alpha;\, c,\, rc(n-1)}$ or $q_{\alpha;\, r,\, rc(n-1)}$ by $\sqrt{2}\, t_{\alpha/2K;\, rc(n-1)}$ in which K is the number of hypotheses or comparisons in the family. In the present case of paired comparisons K is $r(r - 1)/2$ or $c(c - 1)/2$ for rows and columns, respectively.

Example 9-6

Let us return to Example 9-3 and assess the illumination differences of the table lamp settings by the Tukey method. The critical values of the studentized range distribution are

$$q_{0.05;\, 4,\, 12} = 4.20$$

$$q_{0.01;\, 4,\, 12} = 5.50$$

The three estimated successive contrasts and their studentized range statistics (5) had these values:

Contrast	$\hat{\psi}_{i+1,i}$	$q_{i+1,i}$
50–0	1.41	12.65
100–50	0.64	5.74
150–100	0.66	5.92

We see that all four treatments are different in the multiple testing sense at the 0.01 level. Because the other three estimated contrasts are even larger they would also be significantly different from zero at any reasonable level.

We may carry out Scheffé multiple tests on all contrasts

$$\psi_R = \sum_{i=1}^{r} c_i \alpha_i \tag{7}$$

of the row effects by referring

$$t = \frac{\hat{\psi}_R}{\hat{\sigma}_{\hat{\psi}_R}} = \frac{\left(\sum_{i=1}^{r} c_i \bar{x}_{i..}\right)\sqrt{nc}}{\hat{\sigma}\sqrt{\sum c_i^2}} \tag{8}$$

to the Scheffé critical value

$$\sqrt{(r-1)F_{\alpha;\, r-1, rc(n-1)}} \tag{9}$$

If $|t|$ exceeds the critical value, we should reject

$$H_0: \quad \psi_R = 0$$

in the multiple testing sense at the α level. Alternatively, we may construct the $100(1-\alpha)\%$ simultaneous confidence interval for ψ_R:

$$\sum_{i=1}^{r} c_i \bar{x}_{i..} - \sqrt{\hat{\sigma}^2 \frac{1}{nc}(r-1)F_{\alpha;\, r-1, rc(n-1)} \sum c_i^2} \le \sum_{i=1}^{r} c_i \alpha_i \le \sum_{i=1}^{r} c_i \bar{x}_{i..} + \sqrt{\hat{\sigma}^2 \frac{1}{nc}(r-1)F_{\alpha;\, r-1, rc(n-1)} \sum c_i^2} \tag{10}$$

For multiple tests of the hypothesis

$$H_0: \quad \psi_C = \sum_{j=1}^{c} c_j \beta_j = 0 \tag{11}$$

on contrasts of the column effects we may compute the estimator and its standard deviation,

$$\hat{\psi}_C = \sum_{j=1}^{c} c_j \bar{x}_{.j.} \qquad \hat{\sigma}_{\hat{\psi}_C} = \hat{\sigma}\sqrt{\frac{\sum c_j^2}{rn}} \tag{12}$$

and refer the statistic

$$t = \frac{\hat{\psi}_C}{\hat{\sigma}_{\hat{\psi}_C}} \tag{13}$$

to the Scheffé critical value

$$\sqrt{(c-1)F_{\alpha;\, c-1, rc(n-1)}} \tag{14}$$

If t exceeds (14), we should conclude that the hypothesis $\psi_C = 0$ is not tenable in the simultaneous sense at the α level. The contrasts may be paired comparisons of the form

$$\psi_C = \beta_1 - \beta_2 \qquad \psi_C = \beta_3 - \beta_2$$

or comparisons of one treatment with an average of a set of others, for example,

$$\psi_C = \beta_2 - \tfrac{1}{3}(\beta_1 + \beta_3 + \beta_5)$$

The $100(1-\alpha)\%$ simultaneous confidence interval for the general contrast

$\psi_C = \sum c_j\beta_j$ is

$$\sum_{j=1}^{c} c_j\bar{x}_{.j.} - \sqrt{\hat{\sigma}^2 \frac{1}{nr}(c-1)F_{\alpha;\,c-1,rc(n-1)} \sum c_j^2}$$

$$\leq \sum_{j=1}^{c} c_j\beta_j \leq \sum_{j=1}^{c} c_j\bar{x}_{.j.} + \sqrt{\hat{\sigma}^2 \frac{1}{nr}(c-1)F_{\alpha;\,c-1,rc(n-1)} \sum c_j^2} \quad (15)$$

Although the confidence intervals are more laborious to compute, they are more informative than the "yea/nay" answers provided by the multiple hypothesis tests.

Example 9-7

Let us suppose that a consumer research group wishes to compare the gasoline mileage of a certain set of four compact car models A, B, C, and D. Each car is available with a four-, six-, or eight-cylinder engine. The cars are to be compared in a 4×3 cross-classified layout. $n = 2$ cars have been selected at random from the 12 model-engine size populations available to the group. Time and cost constraints have dictated the simple two-way layout, although in practice we probably would use a more complicated design with provision for driver effects and controls for engine condition, age, total mileage, and other performance factors. In the present case we merely assume that those variables have been held constant in the choice of the cars. These values in miles per gallon were obtained for a standard route traversed by the same professional chauffeur:

	Number of cylinders			
Model	4	6	8	Mean
A	23.0	17.2	16.3	
	20.2	16.6	15.5	18.13
B	21.7	15.1	13.0	
	21.3	16.5	12.4	16.67
C	20.3	18.2	16.1	
	19.7	16.8	14.9	17.67
D	23.8	16.5	13.4	
	22.6	17.9	13.8	18.00
Mean	21.575	16.85	14.425	17.617

This analysis of variance was computed from the data:

Source	Sum of squares	d.f.	Mean square	F
Model	7.9133	3	2.6378	3.40
Cylinders	211.5433	2	105.7717	136.19
Interaction	19.6167	6	3.2694	4.21
Error	9.3200	12	0.7767	—
Total	248.3933	23	—	—

We note immediately that the interaction F-ratio exceeds the critical value $F_{0.025;\,6,\,12} = 3.73$. An additive model of miles per gallon as the sum of model and engine size effects alone is not supported by the data. We must carry out separate analyses of variance for each row and column of the table. Nevertheless, before doing so let us consider the model and cylinder F-ratios in the complete analysis. The cylinder means are different at any sort of common significance level, and it would appear that Tukey or Scheffé multiple comparison procedures would declare the three paired differences significant, too. The means of the four models are only different at the 0.10 level rather than the conventional 0.05 value. Largely as an illustration of the Scheffé procedure we shall estimate a number of contrasts ψ_i suggested by the row means and test the hypothesis $H_0: \psi_i = 0$ in each case. The contrasts, their estimates, and the statistics (8) are shown in this table:

i	Contrast ψ_i	$\hat{\psi}_i$	t_i
1	$\alpha_1 - \alpha_2$	1.47	2.88
2	$\alpha_4 - \alpha_2$	1.33	2.61
3	$\alpha_3 - \alpha_2$	1.00	1.97
4	$\alpha_1 - \alpha_3$	0.47	0.92
5	$\alpha_4 - \alpha_3$	0.33	0.66
6	$\frac{1}{2}(\alpha_1 + \alpha_4) - \alpha_2$	1.40	3.18
7	$\alpha_1 - \frac{1}{2}(\alpha_2 + \alpha_3)$	0.97	2.19

We note that the standard deviations of the first five estimated contrasts are each equal to 0.5088, whereas those of $\hat{\psi}_6$ and $\hat{\psi}_7$ are equal to 0.4407. The Scheffé $\alpha = 0.10$ critical value is

$$\sqrt{3F_{0.10;\,3,\,12}} = 2.80$$

We note that model B, or the row 2 effect, appears to be different from model A. Similarly, the effect of model B is different from the average of models A and D. Model B is almost different from model D at the 0.10 level. The other paired differences and the seventh contrast are not significantly different from zero.

Now let us return to the two-way design and confront its analysis in the context of the significant interaction effect. It will be helpful to begin with the tables of cell means and within-cells sums of squares:

	Number of cylinders					
	4		6		8	
Model	$\bar{x}_{i1.}$	SSE_{i1}	$\bar{x}_{i2.}$	SSE_{i2}	$\bar{x}_{i3.}$	SSE_{i3}
A	21.6	3.92	16.9	0.18	15.9	0.32
B	21.5	0.08	15.8	0.98	12.7	0.18
C	20.0	0.18	17.5	0.98	15.5	0.72
D	23.2	0.72	17.2	0.98	13.6	0.08

The means are shown in Figure 9-6. Models A and C seem to be the source of the interaction effect. With the possible exception of model A, the relation of decreasing mileage

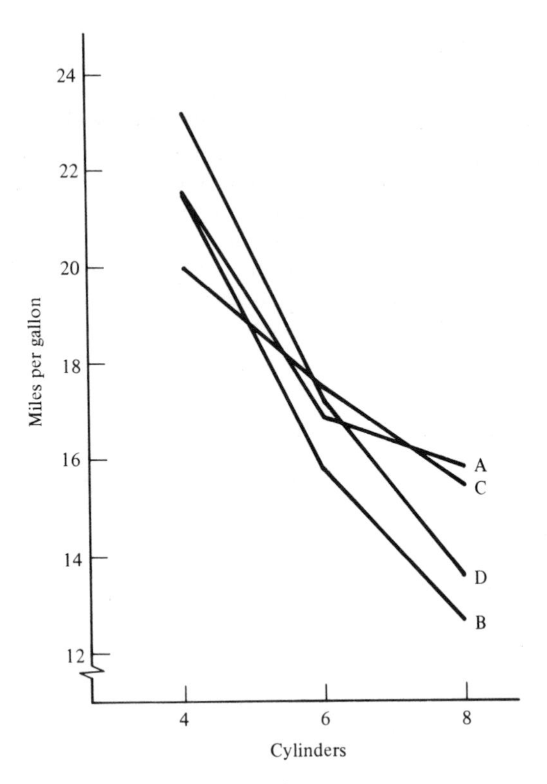

Figure 9-6 Mean miles per gallon.

to a larger engine seems clear and in little need of statistical support. Let us then concentrate on model differences for the three engine sizes. We may test the hypotheses of equal means in two ways. In the first we merely carry out three separate one-way analyses of variance. By so doing we have minimized the effect of the unequal within-cell variances, but our error degrees of freedom are only four in each case. The second approach uses the pooled error mean squares of the initial two-way analysis of variance. Because its error degrees of freedom parameter is 12 it is a more sensitive test. The six F-statistics for both methods are summarized in Table 9-8. The four-cylinder models are not significantly different by the first F-ratio but are different at the 0.05 level when the pooled estimate of the error variance is used. The six-cylinder models are not different by either approach. The eight-cylinder models are different at the 0.05 level when only their within-cells error mean square is used but differ at the 0.01 level when the F-statistic is computed from the pooled error mean square. Finally, we note that the total of the three between-models sums of squares is equal to the sum of the model and interaction sums of squares in the initial two-way analysis of variance.

Let us conclude our analysis by determining which models are different within the four- and eight-cylinder groups. We shall use the Scheffé method and the pooled estimate of σ^2, for its sensitivity appears to have been vindicated by the preceding analysis of variance. The 5% Scheffé critical value is $\sqrt{3F_{0.05;\,3,12}} = 3.24$. The 95% simultaneous confidence interval for the difference of the model D and C effects with four cylinders is

TABLE 9-8 Analysis of Variance for Model Differences

Source	Sum of squares	d.f.	Mean square	F-Ratios: Separate	F-Ratios: Pooled
Between models					
Four-cylinder	10.255	3	3.4183	2.79	4.40*
Six-cylinder	3.3	3	1.1	1.41	1.42
Eight-cylinder	13.975	3	4.658	14.33*	6.00**
Total	27.53	9	3.059	—	3.94
Within models					
Four-cylinder	4.90	4	1.225	—	—
Six-cylinder	3.12	4	0.78	—	—
Eight-cylinder	1.30	4	0.325	—	—
Total (pooled)	9.32	12	0.7767		

*significant at the 0.05 level.
**significant at the 0.01 level.

$$0.34 \leq \alpha_4 - \alpha_3 \leq 6.06$$

Since the interval does not contain zero we should conclude that models D and C are different. The other paired comparison differences are not significant in the Scheffé sense, and other contrasts of the row effects would not appear to have any meaning in the substantive sense. The only 95% Scheffé interval of the eight-cylinder model means that does not contain zero is that for models A and B:

$$0.34 \leq \alpha_1 - \alpha_2 \leq 6.06$$

However, we should note that the Tukey 95% confidence interval for the difference of the effects of rows 3 and 2 also does not contain zero:

$$0.18 \leq \alpha_3 - \alpha_2 \leq 5.42$$

In that sense models B and C are also different.

Simultaneous inference about interactions. The Scheffé method can be applied to contrasts

$$\psi = \sum_{i=1}^{r} \sum_{j=1}^{c} c_{ij}\gamma_{ij} \tag{16}$$

of the interaction parameters. We require that

$$\sum_{i=1}^{r} c_{ij} = \sum_{j=1}^{c} c_{ij} = 0 \tag{17}$$

Then ψ is estimable with estimator

$$\hat{\psi} = \sum_{i=1}^{r} \sum_{j=1}^{c} c_{ij}\bar{x}_{ij.} \tag{18}$$

for

$$E(\hat{\psi}) = \sum_{i=1}^{r} \sum_{j=1}^{c} c_{ij} E(\bar{x}_{ij.})$$

$$= \sum_{i=1}^{r} \sum_{j=1}^{c} c_{ij}(\mu + \alpha_i + \beta_j + \gamma_{ij})$$

$$= \sum_{i=1}^{r} \sum_{j=1}^{c} c_{ij}\gamma_{ij}$$

$$= \psi \tag{19}$$

as a consequence of the constraints (17). The $100(1 - \alpha)\%$ simultaneous confidence interval for ψ is

$$\hat{\psi} - \hat{\sigma}\sqrt{(r-1)(c-1)F_{\alpha;\,(r-1)(c-1),\,rc(n-1)} \frac{\sum\sum c_{ij}^2}{n}}$$

$$\leq \psi \leq \hat{\psi} + \hat{\sigma}\sqrt{(r-1)(c-1)F_{\alpha;\,(r-1)(c-1),\,rc(n-1)} \frac{\sum\sum c_{ij}^2}{n}} \tag{20}$$

If the interval contains zero, the hypothesis

$$H_0\colon \quad \psi = \sum_{i=1}^{r} \sum_{j=1}^{c} c_{ij}\gamma_{ij} = 0 \tag{21}$$

should not be rejected. As in the row and column simultaneous tests we may also compute

$$t = \frac{\hat{\psi}}{\hat{\sigma}\sqrt{\sum\sum c_{ij}^2/n}} \tag{22}$$

If

$$|t| > \sqrt{(r-1)(c-1)F_{\alpha;\,(r-1)(c-1),\,rc(n-1)}} \tag{23}$$

we should reject (21).

Example 9-8

We shall use the Scheffé method to determine the source of the significant interaction in the gas mileage data of Example 9-7. The $\alpha = 0.05$ critical value is

$$\sqrt{6F_{0.05;\,6,\,12}} = 4.2426$$

We begin by searching Figure 9-6 for differences in the slopes of the profile segments of the four models. Model B appears to have a greater decrease in mileage from four to six cylinders than Model C. The contrast measuring that difference is

$$\hat{\psi}_1 = \bar{x}_{21.} - \bar{x}_{22.} - (\bar{x}_{31.} - \bar{x}_{32.}) = 3.2$$

Its estimated standard deviation is

$$\hat{\sigma}_{\hat{\psi}_1} = \sqrt{\frac{4\hat{\sigma}^2}{n}} = 1.2464$$

and the standardized contrast has the value $t_1 = 2.57$. Because t_1 is less than the critical value we should not conclude that the interaction is due to the contrast measured by $\hat{\psi}_1$. Similar contrasts comparing the four- and six-cylinder differences of C with the average of B and D, and C with the average of A, B, and D did not yield statistical significance. However, let us compare the average slope of C from four to eight cylinders with the average slopes of B and D between those engine sizes. The contrast estimate is

$$\hat{\psi}_2 = \tfrac{1}{2}(\bar{x}_{21.} + \bar{x}_{41.} - \bar{x}_{23.} - \bar{x}_{43.}) - (\bar{x}_{31.} - \bar{x}_{33.})$$
$$= 4.7$$

and its estimated standard deviation is

$$\hat{\sigma}_{\psi_2} = \sqrt{\hat{\sigma}^2[(1/4)(4/2) + 2(1/2)]} = 1.0794$$

The test statistic (22) has the value 4.35. The interaction appears to be due at least in part to the fuel consumption differences of model C and the average of models B and D. Because such an average over distinct models has little practical meaning we should see whether the slope of C is different from that of D alone. The estimated contrast is

$$\hat{\psi}_3 = \bar{x}_{41.} - \bar{x}_{43.} - (\bar{x}_{31.} - \bar{x}_{33.}) = 5.1$$

Its standard deviation is 1.2464, and the resulting test statistic is $t_3 = 4.09$. t_3 falls just short of the critical value. It would appear that the contrasts causing the significant interaction may be artifactual combinations of the models. In any case the individual analyses of variance at the three engine sizes are still the most valid and conservative way of analyzing these data.

Some further aspects of multiple comparisons in the two-way layout have been discussed by Chew (1977).

9-6 A TEST FOR THE ADDITIVE MODEL

In the examples of Sections 9-2 to 9-5 we have seen the importance of the assumption of additivity of the row and column effects for the interpretation of the two-way analysis of variance. The failure of the additive model dictates that one-way analyses must be made at each level of the row and column treatments to elicit differences that may have been masked in the row and column means by interactions. Of course, the interaction F-ratio was based on the variation within the cells of the layout. What are we to do when each cell contains only one observation?

Fortunately, a procedure is available for testing the assumption of an additive model against a particular kind of nonadditivity. It is due to Tukey (1949) and is sometimes referred to as "Tukey's single degree of freedom for nonadditivity." For the test we postulate the additive model

$$x_{ij} = \mu + \alpha_i + \beta_j + e_{ij} \qquad i = 1, \ldots, r; \quad j = 1, \ldots, c \tag{1}$$

and for its alternative the particular nonadditive model

$$x_{ij} = \mu + \alpha_i + \beta_j + \gamma\alpha_i\beta_j + e_{ij} \tag{2}$$

The usual independence, common variance, and normality assumptions hold for the e_{ij}. The additive model is expressed by the hypothesis

$$H_0: \quad \gamma = 0 \tag{3}$$

The alternative is $H_1: \gamma \neq 0$, so that the special nonadditive model (2) holds for the layout. Scheffé (1959) has pointed out that a second-degree polynomial model in the row and column effects with cross-product term $\gamma\alpha_i\beta_j$ and the usual zero-sum restrictions on the parameters will reduce to the nonadditive alternative (2). To test the hypothesis (3) we first compute the sum of squares for nonadditivity,

$$\text{SSN} = \frac{\left[\sum_{i=1}^{r}\sum_{j=1}^{c}(\bar{x}_{i.} - \bar{x}_{..})(\bar{x}_{.j} - \bar{x}_{..})x_{ij}\right]^2}{\left[\sum_{i=1}^{r}(\bar{x}_{i.} - \bar{x}_{..})^2\right]\left[\sum_{j=1}^{c}(\bar{x}_{.j} - \bar{x}_{..})^2\right]} \tag{4}$$

When $\gamma = 0$ SSN$/\sigma^2$ has the chi-squared distribution with one degree of freedom. Recall that the error or interaction sum of squares

$$\text{SSI} = \sum_{i=1}^{r}\sum_{j=1}^{c}(x_{ij} - \bar{x}_{i.} - \bar{x}_{.j} + \bar{x}_{..})^2 \tag{5}$$

has the $\sigma^2 x^2$ distribution with $(r - 1)(c - 1)$ degrees of freedom. Tukey has shown that SSN and SSI − SSN are independently distributed, so that

$$F = \frac{(rc - r - c)\text{SSN}}{\text{SSI} - \text{SSN}} \tag{6}$$

has the F-distribution with 1 and $rc - r - c$ degrees of freedom when $H_0: \gamma = 0$ is true. We note that the interaction sum of squares has lost a single degree of freedom to provide the test for nonadditivity: hence the name of Tukey's test. The derivation of the test statistic and its distribution follows from the theory of least squares and quadratic forms. It is beyond the mathematical level we have chosen and will be omitted. Developments of the test can be found in the original paper by Tukey (1949) and in the texts of Rao (1965) and Scheffé (1959).

Example 9-9

We illustrate the test for nonadditivity by its application to the manuscript submission data of Example 9-5 and Exercise 2, Section 4-6. The necessary sums of squares and products are

$$\sum_{i=1}^{4}(\bar{x}_{i.} - \bar{x}_{..})^2 = 65.0763 \qquad \sum_{j=1}^{6}(\bar{x}_{.j} - \bar{x}_{..})^2 = 86.21875$$

$$\sum_{i=1}^{4}\sum_{j=1}^{6}(\bar{x}_{i.} - \bar{x}_{..})(\bar{x}_{.j} - \bar{x}_{..})x_{ij} = -97.1390$$

$$\text{SSN} = 1.6818 \qquad \text{SSI} - \text{SSN} = 732.6102$$

$$F = \frac{14(1.6818)}{732.6102} = 0.032$$

The F-statistic is miniscule: The hypothesis of an additive model is certainly tenable.

Example 9-10

Let us apply the Tukey test to the cell means of the gas mileage data in Example 9-7. We wish to compare the result of the nonadditivity test with the significant interaction found by the usual F-ratio based on the within-cell variation. We shall treat the cell means as if they are single observations. The required statistics are

$$\sum_{i=1}^{4}(\bar{x}_{i..} - \bar{x}_{...})^2 = 1.3189 \qquad \sum_{j=1}^{3}(\bar{x}_{.j.} - \bar{x}_{...})^2 = 26.4427$$

$$\sum_{i=1}^{4}\sum_{j=1}^{3}(\bar{x}_{i..} - \bar{x}_{...})(\bar{x}_{.j.} - \bar{x}_{...})\bar{x}_{ij.} = -5.2466$$

$$\text{SSI} = \sum_{i=1}^{4} \sum_{j=1}^{3} (\bar{x}_{ij.} - \bar{x}_{i..} - \bar{x}_{.j.} + \bar{x}_{...})^2 = 9.8083$$

$$\text{SSN} = 0.7893 \qquad \text{SSI} - \text{SSN} = 9.0190$$

$$F = 5\left(\frac{0.7893}{9.0190}\right) = 0.44$$

The F-statistic does not even approach significance. Contrary to our finding of an interaction in Example 9-7, the additive model for the cell means is tenable.

9-7 THREE-WAY LAYOUTS

The two-way layout introduced in Section 9-3 can be extended to any number of dimensions. We treat the case of data cross-classified by three kinds of treatments in this section. The layout is balanced: Each cell in the table contains exactly n independent observations. The analysis of variance, contrasts, and multiple comparisons can be carried out as in the two-way layout with changes in notation and degrees of freedom, and some necessary extensions in the model and methods. We shall indicate how the analyses can be generalized to more dimensions. Explicit results for the special case of two treatments in each dimension, or a factorial layout, are given in Section 9-8.

Let us begin by describing a three-way experimental design. In Example 9-2 we considered the crushing strength of mortar cubes cross-classified by laboratory and the consistency of the mix. Now let us add another dimension after Wilsdon (1934): curing time of the mortar. The table for the data might be as shown in Table 9-9 for four curing times, three consistencies (8, 10, and 12.5% water), and three laboratories. We would assume that each of the 36 cells of the table contains exactly n crushing strength measurements. We prefer that n be at least equal to two, so that a "pure," or within-cell, estimate of the error variance can be computed. From the data we would test the three hypotheses of equal curing time effects, equal consistency effects, and equal laboratory means. In addition, we wish to test for the presence of interactions of the three treatments in their various combinations. For example, are the laboratory

TABLE 9-9 Three-Way Layout

	Curing time (days)											
	1			3			7			28		
Consistency / Laboratory	8	10	12.5	8	10	12.5	8	10	12.5	8	10	12.5
R												
S												
T												

differences the same for all three consistencies when the observations have been averaged over curing times? These tests for the three main effects, three paired-treatment interactions, and the single three-treatment interaction may be made by a single analysis of variance.

As in any application of linear statistical inference let us begin with the additive model for the hth observation in the ith row, jth column, and kth layer of a general $r \times c \times d$ layout:

$$x_{ijkh} = \mu + \alpha_i + \beta_j + \gamma_k + (\alpha\beta)_{ij} + (\alpha\gamma)_{ik} + (\beta\gamma)_{jk} + (\alpha\beta\gamma)_{ijk} + e_{ijkh}$$
$$i = 1, \ldots, r; \quad j = 1, \ldots, c; \quad k = 1, \ldots, d; \quad h = 1, \ldots, n \qquad (1)$$

μ is the usual general effect. α_i, β_j, and γ_k are the row, column, and layer effects, respectively. To avoid running out of Greek letters we have denoted the interaction parameters thus:

$(\alpha\beta)_{ij}$: row i and column j interaction

$(\alpha\gamma)_{ik}$: row i and layer k interaction

$(\beta\gamma)_{jk}$: column j and layer k interaction

$(\alpha\beta\gamma)_{ijk}$: row i, column j, and layer k interaction

The disturbance terms e_{ijkh} are independently distributed normal random variables with mean zero and a common variance σ^2. If $n = 1$ the second-order interaction term $(\alpha\beta\gamma)_{ijk}$ and the disturbance term are indistinguishable: In that case the error variance for the forthcoming tests must be estimated by the interaction mean square.

To obtain explicit estimators of the main-effect parameters we must impose certain constraints on the elements of the three-way model (1). We shall require that all the main effect and interaction parameters sum to zero over all combinations of their subscripts. Those constraints are rather numerous:

$$\sum_{i=1}^{r} \alpha_i = \sum_{j=1}^{c} \beta_j = \sum_{k=1}^{d} \gamma_k = 0$$

$$\sum_{i=1}^{r} \sum_{j=1}^{c} (\alpha\beta)_{ij} = \sum_{i=1}^{r} \sum_{k=1}^{d} (\alpha\gamma)_{ik} = \sum_{j=1}^{c} \sum_{k=1}^{d} (\beta\gamma)_{jk} = \sum_{i=1}^{r} \sum_{j=1}^{c} \sum_{k=1}^{d} (\alpha\beta\gamma)_{ijk} = 0$$

$$\sum_{i=1}^{r} (\alpha\beta)_{ij} = \sum_{j=1}^{c} (\alpha\beta)_{ij} = \sum_{i=1}^{r} (\alpha\gamma)_{ik} = \sum_{k=1}^{d} (\alpha\gamma)_{ik} = \sum_{j=1}^{c} (\beta\gamma)_{jk} = \sum_{k=1}^{d} (\beta\gamma)_{jk} = 0 \qquad (2)$$

$$\sum_{i=1}^{r} \sum_{j=1}^{c} (\alpha\beta\gamma)_{ijk} = \sum_{i=1}^{r} \sum_{k=1}^{d} (\alpha\beta\gamma)_{ijk} = \sum_{j=1}^{c} \sum_{k=1}^{d} (\alpha\beta\gamma)_{ijk} = 0$$

Then the estimators of the main effects are

$$\hat{\alpha}_i = \bar{x}_{i...} - \bar{x}_{....} \qquad i = 1, \ldots, r$$
$$\hat{\beta}_j = \bar{x}_{.j..} - \bar{x}_{....} \qquad j = 1, \ldots, c \qquad (3)$$
$$\hat{\gamma}_k = \bar{x}_{..k.} - \bar{x}_{....} \qquad k = 1, \ldots, d$$

in which the means

$$\bar{x}_{i...} = \frac{1}{cdn} \sum_{j=1}^{c} \sum_{k=1}^{d} \sum_{h=1}^{n} x_{ijkh}$$

$$\bar{x}_{.j..} = \frac{1}{rdn} \sum_{i=1}^{r} \sum_{k=1}^{d} \sum_{h=1}^{n} x_{ijkh}$$

$$\bar{x}_{..k.} = \frac{1}{rcn} \sum_{i=1}^{r} \sum_{j=1}^{c} \sum_{h=1}^{n} x_{ijkh} \tag{4}$$

$$\bar{x}_{....} = \frac{1}{rcdn} \sum_{i=1}^{r} \sum_{j=1}^{c} \sum_{k=1}^{d} \sum_{h=1}^{n} x_{ijkh}$$

are the respective averages of the observations in the ith row, jth column, kth layer, and the entire table. Under the constraints the estimator of the general effects is the grand mean

$$\hat{\mu} = \bar{x}_{....} \tag{5}$$

Because the main-effect estimators are deviations from the grand mean their sums are zero, and the constraints in the first line of (2) are satisfied. In applications we shall usually be more concerned with the estimation of contrasts for a particular dimension. The contrasts, their estimators, and their variances are shown in Table 9-10. The grand mean has vanished in each estimator from the

TABLE 9-10 Contrasts of Main Effects

Dimension	Contrast	Estimator	Variance
Row	$\psi_R = \sum_{i=1}^{r} c_i \alpha_i$	$\hat{\psi}_R = \sum_{i=1}^{r} c_i \bar{x}_{i...}$	$\text{Var}(\hat{\psi}_R) = \frac{\sigma^2}{cdn} \sum_{i=1}^{r} c_i^2$
Column	$\psi_C = \sum_{j=1}^{c} c_j \beta_j$	$\hat{\psi}_C = \sum_{j=1}^{c} c_j \bar{x}_{.j..}$	$\text{Var}(\hat{\psi}_C) = \frac{\sigma^2}{rdn} \sum_{j=1}^{c} c_j^2$
Layer	$\psi_D = \sum_{k=1}^{d} c_k \gamma_k$	$\hat{\psi}_D = \sum_{k=1}^{d} c_k \bar{x}_{..k.}$	$\text{Var}(\hat{\psi}_D) = \frac{\sigma^2}{rcn} \sum_{k=1}^{d} c_k^2$

contrast property of a coefficient zero sum. Our inferences about the individual effects will be in terms of these contrasts.

The estimators of the two-factor interaction parameters are

$$\begin{aligned} \widehat{(\alpha\beta)}_{ij} &= x_{ij..} - \hat{\alpha}_i - \hat{\beta}_j - \hat{\mu} \\ &= \bar{x}_{ij..} - \bar{x}_{i...} - \bar{x}_{.j..} + \bar{x}_{....} \\ \widehat{(\alpha\gamma)}_{ik} &= \bar{x}_{i.k.} - \bar{x}_{i...} - \bar{x}_{..k.} + \bar{x}_{....} \\ \widehat{(\beta\gamma)}_{jk} &= \bar{x}_{.jk.} - \bar{x}_{.j..} - \bar{x}_{..k.} + \bar{x}_{....} \end{aligned} \tag{6}$$

The combination means are defined by

$$\begin{aligned} \bar{x}_{ij..} &= \frac{1}{nd} \sum_{k=1}^{d} \sum_{h=1}^{n} x_{ijkh} \\ \bar{x}_{i.k.} &= \frac{1}{nc} \sum_{j=1}^{c} \sum_{h=1}^{n} x_{ijkh} \\ \bar{x}_{.jk.} &= \frac{1}{nr} \sum_{i=1}^{r} \sum_{h=1}^{n} x_{ijkh} \end{aligned} \tag{7}$$

Of course, the subscripts take on the various combinations of these row, column, and layer indices. One should verify that the appropriate constraints of (2) are satisfied by the estimators. For example,

$$\begin{aligned}\sum_{i=1}^{r}\sum_{k=1}^{d}(\alpha\gamma)_{ik} &= \sum_{i=1}^{r}\sum_{k=1}^{d}(\bar{x}_{i.k} - \bar{x}_{i...} - \bar{x}_{..k.} + \bar{x}_{....}) \\ &= rd(\bar{x}_{....} - \bar{x}_{....} - \bar{x}_{....} + \bar{x}_{....}) \\ &= 0 \end{aligned} \tag{8}$$

The three-factor interactions are estimated from the residuals of the cell means

$$\bar{x}_{ijk.} = \frac{1}{n}\sum_{h=1}^{n} x_{ijkh} \tag{9}$$

and their predictions by the main effects and the two-factor interactions:

$$\begin{aligned}\widehat{(\alpha\beta\gamma)}_{ijk} &= \bar{x}_{ijk} - [\hat{\mu} + \hat{\alpha}_i + \hat{\beta}_j + \hat{\gamma}_k + \widehat{(\alpha\beta)}_{ij} + \widehat{(\alpha\gamma)}_{ik} + \widehat{(\beta\gamma)}_{jk}] \\ &= \bar{x}_{ijk.} - \bar{x}_{ij..} - \bar{x}_{i.k.} - \bar{x}_{.jk.} + \bar{x}_{i...} + \bar{x}_{.j..} + \bar{x}_{..k.} - \bar{x}_{....} \\ &\qquad i = 1, \ldots, r; \quad j = 1, \ldots, c; \quad k = 1, \ldots, d \end{aligned} \tag{10}$$

Again we can verify that the estimators satisfy the proper conditions of (2).

With the estimators stored away for some future uses let us now turn to the analysis of variance for the three-way layout. The hypotheses of equal row, column, and layer effects are, respectively,

$$\begin{aligned} H_0&: \ \alpha_1 = \cdots = \alpha_r \\ H_0&: \ \beta_1 = \cdots = \beta_c \\ H_0&: \ \gamma_1 = \cdots = \gamma_d \end{aligned} \tag{11a}$$

Under the constraints (2) the hypotheses become

$$\begin{aligned} H_0&: \ \alpha_i = 0 \qquad i = 1, \ldots, r \\ H_0&: \ \beta_j = 0 \qquad j = 1, \ldots, c \\ H_0&: \ \gamma_k = 0 \qquad k = 1, \ldots, d \end{aligned} \tag{11b}$$

The alternative hypotheses

$$\begin{aligned} H_1&: \ \sum_{i=1}^{r} c_i\alpha_i \neq 0 \\ H_1&: \ \sum_{j=1}^{c} d_j\beta_j \neq 0 \\ H_1&: \ \sum_{k=1}^{d} f_k\gamma_k \neq 0 \end{aligned} \tag{12}$$

merely state that contrasts of the particular treatment effects exist that are not equal to zero. For most of our purposes the imprecise alternative "some of the [row, column, layer] effects are unequal" will suffice. The hypotheses of no interaction are these:

$$\begin{aligned} &\text{Rows} \times \text{columns} && H_0: \ (\alpha\beta)_{ij} = 0 \quad \text{all } i, j \\ &\text{Rows} \times \text{layers} && H_0: \ (\alpha\gamma)_{ik} = 0 \quad \text{all } i, k \\ &\text{Columns} \times \text{layers} && H_0: \ (\beta\gamma)_{jk} = 0 \quad \text{all } j, k \\ &\text{Rows} \times \text{columns} \times \text{layers} && H_0: \ (\alpha\beta\gamma)_{ijk} = 0 \quad \text{all } i, j, k \end{aligned} \tag{13}$$

Under the zero-sum constraints on the interaction parameters the alternatives to those hypotheses are merely that some of the parameters are different from zero. The analysis of variance for testing the seven hypotheses is given in Table 9-11. The sums of squares are computed by these formulas:

$$\begin{aligned}
\text{Rows:}\quad \text{SSR} &= ncd \sum_{i=1}^{r} (\bar{x}_{i...} - \bar{x}_{....})^2 \\
&= ncd \sum_{i=1}^{r} \bar{x}_{i...}^2 - \frac{G^2}{rcdn} \\
\text{Columns:}\quad \text{SSC} &= nrd \sum_{j=1}^{c} (\bar{x}_{.j..} - \bar{x}_{....})^2 \\
&= nrd \sum_{j=1}^{c} \bar{x}_{.j..}^2 - \frac{G^2}{rcdn} \\
\text{Layers:}\quad \text{SSL} &= nrc \sum_{k=1}^{d} (\bar{x}_{..k.} - \bar{x}_{....})^2 \\
&= nrc \sum_{k=1}^{d} \bar{x}_{..k.}^2 - \frac{G^2}{rcdn} \\
\text{Rows} \times \text{columns:}\quad \text{SSRC} &= nd \sum_{i=1}^{r} \sum_{j=1}^{c} (\bar{x}_{ij..} - \bar{x}_{i...} - \bar{x}_{.j..} + \bar{x}_{....})^2 \\
&= nd \sum_{i=1}^{r} \sum_{j=1}^{c} \bar{x}_{ij..}^2 - \frac{G^2}{rcdn} - \text{SSR} - \text{SSC} \\
\text{Rows} \times \text{layers:}\quad \text{SSRL} &= nc \sum_{i=1}^{r} \sum_{k=1}^{d} (\bar{x}_{i.k.} - \bar{x}_{i...} - \bar{x}_{..k.} + \bar{x}_{....})^2 \\
&= nc \sum_{i=1}^{r} \sum_{k=1}^{d} \bar{x}_{i.k.}^2 - \frac{G^2}{rcdn} - \text{SSR} - \text{SSL} \\
\text{Columns} \times \text{layers:}\quad \text{SSCL} &= nr \sum_{j=1}^{c} \sum_{k=1}^{d} (\bar{x}_{.jk.} - \bar{x}_{.j..} - \bar{x}_{..k.} + \bar{x}_{....})^2 \\
&= nr \sum_{j=1}^{c} \sum_{k=1}^{d} \bar{x}_{.jk.}^2 - \frac{G^2}{rcdn} - \text{SSC} - \text{SSL}
\end{aligned} \tag{14}$$

Rows $\times$ columns $\times$ layers:

$$\begin{aligned}
\text{SSRCL} &= n \sum_{i=1}^{r} \sum_{j=1}^{c} \sum_{k=1}^{d} (\bar{x}_{ijk.} - \bar{x}_{ij..} - \bar{x}_{i.k.} - \bar{x}_{.jk.} + \bar{x}_{i...} \\
&\qquad + \bar{x}_{.j..} + \bar{x}_{..k.} - \bar{x}_{....})^2 \\
&= n \sum_{i=1}^{r} \sum_{j=1}^{c} \sum_{k=1}^{d} \bar{x}_{ijk.}^2 - \frac{G^2}{rcdn} - \text{SSRC} - \text{SSRL} - \text{SSCL} \\
&\qquad - \text{SSR} - \text{SSC} - \text{SSL}
\end{aligned}$$

$$\text{Within cells (error):}\quad \text{SSE} = \sum_{i=1}^{r} \sum_{j=1}^{c} \sum_{k=1}^{d} \sum_{h=1}^{n} (x_{ijkh} - \bar{x}_{ijk.})^2$$

$$\text{Total:}\quad \text{SST} = \sum_{i=1}^{r} \sum_{j=1}^{c} \sum_{k=1}^{d} \sum_{h=1}^{n} x_{ijkh}^2 - \frac{G^2}{rcdn}$$

The sums of squares for the main effects, interactions, and error (within-cells) are independently distributed. If the hypothesis about the parameters of a particular effect is true, its sum of squares scaled by σ^2 will have the chi-squared

TABLE 9-11 Analysis of Variance for the Three-Way Layout

Source	Sum of squares	d.f.	Mean square	F
Rows	SSR	$r-1$	$\text{MSR} = \dfrac{\text{SSR}}{r-1}$	$F_R = \dfrac{\text{MSR}}{\text{MSE}}$
Columns	SSC	$c-1$	$\text{MSC} = \dfrac{\text{SSC}}{c-1}$	$F_C = \dfrac{\text{MSC}}{\text{MSE}}$
Layers	SSL	$d-1$	$\text{MSL} = \dfrac{\text{SSL}}{d-1}$	$F_L = \dfrac{\text{MSL}}{\text{MSE}}$
Rows × columns	SSRC	$(r-1)(c-1)$	$\text{MSRC} = \dfrac{\text{SSRC}}{(r-1)(c-1)}$	$F_{RC} = \dfrac{\text{MSRC}}{\text{MSE}}$
Rows × layers	SSRL	$(r-1)(d-1)$	$\text{MSRL} = \dfrac{\text{SSRL}}{(r-1)(d-1)}$	$F_{RL} = \dfrac{\text{MSRL}}{\text{MSE}}$
Columns × layers	SSCL	$(c-1)(d-1)$	$\text{MSCL} = \dfrac{\text{SSCL}}{(c-1)(d-1)}$	$F_{CL} = \dfrac{\text{MSCL}}{\text{MSE}}$
Rows × columns × layers	SSRCL	$(r-1)(c-1)(d-1)$	$\text{MSRCL} = \dfrac{\text{SSRCL}}{(r-1)(c-1)(d-1)}$	$F_{RCL} = \dfrac{\text{MSRCL}}{\text{MSE}}$
Within cells	SSE	$rcd(n-1)$	$\text{MSE} = \dfrac{\text{SSE}}{rcd(n-1)}$	—
Total	SST	$rcd-1$	—	—

distribution with degrees of freedom given by the third column of Table 9-11. Similarly, SSE/σ^2 has the chi-squared distribution with $rcd(n-1)$ degrees of freedom, or $n-1$ degrees of freedom from each of the rcd cells. The ratios of the hypothesis and error mean squares have F-distributions with appropriate degrees of freedom found from the third column of the table. If the computed F-statistic exceeds the $100\alpha\%$ critical value of that F-distribution, we would reject the hypothesis. For example, if

$$F_C > F_{\alpha;\, c-1,\, rcd(n-1)} \tag{15}$$

we should conclude that the column effects are different at the α level. If

$$F_{RL} > F_{\alpha;\, (r-1)(d-1),\, rcd(n-1)} \tag{16}$$

the rows by layers interaction is significant at the α level. Similarly,

$$F_{RCL} > F_{\alpha;\, (r-1)(c-1)(d-1),\, rcd(n-1)} \tag{17}$$

would imply that the rows $\times$ columns $\times$ layers interaction effects was different from zero at the α level.

Example 9-11

As an illustration of a three-way layout with two observations per cell we use an extension of the illumination experiment described in Example 9-3. In the present case the row treatments are the four settings 0, 50, 100, and 150 watts of the table lamp, columns correspond to the levels of no light, one 60-watt lamp, and two 60-watt lamps in a pole lamp, and the layer treatments are the conditions of no light and 40-watt illumination of a fluorescent desk lamp to the side of the observer. As in Example 9-3, illumination has been measured in light units on a photographic light meter. A complete set of 24 readings was recorded as the first observation in each cell. Then a second replication of the experiment was made, and its readings recorded as the second observation of the cell pairs. We assume that the replications are independent and that the experimental conditions did not change from one replication to another.

The observed light values are shown in Table 9-12. We have omitted the cell

TABLE 9-12 Light Values

	Pole lamp (watts)						
	0		60		120		
Table \ Desk	0	40	0	40	0	40	Mean
0	6.3	7.1	7.4	7.2	7.6	7.3	7.0167
	6.7	6.5	6.8	7.1	7.0	7.2	
50	8.0	7.9	8.1	7.9	8.0	8.5	8.0167
	7.8	7.9	7.9	8.1	8.0	8.1	
100	9.0	9.1	9.0	9.1	9.1	9.1	8.90
	8.7	8.3	8.8	8.9	8.9	8.8	
150	9.1	9.2	9.5	9.6	9.4	9.4	9.30
	9.0	9.1	9.2	9.2	9.5	9.4	
Mean	8.075	8.1375	8.3375	8.3875	8.4375	8.475	8.3083
Pole mean	8.1062		8.3625		8.4562		

means and individual within-cell sums of squares to conserve space. We shall need these tables of means for the two-factor interactions:

Pole and Table Lamp Means

Table	Pole 0	Pole 60	Pole 120
0	6.65	7.125	7.275
50	7.90	8.00	8.15
100	8.775	8.95	8.975
150	9.10	9.375	9.425

Desk and Table Lamp Means

Table	Desk 0	Desk 40
0	6.9667	7.0667
50	7.9667	8.0667
100	8.9167	8.8833
150	9.2833	9.3167
Mean	8.2833	8.3333

The analysis of variance is given in Table 9-13. As in Example 9-3 the desk lamp had no significant effect on the light values. None of the four interaction effects even

TABLE 9-13 Light Values Analysis of Variance

Source	Sum of squares	d.f.	Mean square	F
Table	37.0433	3	12.3478	215
Pole	1.0504	2	0.5252	9.13
Desk	0.0300	1	0.03	0.52
T × P	0.2679	6	0.04465	0.78
T × D	0.0367	3	0.01222	0.21
P × D	0.00125	2	0.00062	0.01
T × P × D	0.1671	6	0.02785	0.48
Within cells	1.3800	24	0.0575	—
Total	39.9767	47	—	—

approaches statistical significance: that should not be surprising if we think of the illumination as the sum of the individual sources. The pole and table lamps have obvious effects. The question of which of their level means are different seems academic. A less evident matter is that of the second-order differences among those means. Its answer is left as an exercise.

Because the desk lamp and interaction effects were not significant we might ignore the third dimension of the layout and treat the data as coming from a 4 × 3 experimental design with $n = 4$ observations in each cell. This analysis of variance would be obtained:

Source	Sum of squares	d.f.	Mean square	F
Table	37.0433	3	12.3478	280
Pole	1.0504	2	0.5252	11.9
T × P	0.2679	6	0.04465	1.01
Within cells	1.5850	36	0.04403	—
Total	39.9767	47	—	—

Such a "pooled" analysis would probably be preferable in reporting the results of the experiment. Its conclusions are unchanged, although the error mean square is slightly smaller than that in Table 9-13.

Example 9-12

In the preceding example we had two observations in each cell of the experimental plan. As long as the values in a pair were obtained under the same conditions they would provide a "pure" estimate of measurement, sampling, or experimental error variance. If a single observation has been obtained in each cell we must use the highest-order interaction, or that of rows by columns by layers, to estimate the disturbance variance σ^2. As we indicated in the preceding example we may improve the estimate by including any insignificant interactions in its sum of squares.

Let us illustrate the single-observation layout with data collected by Wilsdon (1934) on tensile strengths of mortar samples. The rows of the design are $r = 10$ testing laboratories, the columns are $c = 4$ curing times of 1, 3, 7, and 28 days, and the layers are the $d = 3$ water percentages 8, 10, and 12.5% in the mix. The data and certain means are shown in Table 9-14. The two-way tables of means of treatment combinations

TABLE 9-14 Tensile Strengths of Mortar Samples

	Percentage of water in mix											
	1 day			3 days			7 days			28 days		
Laboratory	8	10	12.5	8	10	12.5	8	10	12.5	8	10	12.5
R	457	337	214	559	508	367	611	539	380	700	632	537
S	500	281	150	636	521	361	689	589	450	721	669	538
T	541	299	140	644	488	330	710	542	435	747	610	505
U	435	265	132	621	498	338	726	610	449	762	683	495
V	642	322	155	838	569	378	878	639	465	934	690	518
W	430	345	187	627	520	348	697	580	400	723	667	465
X	531	276	141	679	531	348	708	555	413	740	612	460
Y	501	349	163	658	572	338	705	621	427	761	652	448
Z	390	278	123	600	507	348	596	522	429	663	673	513
A	382	229	147	451	486	323	522	523	405	546	609	495
Mean	481	298	155	631	520	348	684	572	425	730	650	497

SOURCE: Data reproduced with the permission of the Royal Statistical Society.

TABLE 9-15 Mean Tensile Strengths

a. Laboratories × Curing Time

Laboratory	Days 1	3	7	28	Mean
R	336	478	510	623	486.75
S	310.33	506	576	642.67	508.75
T	326.67	487.33	562.33	620.67	499.25
U	277.33	485.67	595	646.67	501.17
V	373	595	660.67	714	585.67
W	320.67	498.33	559	618.33	499.08
X	316	519.33	558.67	604	499.5
Y	337.67	522.67	584.33	620.33	516.25
Z	263.67	485	515.67	616.33	470.17
A	252.67	420	483.33	550	426.5
Mean	311.33	499.67	560.5	625.6	499.3

b. Laboratories × Percent Water

Laboratory	Percent water 8	10	12.5	Mean
R	581.75	504	379.5	486.75
S	636.5	515	374.75	508.75
T	660.5	484.75	352.5	499.25
U	636	514	353.5	501.17
V	823	555	379	585.67
W	619.25	528	350	499.08
X	664.5	493.5	340.5	499.5
Y	656.25	548.5	344	516.25
Z	562.25	495	353.25	470.17
A	475.25	461.75	342.5	426.5
Mean	631.525	509.95	356.45	499.33

c. Curing Time × Percent Water

Days	Percent water 8	10	12.5	Mean
1	480.9	298.1	155.2	311.4
3	631.3	520.0	347.9	499.67
7	684.2	572	425.3	560.5
28	729.7	649.7	497.4	625.6
Mean	631.525	509.95	356.45	499.3

are given in Table 9-15. From these we may compute the analysis of variance:

Source	Sum of squares	d.f.	Mean square	F
Laboratories	169,746	9	18,861	45
Curing time	1,650,112	3	550,037	1,306
Percent water	1,520,119	2	760,060	1,804
L × CT	31,944	27	1,183	2.81
L × PW	151,515	18	8,418	19.98
CT × PW	36,333	6	6,056	14.38
L × CT × PW	22,744	54	421	—
Total	3,582,514	119	—	—

All of the effects are highly significant. Because of the interaction effects we should make separate analyses of variance for each of the two-way layouts at the four curing times and the three percentages of water. Because the time, laboratory, and water effects are so great we undoubtedly would reach the same conclusions of the complete three-way analysis. The two laboratories with extremely low and high means might be eliminated and the analysis of variance recomputed. It is likely that the more homogeneous observations would yield even greater significance of the main effects. Finally, one might fit polynomials to the curing time and water percentage means, although the unequally spaced points would preclude orthogonal polynomials as we have treated them. Alternatively, we might search for the source of the interaction effects by means of contrasts and appropriate Scheffé simultaneous critical values.

Three-way layouts can be extended to an indefinite number of dimensions. The formulas for the main effects follow from those in (14) with appropriate multipliers and additional subscripts. When the treatments have only two levels, for example, presence or absence of the treatment, we may take advantage of many simplifications in the analysis of the data. Such experimental plans are called *factorial designs*, and a factorial layout with p treatments applied in all combinations of two levels is called a 2^p design. We consider the special properties of factorial plans in Section 9-8.

9-8 FACTORIAL DESIGNS

When the treatments of higher-way layouts consist of some experimental conditions applied at different degrees the design is called a *factorial* plan. The conditions, or the ways of classification of the design, are called the *factors*, and their individual treatments are their *levels*. Our examples of two- and three-way designs in the preceding sections involved factorial plans. Lamps were extinguished or illuminated at different wattages, and mortar or concrete was mixed with varying proportions of water. Conversely, the individual laboratories in the concrete strength examples cannot be construed as levels of a condition.

Factorial designs have useful properties that are not shared by more general cross-classified layouts. For example, we may fit polynomial models to the response means at the different levels of a treatment or fit polynomial surfaces to the cell means of treatment combinations. When each factor occurs at two or three levels it is possible to take advantage of certain simplifications in the construction and analysis of the experiment. In this section we discuss 2^p-factorials, or designs whose p ways of classification each have two levels. Not only are those factorials very useful for agricultural, engineering, or behavioral sciences experiments whose treatments have presence/absence or other dyadic characteristics, but their statistical analysis has this useful property: Because each main effect and interaction has a single degree of freedom their sums of squares may be expressed as the squared values of contrasts of the cell means of the data. In a 2^p-factorial with n observations in each cell the estimator of the effect of the ith factor's presence at level 2 and absence at Level 1 would have this general form:

Estimated effect of factor i

$$= \frac{1}{2^{p-1}n}\begin{pmatrix}\text{sum of all observations with factor } i \text{ at level 2} - \text{sum} \\ \text{of all observations with factor } i \text{ at level 1}\end{pmatrix}$$

$$i = 1, \ldots, p \tag{1}$$

We divided by the number of terms in each sum to make the estimator commensurate with the original observations. In our formal treatment of the factorial effects and interactions we shall omit the divisor but introduce a scale factor in the sums of squares or t-statistics computed from the contrasts.

The linear model for the observations in the 2^p factorial design is equivalent to the p-way generalization of the fixed-effects models in Sections 9-6 and 9-7, except that the level subscripts only assume the values 1 and 2 for each factor. For example, the 2^3-model with n observations in each of the eight cells is

$$x_{ijkh} = \mu + \alpha_i + \beta_j + \gamma_k + (\alpha\beta)_{ij} + (\alpha\gamma)_{ik} + (\beta\gamma)_{jk} + (\alpha\beta\gamma)_{ijk} + e_{ijkh}$$
$$i = 1, 2; \quad j = 1, 2; \quad k = 1, 2; \quad h = 1, \ldots, n \tag{2}$$

Let the sums of the n observations in the ijkth cell be denoted by

$$x_{ijk.} = \sum_{h=1}^{n} x_{ijkh} \tag{3}$$

where the dot indicates that the summation has been made over the values of the last subscript. The eight sums of the layout can be summarized as in Table 9-16. We note that the effect of factor A can be estimated by the contrast

$$\hat{\psi}_A = (x_{211.} + x_{221.} + x_{212.} + x_{222.}) - (x_{111.} + x_{121.} + x_{112.} + x_{122.}) \tag{4}$$

The terms with positive signs are the sums from the cells in which factor A had the subscript 2, or was "present" in our presence/absence usage. From those terms the sums of the cells for which factor A was "absent" have been subtracted. The contrasts for factors B and C could be formed in the same way.

We can measure interactions by similar contrasts of the cell sums. The $A \times$ B interaction is measured by the difference of the changes from level 1 to level 2 of the A factor at levels 1 and 2 of B. In computing that second difference

TABLE 9-16 Cell Totals of the 2^3-Layout

Factor A	Factor C	Factor B Level 1	Factor B Level 2
Level 1	Level 1	$x_{111.}$	$x_{121.}$
	Level 2	$x_{112.}$	$x_{122.}$
Level 2	Level 1	$x_{211.}$	$x_{221.}$
	Level 2	$x_{212.}$	$x_{222.}$

we must average over the two levels of the third factor, C. The interaction is then measured by the contrast

$$\begin{aligned}\hat{\psi}_{AB} &= (x_{221.} + x_{222.}) - (x_{211.} + x_{212.}) \\ &\quad - [(x_{121.} + x_{122.}) - (x_{111.} + x_{112.})] \\ &= x_{221.} + x_{222.} - x_{211.} - x_{212.} - x_{121.} - x_{122.} + x_{111.} + x_{112.} \end{aligned} \tag{5}$$

The three-factor interaction of A, B, and C would be measured by the difference of the A $\times$ B contrasts at levels 1 and 2 of C. The coefficients of the contrasts for the seven main effects and interactions are shown in Table 9-17. The cells of

TABLE 9-17 Contrast Coefficients for Main Effects and Interactions in the 2^3-Layout

Effect	(1) ABC 111	*a* ABC 211	*b* ABC 121	*c* ABC 112	*ab* ABC 221	*ac* ABC 212	*bc* ABC 122	*abc* ABC 222
A	−1	1	−1	−1	1	1	−1	1
B	−1	−1	1	−1	1	−1	1	1
C	−1	−1	−1	1	−1	1	1	1
A × B	1	−1	−1	1	1	−1	−1	1
A × C	1	−1	1	−1	−1	1	−1	1
B × C	1	1	−1	−1	−1	−1	1	1
A × B × C	−1	1	1	1	−1	−1	−1	1

(Column headings are grouped under "Cell and level subscripts".)

the layout have been designated by the level subscripts of the three factors and by the following notation due to Yates (1937): the *ijk*th cell is represented by the symbol

$$a^{i-1}b^{j-1}c^{k-1} \tag{6}$$

for the eight combinations of i, j, and k. When $i = j = k = 1$, we shall denote the cell by the symbol (1) to avoid confusion with the integer 1. The Yates notation is consistent with the "presence" of the factor in the cells with subscript 2 and its "absence" for subscript 1. If the factor is present its lowercase letter appears in the product and does not if it is absent. For example, a denotes the combination of level 2 of A and level 1 of B and C. In the *abc* cell all factors occur at the second level.

We note that the seven contrasts are orthogonal: The sum of the products of corresponding coefficients in each pair of rows of Table 9-17 is zero. Each contrast is normally distributed with variance

$$\operatorname{Var}(\hat{\psi}) = 8n\sigma^2 \tag{7}$$

because the sum of the squares of the coefficients in each line of the table is eight. If the main effect or interaction parameter is zero, the respective normal distribution of its contrast estimator has mean zero. We estimate σ^2 by the usual within-cells mean square:

$$\hat{\sigma}^2 = \frac{\sum_{i=1}^{2}\sum_{j=1}^{2}\sum_{k=1}^{2}\sum_{h=1}^{n}(x_{ijkh} - \bar{x}_{ijk.})^2}{8(n-1)} = \frac{\text{SSE}}{8(n-1)} \tag{8}$$

If $n = 1$ we must use the highest-order interaction (for example, $\text{A} \times \text{B} \times \text{C}$) or sums of the two-way and three-way interactions, for the estimation of σ^2:

$$\hat{\sigma}^2 = \frac{\hat{\psi}^2_{\text{ABC}}}{8} \tag{9}$$

where $\hat{\psi}_{\text{ABC}}$ is the last contrast in Table 9-17.

To test the hypothesis

$$H_0:\quad \alpha_1 = \alpha_2 \tag{10}$$

of no row effect we refer

$$t = \frac{\hat{\psi}_{\text{A}}}{\sqrt{8n\hat{\sigma}^2}} \tag{11}$$

to the $100\alpha\%$ critical value of the t-distribution with $8(n - 1)$ degrees of freedom. The test may be one- or two-sided—an advantage not shared with the analysis of variance tests for treatments with more than two levels. Similarly, we test the column and layer hypotheses

$$H_0:\quad \beta_1 = \beta_2 \qquad H_0:\quad \gamma_1 = \gamma_2 \tag{12}$$

by forming their respective contrasts from the coefficients in the second and third lines of Table 9-17 and referring their t-statistics of the sort defined by (11) to appropriate critical values. The same procedure is used for testing the hypotheses of no interactions by means of the contrasts in the table and the same t-statistics with $8(n - 1)$ degrees of freedom.

Alternatively, we may prepare an analysis of variance table from the contrasts and test the seven hypotheses by referring their F-ratios to critical values of the F-distribution with one and $8(n - 1)$ degrees of freedom. The analysis of variance is shown in Table 9-18. The error mean square $\hat{\sigma}^2$ is defined by (8). If $n = 1$ the three-factor interaction mean square or perhaps some pooled combination of the interaction sums of squares must be used as the divisor for the F-ratios.

TABLE 9-18 Analysis of Variance for the 2^3 Factorial Design

Effect	Sum of squares	d.f.	Mean square
A	$\frac{\hat{\psi}_A^2}{8n}$	1	$\frac{\hat{\psi}_A^2}{8n}$
B	$\frac{\hat{\psi}_B^2}{8n}$	1	$\frac{\hat{\psi}_B^2}{8n}$
C	$\frac{\hat{\psi}_C^2}{8n}$	1	$\frac{\hat{\psi}_C^2}{8n}$
A × B	$\frac{\hat{\psi}_{AB}^2}{8n}$	1	$\frac{\hat{\psi}_{AB}^2}{8n}$
A × C	$\frac{\hat{\psi}_{AC}^2}{8n}$	1	$\frac{\hat{\psi}_{AC}^2}{8n}$
B × C	$\frac{\hat{\psi}_{BC}^2}{8n}$	1	$\frac{\hat{\psi}_{BC}^2}{8n}$
A × B × C	$\frac{\hat{\psi}_{ABC}^2}{8n}$	1	$\frac{\hat{\psi}_{ABC}^2}{8n}$
Within cells	SSE	$8(n-1)$	$\hat{\sigma}^2 = \frac{\text{SSE}}{8(n-1)}$
Total	$\sum\sum\sum\sum x_{ijkh}^2 - \frac{G^2}{8n}$	$8n-1$	—

Example 9-13

We shall illustrate a 2 × 2 factorial with some fictitious, though realistic, data. Size C flashlight batteries can be purchased from brand A or brand B. Each brand includes several types of batteries,* including alkaline (A) and standard (S). Four batteries were selected at random from each combination of brand and type on sale in retail stores. The voltages, means, and within-cell sums of squares are shown in Table 9-19.

TABLE 9-19 Flashlight Battery Voltages

	Brand		
Type	A	B	Mean
Alkaline			
Volts	1.42, 1.45, 1.46, 1.47	1.51, 1.52, 1.54, 1.55	1.49
Mean	1.45	1.53	
SSE	0.0014	0.001	
Standard			
Volts	1.40, 1.41, 1.43, 1.44	1.51, 1.53, 1.53, 1.55	1.475
Mean	1.42	1.53	
SSE	0.001	0.0008	
Mean	1.435	1.53	1.4825

*For a semitechnical description of the properties of small batteries, see the article by Stone (1979).

The error sum of squares is merely

$$\text{SSE} = \text{SSE}_{11} + \text{SSE}_{12} + \text{SSE}_{21} + \text{SSE}_{22} = 0.0042$$

The treatment sums of squares can be computed most conveniently by these formulas in terms of the cell means:

$$\text{SS}_{\text{brand}} = n(\bar{x}_{.1.} - \bar{x}_{.2.})^2 = 0.0361$$

$$\text{SS}_{\text{type}} = n(\bar{x}_{1..} - \bar{x}_{2..})^2 = 0.0009$$

$$\text{SS}_{\text{B} \times \text{T}} = \frac{n}{4}(\bar{x}_{11.} - \bar{x}_{12.} - \bar{x}_{21.} + \bar{x}_{22.})^2 = 0.0009$$

The analysis of variance is shown in the table:

Analysis of Variance

Source	Sum of squares	d.f.	Mean square	F
Brands	0.0361	1	0.0361	103.14
Type	0.0009	1	0.0009	2.57
B × T	0.0009	1	0.0009	2.57
Within	0.0042	12	0.00035	—
Total	0.0421	15	—	—

The difference in the brand means is clearly significant at any conventional level. The difference between the alkaline and standard means, and the interaction sum of squares, are not significant even at the 0.10 level. Nevertheless, we might still test the hypothesis that the alkaline and standard means are the same for brand A. We may do so by referring the t-statistic

$$t = \frac{(\bar{x}_{11.} - \bar{x}_{21.})\sqrt{n/2}}{\hat{\sigma}} = 2.27$$

to the critical value $t_{0.025;\,12} = 2.18$. We should conclude that the alkaline brand A batteries may have a higher mean voltage than the standard brand A batteries. However, we incur a risk of being wrong in that conclusion nearly 5% of the time. We should conduct a second experiment with larger sample sizes of both types of brand A batteries. We also should find a way to control for the shelf life, or time since manufacture, of each battery. We shall address that problem in our treatment of the analysis of covariance for concomitant variables in Chapter 10.

Contrasts and sums of squares for higher-way factorial designs. We know that in any two-level factorial layout the effect of a given factor can be estimated by comparing the sum of all observations at one level with those at the other. Similarly, the interaction of two factors can be estimated by the usual two-way layout formula applied to sums or means of the observations taken over all levels of the remaining $p - 2$ factors not in the interaction. Although we might proceed in that fashion to develop the contrasts and sums of squares for the higher-order interactions, we would prefer some systematic rule or algorithm. Such a method is provided by the Yates symbolic notation introduced earlier in expression (6) of this section.

To use the Yates notation we assign each factor a letter, for example, a, b, c, d for a 2^4-layout. The levels of the factor will be denoted by

$$\begin{aligned} a^0 &= 1 \text{ (absence of factor A)} \\ a &= \text{(presence of factor A)} \end{aligned} \tag{13}$$

and the various combinations of levels of the factors will be shown by products of the letters. For example, in the 2^4 design we could represent the treatment combinations in this way:

Symbol product	Treatment combination
(1)	All factors at first level
a	Factor A at level 2; B, C, D at level 1
bc	Factors B and C at level 2; A and D at level 1
$abcd$	All factors at level 2

We note that the product

$$\begin{aligned} (a+1)(b+1)(c+1)(d+1) = {} & (1) + a + b + c + d \\ & + ab + ac + ad + bc + bd + cd \\ & + abc + abd + acd + bcd \\ & + abcd \end{aligned} \tag{14}$$

contains all $2^4 = 16$ treatment combinations and represents the sum of the cell means or totals. Similar products can be formed to give the contrasts measuring the main effects and interactions. The effect of factor C can be represented by the product

$$\begin{aligned} & (c-1)(a+1)(b+1)(d+1) \\ & \quad = (c-1)(1 + a + b + d + ab + ad + bd + abd) \\ & \quad = c + ac + bc + cd + abc + acd + bcd + abcd \\ & \qquad - (1 + a + b + d + ab + ad + bd + abd) \end{aligned} \tag{15}$$

The first group of treatment combinations in the last line contains factor C, whereas that factor is absent, or at level one, in the combinations in the second group. The symbolic product has generated the signs in the factor C contrast for the cell means or totals. We may repeat this process for the other effects. For example, the factor A effect can be estimated by the contrast

$$\begin{aligned} & (a-1)(b+1)(c+1)(d+1) \\ & \qquad = a + ab + ac + ad + abc + abd + acd + abcd \\ & \qquad\quad - (1 + b + c + d + bc + bd + cd + bcd) \end{aligned} \tag{16}$$

and so on for factors B and D.

The products generating the interaction contrasts contain terms in which the symbol for an interacting factor is followed by -1. Factors excluded from the interaction appear in the product as $(a+1)$, $(b+1)$, $(c+1)$, and so on.

For example, the A $\times$ B interaction in the 2^4-factorial can be estimated from the symbolic contrast

$$\begin{aligned}(a-1)(b-1)(c+1)(d+1) \\ &= (ab-a-b+1)(cd+c+d+1) \\ &= 1-a-b+c+d+ab-ac-ad-bc-bd+cd \\ &\quad + abc+abd-acd-bcd+abcd \qquad (17)\end{aligned}$$

The A $\times$ B $\times$ C interaction would require the contrast

$$\begin{aligned}(a-1)(b-1)(c-1)(d+1) \\ &= -1+a+b+c-d+ab-ac+ad-bc+bd+cd \\ &\quad + abc-bcd-acd-abd+abcd \qquad (18)\end{aligned}$$

Finally, the four-factor interaction A $\times$ B $\times$ C $\times$ D is estimated by the symbolic contrast

$$\begin{aligned}(a-1)(b-1)(c-1)(d-1) \\ &= 1-a-b-c-d+ab+ac+ad+bc+bd+cd \\ &\quad - abc-abd-acd-bcd+abcd \qquad (19)\end{aligned}$$

The signs of the coefficients for the 15 main effect and interaction contrasts are given in Table 9-20. One should verify that these are orthogonal. The variance of any contrast $\hat{\psi}$ under the usual assumptions of the fixed-effects model with n observations in each cell is

$$\text{Var}(\hat{\psi}) = \frac{16\sigma^2}{n} \qquad (20)$$

We may test the hypothesis

$$H_0: \quad E(\hat{\psi}) = 0 \qquad (21)$$

of no main effect or interaction in the population against the two-sided alternative

$$H_1: \quad E(\hat{\psi}) \neq 0$$

by referring

$$t = \frac{\hat{\psi}\sqrt{n}}{4\hat{\sigma}} \qquad (22)$$

to the critical value $t_{\alpha/2;\,16(n-1)}$ of the t-distribution with $16(n-1)$ degrees of freedom. Alternatively, we may form an analysis of variance table by computing the effect sum of squares

$$\text{SS}_\psi = \frac{n\hat{\psi}^2}{16} \qquad (23)$$

When the hypothesis (21) is true for the particular contrast ψ, $\text{SS}_\psi/\hat{\sigma}^2$ will have the chi-squared distribution with one degree of freedom. We should reject the null hypothesis if

$$F = \frac{\text{SS}_\psi}{\hat{\sigma}^2} \qquad (24)$$

TABLE 9-20 Contrast Coefficient Signs for the 2^4-Factorial Layout

	Treatment combination															
Effect	(1)	*a*	*b*	*c*	*d*	*ab*	*ac*	*ad*	*bc*	*bd*	*cd*	*abc*	*abd*	*acd*	*bcd*	*abcd*
A	−	+	−	−	−	+	+	+	−	−	−	+	+	+	−	+
B	−	−	+	−	−	+	−	−	+	+	−	+	+	−	+	+
C	−	−	−	+	−	−	+	−	+	−	+	+	−	+	+	+
D	−	−	−	−	+	−	−	+	−	+	+	−	+	+	+	+
AB	+	−	−	+	+	+	−	−	−	−	+	+	+	−	−	+
AC	+	−	+	−	+	−	+	−	−	+	−	+	−	+	−	+
AD	+	−	+	+	−	−	−	+	+	−	−	−	+	+	−	+
BC	+	+	−	−	+	−	−	+	+	−	−	+	−	−	+	+
BD	+	+	−	+	−	−	+	−	−	+	−	−	+	−	+	+
CD	+	+	+	−	−	+	−	−	−	−	+	−	−	+	+	+
ABC	−	+	+	+	−	−	−	+	−	+	+	+	−	−	−	+
ABD	−	+	+	−	+	−	+	−	+	−	+	−	+	−	−	+
ACD	−	+	−	+	+	+	−	−	+	+	−	−	−	+	−	+
BCD	−	−	+	+	+	+	+	+	−	−	−	−	−	−	+	+
ABCD	+	−	−	−	−	+	+	+	+	+	+	−	−	−	−	+

exceeds the upper $100\alpha\%$ critical value $F_{\alpha;\,1,\,16(n-1)}$ of the F-distribution with one and $16(n-1)$ degrees of freedom.

The symbolic notation for generating interaction contrasts provides another interpretation of higher-order interactions. The three-factor interaction in a 2^3-layout with factors A, B, and C can be written symbolically as

$$\begin{aligned}(a-1)(b-1)(c-1) &= c(a-1)(b-1)-(a-1)(b-1)\\ &= b(a-1)(c-1)-(a-1)(c-1)\\ &= a(b-1)(c-1)-(b-1)(c-1)\end{aligned} \tag{25}$$

In the first of the right-hand lines we are comparing the AB interaction at level 2 (presence) of C with the AB interaction at level 1 (absence) of the C factor. In that sense the ABC interaction is an "interaction" of the AB contrasts at the two levels of C. Similarly, the second line compares the AC interaction at the two levels of B, and the third expression compares the BC contrasts at the A levels. This interpretation of higher-order interactions as "interactions of interactions of lower order" is a convenient computational and mnemonic device for calculating interaction contrasts and sums of squares.

Example 9-14

This 2^4-factorial experiment was carried out to measure the voltages of four AA penlight batteries. All 15 combinations of the batteries singly or in series were measured on a voltmeter. The batteries A, B, C, and D were the factors, and the levels consisted of the presence or absence of a battery in the circuit. The sixteenth observation was the zero voltage when all batteries were absent. Such an experiment is an example of a "weighing" design (Yates, 1935; Kempthorne, 1952), for apart from measurement biases in the meter at different voltages, internal resistance in the batteries, and perhaps other nonlinear disturbances, we should not have interaction terms in the factorial model. A single voltage reading was made at each combination, so that $n = 1$, and no within-cell measure of experimental error was available. If the battery combinations are designated in the Yates notation, the voltage readings may be listed in this fashion:

(1)	*a*	*b*	*c*	*d*	*ab*	*ac*	*ad*
0	1.45	1.40	1.55	1.56	2.85	3.04	3.00

bc	*bd*	*cd*	*abc*	*abd*	*acd*	*bcd*	*abcd*
2.95	2.92	3.10	4.45	4.42	4.56	4.52	6.04

The sample contrasts and some functions of them are shown in Table 9-21. One should verify that the sum of the squared contrasts divided by 16 is equal to the total sum of squares

$$\sum\sum\sum\sum (x_{ijkh} - \bar{x}_{....})^2 = 36.2838$$

What conclusions can we draw from the sample contrasts? First, any question of testing for the main effects, or positive voltages, of the four batteries is beneath

TABLE 9-21 Contrasts from the Voltage Experiment

Effect	Contrast	Contrast/8	(Contrast)^2/16
A	11.81	1.476	8.72
B	11.29	1.411	7.97
C	12.61	1.576	9.94
D	12.43	1.554	9.66
AB	0.13	0.01625	0.001056
AC	0.13	0.01625	0.001056
AD	0.03	0.00375	0.000056
BC	0.13	0.01625	0.001056
BD	0.07	0.00875	0.000306
CD	0.03	0.00375	0.000056
ABC	0.01	0.00125	0.000006
ABD	0.11	0.01375	0.000756
ACD	−0.05	−0.00625	0.000156
BCD	0.11	0.01375	0.000756
ABCD	−0.01	−0.00125	0.00000625

consideration. Those estimates are two orders of magnitude greater than the interactions. Can we discard the interaction terms in the factorial model in favor of one of series voltage as a purely additive function? To make those tests we must have an estimate of σ^2, and it is here that we must make some subjective choices. The usual estimate based on the highest-order interaction would be

$$\hat{\sigma}^2 = 0.0000625$$

with but a single degree of freedom. The resulting F-statistic for the AB, AC, and BC interactions would be equal to 169, that is, barely greater than $F_{0.05;\,1,1} = 161$. Although the test is insensitive because of its second single degree of freedom, the use of the highest-order interaction in a model where the existence of all interactions is suspect may cause an upward bias in the F-statistic. We might consider these estimates of the error variance:

Estimate	$\hat{\sigma}^2$	d.f.
1. Pool the three- and four-factor interactions	0.00033625	5
2. Pool the three- and four-factor interactions and the smallest two-factor interactions AD and CD	0.00025625	7
3. Use the interactions AD, BD, CD, ABC, ACD, ABCD	0.00009792	6

With the first estimate as denominator the F-statistics for AB, AC, and BC do not exceed even the 0.10 critical value. With the second error mean square those F-statistics exceed the 0.10 critical value. The third estimate of σ^2 gives F-statistics for AB, AC, and BC significant at the 0.025 level, whereas those in the ABD and BCD interactions are significant at the 0.05 level. How should we resolve these subjective and contradictory results? A conservative conclusion might be that there is a suggestion of pairwise interactions among batteries A, B, and C, but their existence cannot be confirmed by a statistical test with a small and well-determined Type I error probability.

Contrasts and confidence intervals for factorial effects. In Example 9-14 we saw that the main effects, or the four battery voltages, were overwhelmingly different from zero. Although the question of positive main effects has been answered decisively, we might still wish confidence intervals for the four effects as measures of the experimental errors of the voltages. The $100(1-\alpha)\%$ confidence interval for the A-factor effect is

$$\begin{aligned} \hat{\psi}_A - t_{\alpha/2;\,16(n-1)}\frac{\hat{\sigma}}{2\sqrt{n}} &\leq \psi_A \\ &\leq \hat{\psi}_A + t_{\alpha/2;\,16(n-1)}\frac{\hat{\sigma}}{2\sqrt{n}} \end{aligned} \tag{26}$$

in which we have represented the effect and its estimate by the contrasts

$$\begin{aligned} \psi_{\mathrm{A}} &= \alpha_2 - \alpha_1 \\ \hat{\psi}_{\mathrm{A}} &= \bar{x}_{2\cdots\cdot} - \bar{x}_{1\cdots\cdot} \end{aligned} \tag{27}$$

The standard deviation $\hat{\sigma}$ is the square root of the error mean square computed from the sums of squares within the 16 cells. If $n = 1$ we must replace $\hat{\sigma}$ by an estimate computed from the interaction sums of squares. The confidence intervals for the other three main effects

$$\psi_{\mathrm{B}} = \beta_2 - \beta_1 \qquad \psi_{\mathrm{C}} = \gamma_2 - \gamma_1 \qquad \psi_{\mathrm{D}} = \delta_2 - \delta_1 \tag{28}$$

follow by replacing $\hat{\psi}_{\mathrm{B}}$ by the respective estimators

$$\hat{\psi}_{\mathrm{B}} = \bar{x}_{\cdot 2\cdots} - \bar{x}_{\cdot 1\cdots} \qquad \hat{\psi}_{\mathrm{C}} = \bar{x}_{\cdot\cdot 2\cdot\cdot} - \bar{x}_{\cdot\cdot 1\cdot\cdot} \qquad \hat{\psi}_{\mathrm{D}} = \bar{x}_{\cdots 2\cdot} - \bar{x}_{\cdots 1\cdot} \tag{29}$$

We note that the variance of each estimator is

$$\text{Var (average at level 2 − average at level 1)} = \sigma^2\left(\frac{1}{8n} + \frac{1}{8n}\right) = \frac{\sigma^2}{4n} \tag{30}$$

The divisor of $\hat{\sigma}$ in (26) follows from that result.

Occasionally, contrasts of the main effects may be of interest in analyzing the results of a factorial experiment. For example, in the previous 2^4-design for measuring battery voltage we might wish to test the hypothesis that the effects (voltage) of batteries A and B or C and D are equal or construct a confidence interval for their difference. We might also wish to make similar tests or inferences about the differences of the interaction parameters. For the test of the single hypothesis

$$H_0\colon\ \alpha_2 - \alpha_1 = \beta_2 - \beta_1 \tag{31}$$

of equal A and B effects we begin by computing the variance of the difference of their estimators:

$$\operatorname{Var}(\hat{\psi}_{\mathrm{A}} - \hat{\psi}_{\mathrm{B}}) = \sigma^2\left(\frac{1}{4n} + \frac{1}{4n}\right) = \frac{\sigma^2}{2n} \tag{32}$$

Because the estimators are orthogonal contrasts, and thus independent, the right-hand side contains no covariance term. We should reject the hypothesis

(31) in favor of a two-sided alternative if the t-statistic

$$t = \frac{(\hat{\psi}_A - \hat{\psi}_B)\sqrt{2n}}{\hat{\sigma}} \tag{33}$$

exceeds the usual critical value $t_{\alpha/2;\,16(n-1)}$. As in all of our previous methods, $n = 1$ requires that we estimate σ^2 from the interaction sums of squares, and take as the degrees of freedom of t the number of interactions used in the mean square for the estimator.

It is essential to remember that the preceding tests and confidence statements refer to *single* functions of the parameters chosen before an examination of the data. If we wish protection for a family of tests or intervals we must replace the t critical value by one from the Bonferroni, Scheffé, or Tukey multiple comparison methods.

Example 9-15

Now we illustrate the tests and confidence intervals with the factorial data of Example 9-14. First we construct 99% confidence intervals for the four main effects. To do so we must decide on an estimate of σ^2 based on the interaction sums of squares because $n = 1$. Interaction effects would seem to be ruled out by the additive model for series voltage without regard for various resistances, and one estimate of σ^2 might use all the interaction sums of squares in Table 9-21:

$$\hat{\sigma}^2 = \frac{\text{SSAB} + \text{SSAC} + \text{SSBC} + \cdots + \text{SSABCD}}{11}$$

$$= \frac{0.00526625}{11} = 0.00047875$$

and $\hat{\sigma} = 0.02188$. The estimated variance is larger than those in Example 9-14 based on the smaller interaction sums of squares and in that sense will lead to more conservative (wider) confidence intervals. The two-sided critical value is $t_{0.005;\,11} = 3.106$. These confidence intervals were obtained for the main effects:

Battery	$\hat{\psi}$	Confidence interval for effect (voltage)
A	1.476	$1.442 \leq \psi_A \leq 1.510$
B	1.411	$1.377 \leq \psi_B \leq 1.445$
C	1.576	$1.542 \leq \psi_C \leq 1.610$
D	1.554	$1.520 \leq \psi_D \leq 1.588$

Next we shall test the hypothesis (31) that the A and B effects are equal, that is, that the true voltages of batteries A and B are the same. To do so we compute the statistic (33):

$$t = \frac{(1.476 - 1.411)\sqrt{2}}{0.02188} = 4.20$$

Because t exceeds the $\alpha = 0.01$ critical value of the t-distribution with 11 degrees of freedom we should reject the hypothesis of equal voltages. We might test the hypothesis

$$H_0\colon \psi_C = \psi_D$$

of equal voltage effects for batteries C and D. However, $t = 1.42$ in that case, and the null hypothesis cannot be rejected at any conventional level. The other four paired comparisons would yield t-statistics greater than the individual or Bonferroni t critical values. It appears that only C and D may have the same voltage, whereas A and B are different from each other and C and D.

Example 9-16

Let us consider a second example of a large factorial experiment described by Samuel et al. (1976). Unlike the preceding factorial each cell contained $n = 13$ independent observations, so that an estimate of pure subject variance could be obtained without resorting to the higher-order interactions. The purpose of the experiment was to determine whether test environment, race of tester, tester expectation, and sex and race of the subject affected the IQ scores derived from the performance scales of the Wechsler Intelligence Scale for Children. The experimental plan was a 2^5-factorial, in which 208 black and 208 white junior and senior high school students equally divided by sex were administered the Wechsler test. The factors in the experiment were these:

A: sex (male/female)
B: atmosphere (evaluative or gamelike)
C: expectation of tester (high or low)
D: race of tester (black or white)
E: race of subject (black or white)

The mean performance IQ scores are shown in Table 9-22. The authors also gave the standard deviations of the 13 observations within each cell. From those statistics we may compute the error sum of squares as

$$\text{SSE} = 12 \sum_{\text{all 32 cells}} (\text{SD})^2 = 70{,}587.61$$

TABLE 9-22 Performance IQ Means

a. Cell means of the 2^5-layout

			Subject race			
			Black		White	
			Tester race		Tester race	
Subject sex	Atmosphere	Expectation	Black	White	Black	White
Male	Evaluative	High	98.54	97.00	111.08	121.46
		Low	91.69	102.00	109.38	110.31
	Gamelike	High	93.23	96.46	107.15	109.46
		Low	96.23	105.62	109.15	118.46
Female	Evaluative	High	96.69	97.38	106.76	111.00
		Low	91.69	94.69	104.76	117.53
	Gamelike	High	89.30	101.08	103.15	112.30
		Low	97.69	97.46	110.15	115.92

TABLE 9-22 (*cont.*)

b. Main effect means

Effect	Levels	
A. Sex of subject	Male	Female
	104.83	102.97
B. Atmosphere	Evaluative	Gamelike
	103.87	103.93
C. Expectations	High	Low
	103.25	104.55
D. Race of tester	Black	White
	101.04	106.76
E. Race of subject	Black	White
	96.67	111.13

The error degrees of freedom are (32)(13 − 1) = 384, so that the error mean square is

$$\hat{\sigma}^2 = \text{MSE} = \frac{\text{SSE}}{384} = 183.8219$$

We may begin our analysis of variance in Table 9-23 with that last entry.

TABLE 9-23 Selected Lines of the 2^5-Layout Analysis of Variance

Source	Sum of squares	d.f.	Mean square	F
A. Sex	357.62	1	357.62	1.95
B. Atmosphere	0.29	1	0.29	0.00
C. Expectation	173.91	1	173.91	0.95
D. Tester race	3,401.23	1	3,401.23	18.50
E. Subject race	21,728.61	1	21,728.61	118.20
BC	1,292.70	1	1,292.70	7.03
ACE	371.25	1	371.25	2.02
ABCD	555.86	1	555.86	3.02
ACDE	506.51	1	506.51	2.76
Within cells	70,587.61	384	183.82	—

The five main-effect sums of squares can be computed conveniently from the treatment means at the two levels of each factor. Let

$$\bar{x}_{1.....} = \frac{1}{208} \sum_{j,k,h,l,m} x_{1jkhlm} = 104.82625$$

$$\bar{x}_{2.....} = \frac{1}{208} \sum_{j,k,h,l,m} x_{2jkhlm} = 102.9719$$

denote the means of the scores of male and female subjects, respectively. Then the sum of squares for the sex (A) factor is

$$\text{SSA} = 104(\bar{x}_{1.....} - \bar{x}_{2.....})^2 = 357.62$$

The other main-effect sums of squares are computed by the same formula with appropriate changes in the mean subscripts. We note from the first row of Table 9-23 that the sex F-ratio does not exceed the nearest conservative critical value $F_{0.10;\,1,120} = 2.75$. The kind of testing atmosphere and the level of the tester's expectations of the subject do not appear to have any significance as main effects. The last factors D (tester race) and E (subject race) are highly significant.

Because the A, B, and C effects were not significant one might be tempted to average over them and treat the experiment as a 2^2-layout with 104 observations in each cell as opposed to calculating all of the higher-order interactions. Rather than follow that tack directly we shall first calculate the B $\times$ C interaction sum of squares by the formula

$$\begin{aligned} \text{SSBC} &= 2n(\bar{x}_{.11...} - \bar{x}_{.12...} - \bar{x}_{.21...} + \bar{x}_{.22...})^2 \\ &= 1292.70 \end{aligned}$$

The resulting test statistic

$$F = \frac{1292.70}{183.82} = 7.03$$

exceeds the large-sample 0.01 critical value of the F-distribution with 1 and 384 degrees of freedom. Apparently, the effects of the kind of testing atmosphere and the tester's indications that he expected a high or low performance of the subject on the test were not additive. We might examine the nonadditivity with the aid of Figure 9.7. Those

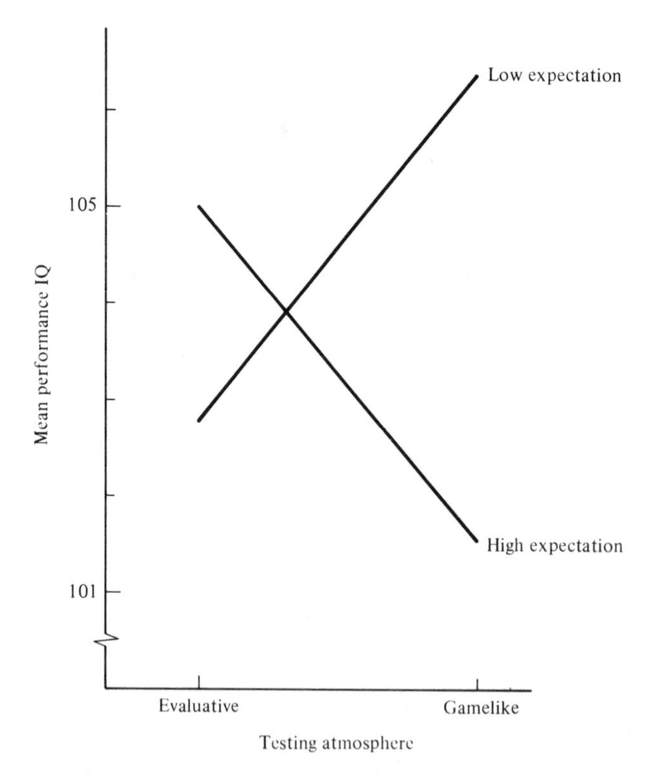

Figure 9-7 Atmosphere and expectation interaction.

subjects who were told that they should perform well (high expectation) performed better in the competitive, ego-threatening "evaluative" setting for the examinations, whereas the subjects who were informed they probably would not score highly on the test performed better in the relaxed "gamelike" testing setting.

All of the other two-factor interactions can be shown to have F-statistics less than 1, and their sums of squares will not be given. The statistics for the sex $\times$ expectation $\times$ subject race (ACE) interaction are given in Table 9-23; the other three-factor interactions had F-ratios less than 1, and have been omitted. The ABCD and ACDE interactions were significant at the 0.10 level; the other four-factor interactions and the single interaction of all five factors had sums of squares too small to merit consideration. The investigators concluded that much of the interracial differences in mean IQ could be attributed to the social psychological characteristics of the testing situation. Further information and a second factorial analysis also implicated the socioeconomic status of the subject as an important determinant of IQ performance.

Some further topics. Large factorial designs may require more experimental units than can be found in a single homogeneous block of land, raw material, or similar subjects. In that case we may use *confounded* or *fractional factorials*. A confounded design has one or more effect contrasts equivalent to, or confounded with, block differences. Usually, the confounded effects are the highest-order interactions or other comparisons considered a priori to be insignificant or of little interest to the investigator. For example, suppose that the 2^3-factorial experiment with factors A, B, and C can be run with at most four treatment combinations under the same conditions. Those four units will constitute a *block*. We might conduct the experiment as in Table 9-24. The

TABLE 9-24 Confounded 2^3-Factorial Design

Block 1		Block 2	
(1)	*ab*	*a*	*b*
ac	*bc*	*c*	*abc*

contrast estimating the block effect is in symbolic form

$$\begin{aligned}\hat{\psi}_{\text{block}} &= a + b + c + abc - ab - ac - bc - (1) \\ &= (a-1)(b-1)(c-1) \\ &= \hat{\psi}_{\text{ABC}} \end{aligned} \tag{34}$$

or, as we may verify from the last line of Table 9-17, the contrast measuring the three-factor interaction of A, B, and C. In all other main-effect and two-factor interaction contrasts the two block effects would cancel; the usual contrasts of the means are unbiased estimators of the six parameters. The construction and properties of confounded factorials have been described by Yates (1937), Kempthorne (1952), Cochran and Cox (1957), and John (1971). The algebraic properties of the treatment combinations in the design provide some interesting and useful ways of confounding certain effects for blocks of given size.

A fractional factorial consists of one-half, one-quarter, or another part of a factorial layout. Half-replicates are denoted by 2^{p-1}, or one-half of the full 2^p-design, quarter replicates by 2^{p-2}, and three-quarter replicates by $3(2^{p-2})$. For example, blocks 1 and 2 in Table 9-24 are the one-half-replicate fractions of the 2^3-design. As in the case of confounded factorials, it may not be feasible to collect all 2^p observations on many complex agricultural or chemical processes. Furthermore, only p of the 2^{p-1} degrees of freedom about the grand mean are used for estimation of the main effects. The remainder are needed only for inferences about the interactions. If we are only interested in the main effects and can validly assume that the interactions in the model are negligible, a fractional factorial will provide main-effect estimates with a considerable reduction in experimental units or replications. For example, the symbolic notation contrasts

$$\hat{\psi}_1 = \frac{abc + a - b - c}{4}$$
$$\hat{\psi}_2 = \frac{abc + b - a - c}{4} \tag{35}$$
$$\hat{\psi}_3 = \frac{abc + c - a - b}{4}$$

of the treatment combinations of block 2 of Table 9-24 provide estimators of the parametric functions

$$\psi_1 = \alpha_2 + (\beta\gamma)_{22}$$
$$\psi_2 = \beta_2 + (\alpha\gamma)_{22} \tag{36}$$
$$\psi_3 = \gamma_2 + (\alpha\beta)_{22}$$

We have imposed the usual zero-sum constraints of the type

$$\alpha_1 + \alpha_2 = 0 \qquad \sum_{i=1}^{2} \sum_{j=1}^{2} (\alpha\beta)_{ij} = 0 \qquad \sum\sum\sum (\alpha\beta\gamma)_{ijk} = 0$$

on the parameters of the original linear model (2) to simplify the expected values. The three estimators are unbiased for the main effects only if the two-factor interactions vanish.

The fractional factorial was introduced by Finney (1945). Construction of the designs leads to interesting combinatorial and algebraic problems; these have been discussed by Kempthorne (1952). Cochran and Cox (1957), Cox (1958), Davies (1963), and John (1971) have given examples of fractional factorial designs.

The three-level factorial has also been developed intensively. The three points permit the fitting of quadratic functions to the various means, estimation of a minimum or maximum in the response variable, and the decomposition of the main effects and interactions into linear and quadratic components. A readable first source with numerical examples is the original Yates memoir (1937). The construction, confounding, fractional replication, and analysis of 3^p-factorials have been treated in the texts by Cochran and Cox, Kempthorne, and John.

9-9 UNBALANCED TWO-WAY LAYOUTS

Our preceding methods for cross-classified data have always assumed that each cell of the layout contained the same number of independent observations. In the design of the experiment exactly n subjects were chosen randomly for each treatment combination, and in truth an observation was actually collected from each subject. No testing appointments were overlooked, no rats or mice died from other causes before their sacrifice for the experimental assay, or no new factors or levels were added after the experiment was underway.

If the assumption of equal cell sizes, or a *balanced* layout, may be difficult to attain even in a carefully conducted investigation in the laboratory, it simply will not hold for survey or observational data. When the total sample has been cross-classified according to sex, race, demographic characteristics, or other categorical variables the cell sizes will probably vary greatly, especially if the classification factors are correlated variables. When the data are unbalanced the earlier sums of squares formulas for main effects and interactions are no longer valid, and we must obtain new expressions that take into account the unequal cell numbers. The unbalanced data, and the sums of squares for their effects, are said to be *nonorthogonal*, for their total sum of squares cannot be partitioned by a single analysis of variance into independent components due to row, column, and interaction effects.

The analysis of unbalanced data is fraught with unresolved controversies. Some of those problems will be described briefly at the end of the section. Within the constraints of mathematical level and space, we have chosen a simple method for row, column, and interaction tests due to Yates known as the *weighted squares of means method.* Unlike Yates's method of unweighted means the weighted squares of means leads to variance ratios with exact F-distributions. The Yates analyses do not require inversion of a matrix or solution of estimation equations, but they do suffer from this limitation: *All cells in the layout must contain at least one observation.* Empty cells require the use of other more complicated techniques.

For the weighted squares of means method we begin with the linear model

$$x_{ijh} = \mu + \alpha_i + \beta_j + \gamma_{ij} + e_{ijh} \qquad i = 1, \ldots, r; \quad j = 1, \ldots, c; \quad h = 1, \ldots, N_{ij} \tag{1}$$

in which α_i, β_j, and γ_{ij} are the respective row, column, and interaction parameters. To keep the hypotheses on row and column effects free of the interaction terms we shall impose the constraints

$$\begin{aligned} \sum_{j=1}^{c} \gamma_{1j} &= \cdots = \sum_{j=1}^{c} \gamma_{rj} = 0 \\ \sum_{i=1}^{r} \gamma_{i1} &= \cdots = \sum_{i=1}^{r} \gamma_{ic} = 0 \end{aligned} \tag{2}$$

The e_{ijh} are independently and normally distributed with zero mean and a common variance σ^2 for all combinations of i, j, and h. The number of observations N_{ij} in the ijth cell must be one or more: The layout contains no empty cells. The cell means are

$$\bar{x}_{ij.} = \frac{1}{N_{ij}} \sum_{h=1}^{N_{ij}} x_{ijh} \tag{3}$$

From those we must calculate the *average* cell mean for each row and column:

$$\begin{aligned} \bar{u}_{i.} &= \frac{1}{c} \sum_{j=1}^{c} \bar{x}_{ij.} \qquad i = 1, \ldots, r \\ \bar{u}_{.j} &= \frac{1}{r} \sum_{i=1}^{r} \bar{x}_{ij.} \qquad j = 1, \ldots, c \end{aligned} \tag{4}$$

Note that those averages give each cell the same weight. They are not necessarily equal to the usual row and column means found by averaging all individual observations in a row or column. We need some row and column weights

$$\begin{aligned} w_i &= c^2 \left(\sum_{j=1}^{c} \frac{1}{N_{ij}} \right)^{-1} \qquad i = 1, \ldots, r \\ v_j &= r^2 \left(\sum_{i=1}^{r} \frac{1}{N_{ij}} \right)^{-1} \qquad j = 1, \ldots, c \end{aligned} \tag{5}$$

for the computation of the weighted row and column grand means:

$$\begin{aligned} \bar{u}_{\text{RW}} &= \frac{1}{\sum_{i=1}^{r} w_i} \sum_{i=1}^{r} w_i \bar{u}_{i.} \\ \bar{u}_{\text{CW}} &= \frac{1}{\sum_{j=1}^{c} v_j} \sum_{j=1}^{c} v_j \bar{u}_{.j} \end{aligned} \tag{6}$$

Then the weighted sums of squares for the rows and columns are

$$\begin{aligned} \text{SSR} &= \sum_{i=1}^{r} w_i (\bar{u}_{i.} - \bar{u}_{\text{RW}})^2 \\ \text{SSC} &= \sum_{j=1}^{c} v_j (\bar{u}_{.j} - \bar{u}_{\text{CW}})^2 \end{aligned} \tag{7}$$

If the hypothesis

$$H_0\colon \alpha_1 = \cdots = \alpha_r \tag{8}$$

of equal row treatment effects is true, SSR/σ^2 will have the chi-squared distribution with $r - 1$ degrees of freedom. Similarly, if the hypothesis

$$H_0\colon \beta_1 = \cdots = \beta_c \tag{9}$$

of equal column effects holds, SSC/σ^2 is a chi-squared random variable with $c - 1$ degrees of freedom. SSR and SSC are not independently distributed. In the special case of the 2^2-layout the sums of squares can be written as

$$\begin{aligned} \text{SSR} &= \frac{w_1 w_2}{4(w_1 + w_2)} (\bar{x}_{11.} + \bar{x}_{12.} - \bar{x}_{21.} - \bar{x}_{22.})^2 \\ &= \left(\frac{1}{N_{11}} + \frac{1}{N_{12}} + \frac{1}{N_{21}} + \frac{1}{N_{22}} \right)^{-1} (\bar{x}_{11.} + \bar{x}_{12.} - \bar{x}_{21.} - \bar{x}_{22.})^2 \end{aligned} \tag{10}$$

$$\text{SSC} = \left(\frac{1}{N_{11}} + \frac{1}{N_{12}} + \frac{1}{N_{21}} + \frac{1}{N_{22}} \right)^{-1} (\bar{x}_{11.} - \bar{x}_{12.} + \bar{x}_{21.} - \bar{x}_{22.})^2 \tag{11}$$

or the squares of the unweighted sample row and column contrasts scaled proportionately to their variances.

The sum of squares for testing the hypothesis

$$H_0: \gamma_{ij} = 0 \qquad \text{all } i, j \tag{12}$$

of no row by column interaction effect is

$$\begin{aligned} \text{SSI} &= N_H \sum_{i=1}^{r} \sum_{j=1}^{c} (\bar{x}_{ij.} - \bar{u}_{i.} - \bar{u}_{.j} + \bar{u}_{..})^2 \\ &= N_H \left(\sum_{i=1}^{r} \sum_{j=1}^{c} \bar{x}_{ij.}^2 - c \sum_{i=1}^{r} \bar{u}_{i.}^2 - r \sum_{j=1}^{c} \bar{u}_{.j}^2 + rc\bar{u}_{..}^2 \right) \end{aligned} \tag{13}$$

in which

$$N_H = rc \left(\sum_{i=1}^{r} \sum_{j=1}^{c} \frac{1}{N_{ij}} \right)^{-1} \tag{14}$$

is the *harmonic mean* of the cell sizes, and

$$\bar{u}_{..} = \frac{1}{r} \sum_{i=1}^{r} \bar{u}_{i.} = \frac{1}{c} \sum_{j=1}^{c} \bar{u}_{.j} = \frac{1}{rc} \sum_{i=1}^{r} \sum_{j=1}^{c} \bar{x}_{ij.} \tag{15}$$

is the unweighted average of all cell means. If the hypothesis (12) is true, SSI/σ^2 has the chi-squared distribution with $(r - 1)(c - 1)$ degrees of freedom. In the special case of a 2×2 layout,

$$\text{SSI} = \left(\frac{1}{N_{11}} + \frac{1}{N_{12}} + \frac{1}{N_{21}} + \frac{1}{N_{22}} \right)^{-1} (\bar{x}_{11.} - \bar{x}_{12.} - \bar{x}_{21.} + \bar{x}_{22.})^2 \tag{16}$$

The error sum of squares is computed from the variation within the cells about their means:

$$\text{SSE} = \sum_{i=1}^{r} \sum_{j=1}^{c} \sum_{h=1}^{N_{ij}} (x_{ijh} - \bar{x}_{ij.})^2 \tag{17}$$

The estimator of the disturbance term variance is the error mean square

$$\hat{\sigma}^2 = \text{MSE} = \frac{\text{SSE}}{N - rc} \tag{18}$$

where for convenience

$$N = \sum_{i=1}^{r} \sum_{j=1}^{c} N_{ij} \tag{19}$$

When the assumption of independent and normal disturbance terms is true SSE/σ^2 has a chi-squared distribution with $N - rc$ degrees of freedom and is distributed independently of the three sums of squares for row, column, and interaction effects.

Now we are ready to make the hypothesis tests. For the test of equal row effects we refer

$$F = \frac{\text{SSR}/(r - 1)}{\hat{\sigma}^2} \tag{20}$$

to the critical value $F_{\alpha; r-1, N-rc}$ of the F-tables, and reject the hypothesis (8) if (20) exceeds that number. Similarly, we reject the hypothesis (9) of equal column effects if

$$F = \frac{\text{SSC}/(c - 1)}{\hat{\sigma}^2} \tag{21}$$

exceeds $F_{\alpha;c-1,N-rc}$. To test that the row and column effects are additive, or that their treatments do not interact, we compute the third statistic

$$F = \frac{\text{SSI}/(r-1)(c-1)}{\hat{\sigma}^2} \tag{22}$$

and accept the hypothesis if $F \leq F_{\alpha;(r-1)(c-1),N-rc}$. If the alternative of an interaction effect is indicated, we should omit the row and column tests (20) and (21) in favor of separate one-way analyses of variance for each row and column. By so doing we can also avoid some of the questions of the definitions of the row and column hypotheses with unbalanced data and interaction effects.

Example 9-17

In Exercise 11, Section 8-10, we considered statistics computed by Lansdell (1968) for a 2 × 2 cross-classification of patients who underwent left or right temporal lobe ablation for relief of epilepsy. These vocabulary test mean scores were given for the four cells:

Temporal lobe	Male	Female
Left	55.5 ($N_{11} = 11$)	47.6 ($N_{12} = 8$)
Right	65.2 ($N_{21} = 13$)	64.1 ($N_{22} = 20$)

The unweighted row and column means are

$$\bar{u}_{1.} = 51.55 \qquad \bar{u}_{2.} = 64.65$$
$$\bar{u}_{.1} = 60.35 \qquad \bar{u}_{.2} = 55.85$$

respectively, and the unweighted grand mean is $\bar{u}_{..} = 58.1$. The row weights are

$$w_1 = 4(\tfrac{1}{11} + \tfrac{1}{8})^{-1} = 18.526$$
$$w_2 = 4(\tfrac{1}{13} + \tfrac{1}{20})^{-1} = 31.515$$

and the column weights are

$$v_1 = 4(\tfrac{1}{11} + \tfrac{1}{13})^{-1} = 23.833$$
$$v_2 = 4(\tfrac{1}{8} + \tfrac{1}{20})^{-1} = 22.857$$

The weighted row and column grand means are

$$\bar{u}_{\text{RW}} = 59.800 \qquad \bar{u}_{\text{CW}} = 58.147$$

and the row and column sums of squares are

$$\begin{aligned}
\text{SSR} &= 18.526(51.55 - 59.80)^2 + 31.515(64.65 - 59.80)^2 \\
&= 2002.263 \\
\text{SSC} &= 23.833(60.35 - 58.147)^2 + 22.857(55.85 - 58.147)^2 \\
&= 236.27
\end{aligned}$$

The interaction sum of squares is

$$\text{SSI} = 11.6(13{,}705.86 - 13{,}522.69 - 13{,}674.05 + 13{,}502.44)$$
$$= 134.88$$

The within-cell mean square was given by Lansdell as SSE/48 = 283.3; that value is within round-off error of the value calculated from the individual standard deviations of the scores in the four cells. We may summarize the preceding statistics in this analysis of variance table:

Source	Sum of squares	d.f.	Mean square	F
Side (rows)	2,002.26	1	2,002.26	7.07
Sex (columns)	236.27	1	236.27	0.83
Interaction	134.88	1	134.88	0.48
Error	13,598.4	48	283.3	—

The F-statistic for the side effect exceeds the 0.025 critical value, but is not quite larger than the interpolated 0.01 value. The sex and sex by side interaction effects do not even approach significance.

Example 9-18

As a second illustration we shall use the blood lead levels of Example 8-4. For convenience we shall repeat Table 8-5, given here as Table 9-25, although with its cell variances replaced by sums of squares about the cell means. The row and column

TABLE 9-25 Blood Lead Levels

	Sex	
Ethnic group	Male	Female
White		
Sample size	151	139
Mean	21.40	20.39
$\sum_{h=1}^{N_{1j}} (x_{1jh} - \bar{x}_{1j.})^2$	7298.25	4899
Nonwhite		
Sample size	60	57
Mean	23.92	25.70
$\sum_{h=1}^{N_{2j}} (x_{2jh} - \bar{x}_{2j.})^2$	2944.60	3583.94

weights and the harmonic mean of the cell sizes are

$$w_1 = 289.50 \qquad w_2 = 116.92$$
$$v_1 = 171.75 \qquad v_2 = 161.69$$
$$N_H = 83.29$$

The averages of the row and column means are

$$\bar{u}_{1.} = 20.895 \qquad \bar{u}_{2.} = 24.81$$
$$\bar{u}_{.1} = 22.66 \qquad \bar{u}_{.2} = 23.045$$
$$\bar{u}_{..} = 22.852$$

and the weighted row and column grand means are

$$\bar{u}_{\text{RW}} = 22.02 \qquad \bar{u}_{\text{CW}} = 22.85$$

The sum of squares within the four cells is

$$\text{SSE} = 18{,}725.79$$

and the error mean square is

$$\hat{\sigma}^2 = \frac{\text{SSE}}{403} = 46.466$$

The sum of squares for the ethnic difference is

$$\begin{aligned}\text{SSR} &= 289.5034(20.895 - 22.02)^2 + 116.9231(24.81 - 22.02)^2 \\ &= 1276.54\end{aligned}$$

and its test statistic is

$$F = \frac{1276.54}{46.466} = 27.47$$

Because F exceeds any conventional critical value with one and 120 or more degrees of freedom we should conclude that the ethnic groups have different mean blood lead levels in their populations. We note that the F-statistic is close to the square of the t-value for the weighted means contrast in Example 8-4. The columns, or sex, sum of squares had the value

$$\begin{aligned}\text{SSC} &= 171.7536(22.66 - 22.85)^2 + 161.6939(23.045 - 22.85)^2 \\ &= 12.34\end{aligned}$$

Its F-statistic is only equal to 0.27; the data do not support a blood lead mean difference for male and female children.

We may compute the interaction sum of squares by the 2×2 layout formula (16):

$$\begin{aligned}\text{SSI} &= \frac{83.29}{4}(21.40 - 20.39 - 23.92 + 25.70)^2 \\ &= 162.08\end{aligned}$$

The associated F-ratio is equal to 3.49, or the square of the t-statistic for the interaction contrast in Example 8-4. The interaction is significant at the 0.10 level. We probably should examine the ethnic mean differences separately for males and females. The t-statistic for the mean difference of white and nonwhite males is equal to 2.42; that exceeds the 0.02 two-sided t critical value. Similarly, the t-statistic for females is equal to 5.31. The ethnic difference is more pronounced for female children, although an examination of the original blood lead levels indicates that the elevated mean for nonwhite females may be due to a small number of extreme values in one dwelling unit—a fact reflected in the large variance of the data in the nonwhite female cell of Table 9-25.

Some further remarks. The Yates unweighted means method (Yates, 1934) is computationally simple and more direct than the preceding weighted squares of means analysis but does not lead to ratios with exact F-distributions. Gosslee and Lucas (1965) have found approximate F-distributions with modified degrees of freedom for the statistics. Speed and Monlezun (1979) have pointed out that both methods are equivalent, and hence lead to exact F-distributions, in the case of the 2^p- factorial layout.

One special pattern of unequal cell sizes will permit an exact analysis of variance of the three effects. If

$$N_{ij} = \frac{N_{i.}N_{.j}}{N} \tag{23}$$

for all combinations of i and j the cell sizes are said to be *proportional*. That is, in every row of the layout the cell numbers are the proportions

$$\frac{N_{.1}}{N}, \ldots, \frac{N_{.c}}{N} \tag{24}$$

of the row totals. A similar relation holds for the proportions of the column totals. The pattern is the same as that for perfect independence of two cross-classified categorical variates: The uncorrected chi-squared statistic would equal zero. The sums of squares formulas for row, column, and interaction effects are

$$\begin{aligned} \text{SSR} &= \sum_{i=1}^{r} N_{i.}(\bar{x}_{i..} - \bar{x}_{...})^2 \\ \text{SSC} &= \sum_{j=1}^{c} N_{.j}(\bar{x}_{.j.} - \bar{x}_{...})^2 \\ \text{SSI} &= \sum_{i=1}^{r} \sum_{j=1}^{c} N_{ij}(\bar{x}_{ij.} - \bar{x}_{i..} - \bar{x}_{.j.} + \bar{x}_{...})^2 \end{aligned} \tag{25}$$

The error sum of squares is still given by (17). The mean squares, F-ratios, and decision rules for the hypotheses on the effect parameters are unchanged from expressions (20)–(22).

Other analyses for unbalanced data can be obtained from regression models with dummy variables for the row, column, and interaction effects and the reduction formulas of Section 3-3 for hypotheses on the appropriate subsets of the parameters. However, the conclusions may depend on the order in which the effects are eliminated and can be misleading. Those shortcomings have been described by Appelbaum and Cramer (1974). Searle (1971) has given a comprehensive treatment of unbalanced two-way layouts by converting the linear model (1) to full rank, and employing the reduction formula notation for the various sums of squares. He has described the difficulty of reaching conclusions about the main effects from the different F-statistics. Kutner (1974) and Speed and Hocking (1976) have contended that the reduction notation does not indicate the true hypothesis being tested and can be misapplied in some cases. Heiberger and Laster (1977, 1980) have compared several of the unbalanced data techniques, particularly for the 2×2 layout, and have offered suggestions for tests with maximum power.

9-10 NESTED LAYOUTS

In all of the preceding layouts of this chapter the data have been cross-classified by types of treatments. Every treatment combination in the two-way or higher-way table gave at least one observation, although a more general and advanced model for unbalanced data would have allowed for empty cells in the table. Now we shall describe a simple kind of *nested* layout, in which each treatment of a one-way layout contains in turn several variants of another class of conditions. Nested designs are also called *hierarchical*, for the treatments form stages or a tree rather than cross-tabulations.

Let us begin by introducing some nested data. A large introductory statistics course has two teaching assistants, A and B, who conduct recitation sessions of the course each week. A and B each have three recitation sessions. At the end of the term students in the course were asked for an overall evaluation of the instructors on the usual 1 through 5 scale described in earlier examples. The frequencies of the first three scale points for the six sections are shown in Table 9-26; no "4" or "5" ratings were reported. We would like to answer

TABLE 9-26 Teaching Assistant Ratings

	Teaching assistant								
	A				B				
	Section				Section				
Rating	1	2	3	Total	4	5	6	Total	Grand total
1	5	8	10	23	2	6	3	11	34
2	4	3	4	11	10	2	3	15	26
3	0	0	1	1	2	0	2	4	5
Responses	9	11	15	35	14	8	8	30	65
Mean	1.44	1.27	1.40	1.37	2.00	1.25	1.88	1.77	1.55
Sum of squares about mean	2.22	2.18	5.6	10.17	4.00	1.50	4.88	13.37	26.06

these questions about the two teaching assistants and the recitation sections:

1. Are the mean evaluations of the two teaching assistants significantly different?
2. Do the sections differ significantly in their evaluations of a particular instructor?

We might provide answers by carrying out a one-way analysis of variance on the ratings from the six sections and then expressing questions 1 and 2 in terms of appropriate contrasts of the six averages. Instead, we decompose the sum of squares for the six sections into independent sums of squares associated with instructors and sections-within-instructor differences. The first analysis of variance with the sections as treatments had this form:

Source	Sum of squares	d.f.	Mean square	F
Sections 1–6	5.6825	5	1.1365	3.29
Within sections	20.3790	59	0.3454	—
Total	26.0615	64	—	—

The error sum of squares is the total of the six within-section sums of squares in the last line of Table 9-26. The total sum of squares is the last entry in that line, and of course the sections 1 through 6 term follows by subtraction. The F-ratio nearly exceeds the closest 1% critical value $F_{0.01;5,60} = 3.34$, and we should conclude that the means of the six normal populations of the sections are different. We may proceed with confidence that question 1 or 2 will be answered affirmatively.

For the test of the first hypothesis of equal instructor mean ratings we decompose the sections 1 through 6 sum of squares into components due to the difference of the instructors' means and the variation among the section means for each instructor:

$$5.6825 = \frac{(35)(30)}{65}(1.3714 - 1.7667)^2 + (0.1679 + 2.9917)$$
$$= 2.5230 + 3.1595 \tag{1}$$

To test the hypothesis we compute

$$F = \frac{\text{instructors' mean square}}{\text{error mean square}}$$
$$= \frac{2.5230/1}{0.3454}$$
$$= 7.30 \tag{2}$$

and refer it to a critical value of the F-distribution with 1 and 59 degrees of freedom; that is, $F_{0.01;\,1,59} = 7.08$. The instructors' mean ratings are indeed different.

Now we wish to test whether the section means within each instructor are significantly different. Recall that the sums of squares for the sections of instructors A and B were

$$SSS_A = 0.1679 \qquad SSS_B = 2.9917 \tag{3}$$

and their total was

$$SSS_{WI} = 3.1595 \tag{4}$$

We may answer question 2 by computing

$$F = \frac{\text{SSS}_{\text{WI}}/4}{\text{MSE}}$$

$$= \frac{3.1595/4}{0.3454}$$

$$= 2.29 \tag{5}$$

That ratio is greater than the 0.10 critical value but less than the 0.05 point. Before concluding that the differences among the section means are only of marginal significance we should compute the F-statistics for the individual instructors' sections:

$$F_{\text{A}} = \frac{0.1679/2}{0.3454} = 0.24$$

$$F_{\text{B}} = \frac{2.9917/2}{0.3454} = 4.33 \tag{6}$$

Clearly, the section mean ratings for A are not different, but those for B are significantly so at the 0.025 level.

We may summarize our statistics in the analysis of variance of Table 9-27. Note the nesting of the sums of squares and degrees of freedom for the section and instructor effects. From the analysis we have determined these answers to the original questions:

TABLE 9-27 Hierarchical Analysis of Variance for the Teaching Ratings

Source	Sum of squares	d.f.	Mean square	F
Sections 1–6	5.6825	5	1.1365	3.29
Instructors	2.5230	1	2.5230	7.30
Sections within instructors	3.1595	4	0.7899	2.29
A. Sections 1–3	0.1679	2	0.08395	0.24
B. Sections 4–6	2.9917	2	1.4958	4.33
Within sections	20.379	59	0.3454	—
Total	26.0615	65	—	—

1. The teaching assistant mean ratings are different.
2. The section mean ratings are not different for instructor A, but are for B.

Now let us turn to a mathematical model and general formulas for a nested two-way layout. The upper, or primary, way of classification consists of k treatments. The primary treatments contain $m_1, \ldots, m_k$ nested, or secondary, treatments, respectively. The jth nested treatment in the ith primary condition has N_{ij} independent observations. This general nested layout has the appearance of Table 9-28. The hth observation under the jth treatment

TABLE 9-28 Two-Way Nested Layout

	Primary treatment							
	1 Nested treatments			$\cdots$	k Nested treatments		
	1	$\cdots$	m_1		1	$\cdots$	m_k
	x_{111}	$\cdots$	x_{1m_11}	$\cdots$	x_{k11}	$\cdots$	x_{km_k1}
	$\cdot$		$\cdot$		$\cdot$		$\cdot$
	$\cdot$		$\cdot$		$\cdot$		$\cdot$
	$\cdot$		$\cdot$		$\cdot$		$\cdot$
	$x_{11N_{11}}$	$\cdots$	$x_{1m_1N_{1m_1}}$	$\cdots$	$x_{k1N_{k1}}$	$\cdots$	$x_{km_kN_{km_k}}$
Nested treatments							
Sample size	N_{11}	$\cdots$	N_{1m_1}	$\cdots$	N_{k1}	$\cdots$	N_{km_k}
Mean	$\bar{x}_{11.}$	$\cdots$	$\bar{x}_{1m_1.}$	$\cdots$	$\bar{x}_{k1.}$	$\cdots$	$\bar{x}_{km_k.}$
Primary treatments							
Sample size		N_1		$\cdots$		N_k	
Mean		$\bar{x}_{1..}$		$\cdots$		$\bar{x}_{k..}$	

nested within the ith primary condition has the linear model

$$x_{ijh} = \mu + \alpha_i + \beta_{ij} + e_{ijh} \qquad \begin{array}{l} i = 1, \ldots, k \\ j = 1, \ldots, m_i \\ h = 1, \ldots, N_{ij} \end{array} \tag{7}$$

in which μ = usual general effect

α_i = effect of the ith primary treatment

β_{ij} = effect of the jth nested treatment under condition i

and the e_{ijh} are independently and normally distributed random variables with mean zero and a common variance σ^2 for all combinations of $i, j,$ and h. In order that the tests on the primary treatment effects are unaffected by the nested treatment parameters β_{ij} we shall impose these constraints:

$$\sum_{j=1}^{m_i} N_{ij}\beta_{ij} = 0 \qquad i = 1, \ldots, k \tag{8}$$

In addition, we shall assume that

$$\sum_{i=1}^{k} N_i\alpha_i = 0 \tag{9}$$

The numbers of observations in the layout will be denoted by these symbols:

1. *Primary treatment sample sizes*

$$\begin{array}{l} N_1 = N_{11} + \cdots + N_{1m_1} \\ \quad\vdots \\ N_k = N_{k1} + \cdots + N_{km_k} \end{array} \tag{10a}$$

2. *Grand total*

$$N = N_1 + \cdots + N_k \tag{10b}$$

3. *Total number of nested treatments*

$$m = m_1 + \cdots + m_k \tag{10c}$$

We shall need these averages from Table 9-28:

1. *Primary treatment i, nested treatment j*

$$\bar{x}_{ij.} = \frac{1}{N_{ij}} \sum_{h=1}^{N_{ij}} x_{ijh} \qquad i = 1, \ldots, k; \quad j = 1, \ldots, m_i \tag{11a}$$

2. *Primary treatment i*

$$\begin{aligned} \bar{x}_{i..} &= \frac{1}{N_i} \sum_{j=1}^{m_i} \sum_{h=1}^{N_{ij}} x_{ijh} \\ &= \frac{1}{N_i} \sum_{j=1}^{m_i} N_{ij} \bar{x}_{ij.} \end{aligned} \tag{11b}$$

3. *Grand mean*

$$\begin{aligned} \bar{x}_{...} &= \frac{1}{N} \sum_{i=1}^{k} \sum_{j=1}^{m_i} \sum_{h=1}^{N_{ij}} x_{ijh} \\ &= \frac{1}{N} \sum_{i=1}^{k} N_i \bar{x}_{i..} \end{aligned} \tag{11c}$$

When the linear model (7) and the constraints (8) and (9) hold the estimators of the nested treatment parameters are

$$\hat{\beta}_{ij} = \bar{x}_{ij.} - \bar{x}_{i..} \qquad j = 1, \ldots, m_i; \quad i = 1, \ldots, k \tag{12}$$

and

$$\hat{\alpha}_i = \bar{x}_{i..} - \bar{x}_{...} \qquad i = 1, \ldots, k \tag{13}$$

The development of the estimators, constraints, and more general estimable functions has been given by Searle (1971). In practice we shall be concerned principally with the estimators of the difference of pairs of nested treatment effects, or

$$\widehat{\beta_{ij} - \beta_{il}} = \bar{x}_{ij.} - \bar{x}_{il.} \tag{14}$$

and the differences

$$\widehat{\alpha_i - \alpha_g} = \bar{x}_{i..} - \bar{x}_{g..} \tag{15}$$

of the primary treatment means.

Now we shall obtain an analysis of variance to test the hypothesis

$$H_{01}: \alpha_1 = \cdots = \alpha_k \tag{16}$$

of the equality of the primary effects, the general hypothesis

$$H_{02}: \begin{array}{c} \beta_{11} = \cdots = \beta_{1m_1} \\ \vdots \\ \beta_{k1} = \cdots = \beta_{km_k} \end{array} \tag{17}$$

of the same nested treated effects *within* the primary treatments, and the equality of the nested treatments within an individual primary condition:

$$H_{03}\colon\ \beta_{i1} = \cdots = \beta_{im_i} \qquad i = 1, \ldots, k \tag{18}$$

Let us begin with the estimator of the common variance σ^2 of the random disturbance term in the model. That source of variation is measured by the sums of squares about each nested treatment mean:

$$\text{SSE}_{ij} = \sum_{h=1}^{N_{ij}} (x_{ijh} - \bar{x}_{ij.})^2 \tag{19}$$

SSE_{ij}/σ^2 is a chi-squared variate with $N_{ij} - 1$ degrees of freedom, so that individual estimators of σ^2 could be obtained from the nested data as

$$\hat{\sigma}^2 = \frac{\text{SSE}_{ij}}{N_{ij} - 1} \tag{20}$$

The proper estimator of the common variance would be obtained from the pooled within-treatments sum of squares

$$\text{SSE} = \sum_{i=1}^{k} \sum_{j=1}^{m_i} \text{SSE}_{ij}$$

as its mean square

$$\hat{\sigma}^2 = \text{MSE} = \frac{\text{SSE}}{N - m} \tag{21}$$

N and m were defined by (10b) and (10c). That mean square will be the divisor for all of the F-statistics for the tests of the hypotheses (16)–(18).

Now let us develop the sums of squares for the hypotheses or main effects of the nested layout. We begin with the sum of squares for all $m = \sum m_i$ nested treatments without the primary treatment distinction:

$$\text{SST} = \sum_{i=1}^{k} \sum_{j=1}^{m_i} N_{ij}(\bar{x}_{ij.} - \bar{x}_{...})^2 \tag{22}$$

We can decompose SST into the sum of the independent components

$$\text{SSP} = \sum_{i=1}^{k} N_i(\bar{x}_{i..} - \bar{x}_{...})^2 \tag{23}$$

$$\text{SSN} = \sum_{i=1}^{k} \sum_{j=1}^{m_i} N_{ij}(\bar{x}_{ij.} - \bar{x}_{i..})^2 \tag{24}$$

by writing

$$\text{SST} = \sum_{i=1}^{k} \sum_{j=1}^{m_i} N_{ij}[\bar{x}_{ij.} - \bar{x}_{i..} + (\bar{x}_{i..} - \bar{x}_{...})]^2$$

and noting that the sum of products vanishes when the quadratic is expanded. When the hypothesis (16) of no primary treatment difference is true the ratio

$$F = \frac{\text{SSP}/(k-1)}{\hat{\sigma}^2} \tag{25}$$

has the F-distribution with $k - 1$ and $N - m$ degrees of freedom. We should reject the hypothesis if

$$F > F_{\alpha; k-1, N-m} \tag{26}$$

To test the hypothesis (17) that the nested treatments within each primary condition have the same effect we compute the ratio

$$F = \frac{\text{SSN}/(m-k)}{\hat{\sigma}^2} \tag{27}$$

and refer it to some critical value of the F-distribution with $m-k$ and $N-m$ degrees of freedom. If

$$F > F_{\alpha; m-k, N-m} \tag{28}$$

we should reject the hypothesis in favor of the alternative that some of the nested treatments differ in their effects.

Finally, we may test for the equality of the treatments nested within one particular primary treatment. If that hypothesis (18) is true for the ith primary treatment, then

$$\frac{\text{SSN}_i}{\sigma^2} = \frac{\sum_{j=1}^{m_i} N_{ij}(\bar{x}_{ij.} - \bar{x}_{i..})^2}{\sigma^2} \tag{29}$$

is a chi-squared random variable with $m_i - 1$ degrees of freedom. To test (18) we compute

$$F = \frac{\text{SSN}_i/(m_i - 1)}{\hat{\sigma}^2} \tag{30}$$

If

$$F > F_{\alpha; m_i-1, N-m} \tag{31}$$

we should reject the hypothesis of equal treatment effects nested within the ith primary condition.

The two principal hypothesis tests can be summarized in the analysis of variance of Table 9-29. For simplicity we have omitted the breakdown

$$\text{SSN} = \text{SSN}_1 + \cdots + \text{SSN}_k \tag{32}$$

in the individual nested sums of squares for the test of (18) and the statistics given by (30). We shall not pursue the powers of the analysis of variance tests,

TABLE 9-29 Analysis of Variance for the Two-Way Nested Layout

Source	Sum of squares	d.f.	Mean square	F
All nested treatments	SST	$m-1$	$\frac{\text{SST}}{m-1}$	$\frac{\text{SST}}{(m-1)\hat{\sigma}^2}$
Primary treatments	SSP	$k-1$	$\frac{\text{SSP}}{k-1}$	$\frac{\text{SSP}}{(k-1)\hat{\sigma}^2}$
Nested treatments	SSN	$m-k$	$\frac{\text{SSN}}{m-k}$	$\frac{\text{SSN}}{(m-k)\hat{\sigma}^2}$
Within treatments	SSE	$N-m$	$\hat{\sigma}^2 = \frac{\text{SSE}}{N-m}$	—
Total	$\sum_{i=1}^{k}\sum_{j=1}^{m_i}\sum_{h=1}^{N_{ij}}(x_{ijh} - \bar{x}_{...})^2$	$N-1$	—	—

although they can be obtained by the usual noncentral F-distributions with appropriate noncentrality parameters. Multiple comparisons of the primary or nested treatments can be made by the Bonferroni or Scheffé methods, or by the Tukey procedure in the event of balanced data, that is, a common sample size N_{ij} for all treatments. We shall show some multiple comparisons in the next example.

Example 9-19

Let us illustrate the computational formulas with the nested analysis of variance for another set of teaching evaluations. The primary "treatments" are five teaching assistants (TAs) in a large undergraduate statistics course. The nested treatments are the three recitation sections taught by each assistant. The data were adapted from actual evaluations to preserve confidentiality. Only the summary statistics for the nested layout will be given in Table 9-30. In addition, we shall need

$$G = \sum\sum\sum x_{ijh} = 403 \qquad \sum\sum\sum x_{ijh}^2 = 951$$

$$N = 204 \qquad m = 15$$

$$\frac{G^2}{N} = 796.1225$$

TABLE 9-30 Teaching Evaluation Summary Statistics

a. Nested sections

Assistant	Section	N_{ij}	Sum of ratings	Sum of squared ratings	Mean
D	1	15	22	40	1.47
	2	10	14	22	1.40
	3	10	18	36	1.80
E	4	10	15	25	1.50
	5	12	23	51	1.92
	6	15	30	72	2.00
F	7	10	17	33	1.70
	8	20	43	109	2.15
	9	11	26	78	2.36
G	10	18	36	80	2.00
	11	15	32	76	2.13
	12	13	32	94	2.46
H	13	15	28	58	1.87
	14	14	28	66	2.00
	15	16	39	111	2.44

b. Teaching assistants

Assistant	N_i	Sum of ratings	Mean	Sum of squares about mean
D	35	54	1.54	14.6857
E	37	68	1.84	23.0270
F	41	86	2.10	39.6098
G	46	100	2.17	32.6087
H	45	95	2.11	34.4444

The sum of squares between the means of the $m = 15$ nested sections without regard for their five teaching assistants is

$$\text{SST} = \sum_{i=1}^{5} \sum_{j=1}^{3} \frac{(\sum x_{ijh})^2}{N_{ij}} - \frac{G^2}{N}$$

$$= \frac{22^2}{15} + \frac{14^2}{10} + \cdots + \frac{39^2}{16} - 796.1225$$

$$= 19.8970$$

The primary treatments, or teaching assistants, sum of squares is

$$\text{SSP} = \sum_{i=1}^{5} \frac{(\sum\sum x_{ijh})^2}{N_i} - \frac{G^2}{N}$$

$$= \frac{54^2}{35} + \frac{68^2}{37} + \frac{86^2}{41} + \frac{100^2}{46} + \frac{95^2}{45} - 796.1225$$

$$= 10.5018$$

The nested treatments sum of squares is composed of the sums of squares between sections for each TA. For example, the component due to TA D is

$$\text{SSN}_\text{D} = \frac{22^2}{15} + \frac{14^2}{10} + \frac{18^2}{10} - \frac{54^2}{35}$$

$$= 0.9524$$

We repeat that calculation for the remaining four sections and sum the components to obtain the sum of squares for sections nested within TAs:

$$\text{SSN} = \text{SSN}_\text{D} + \text{SSN}_\text{E} + \text{SSN}_\text{F} + \text{SSN}_\text{G} + \text{SSN}_\text{H}$$

$$= 9.3952$$

Finally, the within-sections sum of squares can be computed by subtraction, or as a check on the calculations, from the sums of squares (19)

$$\text{SSE}_{ij} = \sum_h x_{ijh}^2 - \frac{(\sum_h x_{ijh})^2}{N_{ij}}$$

and summed over all sections.

The sums of squares and F-statistics are given in the analysis of variance of Table 9-31. The F-ratio for teaching assistants exceeds the nearest 1% critical value, and we

TABLE 9-31 Analysis of Variance for the TA Evaluations

Source	Sum of squares	d.f.	Mean square	F
All sections	19.8971	14	1.4212	1.99
TAs	10.5018	4	2.6254	3.68
Sections within TAs	9.3952	10	0.9395	1.32
D	0.9524	2	0.4762	0.67
E	1.6104	2	0.8052	1.13
F	2.4143	2	1.2072	1.69
G	1.6446	2	0.8223	1.15
H	2.7736	2	1.3868	1.94
Within sections	134.9804	189	0.7142	—
Total	154.8775	203	—	—

should conclude that the TA differences in mean ratings are greater than those due to random or sampling variation. The F-statistic for sections nested by TAs does not ever exceed the conservative critical value $F_{0.10;\,10,120} = 1.65$ for $\alpha = 0.10$, and we should conclude that the within-section mean differences are not significant at any conventional level. The individual TA section differences are even less significant: The F-ratios for TAs D, E, and G do not exceed even the 25% critical value. Differences in the mean ratings appear to be associated with instructors rather than the characteristics of particular sections.

Let us use the Bonferroni multiple comparison procedure to determine which differences among the TA means are statistically significant. We have chosen the Bonferroni method because we are only interested in the $\binom{5}{2} = 10$ paired comparisons; general contrasts of the instructor means would appear to have little meaning. The critical value for the family of 10 tests with an overall error rate $\alpha = 0.05$ is approximately 2.87, if we adopt the conservative practice of using 120 degrees of freedom rather than the limiting normal distribution. The statistic for comparing the rth and sth instructors is

$$t_{rs} = \frac{|x_{r..} - x_{s..}|}{\hat{\sigma}\sqrt{1/N_r + 1/N_s}}$$

Some values of t_{rs} are shown in the table:

Pair	$\bar{x}_{r..} - \bar{x}_{s..}$	t_{rs}
D, E	0.30	1.51
D, F	0.56	2.88
D, H	0.63	3.32
D, G	0.57	2.99
E, F	0.26	1.36
E, H	0.33	1.77

The remaining pairs can be seen by inspection to be not significantly different. Teaching assistant D has a mean rating significantly lower (that is, better) than F, H, and G. E is not different from the remaining TAs, and F, G, and H are scarcely different. The overall difference among the instructors appears to be due to the significantly lower score of D.

Similar multiple comparisons can be made among the nested means, but because the analysis of variance tests of the nested parameters were not significant we shall not pursue the comparisons.

9-11 THE RANDOM-EFFECTS MODEL

The inferences we have made in the preceding experimental designs were upon the particular treatments and interactions comprising the layout. Each condition or level was included so that we might estimate its own effect upon the experimental units. Those effects were represented by unknown parameters in the linear model for the observations. With a sufficiently large number of observations we could estimate the parameters or at least contrasts of them to a desired degree of accuracy. Because our intent was inference about fixed parameters we referred to the models as containing *fixed* main and interaction effects.

Now we shall extend the one-way model of Section 8-8 with random effects to the two-way layout. As in the earlier model we begin by assuming that the row and column treatments were drawn at random from two very large populations of the experimental conditions. A good illustration should be concrete, so we shall draw an example from the mortar crushing strength experiments of Sections 9-3 and 9-7. A batching plant prepares large amounts of concrete mix for a highway project. The concrete is mixed in trucks on the way to the pouring site. Samples are taken at the site, cast in cubes, and tested for crushing strength after curing for 10 days. We wish to determine if the daily batches differ in crushing strength, and at the same time, find whether the mixing trucks give cubes with different crushing strengths. Because the number of daily batches is large we shall draw a random sample of r batches from the available daily productions. Similarly, our limited number of technicians and the available laboratory facilities will not permit strength measurements on specimens from all transit-mix trucks: We shall instead choose c trucks at random from the fleet and draw two samples of concrete from each truck on each day. The cross-classified layout will appear as in Table 9-32.

TABLE 9-32 Concrete Specimen Layout

		Truck		
Batch	Specimen	1	...	c
1	1	x_{111}	...	x_{1c1}
	2	x_{112}	...	x_{1c2}
⋮	⋮	....		
r	1	x_{r11}	...	x_{rc1}
	2	x_{r12}	...	x_{rc2}

Because the batches and trucks were drawn at random their effects on crushing strength of the cubes will be random variables. Let us denote the batch effect by the random variable u_i. If the batches do not differ in their effects, the u_i will not vary among themselves, and

$$\text{Var}(u_i) = \sigma_R^2 = 0 \tag{1}$$

Our test for no batch, or row, effects in the random model is equivalent to testing the hypothesis

$$H_0: \sigma_R^2 = 0 \tag{2}$$

as opposed to the alternative

$$H_1: \sigma_R^2 > 0 \tag{3}$$

that the row effects do vary from batch to batch. We refer to σ_R^2 as the batch (row) *component of variance*. Similarly, the truck effects are described by the random variables $v_1, \ldots, v_c$. If the trucks are not different,

$$\text{Var}(v_j) = \sigma_C^2 = 0 \tag{4}$$

and our hypothesis of no truck, or column, differences is (4), or

$$H_0: \sigma_C^2 = 0 \tag{5}$$

The alternative is a positive variance for the column effects. The row and column variates enter linearly into this model for the hth observation in the cell formed by row i and column j:

$$x_{ijh} = \mu + u_i + v_j + w_{ij} + e_{ijh}$$
$$i = 1, \ldots, r; \quad j = 1, \ldots, c; \quad h = 1, \ldots, n \tag{6}$$

μ is the usual general effect common to all observations. The row effects u_i are normally distributed random variables with mean zero and variance σ_R^2. The column effects v_j are also normally distributed with zero mean and variance σ_C^2. The w_{ij} are normal random variables measuring the interaction between the row and column treatments; they have mean zero and a common variance σ_I^2 for all combinations of i and j. The hypothesis of no row by column interaction, or an additive model, is

$$H_0: \sigma_I^2 = 0 \tag{7}$$

Finally, the e_{ijh} are the usual disturbance terms with mean zero and a common variance σ^2 for all observations. The four kinds of random variables in the model are independent for each observation. Under those assumptions we can write the variance of the ijhth observation as

$$\text{Var}(x_{ijh}) = \sigma_R^2 + \sigma_C^2 + \sigma_I^2 + \sigma^2 \tag{8}$$

or as the sum of variances due to the row, column, interaction, and disturbance terms. Because of that decomposition the random-effects model is often called the *variance components* model.

To test the hypotheses (2), (5), and (7) on the three treatment variance components we begin by computing the usual sums of squares for rows, columns, and interactions given by expressions (10), (15), and (26) of Section 9-3:

$$\text{SSR} = nc \sum_{i=1}^{r} (\bar{x}_{i..} - \bar{x}_{...})^2 \tag{9a}$$

$$\text{SSC} = nr \sum_{j=1}^{c} (\bar{x}_{.j.} - \bar{x}_{...})^2 \tag{9b}$$

$$\text{SSI} = n \sum_{i=1}^{r} \sum_{j=1}^{c} (\bar{x}_{ij.} - \bar{x}_{i..} - \bar{x}_{.j.} + \bar{x}_{...})^2 \tag{9c}$$

We shall also need the within-cells sum of squares (5) of that section:

$$\text{SSE} = \sum_{i=1}^{r} \sum_{j=1}^{c} \sum_{h=1}^{n} (x_{ijh} - \bar{x}_{ij.})^2 \tag{10}$$

Then, it is possible to show that SSR, SSC, SSI, and SSE are independently distributed and, when scaled by appropriate constants, have chi-squared distributions with the same degrees of freedom parameters as the fixed-effects analysis of variance of Table 9-4 [for the details and derivations see, for example, Scheffé (1959, Section 7.4)]. The distinction between the two analyses lies in the expectations of the row and column mean squares shown in Table 9-33. Each expectation contains the term $n\sigma_I^2$: The presence of interaction effects in the two-way layout serves to increase the average row and column mean squares, even when the variance components due to the main effects are zero. The proper divisor for the row and column F-statistics is the interaction mean square MSI. If

$$F = \frac{\text{MSR}}{\text{MSI}} > F_{\alpha; r-1, (r-1)(c-1)} \tag{11}$$

we should conclude that the hypothesis (2) of no variation in the row effects is untenable at the α level. Similarly, if

$$F = \frac{\text{MSC}}{\text{MSI}} > F_{\alpha; c-1, (r-1)(c-1)} \tag{12}$$

the hypothesis (5) of no differences among the column effects should be rejected at the α level. The statistic for testing the hypothesis (7) of no interaction is still

$$F = \frac{\text{MSI}}{\hat{\sigma}^2} \tag{13}$$

as in the fixed-effects analysis.

Simple estimators of the variance components can be obtained by equating the mean squares to their expectations and solving for the components. The estimator of σ^2 is unchanged from the fixed-effects analysis:

$$\hat{\sigma}^2 = \frac{\text{SSE}}{rc(n-1)} \tag{14}$$

The estimator of the interaction variance component is

$$\hat{\sigma}_I^2 = \frac{\text{MSI} - \hat{\sigma}^2}{n} \tag{15}$$

whereas those of the row and column variates are

$$\begin{aligned} \hat{\sigma}_R^2 &= \frac{\text{MSR} - \text{MSI}}{nc} \\ \hat{\sigma}_C^2 &= \frac{\text{MSC} - \text{MSI}}{nr} \end{aligned} \tag{16}$$

respectively. As in the one-way analysis the likelihood of negative variance estimates is high.

We may compute intraclass correlation coefficients from the variance components of the two-way model as in the one-way layout of Section 8-8. The intraclass correlation of elements in the same row and different columns is the variance ratio

TABLE 9-33 Analysis of Variance for the Two-Way Random-Effects Model

Source	Sum of squares	d.f.	Mean square	Expected mean square	F
Rows	SSR	$r-1$	$\text{MSR} = \frac{\text{SSR}}{r-1}$	$\sigma^2 + n\sigma_I^2 + nc\sigma_R^2$	$\frac{\text{MSR}}{\text{MSI}}$
Columns	SSC	$c-1$	$\text{MSC} = \frac{\text{SSC}}{c-1}$	$\sigma^2 + n\sigma_I^2 + nr\sigma_C^2$	$\frac{\text{MSC}}{\text{MSI}}$
Interaction	SSI	$(r-1)(c-1)$	$\text{MSI} = \frac{\text{SSI}}{(r-1)(c-1)}$	$\sigma^2 + n\sigma_I^2$	$\frac{\text{MSI}}{\hat{\sigma}^2}$
Error	SSE	$rc(n-1)$	$\hat{\sigma}^2 = \frac{\text{SSE}}{rc(n-1)}$	σ^2	—
Total	$\sum\sum\sum (x_{ijh} - \bar{x}_{...})^2$	$rcn-1$	—	—	—

$$\rho_R = \frac{\text{Cov}(x_{ijh}, x_{irs})}{\text{Var}(x_{ijh})} \qquad j \neq r$$

$$= \frac{\text{Cov}(\mu + u_i + v_j + w_{ij} + e_{ijh}, \mu + u_i + v_r + w_{ir} + e_{irs})}{\text{Var}(\mu + u_i + v_j + w_{ij} + e_{ijh})}$$

$$= \frac{\sigma_R^2}{\sigma_R^2 + \sigma_C^2 + \sigma_I^2 + \sigma^2} \tag{17}$$

To estimate ρ_R we replace the variance components by their estimators (14)–(16):

$$\hat{\rho}_R = \frac{\text{MSR} - \text{MSI}}{\text{MSR} - \text{MSI} + (c/r)(\text{MSC} - \text{MSI}) + c(\text{MSI} - \hat{\sigma}^2) + nc\hat{\sigma}^2} \tag{18}$$

Similarly, the intraclass correlation of observations in the same column but different rows is

$$\rho_C = \frac{\sigma_C^2}{\sigma_R^2 + \sigma_C^2 + \sigma_I^2 + \sigma^2} \tag{19}$$

and its estimator is

$$\hat{\rho}_C = \frac{\text{MSC} - \text{MSI}}{\text{MSC} - \text{MSI} + (r/c)(\text{MSR} - \text{MSI}) + r(\text{MSI} - \hat{\sigma}^2) + nr\hat{\sigma}^2} \tag{20}$$

The intraclass correlation for observations in the same cell of the layout is

$$\rho_W = \frac{\text{Cov}(x_{ijh}, x_{ijr})}{\text{Var}(x_{ijh})}$$

$$= \frac{\sigma_I^2}{\sigma_R^2 + \sigma_C^2 + \sigma_I^2 + \sigma^2} \tag{21}$$

and its estimator would be computed by substitution of the variance component estimators as before. As in the case of the ordinary product moment correlation, the estimators are biased. Because the model contains variance components for rows and columns, the estimators will have different values from those obtained in the simpler one-way layout of Section 8-8.

Example 9-20

A traditional illustration of a two-way layout with random effects has been that of a sample of r machine operators selected randomly from a factory population, and assigned to operate each of c randomly chosen machines n times. In a military setting the "machines" might be weapons drawn at random from a production lot, the operators a random sample of marksmen from a certain class of proficiency, and the n observations the radial distance of each shot from the center of the target. Not having an arsenal we generated some similar data with a child's dartboard and four plastic balls covered with an adhesive fabric. The balls are the column treatments and will be considered a random sample from the population of balls manufactured for the game. The row treatments are the three players N, D, and P: They are taken as a random sample from the universe of players of similar skill. Our inferences will be made about those populations rather than the individual balls and players. Each player will make $n = 3$ tosses. The score for a toss will be the distance from the ball to the target center measured to the nearest quarter-inch. When a ball failed to adhere to the target the toss was repeated; the data are actually truncated to exclude wild scores. The distances are shown in Table 9-34.

TABLE 9-34 **Scores for the Target Game**

Player		Ball 1	Ball 2	Ball 3	Ball 4	Sum
N		4	2.25	3	5.25	
		3	2	3.5	2.25	
		3.5	5.25	0.75	4.75	
	Sum	10.5	9.5	7.25	12.25	39.5
D		2.5	4.75	4	3	
		0.75	1.25	2.75	2.25	
		2	1.50	2.25	0.5	
	Sum	5.25	7.5	9.0	5.75	27.5
P		2	2.5	1.75	1.5	
		0.75	3	1	2.5	
		3.75	1.5	3.75	2.5	
	Sum	6.5	7	6.5	6.5	26.5
	Sum	22.25	24.0	22.75	24.5	

A visual examination of the data indicates that players D and P have nearly equal average scores, although that of N is higher. The averages for the four balls are close. We should expect little significance in the analysis of variance, and indeed that impression is supported by the mean squares and F-ratios:

Source	Sum of squares	d.f.	Mean square	F
Players	8.72	2	4.36	3.75
Balls	0.37	3	0.12	0.11
Interaction	6.99	6	1.16	0.68
Error	41.14	24	1.71	—
Total	57.21	35	—	—

The differences among the players are significant at the 0.05 level: We should conclude that the player variance component σ_R^2 is positive. Its estimate is

$$\hat{\sigma}_R^2 = \frac{4.36 - 1.16}{12} = 0.267$$

Nevertheless, the estimate of the disturbance term variance is $\hat{\sigma}^2 = 1.71$, or more than six times as great. The estimates of the variance components for balls and the interaction are negative, as we might expect from the large within-cell variation of the distances. We should merely combine the scores for each player across the columns and not attempt to estimate those components of variance.

Example 9-21

Let us consider a more serious scientific application of the random-effects model to mean arterial blood pressures of healthy young males recorded at two times. The data were reported by Mangold et al. (1955), as part of an investigation of cerebral circulation and metabolism in normal, fatigued, and sleeping subjects, and are shown in Table 9-35.

TABLE 9-35 Mean Arterial Blood Pressure

	Time			
Subject	1	2	Mean	Difference
1	86	97	91.5	11
2	90	94	92	4
3	82	86	84	4
4	100	103	101.5	3
5	81	102	91.5	21
6	102	104	103	2
7	87	89	88	2
8	95	94	94.5	−1
9	88	100	94	12
10	109	110	109.5	1
11	80	86	83	6
12	89	89	89	0
13	83	94	88.5	11
Mean	90.15	96	93.08	5.85

From the data this analysis of variance was computed:

Source	Sum of squares	d.f.	Mean square	F
Subjects	1390.85	12	115.90	5.92
Times	222.16	1	222.16	11.35
S × T	234.84	12	19.57	—
Total	1847.85	25	—	—

Each F-statistic exceeds its $\alpha = 0.01$ critical value, and we may conclude that the variations due to subjects and measurement times are greater than we might expect from chance fluctuations. The variance component estimates are

$$\hat{\sigma}^2_{\text{subjects}} = \frac{115.90 - 19.57}{2} = 48.17$$

$$\hat{\sigma}^2_{\text{times}} = \frac{222.16 - 19.57}{13} = 15.58$$

As we might expect, subjects are more variable than the repeated measurements of their blood pressures.

Some insight into the analysis of variance tests might be gained by examining the interaction sum of squares for the case of $c = 2$ columns. A little algebra gives

$$\text{SSI} = \sum_{i=1}^{r} \sum_{j=1}^{2} (x_{ij} - \bar{x}_{i.} - \bar{x}_{.j} + \bar{x}_{..})^2$$

$$= \frac{1}{2}\left[\sum_{i=1}^{r} (x_{i1} - x_{i2})^2 - \frac{\left[\sum_{i=1}^{r} (x_{i1} - x_{i2})\right]^2}{r}\right]$$

or one-half the sum of squared deviations of the paired column differences $x_{i1} - x_{i2}$ about their mean. In using the interaction mean square for the F-statistics we are comparing the row and column mean variation to that of the time 2–time 1 differences in blood pressure.

The estimates of the intraclass correlations within rows and columns are

$$\hat{\rho}_R = \frac{115.90 - 19.57}{115.90 - 19.57 + (2/13)(222.16 - 19.57) + 2(19.57)} = 0.58$$

$$\hat{\rho}_C = \frac{222.16 - 19.57}{222.16 - 19.57 + (13/2)(115.9 - 19.57) + 13(19.57)} = 0.19$$

respectively. Because the layout only contains two columns it will be interesting to compare $\hat{\rho}_R$ directly with the ordinary product moment correlation of the pressures at the two times. We compute

$$\sum_{i=1}^{13} (x_{i1} - \bar{x}_{.1})^2 = 953.69 \qquad \sum_{i=1}^{13} (x_{i2} - \bar{x}_{.2})^2 = 672$$

$$\sum_{i=1}^{13} (x_{i1} - \bar{x}_{.1})(x_{i2} - \bar{x}_{.2}) = 578$$

and from them $r = 0.72$. That correlation is larger than $\hat{\rho}_R$ because the denominator of the latter measure contains a variance component due to the times as well as the "error" component represented by the subjects $\times$ times interaction.

Power of the random-effects tests. We have seen that the random-effects model leads to F-ratios for its hypothesis tests, although the denominators are different from those in the fixed model when $n \geq 2$. When the alternative hypotheses of positive variance components are true the statistics are still proportional to central F random variables, and their power probabilities can be calculated easily if adequate tables of the F-distribution are available. To determine the power of the hypothesis test (2) of no variation in row effects we begin by noting that

$$\chi_R^2 = \frac{\text{SSR}}{\sigma^2 + n\sigma_I^2 + nc\sigma_R^2} \tag{22}$$

has the chi-squared distribution with $r - 1$ degrees of freedom, and

$$\chi_I^2 = \frac{\text{SSI}}{\sigma^2 + n\sigma_I^2} \tag{23}$$

is independently distributed as a chi-squared variate with $(r - 1)(c - 1)$ degrees of freedom. The ratio for testing the hypothesis (2) can be represented as

$$F = \frac{(r-1)(c-1)}{r-1} \frac{\text{SSR}}{\text{SSI}}$$

$$= \frac{\sigma^2 + n\sigma_I^2 + nc\sigma_R^2}{\sigma^2 + n\sigma_I^2} \frac{\chi_R^2/(r-1)}{\chi_I^2/(r-1)(c-1)}$$

$$= \frac{\sigma^2 + n\sigma_I^2 + nc\sigma_R^2}{\sigma^2 + n\sigma_I^2} F'_R \tag{24}$$

where F'_R denotes an F random variable with $r-1$ and $(r-1)(c-1)$ degrees of freedom. When $\sigma_R^2 = 0$ the scale factor is 1, and F is identical to F'_R. The power of the test is

$$\text{Power} = P(F > F_{\alpha; r-1, (r-1)(c-1)})$$

$$= P\left(\frac{\sigma^2 + n\sigma_I^2 + nc\sigma_R^2}{\sigma^2 + n\sigma_I^2} F'_R > F_{\alpha; r-1, (r-1)(c-1)}\right)$$

$$= P\left(F'_R > \frac{\sigma^2 + n\sigma_I^2}{\sigma^2 + n\sigma_I^2 + nc\sigma_R^2} F_{\alpha; r-1, (r-1)(c-1)}\right) \tag{25}$$

The power probability only involves the usual central F-distribution and the variance ratio

$$\delta = \frac{nc\sigma_R^2}{\sigma^2 + n\sigma_I^2} \tag{26}$$

As δ increases from its value of zero under the null hypothesis the right-hand constant term in the last line of (25) becomes smaller, and the power will tend toward a probability of 1 of rejecting the hypothesis.

Similarly, the variate

$$x_C^2 = \frac{\text{SSC}}{\sigma^2 + n\sigma_I^2 + nr\sigma_C^2} \tag{27}$$

has the chi-squared distribution with $c-1$ degrees of freedom, and the same argument shows that the power of the test of the hypothesis (5) of no column treatments variation is

$$\text{Power} = P\left(F'_C > \frac{\sigma^2 + n\sigma_I^2}{\sigma^2 + n\sigma_I^2 + nr\sigma_C^2} F_{\alpha; c-1, (r-1)(c-1)}\right) \tag{28}$$

F'_C has the F-distribution with $c-1$ and $(r-1)(c-1)$ degrees of freedom. The power is a function of the ratio

$$\delta = \frac{nr\sigma_C^2}{\sigma^2 + n\sigma_I^2} \tag{29}$$

Again the power probabilities can be computed from more extensive tables of the F-distribution than we have included in this text or from the charts given by Bowker and Lieberman (1972). Finally, for the power of the interaction test (7) we note that

$$\frac{\text{SSI}}{\sigma^2 + n\sigma_I^2} \tag{30}$$

has the chi-squared distribution. The power is

$$\text{Power} = P\left(F'_I > \frac{\sigma^2}{\sigma^2 + n\sigma_I^2} F_{\alpha; (r-1)(c-1), rc(n-1)}\right) \tag{31}$$

where F'_I has the F-distribution with $(r-1)(c-1)$ and $rc(n-1)$ degrees of freedom.

We shall illustrate the power probabilities for a layout with $r = 5$ row treatments, $c = 3$ columns, and $n = 5$ independent observations in each cell. The significance level was chosen to be $\alpha = 0.05$. The probabilities were calculated for each of the three hypotheses by transforming the probability statements (25), (28), and (31) to incomplete beta functions, and employing the extensive tables of that function compiled by Pearson (1968). The powers are shown in Table 9-36. The powers increase rather slowly for larger values of the

TABLE 9-36 Power Probabilities for the Random-Effects Model ($r = 5$, $c = 3$, $n = 5$, $\alpha = 0.05$)

Rows ($H_0: \sigma_R^2 = 0$)		Columns ($H_0: \sigma_C^2 = 0$)		Interaction ($H_0: \sigma_I^2 = 0$)	
$\frac{15\sigma_R^2}{\sigma^2 + 5\sigma_I^2}$	Power	$\frac{25\sigma_C^2}{\sigma^2 + 5\sigma_I^2}$	Power	$\frac{5\sigma_I^2}{\sigma^2}$	Power
0	0.05	0	0.05	0	0.05
1.5	0.28	0.5	0.28	1	0.41
3	0.48	1	0.37	2	0.69
4.5	0.61	2	0.51	3	0.83
6	0.71	4	0.66	4	0.90
9	0.81	8	0.79	5	0.94
15	0.91	16	0.88	6	0.97
18	0.93	32	0.93	—	—
24	0.96	64	0.97	—	—
36	0.98	128	0.98	—	—

variance ratios. Large relative increases in the random-effect variances may raise the power only slightly. Nevertheless, if $\sigma_R^2 = \sigma^2 + 5\sigma_I^2$, or if the row variance component is equal to the expectation of the interaction mean square, the probability of rejecting the null hypothesis $H_0: \sigma_R^2 = 0$ is 0.91. The powers of the column and interaction tests for similar unit variance ratios are also greater than 0.90. The analysis of variance seems sensitive for that degree of departure from the null hypotheses.

Some further topics. The random-effects model for unbalanced two-way layouts leads to difficult problems for the estimation of the variance components. Estimation methods and their properties have been summarized in detail by Searle (1971, Chapters 10 and 11). Nested layouts can be analyzed with random component models, although the primary and nested sample sizes should be balanced to give proper test statistics with exact F-distributions. The analysis of nested layouts has been described by Scheffé (1959, Section 7-6). Rules for extending the analysis of variance to many cross-classified types of treatments with the same number of observations have been given by Scheffé (1959, Chapter 7). Unfortunately, the expected mean squares do not lead to exact F-statistics unless certain of the interaction variance components equal zero. Scheffé has suggested approximate F-tests due to Satterthwaite (1946).

9-12 THE MIXED MODEL

One of the most important models for the data in a two-way layout consists of both fixed and random effects. It is known as the *mixed model*, or *Model* III. The mixed model may be appropriate when the row treatments were drawn randomly from a large population of conditions, but the column treatments had fixed effects on the response variable. For example, the row conditions might consist of subjects chosen randomly from a universe of persons in good health, whereas the column treatments were the two unique states of the subject before and after administration of a drug. The model differs from the one-way layout with two fixed treatments in that *the same subject is observed under the fixed conditions*. Those treatments consist of repeated measurements. For example, the arterial blood pressure measurements of Example 9-20 might be treated as a mixed rather than random-effects model if the two times had been selected with respect to fixed reference points (for example, pre- and post-prandial, measurements before and during a stressful procedure, and so forth). Each cell in the layout may have a single observation or a common number n for balance.

The linear model we shall use for the two-way mixed-effects layout is

$$x_{ijh} = \mu + u_i + \tau_j + w_{ij} + e_{ijh}$$
$$i = 1, \ldots, r; \quad j = 1, \ldots, c; \quad h = 1, \ldots, n \qquad (1)$$

μ is the general parameter common to all observations, u_i is a random variable reflecting the effect of the ith row treatment, τ_j is a parameter measuring the contribution of the jth fixed column treatment, w_{ij} is the random effect of the interaction of the ith row and jth column conditions, and e_{ijh} is the usual sampling or experimental error disturbance accounting for the variation within the ijth cell. The random effects u_i, w_{ij}, and e_{ijh} are independently and normally distributed with zero means and respective variances

$$\begin{aligned} \text{Var}(u_i) &= \sigma_R^2 \\ \text{Var}(w_{ij}) &= \sigma_I^2 \\ \text{Var}(e_{ijh}) &= \sigma^2 \end{aligned} \qquad (2)$$

for all combinations of their subscripts. We have selected this model from the several ones proposed for the mixed layout because of the simplicity of its assumptions on the variances and covariances of the random components. Other models lead to a different test statistic for the random row effect. We shall mention those models briefly later in this section and also discuss a multivariate approach to tests for the fixed effects that purchases generality at the expense of power and sensitivity.

For the fixed column treatments we wish to test the hypothesis

$$H_0: \tau_1 = \cdots = \tau_c \qquad (3)$$

of equal effects, as opposed to the usual alternative that some contrast of the τ_j does not equal zero. To make the test we prepare the usual two-way analysis of variance table and form an F-ratio appropriate to the expected value of the column mean square. Those expected values are shown in Table 9-37. The

TABLE 9-37 Analysis of Variance for the Mixed Model

Source	Sum of squares	d.f.	Mean square	Expected mean square	F
Rows (random)	SSR	$r-1$	$\text{MSR} = \dfrac{\text{SSR}}{r-1}$	$\sigma^2 + n\sigma_I^2 + nc\sigma_R^2$	$\dfrac{\text{MSR}}{\text{MSI}}$
Columns (fixed)	SSC	$c-1$	$\text{MSC} = \dfrac{\text{SSC}}{c-1}$	$\sigma^2 + n\sigma_I^2 + \dfrac{rn}{c-1}\sum_{j=1}^{c}(\tau_j - \bar{\tau})^2$	$\dfrac{\text{MSC}}{\text{MSI}}$
Interaction (random)	SSI	$(r-1)(c-1)$	$\text{MSI} = \dfrac{\text{SSI}}{(r-1)(c-1)}$	$\sigma^2 + n\sigma_I^2$	$\dfrac{\text{MSI}}{\hat{\sigma}^2}$
Within cells	SSE	$rc(n-1)$	$\hat{\sigma}^2 = \dfrac{\text{SSE}}{rc(n-1)}$	σ^2	—
Total	$\sum\sum\sum(x_{ijh} - \bar{x}_{...})^2$	$rcn-1$	—	—	—

sums of squares are calculated by the usual formulas (9) and (10) of Section 9-11 for a balanced layout. Because the expected columns mean square contains the term $n\sigma_I^2$ the interaction mean square is the correct denominator for the F-ratio. We reject the hypothesis (3) if

$$F = \frac{\text{MSC}}{\text{MSI}} > F_{\alpha; c-1, (r-1)(c-1)} \tag{4}$$

for a test at the α level. Similarly, we should reject the hypothesis

$$H_0\colon \sigma_R^2 = 0 \tag{5}$$

of no variation among the random row effects if

$$F = \frac{\text{MSR}}{\text{MSI}} > F_{\alpha; r-1, (r-1)(c-1)} \tag{6}$$

The test of the hypothesis of no interaction, or

$$H_0\colon \sigma_I^2 = 0 \tag{7}$$

follows as in the fixed and random layouts by referring $F = \text{MSI}/\hat{\sigma}^2$ to a critical value of the F-distribution with $(r-1)(c-1)$ and $rc(n-1)$ degrees of freedom.

The important special case of the mixed model with $c = 2$ columns and a single observation in each cell is equivalent to the well-known paired t-test for correlated observations. The data have this appearance:

	Treatment		
Subject	1	2	Difference
1	x_{11}	x_{12}	$d_1 = x_{11} - x_{12}$
$\cdot$	$\cdot$	$\cdot$	$\cdot$
$\cdot$	$\cdot$	$\cdot$	$\cdot$
$\cdot$	$\cdot$	$\cdot$	$\cdot$
r	x_{r1}	x_{r2}	$d_r = x_{r1} - x_{r2}$
Mean	$\bar{x}_{.1}$	$\bar{x}_{.2}$	$\bar{d} = \bar{x}_{.1} - \bar{x}_{.2}$

Since $n = 1$ we have dropped the third subscript. Under the assumption that the pairs of observations were drawn at random from a bivariate normal distribution we wish to test the hypothesis

$$H_0\colon \mu_1 = \mu_2 \tag{8}$$

of equal means for treatments 1 and 2, or its equivalent,

$$H_0\colon E(d_i) = 0 \tag{9}$$

of a zero mean for the pair differences. The paired t-statistic for the test is

$$t = \frac{\bar{d}\sqrt{r}}{s_d} \tag{10}$$

As we showed in Example 9-21, the variance

$$s_d^2 = [1/(r-1)] \sum_{i=1}^{r} (d_i - \bar{d})^2 \tag{11}$$

of the differences is equal to $\frac{1}{2}$ MSI, while for $c = 2$ and $n = 1$,

$$\begin{aligned} \text{MSC} &= r(\bar{x}_{.1} - \bar{x}_{..})^2 + r(\bar{x}_{.2} - \bar{x}_{..})^2 \\ &= \frac{1}{2} r(\bar{x}_{.1} - \bar{x}_{.2})^2 \\ &= \frac{1}{2} r\bar{d}^2 \end{aligned} \tag{12}$$

Then the mixed-model F-ratio for equal column effects is

$$\begin{aligned} F &= \frac{\text{MSC}}{\text{MSI}} \\ &= t^2 \end{aligned} \tag{13}$$

The analysis of variance is a generalization of the paired t approach to several repeated treatments on the same sampling units.

Example 9-22

As a simple illustration of the mixed model let us use the analysis of variance of the mortar crushing strengths in Example 9-2. Recall that the layout consisted of three row treatments for the laboratories T, Y, Z conducting the strength tests, the column effects dry and wet consistency of the mix, and $N = 6$ mortar cubes in each cell. Because the mix consistency is defined as a known percentage of water it would seem proper to treat the column conditions as fixed effects. In Example 9-2 we considered the three laboratories fixed because our inferences were to be made only about them. However, if T, Y, and Z are a random sample from a large population of testing laboratories (or even from the finite set {R, T, V, W, Y, Z}) the mixed model would be more appropriate. The two sets of F-ratios are shown in the table.

Source	Fixed	Mixed or random
Labs	8.52	0.54
Consistency	4,109	259
Interaction	15.85	15.85

The significance of the main effects has clearly decreased with the use of the mixed model. Because the second degrees of freedom of the F- statistics are only two the mixed-model tests based on the interaction mean square are less sensitive than the Model I tests using within-cell variance as a measure of statistical variation. Nevertheless, the consistency F-ratio exceeds the critical value $F_{0.01;\,1,2} = 98.5$, and we should conclude, as before, that the amount of water has a great effect on the mortar strength.

Multiple comparisons of the fixed effects. If the mixed model assumptions hold the column means will have the common variance

$$\text{Var}(\bar{x}_{.j.}) = \frac{n\sigma_R^2 + n\sigma_I^2 + \sigma^2}{nr} \qquad j = 1, \ldots, c \tag{14}$$

and the same covariance

$$\text{Cov}(\bar{x}_{.j.}, \bar{x}_{.k.}) = \frac{\sigma_R^2}{r} \qquad j \neq k \tag{15}$$

for all pairs of columns. This special covariance pattern permits the analysis of variance in Table 9-37 to hold just as in the fixed-effects model with independent column means. It is also a sufficient condition for the validity of the Scheffé and Tukey methods of multiple comparisons (Scheffé, 1959; Hochberg, 1974). If the hypothesis of no row and column interaction is tenable, we can construct Scheffé simultaneous confidence intervals for the general contrast

$$\psi = \sum_{j=1}^{c} b_j \tau_j \tag{16}$$

of the column effects by the formula

$$\hat{\psi} - \sqrt{[(c-1)/nr](\sum b_j^2)\text{MSI}F_{\alpha;c-1,(r-1)(c-1)}} \\ \leq \psi \leq \hat{\psi} + \sqrt{[(c-1)/nr](\sum b_j^2)\text{MSI}F_{\alpha;c-1,(r-1)(c-1)}} \tag{17}$$

in which

$$\hat{\psi} = \sum_{j=1}^{c} b_j \bar{x}_{.j.} \tag{18}$$

is the estimator of ψ. MSI is the interaction mean square from Table 9-37. If the interval includes zero, we should conclude that the hypothesis

$$H_0\colon \sum_{j=1}^{c} b_j \tau_j = 0 \tag{19}$$

is acceptable in the multiple testing sense. Alternatively, we might refer

$$t = \frac{\hat{\psi}\sqrt{nr}}{\sqrt{\text{MSI}(\sum b_j^2)}} \tag{20}$$

to the Scheffé critical value

$$\sqrt{(c-1)F_{\alpha;c-1,(r-1)(c-1)}} \tag{21}$$

If $|t|$ exceeds the critical value, the hypothesis (19) should be rejected at the α level in favor of a nonzero value of the contrast ψ.

The Tukey simultaneous confidence intervals for the paired column effect differences have the general expression

$$\bar{x}_{.j.} - \bar{x}_{.k.} - q_{\alpha;c,(r-1)(c-1)}\sqrt{\frac{\text{MSI}}{nr}} \\ \leq \tau_j - \tau_k \leq \bar{x}_{.j.} - \bar{x}_{.k.} + q_{\alpha;c,(r-1)(c-1)}\sqrt{\frac{\text{MSI}}{nr}} \tag{22}$$

for all j and k. We may test the paired hypotheses

$$H_0\colon \tau_j = \tau_k \tag{23}$$

by ascertaining whether the interval (22) contains zero or directly by comparing the statistic

$$q = \frac{|\bar{x}_{.j.} - \bar{x}_{.k.}|\sqrt{nr}}{\sqrt{\text{MSI}}} \tag{24}$$

to a critical value $q_{\alpha; c, (c-1)(r-1)}$ of the Studentized range distribution from Table 7, Appendix A. Bonferroni tests or confidence intervals for a family of K comparisons can be obtained by replacing the critical values in (21) or (22) by the two-sided t-value $t_{\alpha/2K; (r-1)(c-1)}$. Because the row effects are random variables multiple comparisons among them would not be appropriate.

Example 9-23

Mueller (1962) reported concentrations of plasma free fatty acid (FFA) and blood glucose just before and 15, 30, and 45 minutes after intravenous administration of 0.1 unit of insulin to samples of schizophrenic and normal control subjects. The FFA values (mEq/liter multiplied by 100 to avoid decimals) are shown in Table 9-38. Although one of the purposes of the experiment was the comparison of FFA changes of schizophrenics and normals, we shall only analyze the FFA levels of the normal subjects. The data might be represented by the mixed model with a single observation in each cell, for the column treatments consist of the four distinct observation times, and the subjects are a random sample from a population of healthy male volunteers. Because each subject contributes four repeated measurements the observations in a row are not independent, and the usual one-way fixed effects model or regression analysis with multiple observations at equally spaced points would not be appropriate.

TABLE 9-38 FFA Values for Normal Subjects After Insulin Injection (mEq/liter × 100)

	Minutes after insulin				
Subject	0	15	30	45	Mean
1	30	45	27	30	33
2	41	30	38	30	34.75
3	30	22	19	29	25
4	22	21	26	17	21.5
5	37	34	27	27	31.25
6	34	23	33	24	28.5
7	32	29	17	9	21.75
8	27	21	29	17	23.5
9	32	26	20	20	24.5
10	33	29	24	33	29.75
Mean	31.8	28.0	26.0	23.6	27.35
$\sum (x_{ij} - \bar{x}_{.j})^2$	243.6	494	374	524.4	—

SOURCE: Copyright 1962, American Medical Association.

This analysis of variance was computed from the FFA values:

Source	Sum of squares	d.f.	Mean square	F
Subjects	812.1	9	90.2333	2.96
Times	361.1	3	120.3667	3.94
Interaction	823.9	27	30.5148	—
Total	1997.1	39	—	—

Both F-statistics exceed their $\alpha = 0.05$ critical values: We should reject the hypotheses of no variation in the random row effects and no change in FFA over time. Because we have only one observation in each cell we cannot test for a subject $\times$ time interaction. The nature of the experiment and variable would seem to preclude multiple observations, unless the blood sample had been partitioned to permit a measure of experimental error variance from each cell of the layout.

Let us use the Tukey multiple comparison procedure to determine which of the four FFA means are different. The upper 5% critical value of the studentized range is $q_{0.05;4,27} = 3.87$ by linear interpolation. Then

$$q_{0.05;4,27}\sqrt{\frac{\text{MSI}}{r}} = 6.76$$

Only the 0- to 45-minute mean difference exceeds that value. The significant analysis of variance for times appears to be due to that mean difference, as well as other contrasts among the four means. The differences of successive means are not significant in the multiple testing sense, at least for the present small sample and the amount of variation among its FFA values at each time.

Finally, we might examine some of the random components of the mixed model. The sums of squares about each column mean in the last line of Table 9-38 cast some doubt on the assumption of common variances (14) for each column mean. The estimate of the row variance component can be computed by equating mean squares to their expected values given in Table 9-37:

$$\begin{aligned}\hat{\sigma}_R^2 &= \frac{\text{MSR} - \text{MSI}}{c} \\ &= (90.2333 - 30.5148)/4 \\ &= 14.93\end{aligned}$$

The variance of the subjects random variable is one-half that of the "disturbance," or subjects $\times$ times interaction, variance.

An alternative model: The Hotelling T^2-test. Under our mixed model not only are the column mean variances the same, but all of their covariances are equal to σ_R^2/r. That covariance pattern follows directly from the random effects in the model and is a necessary condition for the fixed-effects statistic to have the F-distribution. However, the sample covariance matrix of the observations may not resemble that pattern, and then the validity of the analysis of variance will be in doubt. Another fixed-effects test is available that makes fewer assumptions on the variances and covariances but suffers less power in return for its generality. The test statistic is known as Hotelling's T^2 and is a generalization of the paired t-ratio to vectors of means. T^2 and its distribution were found by Hotelling (1931); more recent and concise derivations have been given by Anderson (1958) and Rao (1973). The application to the mixed model has been described by Scheffé (1959) and Morrison (1976).

For the T^2-test we shall deal directly with the $r \times c$ table of cell means:

$$\bar{\mathbf{X}} = \begin{bmatrix} \bar{x}_{11.} & \cdots & \bar{x}_{1c.} \\ \cdot & & \cdot \\ \cdot & & \cdot \\ \cdot & & \cdot \\ \bar{x}_{r1.} & \cdots & \bar{x}_{rc.} \end{bmatrix} = \begin{bmatrix} \bar{\mathbf{x}}'_{1.} \\ \cdot \\ \cdot \\ \cdot \\ \bar{\mathbf{x}}'_{r.} \end{bmatrix} \tag{25}$$

We shall assume that the $c \times 1$ mean vectors $\bar{\mathbf{x}}_{1.}, \ldots, \bar{\mathbf{x}}_{r.}$ were independently drawn from a multivariate normal population with mean vector $\boldsymbol{\mu}$ and a positive definite covariance matrix $\boldsymbol{\Sigma}$. We wish to test the hypothesis

$$H_0\colon \mu_1 = \cdots = \mu_c \tag{26}$$

or equivalently,

$$H_0\colon \mu_1 - \mu_2 = 0, \ldots, \mu_{c-1} - \mu_c = 0 \tag{27}$$

We may represent that latter hypothesis in matrix form as

$$H_0\colon \mathbf{C}\boldsymbol{\mu} = \mathbf{0} \tag{28}$$

where $\mathbf{0}$ is the null vector of $c - 1$ zeros and $\mathbf{C}$ is the $(c - 1) \times c$ matrix

$$\mathbf{C} = \begin{bmatrix} 1 & -1 & 0 & \cdots & 0 & 0 \\ 0 & 1 & -1 & \cdots & 0 & 0 \\ \cdot & \cdot & \cdot & & \cdot & \cdot \\ \cdot & \cdot & \cdot & & \cdot & \cdot \\ \cdot & \cdot & \cdot & & \cdot & \cdot \\ 0 & 0 & 0 & \cdots & 1 & -1 \end{bmatrix} \tag{29}$$

that forms the successive differences of the means. Now the estimator of the vector of successive mean differences is

$$\begin{bmatrix} \bar{x}_{.1.} - \bar{x}_{.2.} \\ \cdot \\ \cdot \\ \cdot \\ \bar{x}_{.c-1.} - \bar{x}_{.c.} \end{bmatrix} = \mathbf{C}\bar{\mathbf{x}} \tag{30}$$

in which $\bar{\mathbf{x}}' = [\bar{x}_{.1.}, \ldots, \bar{x}_{.c.}]$ is the vector of column means. From the properties of the original multivariate normal distribution of the independent row mean vectors it is possible to show that $\mathbf{C}\bar{\mathbf{x}}$ is also multinormally distributed. The covariance matrix of the distribution is $(1/r)\mathbf{C}\boldsymbol{\Sigma}\mathbf{C}'$, and when the hypothesis (28) of equal column means is true its mean vector will have zeros in each position. Now the unbiased estimator of the original $c \times c$ covariance matrix $\boldsymbol{\Sigma}$ is the sample covariance matrix

$$\mathbf{S} = \frac{1}{r-1} \sum_{i=1}^{r} (\bar{\mathbf{x}}_{i.} - \bar{\mathbf{x}}_{..})(\bar{\mathbf{x}}_{i.} - \bar{\mathbf{x}}_{..})' \tag{31}$$

and the properties of linear transformations of multinormal random vectors tell us that the estimator of the covariance matrix of the successive differences of the column means is

$$\frac{1}{r}\mathbf{CSC}' \tag{32}$$

To test the hypothesis (26) of equal column effects we compute the value of the quadratic form

$$T^2 = r\bar{\mathbf{x}}'\mathbf{C}'(\mathbf{CSC}')^{-1}\mathbf{C}\bar{\mathbf{x}} \tag{33}$$

and refer its scaled value

$$F = \frac{r-c+1}{(c-1)(r-1)} T^2 \tag{34}$$

to the upper critical values of the F-distribution with $c - 1$ and $r - c + 1$ degrees of freedom. If

$$F > F_{\alpha; c-1, r-c+1} \tag{35}$$

we should reject the hypothesis of equal column effects and conclude that there exists some nonzero contrast of those parameters.

When the alternative hypothesis of a general column mean vector $\boldsymbol{\mu}$ is true, the statistic (34) has the noncentral F-distribution with the same degrees of freedom, and noncentrality parameter

$$\lambda = r\boldsymbol{\mu}'\mathbf{C}'(\mathbf{C}\boldsymbol{\Sigma}\mathbf{C}')^{-1}\mathbf{C}\boldsymbol{\mu} \tag{36}$$

The power of the T^2-test can be obtained by entering the Pearson–Hartley charts with the parameter

$$\phi = \sqrt{\frac{\lambda}{c}} \tag{37}$$

and the values of α and the two degrees of freedom parameters.

Roy and Bose (1953) found the family of simultaneous confidence intervals for contrasts of the column means. The general $100(1 - \alpha)\%$ interval for the contrast

$$\mathbf{b}'\boldsymbol{\mu} = \sum_{j=1}^{c} b_j \mu_j \tag{38}$$

is

$$\mathbf{b}'\bar{\mathbf{x}} - \sqrt{\frac{1}{r}\mathbf{b}'\mathbf{S}\mathbf{b}\frac{(r-1)(c-1)}{r-c+1}F_{\alpha; c-1, r-c+1}} \leq \mathbf{b}'\boldsymbol{\mu}$$

$$< \mathbf{b}'\bar{\mathbf{x}} + \sqrt{\frac{1}{r}\mathbf{b}'\mathbf{S}\mathbf{b}\frac{(r-1)(c-1)}{r-c+1}F_{\alpha; c-1, r-c+1}} \tag{39}$$

Note that the estimated variance $\mathbf{b}'\mathbf{S}\mathbf{b}/r$ of the contrast estimator $\mathbf{b}'\bar{\mathbf{x}}$ has been expressed in terms of the $c \times c$ covariance matrix $\mathbf{S}$ of the cell means $\bar{x}_{ij.}$. If the contrast merely compares two column means, for example, the ith and jth, the interval is

$$\bar{x}_{.i.} - \bar{x}_{.j.} - \sqrt{\frac{1}{r}(s_{ii} + s_{jj} - 2s_{ij})\frac{(r-1)(c-1)}{r-c+1}F_{\alpha;c-1,r-c+1}} \leq \mu_i - \mu_j$$

$$\leq \bar{x}_{.i.} - \bar{x}_{.j.} + \sqrt{\frac{1}{r}(s_{ii} + s_{jj} - 2s_{ij})\frac{(r-1)(c-1)}{r-c+1}F_{\alpha;c-1,r-c+1}} \qquad (40)$$

We should reject the individual hypothesis

$$H_0\colon\ \mu_i = \mu_j$$

if

$$t = \frac{|\bar{x}_{.i.} - \bar{x}_{.j.}|\sqrt{r}}{\sqrt{s_{ii} + s_{jj} - 2s_{ij}}} \qquad (41)$$

exceeds the multiple comparisons critical value

$$\sqrt{\frac{(r-1)(c-1)}{r-c+1}F_{\alpha;c-1,r-c+1}} \qquad (42)$$

Through the confidence intervals we can locate the source of a significant T^2-statistic. If T^2 exceeds its $100\alpha\%$ upper critical value, some contrast must have a simultaneous confidence interval that does not contain zero. Conversely, no interval devoid of zero can be found if T^2 is not significant at the same level α.

Example 9-24

Let us apply the T^2 method to the free fatty acid data of Example 9-23. The sample covariance matrix is

$$\mathbf{S} = \begin{bmatrix} 27.07 & 13.56 & 13.56 & 17.80 \\ 13.56 & 54.89 & 3.78 & 22.44 \\ 13.56 & 3.78 & 41.56 & 18.44 \\ 17.80 & 22.44 & 18.44 & 58.27 \end{bmatrix}$$

and the transformation matrix we shall use is

$$\mathbf{C} = \begin{bmatrix} 1 & -1 & 0 & 0 \\ 0 & 1 & -1 & 0 \\ 0 & 0 & 1 & -1 \end{bmatrix}$$

Then the mean vector of successive differences is

$$\mathbf{C}\bar{\mathbf{x}} = \begin{bmatrix} 3.8 \\ 2 \\ 2.4 \end{bmatrix}$$

The covariance matrix of the successive differences of the $N = 10$ subjects is

$$\mathbf{CSC}' = \begin{bmatrix} 58.84 & -51.11 & 14.42 \\ -51.11 & 88.89 & -41.78 \\ 14.42 & -41.78 & 62.93 \end{bmatrix}$$

and its inverse is

$$(\mathbf{CSC}')^{-1} = \begin{bmatrix} 0.04287 & 0.02912 & 0.00950 \\ 0.02912 & 0.03613 & 0.01731 \\ 0.00950 & 0.01731 & 0.02520 \end{bmatrix}$$

Hotelling's T^2 is equal to 16.91, and its F-statistic is

$$F = \frac{10 - 4 + 1}{3(10 - 1)}(16.91) = 4.38$$

Since $F_{0.05;3,7} = 4.35$, we should reject the hypothesis of a constant mean at the four times, but only barely.

Now we shall use the simultaneous test method of expressions (40)–(42) to determine the source of the significance of the T^2-statistic. For the three contrasts of means at successive times the standard deviations

$$\sqrt{s_{ii} + s_{jj} - 2s_{ij}}$$

are given by the square roots of the diagonal elements of **CSC′**. The critical value (42) is equal to 4.096 for $\alpha = 0.05$. Some of the paired-comparison statistics had these values:

i	j	$\bar{x}_{.i.} - \bar{x}_{.j.}$	t_{ij}
1	2	3.8	1.61
1	3	5.8	2.84
1	4	8.2	3.67
2	4	4.4	1.68

None of the t_{ij} exceeds the Roy–Bose critical value, although that for the comparison of the means at zero and 45 minutes is close. We can see from inspection that the other two paired comparisons would not be significant. However, the relationship of the simultaneous tests to the T^2-statistic states that a data set with a T^2 greater than its $100\alpha\%$ critical value must have *some* contrast of the column effects whose $100(1-\alpha)\%$ simultaneous confidence intervals do *not* include zero, or equivalently, whose statistic t exceeds the critical value (42). It is possible to show [see, for example, Morrison (1976)] that the coefficient vector

$$\mathbf{a} = (\mathbf{CSC}')^{-1}\mathbf{C}\bar{\mathbf{x}}$$

will give the linear compound of the successive mean differences with shortest confidence interval or greatest simultaneous t-statistic. In the present case

$$\mathbf{a}' = [0.24, \quad 0.22, \quad 0.13]$$

The contrast

$$\begin{aligned} \mathbf{a}'\bar{\mathbf{x}} &= 0.24(\bar{x}_{.1.} - \bar{x}_{.2.}) + 0.22(\bar{x}_{.2.} - \bar{x}_{.3.}) + 0.13(\bar{x}_{.3.} - \bar{x}_{.4.}) \\ &= 0.24\bar{x}_{.1.} - 0.02\bar{x}_{.2.} - 0.09\bar{x}_{.3.} - 0.13\bar{x}_{.4.} \end{aligned}$$

has a t-statistic equal to 4.11; this is just greater than the critical value (42). The hypothesis of equal FFA means at the four times is not supported by the data, although the substantive meaning of the contrast $\mathbf{a}'\bar{\mathbf{x}}$ is not clear. The significant difference among the means seems to lie in the comparison of time 0 with the 30- and 45-minute means.

The T-statistic for the contrast $\mathbf{a}'\bar{\mathbf{x}}$ is equal to the square root of Hotelling's T^2 of expression (33). That relation is a consequence of the development of the T^2-test by the *union–intersection principle* (Roy, 1953). Verification of the equality will be left as an exercise.

Some further aspects of mixed models. The simple model we have chosen for the mixed effects layout was first given by Mood (1950) and has been discussed by Scheffé (1956) and Searle (1971). Other mixed models have been proposed by Scheffé (1959), Graybill (1961), and other workers. Those models were obtained by imposing constraints on the random effects, and thereby changing their covariance structure. The assumptions and covariance matrices have been compared and discussed in detail by Hocking (1973). The other Scheffé model leads to an expected mean square for the random effects treatments which implies that the F-ratio for the equality of those effects should be

$$F = \frac{\text{MSR}}{\hat{\sigma}^2} \tag{43}$$

The Graybill model gives the same statistic $F = \text{MSR}/\text{MSI}$ as in Table 9-36. The fixed-effects F-ratios are unchanged. The choice of model should be made on the basis of the sampling scheme used to draw the random effects treatments, the nature of the repeated observations within each cell, and the extent to which the covariance structure of the observations (or at least the fixed effect means) agrees with the covariance matrix pattern implied by the model. Hocking's comparisons are helpful in that choice.

Necessary and sufficient conditions on the covariance structure of repeated measurements for the validity of the mixed-model analysis of variance have been found by Huynh and Feldt (1970); these have been described by Morrison (1976). Mendoza et al. (1976) have extended the conditions to three-way layouts in which two of the factors consist of repeated measurements.

Mixed-model and repeated measurements designs have been extended to several dimensions. Some of these have been treated by Winer (1971). The case of repeated measurements on independent sampling units arranged in several distinct groups, or profile analysis, has been discussed by Morrison (1976).

9-13 REFERENCES

ANDERSON, T. W. (1958): *An Introduction to Multivariate Statistical Analysis*, John Wiley & Sons, Inc., New York.

APPELBAUM, M. I., and E. M. CRAMER (1974): Some problems in the non-orthogonal analysis of variance, *Psychological Bulletin*, vol. 81, pp. 335–343.

BOWKER, A. H., and G. J. LIEBERMAN (1972): *Engineering Statistics*, 2nd ed., Prentice-Hall, Inc., Englewood Cliffs, N.J.

CHEW, V. (1977): *Comparisons Among Treatment Means in an Analysis of Variance*, Agricultural Research Service, U.S. Department of Agriculture, Beltsville, Md.

COCHRAN, W. G., and G. M. COX (1957): *Experimental Designs*, 2nd ed., John Wiley & Sons, Inc., New York.

Cox, D. R. (1958): *Planning of Experiments*, John Wiley & Sons, Inc., New York.

Davies, O. L. (Ed.) (1963): *Design and Analysis of Industrial Experiments*, 2nd ed., Hafner Publishing Company, New York.

Finney, D. J. (1945): The fractional replication of factorial experiments, *Annals of Eugenics*, vol. 12, pp. 291–301.

Gosslee, D. G., and H. L. Lucas (1965): Analysis of disproportionate data when interaction is present, *Biometrics*, vol. 21, pp. 115–133.

Graybill, F. A. (1961): *An Introduction to Linear Statistical Models*, vol. 1, McGraw-Hill Book Company, New York.

Heiberger, R. M., and L. L. Laster (1977): Maximizing power in the unbalanced higher-way ANOVA, *Proceedings of the Statistical Computing Section*, American Statistical Association, pp. 234–239.

Heiberger, R. M., and L. L. Laster (1980): Maximum power tests (Letter to the Editor), *The American Statistician*, vol. 34, p. 62.

Hochberg, Y. (1974): The distribution of the range in general balanced models, *The American Statistician*, vol. 28, pp. 137–138.

Hocking, R. R. (1973): A discussion of the two-way mixed model, *The American Statistician*, vol. 27, pp. 148–152.

Hotelling, H. (1931): The generalization of Student's ratio, *Annals of Mathematical Statistics*, vol. 2, pp. 360–378.

Huynh, H., and L. S. Feldt (1970): Conditions under which mean square ratios in repeated measurements designs have exact F-distributions, *Journal of the American Statistical Association*, vol. 65, pp. 1582–1589.

John, P. W. M. (1971): *Statistical Design and Analysis of Experiments*, Macmillan Publishing Co., Inc., New York.

Kempthorne, O. (1952): *The Design and Analysis of Experiments*, John Wiley & Sons, Inc., New York.

Kimball, A. W. (1951): On dependent tests of significance in the analysis of variance, *Annals of Mathematical Statistics*, vol. 22, pp. 600–602.

Kutner, M. H. (1974): Hypothesis testing in linear models (Eisenhart Model I), *The American Statistician*, vol. 28, pp. 98–100.

Lansdell, H. (1968): Effect of temporal lobe ablations on two lateralized deficits, *Physiology and Behavior*, vol. 3, pp. 271–273.

Mangold, R., L. Sokoloff, E. Conner, J. Kleinerman, P.-O. G. Therman, and S. S. Kety (1955): The effects of sleep and lack of sleep on the cerebral circulation and metabolism of normal young men, *Journal of Clinical Investigation*, vol. 34, pp. 1092–1100.

Mendoza, J. L., L. E. Toothaker, and B. R. Crain (1976): Necessary and sufficient conditions for F ratios in the $L \times J \times K$ factorial design with two repeated factors, *Journal of the American Statistical Association*, vol. 71, pp. 992–993.

Mood, A. M. (1950): *Introduction to the Theory of Statistics*, McGraw-Hill Book Company, New York.

Morrison, D. F. (1976): *Multivariate Statistical Methods*, 2nd ed., McGraw-Hill Book Company, New York.

Mueller, P. S. (1962): Plasma free fatty acid response to insulin in schizophrenia, *Archives of General Psychiatry*, vol. 7, pp. 140–146.

PEARSON, K. (1968): *Tables of the Incomplete Beta-Function*, Cambridge University Press, Cambridge, England.

RAO, C. R. (1973): *Linear Statistical Inference and Its Applications*, 2nd ed., John Wiley & Sons, Inc., New York.

ROY, S. N. (1953): On a heuristic method of test construction and its use in multivariate analysis, *Annals of Mathematical Statistics*, vol. 24, pp. 220–238.

ROY, S. N., and R. C. BOSE (1953): Simultaneous confidence interval estimation, *Annals of Mathematical Statistics*, vol. 24, pp. 513–536.

SAMUEL, W., D. SOTO, M. PARKS, P. NGISSAH, and B. JONES (1976): Motivation, race, social class, and IQ, *Journal of Educational Psychology*, vol. 68, pp. 273–285.

SATTERTHWAITE, F. E. (1946): An approximate distribution of estimates of variance components, *Biometrics Bulletin*, vol. 2, pp. 110–114.

SCHEFFÉ, H. (1956): Alternative models for the analysis of variance, *Annals of Mathematical Statistics*, vol. 27, pp. 251–271.

SCHEFFÉ, H. (1959): *The Analysis of Variance*, John Wiley & Sons, Inc., New York.

SEARLE, S. R. (1971): *Linear Models*, John Wiley & Sons, Inc., New York.

SPEED, F. M., and R. R. Hocking (1976): The use of the $R(\ )$-notation with unbalanced data, *The American Statistician*, vol. 30, pp. 30–33.

SPEED, F. M., and C. J. MONLEZUN (1979): Exact F tests for the method of unweighted means in a 2^k experiment, *The American Statistician*, vol. 33, pp. 15–18.

STONE, G. (1979): Best buys in small batteries, *Popular Science*, vol. 125, no. 2, pp. 79–81.

TUKEY, J. W. (1949): One degree of freedom for non-additivity, *Biometrics*, vol. 5, pp. 232–242.

URBAN, W. D. (1976): *Statistical Analysis of Blood Lead Levels of Children Surveyed in Pittsburgh, Pennsylvania: Analytical Methodology and Summary Results*, NBSIR 76-1024, Institute of Applied Technology, National Bureau of Standards, Washington, D.C.

WILSDON, B. H. (1934): Discrimination by specification statistically considered and illustrated by the standard specification for Portland cement, *Supplement to the Journal of the Royal Statistical Society, Series B*, vol. 1, pp. 152–206.

WINER, B. J. (1971): *Statistical Principles in Experimental Design*, 2nd ed., McGraw-Hill Book Company, New York.

YATES, F. (1934): The analysis of multiple classifications with unequal numbers in the different classes, *Journal of the American Statistical Association*, vol. 29, pp. 51–66.

YATES, F. (1935): Complex experiments, *Journal of the Royal Statistical Society, Supplement*, vol. 2, pp. 181–247.

YATES, F. (1937): *The Design and Analysis of Factorial Experiments*, Imperial Bureau of Soil Science, Technical Communication No. 35, Harpenden, England.

9-14 EXERCISES

1. In a feeding experiment guppy siblings (*Poecilia reticulatia*) were grouped in blocks of six fish with the same initial weight and then assigned randomly to one of six feeding levels. These percentage increases in wet weight were recorded for the blocks and food levels:

Block	Feeding Level 1	2	3	4	5	6	Sum
1	11.25	11.27	9.85	8.94	6.05	3.29	50.65
2	9.81	8.28	7.02	6.36	3.64	2.12	37.23
Sum	21.06	19.55	16.87	15.30	9.69	5.41	87.88

SOURCE: Data reproduced with the kind permission of David Reznick.

(a) Do the relative increases in weight differ among the feeding levels?
(b) Use the Tukey, Scheffé, and Bonferroni multiple comparison methods to determine which feeding level means are significantly different.
(c) Test the hypothesis of an additive model with the Tukey single-degree-of-freedom-for-additivity procedure.

2. The data for Example 9-1 originally included a ninth graduate program with enrollments of 3, 10, and 1 students for years 1, 2, and 3, respectively. The program was omitted from the example because of the extreme fluctuations in its enrollments. Repeat the two-way analysis of variance and simultaneous testing procedures of the earlier examples with the enrollment data for the set of nine programs.

3. Mean offspring weights (milligrams) of $n = 6$ female mosquito fish (*Gambusia affinis*) from male and female parents drawn from populations in North Carolina and Illinois had these values:

	Illinois female: Illinois male	Illinois female: North Carolina male	North Carolina female: Illinois male	North Carolina female: North Carolina male
	1.43	1.25	1.52	1.94
	1.15	1.17	1.82	1.41
	1.10	1.29	1.70	1.59
	1.11	1.09	2.21	1.38
	0.96	1.07	1.58	1.35
	1.01	1.01	1.50	1.65
Sum	6.76	6.88	10.33	9.32
Within-cell sum of squares	0.1337	0.0609	0.3585	0.2521

SOURCE: Data reproduced with the kind permission of David Reznick.

Carry out a two-way fixed-effects analysis of variance for male, female, and interaction effects on offspring weight.

4. This example is grist for many variations on the analysis of variance. In a screening program for vaccines two compounds A and B were known to be effective. The rows way of classification consisted of a control treatment and five dosage levels of each compound. The column conditions were the four investigators who conducted independent and identical replications of the experiment. The measure of the effect of the vaccine was the square root of the number of tail lesions in rats. Although each row and column treatment combination was applied to six rats, only the sums of the square roots of the six counts were used for the analysis of variance. The sums are shown in the table:

		Investigator				
Treatment	Dosage	1	2	3	4	Total
Control	—	19.1	17.0	15.3	16.2	67.6
Compound A	100	13.7	13.3	14.4	11.5	52.9
	50	11.4	14.8	16.3	11.1	53.6
	25	14.1	13.8	16.6	11.5	56.0
	12.5	18.1	11.9	15.5	15.0	60.5
	6.25	15.7	16.4	18.5	12.9	63.5
Compound B	100	15.6	11.8	10.6	9.6	47.6
	50	13.1	12.3	10.8	11.5	48.7
	25	12.1	13.7	10.7	15.9	52.4
	12.5	14.2	9.3	15.8	14.7	54.0
	6.25	15.6	12.2	12.4	14.1	54.3
Total		162.7	146.5	156.9	144.0	611.1

SOURCE: Data kindly provided by S. Michael Free and reproduced with the permission of the SmithKline Beckman Corporation.

(a) First treat the data as an 11×4 fixed-effects layout and make the usual analysis of variance.

(b) Use the Scheffé or Bonferroni multiple comparison methods to determine which investigators or row means are different.

(c) Now we should address the dosage levels. Drop the control row and treat the layout as a $2 \times 5 \times 4$ fixed-effects layout.

(d) Next we should consider regression models for the observations as functions of dosage. Begin with the contrast

$$\mathbf{c}' = \left(\frac{2}{\sqrt{10}}, \frac{1}{\sqrt{10}}, 0, \frac{-1}{\sqrt{10}}, \frac{2}{\sqrt{10}}\right)$$

giving the values of the linear orthogonal polynomial for five equally spaced points, and calculate the sums of squares due to that linear function for compound A, compound B, and the totals of the two compounds.

(e) Use the compound A and compound B contrasts of (d) to test the hypothesis that the linear regression coefficients are the same for each compound. Use the investigators $\times$ groups mean square of (a) or (b) for the t- or F-ratio.

(f) The five dosages are actually in the proportions (1, 2, 4, 8, 16). Convert those values to the linear orthogonal polynomial contrast by subtracting the mean

31/5 from each element, and normalizing:

$$\mathbf{d}' = [0.80, \quad 0.15, \quad -0.18, \quad -0.34, \quad -0.43]$$

Repeat the linear regressions of (d) with these actual dosages and compare the sums of squares due to the straight lines with those found in (d).

(g) Calculate the average response over the five dosage levels for each cell in the 2×4 table formed by compounds A and B and the four investigators, and test for row and column differences. This analysis is equivalent to an analysis of variance on the intercepts of the regression lines fitted to the data of each cell.

5. Verify that the parameters of the two-way linear model (1) of Section 9-2 are estimable when the constraint (2) is imposed on the row and column effects.

6. Find the variances and covariances of the estimators $\hat{\mu}$, $\hat{\alpha}_i$, $\hat{\beta}_j$ in (3), Section 9-2.

7. **(a)** Find the expected value of $\hat{\mu}_{ij.}$, (25), Section 9-3.

(b) Compute Var $(\hat{\mu}_{ij.})$ for the predicted cell mean in (a) when the two-way model does not contain interaction effects.

8. Consider the $k = 15$ nested treatments (sections) of Example 9-19 as a one-way analysis of variance. Use the Bonferroni and Scheffé multiple comparison methods to determine which sections are different. Are your conclusions consistent with those in the example for the five teaching fellows?

9. Treat the data in Example 8-5 as a two-way fixed-effects layout, in which each row corresponds to an experimental condition. Carry out multiple comparisons on the row or column effects as necessary and compare your findings with those of the one-way analysis in the example.

10. Use the Scheffé multiple comparison methods to determine whether the second-order differences of the table and desk lamp means of Example 9-3 are significantly different from zero.

11. Use the Tukey test for additivity to decide if an additive model appears to be appropriate for the graduate program enrollments of Example 9-1.

12. The data of Example 9-3 consist of two observations in each cell of the 3×4 table. The first observations were obtained under the 12 combinations, and then the experiment was repeated for the second set of observations. Treat the data as a $2 \times 3 \times 4$ layout, in which the first way of classification contains the "treatments" replication 1 and replication 2.

(a) Are the replication effects significantly different?

(b) Find the interactions among the three dimensions and attempt to test for the significance of the two-factor interactions.

(c) What effect does the introduction of a third dimension have on the hypothesis tests on the row and column main effects?

13. A 2^3-factorial design has been arranged in the following blocks of two units each:

1	2	3	4
(1)	c	a	b
abc	ab	bc	ac

Show that the three interactions AB, AC, and BC are confounded with the block differences.

14. Let us treat the course evaluations of Exercises 4 and 5, Section 3-9, and Example 8-2 as an unbalanced two-way layout. The following statistics were obtained for five years of ratings:

Course		Year				
		0	1	2	3	4
A	N_{1j}	3	3	4	6	4
	$\bar{x}_{1j.}$	1.33	1.67	1.5	1.67	1.25
	SSE_{1j}	0.67	0.67	1	3.33	0.75
B	N_{2j}	9	11	8	9	13
	$\bar{x}_{2j.}$	2.33	1.45	1.5	1.5	1.38
	SSE_{2j}	10	2.73	2	2	3.08

$$SSE_{ij} = \sum_{h=1}^{N_{ij}} (x_{ijh} - \bar{x}_{ij.})^2$$

(a) Carry out the fixed-effects analysis of variance for courses, years, and interaction effects.

(b) Are any contrasts representing linear or quadratic polynomial coefficients significantly different from zero?

15. In the Pittsburgh blood lead investigation (Urban, 1976) described in other examples this two-way table of blood lead means by sex and the father's educational status was obtained:

Education	Sex	
	Male	Female
Elementary		
N_{1j}	14	15
$\bar{x}_{1j.}$	24.43	26.80
SSE_{ij}	557.44	778.4
High school		
N_{2j}	56	45
$\bar{x}_{2j.}$	21.30	21.67
SSE_{2j}	1951.95	2038.08
College		
N_{3j}	28	25
$\bar{x}_{3j.}$	21.25	18.76
SSE_{3j}	845.37	432.48

As before SSE_{ij} is the within-cell sum of squares about $\bar{x}_{ij.}$. Complete the analysis of variance for the three effects and make any necessary multiple comparison tests.

16. Mangold et al. (1955) obtained these values of cerebral vascular resistance (cc/100 g per minute) in $r = 13$ normal subjects at two times from 45 to 125 minutes apart:

	Time	
Subject	1	2
1	1.4	1.5
2	1.9	1.9
3	1.5	1.6
4	2.0	2.5
5	1.6	2.1
6	0.9	1.6
7	1.6	1.3
8	2.3	2.5
9	1.1	1.4
10	1.1	1.4
11	1.3	1.5
12	1.0	1.1
13	1.3	1.4

Assume that the two-way Model II (random effects) is appropriate, and perform its analysis of variance. Estimate the variance components and the intrasubject correlation.

17. Mangold et al. (1955) also measured blood pH in the femoral artery and internal jugular bulb of $r = 13$ normal young men. Observations were taken at two different times for each location, and gave these values:

	Blood pH			
	Arterial time		Jugular time	
Subject	1	2	1	2
1	7.36	7.38	7.34	7.32
2	7.46	7.47	7.41	7.36
3	7.44	7.43	7.38	7.36
4	7.37	7.37	7.34	7.34
5	7.42	7.40	7.36	7.36
6	7.40	7.40	7.35	7.36
7	7.39	7.34	7.34	7.32
8	7.40	7.39	7.36	7.33
9	7.48	7.38	7.47	7.38
10	7.42	7.41	7.38	7.40
11	7.34	7.36	7.33	7.31
12	7.43	7.44	7.36	7.33
13	7.35	7.33	7.28	7.30

(a) Treat the data in each subject–location cell as multiple observations and use the mixed-model analysis of variance to test for subject, arterial–venous, and interaction effects.

(b) Analyze the data as a mixed-model layout with four repeated measurements for each subject. Use appropriate contrasts to test for time, location, and other differences in pH.

18. Treat the data of Example 9-21 as a one-way random effects layout with $k = 13$ treatments and $n = 2$ observations (times) per treatments. Estimate the subjects and error variance components, and the intraclass correlation coefficient. Compare the coefficient with $\hat{\rho}_R$ and the product moment correlation given in Example 9-21.

19. For the multivariate analysis of the mixed-model by the Hotelling T^2-statistic show that the linear compound vector

$$\mathbf{a} = (\mathbf{CSC}')^{-1}\mathbf{C}\bar{\mathbf{x}}$$

has simultaneous test statistic equal to a square root of T^2.

20. Two different versions of a comprehensive statistics examination covering topics A, B, and C were administered to the same $r = 11$ students at different times several weeks apart. The students will be considered as a random sample from a larger hypothetical population, and the topics as fixed treatments. The pairs of scores from the two versions of the exam will be taken as independent observations on the student–topic combinations. The following data were recorded:

	Topic					
	A Exam		B Exam		C Exam	
Student	1	2	1	2	1	2
1	4	13	12	20	15	13
2	15	30	22	14	16	15
3	20	11	14	21	24	16
4	15	20	13	9	22	16
5	14	15	19	13	24	16
6	9	10	15	19	16	13
7	17	11	16	10	25	9
8	22	21	15	23	17	21
9	5	2	14	15	18	22
10	12	13	16	14	15	16
11	12	13	11	21	23	15

(a) Treat the data as a mixed-model layout with $n = 2$ observations per cell and test the hypothesis of no differences among the three topic means by the mixed-model analysis of variance.

(b) If the topic means are significantly different use an appropriate multiple comparisons method to locate the source of the significance.

(c) Repeat the analyses of (a) and (b) by the Hotelling T^2-statistic and the Roy method of multiple comparisons.

(d) Compute the variances and covariances of the cell totals for topics A, B, and C. Does the assumption of a common variance and equal covariances for the analysis of variance appear to hold?

21. An experiment was conducted to compare the effects of standard and experimental drugs known to reduce blood pressure. Both drugs were administered at three dosages to five dogs. The percentage decreases in blood pressure are shown in the table.

	Standard			Experimental			
Dog	1 mg/kg	3 mg/kg	9 mg/kg	0.1 mg/kg	0.3 mg/kg	0.9 mg/kg	Dog total
1	8	20	29	24	51	42	174
2	29	40	63	41	46	54	273
3	12	9	52	16	41	55	185
4	13	38	65	30	45	57	248
5	6	14	34	17	36	43	150
Mean	13.6	24.2	48.6	25.6	43.8	50.2	

SOURCE: Data kindy provided by S. Michael Free and reproduced with the permission of the SmithKline Beckman Corporation.

Because the data consist of six responses on only five dogs we cannot apply the repeated-measurements T^2 approach to all five basic contrasts. The mixed model analysis of variance seems more appropriate.

(a) Complete the mixed-model analysis of variance with the six dosage levels as the fixed treatments.

(b) Form as many contrasts as you deem necessary to test for standard versus experimental drug differences, linear or quadratic trends in pressure change as functions of dosage, and standard-experimental differences in those trend contrasts. Use the Scheffé multiple comparison critical value with $\alpha = 0.01$.

22. Mueller (1962) reported these values of plasma free fatty acid in the blood of $r = 10$ schizophrenic patients immediately before and at 15, 30, and 45 minutes after injection with insulin:

	Plasma FFA (100 × mEq/liter)			
Patient	0	15	30	45
1	51	29	19	19
2	22	31	28	34
3	38	38	38	41
4	30	18	23	31
5	38	47	38	47
6	50	41	33	50
7	30	18	23	23
8	37	31	26	26
9	26	21	23	34
10	29	29	46	36

SOURCE: Copyright 1962, American Medical Association.

(a) Carry out a mixed-model analysis of variance. If the times are significantly different in mean FFA, use an appropriate multiple comparison method to locate the source of the difference.

(b) Repeat the test for equal time effects by means of the Hotelling T^2-statistic.

23. As the next stage of the experiment described in Exercise 3 the Illinois female mosquito fish were crossed with wild males. Each cross had seven females, and from each female $N_{ij} = 10$ offspring were weighed. The data constitute a nested design with primary treatments the locality of the grandfather (Illinois or North Carolina) of the offspring and the nested treatments the seven females in each cross. Mean weights in milligrams of the 10 offspring in each of the 14 groups and the sums of squares of the weights about those means are shown in the table:

a. Illinois grandfathers

	Litter ($N_{1j} = 10$)						
	1	2	3	4	5	6	7
Mean	1.07	1.0	1.4	1.39	1.12	1.08	1.2
Within-litter sum of squares	0.221	0.20	0.06	0.189	0.136	0.096	0.06
Illinois mean	1.18						
Between-litter sum of squares	1.51						

b. North Carolina grandfathers

	Litter ($N_{2j} = 10$)						
	1	2	3	4	5	6	7
Mean	1.0	1.02	0.96	0.79	1.05	0.73	0.88
Within-litter sum of squares	0.08	0.036	0.264	0.029	0.105	0.161	0.036
North Carolina mean	0.91857						
Between-litter sum of squares	0.8949						

SOURCE: Data reproduced with the kind permission of David Reznick.

(a) Carry out a nested fixed-effects analysis of variance, with multiple comparisons on means as they appear to be necessary.

(b) Criticize the assumption of a fixed-effects linear model for the data.

chapter 10

The Analysis of Covariance

10-1 INTRODUCTION

In Chapter 9 we frequently used multiway layouts to account for different characteristics of the experimental units. In a randomized block design the blocks might represent different fertility levels of plots of land in an agricultural field trial or the litters of rats or other animals used in a drug evaluation. Those designs presupposed that sufficient plots or animals with similar characteristics were available for a two- or higher-way design. Frequently, we do not have that luxury of data, or we might wish to use the actual values of the blocking variables in the analysis of variance. We may introduce those *concomitant variables* or *covariates* into the tests for experimental effects by a method known as the *analysis of covariance*. In this context the term *covariance* does not refer to the correlation or product moment of random variables or their observations but rather to the covariation of the response and concomitant variables.

In Sections 10-2 and 10-3 we describe the analysis of covariance for one-way layouts with fixed treatment effects. We begin with a single concomitant variable and then give the general case of several covariates in Section 10-3.

10-2 CONTROLLING FOR RELATED VARIABLES: THE ANALYSIS OF COVARIANCE

The experimental designs leading to the analyses of variance in Chapters 8 and 9 assumed that the sampling units were drawn at random from homogeneous populations. If the populations were not homogeneous with respect to some important characteristic (for example, weight, age, cognitive ability), then the treatment groups were forced to have similar values, means, and ranges of the concomitant variable by matching units before assigning them randomly to treatments. We wished to avoid a confounding of treatment and concomitant variable effects in the design, so that any differences among the treatment means could be ascribed to the experimental effects rather than undesired differences in the concomitant characteristics of their sampling units.

In many studies the investigator cannot control for the concomitant variable's values. For example, in an experiment conducted to determine the effect of nutrition on the growth of rats we would prefer that the rats in each of the feeding regime groups not only be of the same genetic qualities but have nearly the same birthweights or weights at the beginning of the feeding trials. It is relatively easy to maintain similar genetic characteristics by choosing rats from the same litters or parents, but the requirement of comparable weights might sharply reduce the number of available rats and result in an insensitive experiment. A different situation occurs when a simple random sample of human subjects has been drawn and the persons measured with respect to many variables and cross-classified according to several kinds of categories. Then we might wish to compare the averages of a variable, for example, blood cholesterol or glucose level, for male subjects in each of five occupational groups. It is conceivable that some jobs attract older or younger workers, and since cholesterol and glucose levels both increase with age, occupational differences might really be caused by age variation. We would like to be able to adjust the data for age differences before testing the hypothesis of equal occupational effects.

The analysis of covariance is one method for adjusting for the effect of a concomitant variable. We begin by adding to the usual one- or two-way fixed-effects model a linear function relating the response and concomitant variables. It is necessary that the function be the same, or have the same regression coefficients, in each treatment group of the layout. This is the analysis of covariance model for the ith observation obtained under treatment j of an unbalanced one-way layout:

$$x_{ij} = \mu + \tau_j + \beta(u_{ij} - u_{..}) + e_{ij} \qquad i = 1, \ldots, N_j; \quad j = 1, \ldots, k \qquad (1)$$

As in Section 8-3 μ is a common parameter, τ_j is the effect due to the jth treatment, and the e_{ij} are independently and normally distributed random variables with mean zero and a common variance σ^2. u_{ij} is the value of the single concomitant variable in the ijth sampling unit. As in our earlier development of linear regression we prefer to express the concomitant variable in deviations about its grand mean $\bar{u}_{..}$ for the $N = \sum N_j$ sampling units. The regression coefficient β is the same for each treatment group: The regression lines are parallel, with the difference in their heights due only to the treatment effects.

Let us visually examine some data with a concomitant variable. In Figure 10-1 thyroid gland weights have been plotted against body weights for two samples of laboratory animals. The first sample of $N_1 = 8$ animals received water orally for seven days as a control treatment. The second sample contained

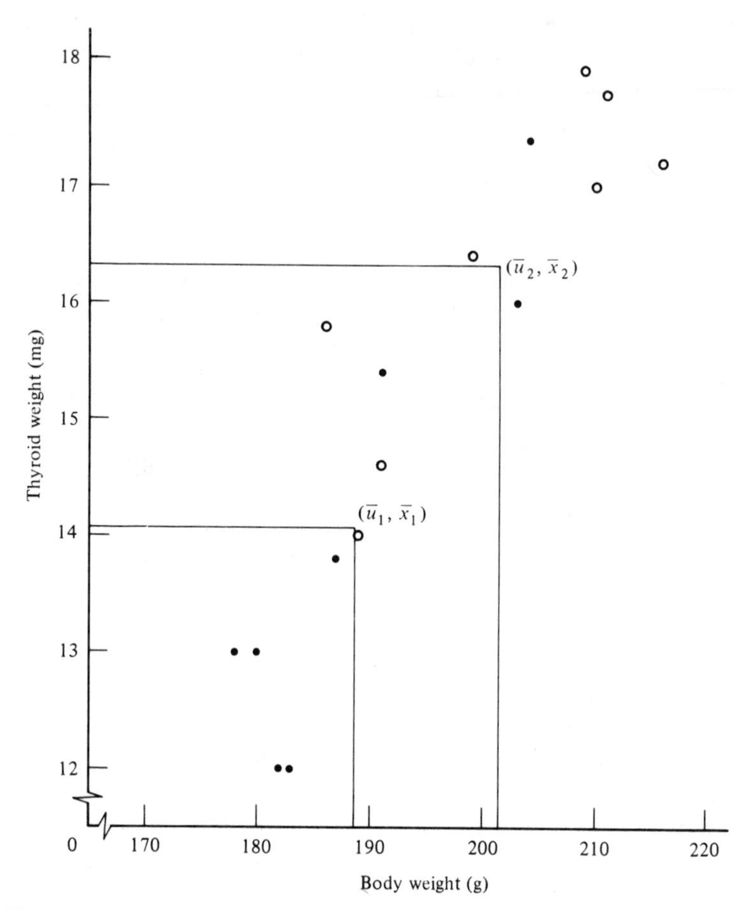

Figure 10-1 Body and thyroid weights. Solid circle, control; open circle, treatment. (Data kindly provided by S. Michael Free and reproduced with the permission of the SmithKline Beckman Corporation.)

$N_2 = 8$ animals who were administered the drug orally for the same period. These averages were computed for the two groups:

Control		Treatment	
Body weight (g)	Thyroid weight (mg)	Body weight (g)	Thyroid weight (mg)
188.5	14.075	201.4	16.35

Clearly, the thyroid mean weights are different, but so are the body weights. A scan of the two scatterplots in the figures indicates that the variables are related in each group. The plots of the two groups overlap each other. The appearance of the plots suggests that the difference in thyroid weights is really due to heavier animals in the drug sample. We shall confirm this impression with the formal analysis of covariance in Example 10-1.

Now let us return to the linear model (1) and describe the analysis of covariance based on it. Our principal purpose is to obtain a test of the hypothesis

$$H_{01}: \quad \tau_1 = \cdots = \tau_k \tag{2}$$

of equal treatment effects when the influence of the concomitant variable has been taken into account by the regression function in the model. As an alternative hypothesis it will suffice to say that some of the τ_j are different. We shall also need a test of the hypothesis

$$H_{02}: \quad \beta = 0 \tag{3}$$

that the response and concomitant variables are unrelated. If we cannot reject (3), we have little justification for the inclusion of the concomitant variable in the model and analysis. Finally, we should have a test of the assumption of a common slope β for the k groups, or the hypothesis

$$H_0: \quad \beta_1 = \cdots = \beta_k \tag{4}$$

that the k within-group regression lines are parallel. All of these tests will be provided by F-ratios from the analysis of covariance.

Let us begin with some estimators of the parameters in the model (1). If we impose the constraint

$$\sum_{j=1}^{k} N_j \hat{\tau}_j = 0 \tag{5}$$

on the least squares equations, the estimator of the jth treatment mean, $\mu + \tau_j$, is

$$\begin{aligned} \widehat{\mu + \tau_j} &= \bar{x}_{.j} - \hat{\beta}(\bar{u}_{.j} - \bar{u}_{..}) \\ &= \bar{x}_{.j(\text{adjusted})} \end{aligned} \tag{6}$$

We shall refer to those estimators as the *adjusted treatment means* for they are the *treatment averages adjusted by the parallel regression lines* to a common average $\bar{u}_{..}$ of the concomitant variable. To estimate the regression line slope β we begin by pooling the sums of products of the two variables within each group:

$$\text{SPE}_{xu} = \sum_{i=1}^{N_1} (x_{i1} - \bar{x}_{.1})(u_{i1} - \bar{u}_{.1}) + \cdots + \sum_{i=1}^{N_k} (x_{ik} - \bar{x}_{.k})(u_{ik} - \bar{u}_{.k}) \tag{7}$$

Similarly, we pool the sums of squares of the concomitant variable about the means of each group:

$$\text{SSE}_{uu} = \sum_{i=1}^{N_1} (u_{i1} - \bar{u}_{.1})^2 + \cdots + \sum_{i=1}^{N_k} (u_{ik} - \bar{u}_{.k})^2 \tag{8}$$

Then the estimator of the common slope is

$$\hat{\beta} = \frac{\text{SPE}_{xu}}{\text{SSE}_{uu}} \tag{9}$$

The unbiased estimator of the variance of the disturbance term e_{ij} is given by the squared residuals of the observations within each group from their predicted values

$$\begin{aligned}\hat{x}_{ij} &= \widehat{\mu + \tau_j} + \hat{\beta}(u_{ij} - \bar{u}_{..}) \\ &= \bar{x}_{.j} + \hat{\beta}[u_{ij} - \bar{u}_{..} - (\bar{u}_{.j} - \bar{u}_{..})] \\ &= \bar{x}_{.j} + \hat{\beta}(u_{ij} - \bar{u}_{.j}) \end{aligned} \tag{10}$$

The estimator is

$$\begin{aligned}\hat{\sigma}^2 &= \frac{1}{N-k-1}\sum_{j=1}^{k}\sum_{i=1}^{N_j}(x_{ij} - \hat{x}_{ij})^2 \\ &= \frac{1}{N-k-1}\sum_{j=1}^{k}\sum_{i=1}^{N_j}[x_{ij} - \bar{x}_{.j} - \hat{\beta}(u_{ij} - \bar{u}_{.j})]^2 \\ &= \frac{1}{N-k-1}\left\{\sum_{j=1}^{k}\sum_{i=1}^{N_j}(x_{ij} - \bar{x}_{.j})^2 - \frac{\left[\sum_{i=1}^{N_j}(x_{ij} - x_{.j})(u_{ij} - \bar{u}_{.j})\right]^2}{\sum_{i=1}^{N_j}(u_{ij} - \bar{u}_{.j})^2}\right\} \\ &= \frac{1}{N-k-1}\left[\mathrm{SSE}_{xx} - \frac{(\mathrm{SPE}_{xu})^2}{\mathrm{SSE}_{uu}}\right] \\ &= \frac{1}{N-k-1}\mathrm{SSE} \end{aligned} \tag{11}$$

We shall test the hypothesis (3) of no relation of the response and concomitant variables first. The common slope estimator $\hat{\beta}$ is a normal random variable with mean β and variance

$$\frac{\sigma^2}{\mathrm{SSE}_{uu}} \tag{12}$$

so that

$$\frac{\hat{\beta}^2 \mathrm{SSE}_{uu}}{\sigma^2} = \frac{(\mathrm{SPE}_{xu})^2}{\sigma^2 \mathrm{SSE}_{uu}} \tag{13}$$

has the chi-squared distribution with one degree of freedom. $\hat{\beta}$ is also distributed independently of $\hat{\sigma}^2$. If H_0: $\beta = 0$ is true,

$$t = \frac{\hat{\beta}\sqrt{\mathrm{SSE}_{uu}}}{\hat{\sigma}} \tag{14}$$

has the t-distribution with $N - k - 1$ degrees of freedom, and we should reject H_0 in favor of the two-sided alternative of an effective linear relation with the covariate if

$$|t| > t_{\alpha/2; N-k-1} \tag{15}$$

We can also make the test by the analysis of variance in Table 10-1. In that table the treatment sum of squares is merely

$$\mathrm{SST}_{xx} = \sum_{j=1}^{k} N_j(\bar{x}_{.j} - \bar{x}_{..})^2 \tag{16}$$

TABLE 10-1 Analysis of Variance for Testing $H_0: \beta = 0$

Source	Sum of squares	d.f.	Mean square
Treatments (unadjusted for covariate)	SST_{xx}	$k - 1$	$\dfrac{\mathrm{SST}_{xx}}{k-1}$
Covariate (within treatments)	$\dfrac{(\mathrm{SPE}_{xu})^2}{\mathrm{SSE}_{uu}}$	1	$\dfrac{(\mathrm{SPE}_{xu})^2}{\mathrm{SSE}_{uu}}$
Error	SSE	$N - k - 1$	$\hat{\sigma}^2$
Total	$\sum\sum (x_{ij} - \bar{x}_{..})^2$	$N - 1$	—

first introduced in expression (11) of Section 8-3. We calculate the ratio

$$F = \frac{(\mathrm{SPE}_{xu})^2}{\hat{\sigma}^2 \mathrm{SSE}_{uu}} \tag{17}$$

and refer it to an upper critical value of the F-distribution with one and $N - k - 1$ degrees of freedom. If

$$F > F_{\alpha;1,N-k-1} \tag{18}$$

we conclude at the α level that the covariate should be retained in the model.

Now with the assurance that β is not zero we can test the hypothesis (1) of equal treatment effects by the second analysis of variance in Table 10-2. The sum of squares for treatments adjusted for the covariate as well as the grand mean is

$$\begin{aligned}\mathrm{SSTA} &= \sum N_j(\bar{x}_{.j} - \bar{x}_{..})^2 + \frac{(\mathrm{SPE}_{xu})^2}{\mathrm{SSE}_{uu}} - \frac{[\sum\sum (x_{ij} - \bar{x}_{..})(u_{ij} - \bar{u}_{..})]^2}{\sum\sum (u_{ij} - \bar{u}_{..})^2} \\ &= \mathrm{SST} + \frac{(\mathrm{SPE}_{xu})^2}{\mathrm{SSE}_{uu}} - \mathrm{SSC}\end{aligned} \tag{19}$$

TABLE 10-2 Analysis of Covariance for the Test of Equal Treatment Effects

Source	Sum of squares	d.f.	Mean square
Treatments (adjusted for covariate)	SSTA	$k - 1$	$\dfrac{\mathrm{SSTA}}{k-1}$
Covariate (unadjusted for treatments)	SSC	1	SSC
Error	SSE	$N - k - 1$	$\hat{\sigma}^2$
Total	$\sum\sum (x_{ij} - \bar{x}_{..})^2$	$N - 1$	—

The adjusted treatment sum of squares can be derived by the methods of Section 3-3; a detailed development in terms of reduction formula notation has been given by Searle (1971, Section 8.2). We note that the sum of squares

$$\text{SSC} = \frac{[\sum\sum (x_{ij} - \bar{x}_{..})(u_{ij} - \bar{u}_{..})]^2}{\sum\sum(u_{ij} - \bar{u}_{..})^2} \tag{20}$$

is merely that due to a single regression line fitted to the N pairs of observations without regard for treatment group distinction. The error sum of squares SSE is the same as in Table 10.1. We test

$$H_0: \quad \tau_1 = \cdots = \tau_k$$

by computing the ratio

$$F = \frac{\text{SSTA}}{\hat{\sigma}^2(k-1)} \tag{21}$$

If

$$F > F_{\alpha; k-1, N-k-1}$$

we should conclude that the treatment effects are different at the α level. Unlike the usual one-way analysis of variance our conclusion refers only to the treatment effects after differences due to the concomitant variable have been removed.

Example 10-1

Now we apply the analysis of covariance to the thyroid and body weight data introduced in Figure 10-1. The observations are shown in Table 10-3. The sums of squares and products of the variables about their group means are given below each column. From

TABLE 10-3 Thyroid and Body Weights

	Control		Drug	
	Thyroid (mg)	Body (g)	Thyroid (mg)	Body (g)
	16.0	203	17.0	210
	13.0	180	16.4	199
	12.0	183	14.0	189
	15.4	191	15.8	186
	17.4	204	17.2	216
	13.0	178	18.0	209
	12.0	182	14.6	191
	13.8	187	17.8	211
Mean	14.075	188.5	16.35	201.4
Sum of squares	27.515	714	14.86	941.875
Sum of products	128.5		100.15	
$\hat{\beta}_j$	0.1800		0.1063	

SOURCE: Data kindly provided by S. Michael Free and reproduced with the permission of the SmithKline Beckman Corporation.

these we may compute the within-treatment regression coefficients

$$\hat{\beta}_1 = 0.1800 \qquad \hat{\beta}_2 = 0.1063$$

and the common, or pooled, estimate

$$\hat{\beta} = \frac{128.5 + 100.15}{714 + 941.875} = 0.1381$$

The control and drug slopes are of the same sign and magnitude, and we shall assume that their population values are equal. Later, in Example 10-2, we shall test that assumption formally. For the hypothesis that the common slope is zero we prepared this analysis of variance:

Source	Sum of squares	d.f.	Mean square	F
Treatments (unadjusted for covariate)	20.7025	1	20.7025	—
Covariate (within treatments)	31.5729	1	31.5729	38.00
Error	10.8021	13	0.83093	—
Total	63.077	15	—	—

The F-statistic exceeds any reasonable critical value with 1 and 13 degrees of freedom. Certainly, body weight is an appropriate and significant concomitant to thyroid weight.

Before carrying out the second analysis of variance adjusted for the covariate let us examine the usual unadjusted one-way analysis of variance. The sum of squares for the two treatments remains the same as in the first line of the preceding table, whereas the error sum of squares can be computed by pooling the sums in the second and third lines:

Source	Sum of squares	d.f.	Mean square	F
Treatments	20.7025	1	20.7025	6.84
Within treatments	42.375	14	3.0268	—
Total	63.077	15	—	—

The difference in thyroid weights is highly significant. Compare that finding with the analysis of covariance of Table 10-2:

Source	Sum of squares	d.f.	Mean square	F
Treatments (adjusted)	0.7060	1	0.7060	0.85
Covariate (unadjusted)	51.5694	1	51.5694	—
Error	10.8021	13	0.83093	—
Total	63.0775	15	—	—

The hypothesis of different control and drug effects on thyroid weight cannot be accepted when the animal's weight is adjusted to a constant mean in the two samples. Our original significant difference in the thyroid means is due to a strong linear relationship of thyroid and body weight.

The adjusted thyroid weights for the control and drug groups are

$$\bar{x}_{.1(\text{adjusted})} = 14.075 - (0.1381)(188.5 - 194.9) = 14.96$$

$$\bar{x}_{.2(\text{adjusted})} = 16.35 - (0.1381)(201.4 - 194.9) = 15.45$$

The parallel regression lines for the control and drug groups are shown in Figure 10-2. The adjustment of the control group's mean thyroid weight amounts to moving up the lower regression line to the point given by the grand mean weight $\bar{u}_{..} = 194.9$. The adjusted mean for the drug group may be found by following the upper line from $\bar{u}_{.2} = 201.4$ back to the grand mean.

When only $k = 2$ treatment groups are present the test of $H_0: \tau_1 = \tau_2$ can be carried out by computing a t-statistic from the difference of the adjusted means

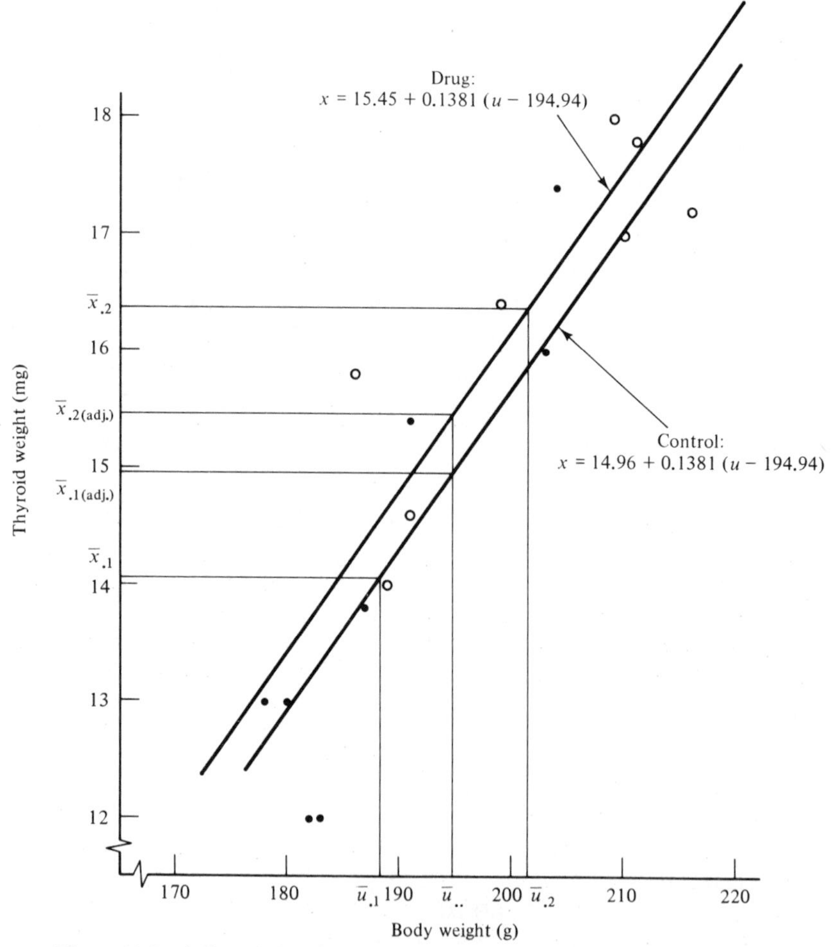

Figure 10-2 Adjusted thyroid mean weights. Solid circle, control; open circle, drug.

$$\bar{x}_{.1(\text{adjusted})} - \bar{x}_{.2(\text{adjusted})} = \bar{x}_{.1} - \bar{x}_{.2} - \hat{\beta}(\bar{u}_{.1} - \bar{u}_{.2})$$

The estimated variance of the difference is

$$\hat{\sigma}^2\left[\frac{1}{N_1} + \frac{1}{N_2} + \frac{(\bar{u}_{.1} - \bar{u}_{.2})^2}{\text{SSE}_{uu}}\right]$$

and is distributed independently of the adjusted mean difference. We may test the hypothesis by computing

$$t = \frac{\bar{x}_{.1(\text{adjusted})} - \bar{x}_{.2(\text{adjusted})}}{\hat{\sigma}\sqrt{\frac{1}{N_1} + \frac{1}{N_2} + \frac{(\bar{u}_{.1} - \bar{u}_{.2})^2}{\text{SSE}_{uu}}}}$$

and referring it to a critical value of the t-distribution with $N_1 + N_2 - 3$ degrees of freedom. In the present case

$$t = \frac{14.964 - 15.461}{\sqrt{0.83093\left[\frac{1}{8} + \frac{1}{8} + \frac{(188.5 - 201.4)^2}{714 + 941.875}\right]}}$$
$$= -0.921$$

or the square root of the original F-statistic. Since $|t|$ is much less than the critical value $t_{0.05;13} = 1.771$ we should not reject the hypothesis of equal treatment effects.

Testing the assumption of parallel regression lines. In Section 3-7 we gave a test of the equality of several regression coefficient vectors. We can use that test for the assumption that a common slope will suffice for the covariate regressions in the k treatment groups. Let the coefficient for the jth treatment be estimated by

$$\hat{\beta}_j = \frac{\sum_{i=1}^{N_j} (x_{ij} - \bar{x}_{.j})(u_{ij} - \bar{u}_{.j})}{\sum_{i=1}^{N_j} (u_{ij} - \bar{u}_{.j})^2} \tag{22}$$

Then the parallelism sum of squares (10) of Section 3-7 can be written as

$$\text{SSP} = \sum_{j=1}^{k} \hat{\beta}_j^2 \sum_{i=1}^{N_j} (u_{ij} - \bar{u}_{.j})^2 - \frac{(\text{SPE}_{xu})^2}{\text{SSE}_{uu}} \tag{23}$$

When

$$H_0\colon\ \beta_1 = \cdots = \beta_k \tag{24}$$

is true SSP/σ^2 will have the chi-squared distribution with $k - 1$ degrees of freedom. SSP is also distributed independently of the error mean square $\hat{\sigma}^2$ defined by (11), so that

$$F = \frac{\text{SSP}}{\hat{\sigma}^2(k-1)} \tag{25}$$

is an F random variable with $k - 1$ and $N - k - 1$ degrees of freedom if the parallelism model holds. We should reject (24) at the α level if $F > F_{\alpha;k-1,N-k-1}$. Searle (1971, Section 8.2) has considered an alternative model for the analysis of covariance with different coefficients in each treatment group.

Example 10-2

Let us test the hypothesis of a common regression coefficient for the data in Example 10-1. Here

$$\text{SSP} = (0.1800)^2(714) + (0.1063)^2(941.875) - 31.5729$$
$$= 2.2036$$
$$F = \frac{2.2036}{0.83093} = 2.65$$

Since $F_{0.10;1,13} = 3.14$, the parallel-line model is tenable for the control and drug groups.

10-3 SEVERAL COVARIATES

Some investigations require more than one concomitant variable. For example, in a comparison of the concentrations of certain biochemicals in the urine of persons classified as normal or suffering from three variants of a metabolic disease we might wish to use both specific gravity and volume of the specimen as covariates. Similarly, the test for significant differences among the overall evaluations of five instructors in a course might use the ratings for three qualities of the course as covariates.

The extension of the analysis of covariance model to multiple covariates is straightforward. We merely increase the number of covariates in (1) of Section 10-2 to p:

$$x_{ij} = \mu + \tau_j + \beta_1(u_{ij1} - \bar{u}_{..1}) + \cdots + \beta_p(u_{ijp} - \bar{u}_{..p}) + e_{ij}$$
$$i = 1, \ldots, N_j; \quad j = 1, \ldots, k \qquad (1)$$

As before x_{ij} is the ith observation under the jth treatment, and $u_{ij1}, \ldots, u_{ijp}$ are the values of its associated concomitant variables. μ and the τ_j are the respective general mean and treatment effects. The regression coefficients

$$\boldsymbol{\beta}' = [\beta_1, \ldots, \beta_p] \qquad (2)$$

are the same in each sampling unit of all treatments: The k regression planes are all parallel. As in our other least squares models the u_{ijh} are fixed, nonrandom variables. The disturbance terms e_{ij} are independently and normally distributed with zero mean and a common variance σ^2.

In our analysis of covariance we wish to test the hypothesis

$$H_0: \quad \tau_1 = \cdots = \tau_k \qquad (3)$$

of equal treatment effects. We also shall need a test of

$$H_0: \quad \boldsymbol{\beta} = \mathbf{0} \qquad (4)$$

or the hypothesis that the response and concomitant variables are unrelated. We should also test the assumption that the regression vector is the same for each treatment group. Let us begin by introducing matrix notation for the variables and the parameters. The response and concomitant observations obtained under the jth treatment will be denoted by

$$\mathbf{x}_j = \begin{bmatrix} x_{1j} \\ \cdot \\ \cdot \\ \cdot \\ x_{N_j j} \end{bmatrix} \qquad \mathbf{U}_{1j} = \begin{bmatrix} u_{1j1} - \bar{u}_{.j1} & \cdots & u_{1jp} - \bar{u}_{.jp} \\ \cdots & \cdots & \cdots \\ u_{N_j j1} - \bar{u}_{.j1} & \cdots & u_{N_j jp} - \bar{u}_{.jp} \end{bmatrix} \tag{5}$$

We have centered the concomitant variables about their averages

$$\bar{u}_{.jh} = \frac{1}{N_j} \sum_{i=1}^{N_j} u_{ijh} \tag{6}$$

under the jth treatment: hence the subscript 1 on $\mathbf{U}_{1j}$. The matrix of sums of squares and products of the concomitant variables within each group will be denoted by

$$\mathbf{E}_{uu} = \sum_{j=1}^{k} \mathbf{U}'_{1j}\mathbf{U}_{1j} \tag{7}$$

The vector of the within-groups sums of products of the concomitant and response variables is

$$\mathbf{E}_{ux} = \sum_{j=1}^{k} \mathbf{U}'_{1j}\mathbf{x}_j \tag{8}$$

These are the extensions of (8) and (7), Section 10-2, to the case of p covariates. Then the estimate of the common regression coefficient vector (2) is

$$\hat{\boldsymbol{\beta}} = \mathbf{E}_{uu}^{-1}\mathbf{E}_{ux} \tag{9}$$

The vector of adjusted treatment means has jth element

$$\bar{x}_{.j(\text{adjusted})} = \bar{x}_{.j} - \hat{\boldsymbol{\beta}}'(\bar{\mathbf{u}}_{.j} - \bar{\mathbf{u}}_{..}) \tag{10}$$

The covariate treatment and grand mean vectors are

$$\bar{\mathbf{u}}_{.j} = \begin{bmatrix} \bar{u}_{.j1} \\ \cdot \\ \cdot \\ \cdot \\ \bar{u}_{.jp} \end{bmatrix} \qquad \bar{\mathbf{u}}_{..} = \begin{bmatrix} \bar{u}_{..1} \\ \cdot \\ \cdot \\ \cdot \\ \bar{u}_{..p} \end{bmatrix} \tag{11}$$

For the estimator of the disturbance term variance we define the sum of squares of residuals about the parallel fitted regression planes within each treatment group:

$$\begin{aligned} \text{SSE} &= \sum_{j=1}^{k} \sum_{i=1}^{N_j} \left[x_{ij} - \bar{x}_{.j} - \sum_{h=1}^{p} \hat{\beta}_h (u_{ijh} - \bar{u}_{.jh}) \right]^2 \\ &= E_{xx} - \mathbf{E}'_{ux}\mathbf{E}_{uu}^{-1}\mathbf{E}_{ux} \end{aligned} \tag{12}$$

where for covenience and consistency we have let

$$E_{xx} = \sum_{j=1}^{k} \sum_{i=1}^{N_j} (x_{ij} - \bar{x}_{.j})^2 \tag{13}$$

The estimator is

$$\hat{\sigma}^2 = \frac{1}{N - k - p} \text{SSE} \tag{14}$$

When the assumptions of the multiple covariate model (1) hold SSE/σ^2 is a chi-squared random variable with $N - k - p$ degrees of freedom.

For our inferences in the multiple covariate model let us begin with the hypothesis (4) of zero regression coefficients. If that is true,

$$\frac{\mathbf{E}'_{ux}\mathbf{E}_{uu}^{-1}\mathbf{E}_{ux}}{\sigma^2} \tag{15}$$

has the chi-squared distribution with p degrees of freedom. To test $H_0\colon \boldsymbol{\beta} = \mathbf{0}$ we compute

$$F = \frac{\mathbf{E}'_{ux}\mathbf{E}_{uu}^{-1}\mathbf{E}_{ux}}{p}\bigg/\frac{\text{SSE}}{N-k-p} \tag{16}$$

and reject the hypothesis if

$$F > F_{\alpha;\, p,\, N-k-p} \tag{17}$$

For the analysis of covariance to be appropriate we should have a significant regression relationship between the response and concomitant variables.

Now we are ready to test the hypothesis

$$H_0\colon \quad \tau_1 = \cdots = \tau_k$$

of no treatment effects after adjustment for the covariates. The analysis of covariance is shown in Table 10-4. The adjusted treatment sum of squares is

$$\text{SSTA} = \sum_{j=1}^{k} N_j(\bar{x}_{.j} - \bar{x}_{..})^2 + \mathbf{E}'_{ux}\mathbf{E}_{uu}^{-1}\mathbf{E}_{ux} - \mathbf{T}'_{ux}\mathbf{T}_{uu}^{-1}\mathbf{T}_{ux} \tag{18}$$

TABLE 10-4 Analysis of Covariance for the Test of Equal Treatment Effects

Source	Sum of squares	d.f.	Mean square
Treatments (adjusted for covariates)	SSTA	$k-1$	$\frac{\text{SSTA}}{k-1}$
Covariates (unadjusted)	SSC	p	$\frac{\text{SSC}}{p}$
Error	SSE	$N-k-p$	$\hat{\sigma}^2$
Total	$\sum\sum (x_{ij} - \bar{x}_{..})^2$	$N-1$	—

The term

$$\text{SSC} = \mathbf{T}'_{ux}\mathbf{T}_{uu}^{-1}\mathbf{T}_{ux} \tag{19}$$

is the generalization of (20), Section 10-2, to p covariates. $\mathbf{T}_{ux}$ is the $p \times 1$ sum of products vector with hth element

$$t_h = \sum_{j=1}^{k}\sum_{i=1}^{N_j} (x_{ij} - \bar{x}_{..})(u_{ijh} - \bar{u}_{..h}) \qquad h = 1, \ldots, p \tag{20}$$

$\mathbf{T}_{uu}$ is the $p \times p$ matrix of sums of squares and products for the concomitant variables. Its general element is

$$t_{hl} = \sum_{j=1}^{k} \sum_{i=1}^{N_j} (u_{ijh} - \bar{u}_{..h})(u_{ijl} - \bar{u}_{..l}) \qquad h, l = 1, \ldots, p \tag{21}$$

We test the hypothesis by referring

$$F = \frac{\text{SSTA}}{\hat{\sigma}^2(k-1)} \tag{22}$$

to the critical value $F_{\alpha; k-1, N-k-p}$ of the F-distribution with $k - 1$ and $N - k - p$ degrees of freedom. If the F-statistic exceeds the critical value we should reject the case of no treatment differences at the $100\alpha\%$ level.

Example 10-3

Averages of student ratings for instructors in five core courses of a Master of Business Administration program were published at the end of a term. The averages were given for faculty members teaching each course; individual scores or their variances and covariances were not available. We wished to see if the average ratings of

$$X = \text{overall evaluation of the instructor}$$

for the courses were significantly different after adjustment for the averages of the concomitant variables

$$U_1 = \text{impact of the course: new skills}$$
$$U_2 = \text{overall evaluation of the course}$$

recorded for each instructor. The average evaluations are shown in Table 10-5. The sums of squares and products of the averages about the course means are given in Table 10-6. The within-course sums of squares and products are found by summing the appropriate elements of those matrices:

$$E_{xx} = 8.7005$$

$$\mathbf{E}_{ux} = \begin{bmatrix} 3.2855 \\ 4.4769 \end{bmatrix} \qquad \mathbf{E}_{uu} = \begin{bmatrix} 2.1119 & 1.9955 \\ 1.9955 & 2.7242 \end{bmatrix}$$

The common regression coefficient vector is

$$\hat{\boldsymbol{\beta}} = \begin{bmatrix} 0.009593 \\ 1.6363 \end{bmatrix}$$

The regression sum of squares is

$$\mathbf{E}'_{ux}\mathbf{E}_{uu}^{-1}\mathbf{E}_{ux} = 7.3572$$

and the error sum of squares for the analysis of covariance model is

$$\text{SSE} = 8.7005 - 7.3572 = 1.3433$$

The estimate of the error variance is

$$\hat{\sigma}^2 = \frac{1.3433}{26 - 5 - 2} = 0.07070$$

We may test for the significance of the concomitant variables by computing the ratio (16):

$$F = \frac{7.3572/2}{0.07070} = 52.03$$

TABLE 10-5 Instructor and Course Average Ratings

	1			2			3			4			5		
	X	U_1	U_2	X	U_1	U_2	X	U_1	U_2	X	U_1	U_2	X	U_1	U_2
	2.14	2.71	2.50	2.77	2.29	2.45	1.11	1.74	1.82	2.19	2.81	2.80	1.50	1.88	1.94
	1.34	2.00	1.95	1.23	1.83	1.64	2.41	2.19	2.54	2.24	2.85	2.97	1.91	1.91	2.33
	2.50	2.66	2.69	1.37	1.78	1.83	1.74	1.40	2.23	3.05	3.15	3.12	2.38	1.76	2.46
	1.40	2.80	2.00	1.52	2.18	2.24	1.15	1.80	1.82	1.91	2.18	2.21			
	1.90	2.38	2.30	1.81	2.14	2.11	1.66	2.17	2.35						
				3.28	2.60	2.76	3.11	2.76	2.88						
				2.00	2.16	2.16	1.45	2.00	2.18						
Mean	1.856	2.51	2.288	1.997	2.14	2.17	1.804	2.08	2.26	2.348	2.748	2.775	1.93	1.85	2.24

TABLE 10-6 Within-Course Sums of Squares and Products

Course		X	U_1	U_2
1	X	0.9715	0.2786	0.6254
	U_1		0.4236	0.1900
	U_2			0.4039
2	X	3.4875	1.1506	1.5709
	U_1		0.4618	0.6027
	U_2			0.8316
3	X	3.1323	1.4007	1.5898
	U_1		0.7154	0.7451
	U_2			0.8654
4	X	0.7213	0.5102	0.4646
	U_1		0.4985	0.4810
	U_2			0.4769
5	X	0.3878	−0.0546	0.2262
	U_1		0.0126	−0.0234
	U_2			0.1465

Since $F_{0.05;2,19} = 5.93$, we should reject the hypothesis of zero coefficients at any reasonable level. Let us proceed with the analysis of covariance; the question of the importance of the individual covariates will be deferred for a moment.

The ordinary one-way analysis of variance without the covariates gave these statistics:

Source	Sum of squares	d.f.	Mean square	F
Courses	0.8363	4	0.2092	0.50
Within courses	8.7005	21	0.4143	—
Total	9.5368	25	—	—

We see that the hypothesis of equal course means for the overall evaluation variable cannot be rejected. Because the error mean square computed from the covariate model is nearly one-sixth that of the ordinary analysis of variance it seems likely that adjustment for the covariates will lead to different results. We begin by computing the total sums of squares and products of the three variables about their grand means. The 2×1 vector $\mathbf{T}_{ux}$ defined by (20) is

$$\mathbf{T}_{ux} = \begin{bmatrix} 4.1136 \\ 5.2323 \end{bmatrix}$$

and the total sums of squares and products matrix defined by (21) is

$$\mathbf{T}_{uu} = \begin{bmatrix} 4.2062 & 3.1384 \\ 3.1384 & 3.7577 \end{bmatrix}$$

Then the total sample regression sum of squares (19) is

$$\text{SSC} = 7.3269$$

The adjusted course sum of squares (18) is

$$\text{SSTA} = 0.8368 + 7.3572 - 7.3269 = 0.8671$$

The analysis of covariance gave this table:

Source	Sum of squares	d.f.	Mean square	F
Courses (adjusted)	0.8671	4	0.2168	3.07
Covariates (after mean)	7.3264	2	3.6632	—
Error	1.3433	19	0.0707	—
Total	9.5368	25	—	—

The critical value for the adjusted F-ratio is $F_{0.05;4,19} = 2.90$. We can reject at the 0.05 level the hypothesis (3) of equal instructor means in the five course populations when adjustment has been made for the variables measuring the impact of the course and its overall evaluation.

To understand the effects of the concomitant variables we should compute the adjusted course means defined by (10). The original and adjusted means are shown in the table. Obviously, course 4, whose instructors had the poorest mean overall rating,

Original and adjusted means		
Course	$\bar{x}_{.j}$	$\bar{x}_{.j\ (\text{adjusted})}$
1	1.856	1.903
2	1.997	2.241
3	1.804	1.901
4	2.348	1.596
5	1.930	2.062

now has the lowest, or best, adjusted mean. The reason is contained in the concomitant variable means for the course in Table 10-5: $\bar{u}_{.41} = 2.348$ and $\bar{u}_{.42} = 2.748$ are higher than most of their counterparts in the other courses and indicate dissatisfaction with course 4 rather than its instructors. When the instructor averages are adjusted for that perception of the course the ratings of the instructors are improved.

Testing for parallel regression planes. The methods of Section 3-7 provide a test of the hypothesis

$$H_0: \begin{bmatrix} \beta_{11} \\ \cdot \\ \cdot \\ \cdot \\ \beta_{p1} \end{bmatrix} = \cdots = \begin{bmatrix} \beta_{1k} \\ \cdot \\ \cdot \\ \cdot \\ \beta_{pk} \end{bmatrix} \tag{23}$$

that a common regression function is appropriate for the k treatments of the one-way layout. For the test we shall need estimates of the group regression vectors and individual sums of squares and products matrices for each group. We shall write the $p \times p$ matrix of sums of squares and products of the concomitant variable in the jth treatment as

$$\mathbf{E}_{uuj} = \mathbf{U}'_{1j}\mathbf{U}_{1j} \qquad j = 1, \ldots, k \tag{24}$$

where $\mathbf{U}_{1j}$ is the $N_j \times p$ matrix of deviations defined in (5). The $\mathbf{E}_{uuj}$ are of course the individual terms in the definition (7) of $\mathbf{E}_{uu}$. Similarly,

$$\mathbf{E}_{uxj} = \mathbf{U}'_{1j}\mathbf{x}_j \qquad j = 1, \ldots, k \tag{25}$$

is the $p \times 1$ vector of sums of products of the response and concomitant observations in the jth group, or the jth term in the formula (8) for $\mathbf{E}_{ux}$. Then the estimate of the jth coefficient vector is

$$\hat{\boldsymbol{\beta}}_j = \mathbf{E}_{uuj}^{-1}\mathbf{E}_{uxj} \tag{26}$$

To test the hypothesis (23) we compute the sum of squares

$$\text{SSP} = \sum_{j=1}^{k} \mathbf{E}'_{uxj}\mathbf{E}_{uuj}^{-1}\mathbf{E}_{uxj} - \mathbf{E}'_{ux}\mathbf{E}_{uu}^{-1}\mathbf{E}_{ux} \tag{27}$$

originally given in (10), Section 3-7. If the hypothesis is true, the statistic

$$F = \frac{\text{SSP}}{p(k-1)\hat{\sigma}_w^2} \tag{28}$$

will have the F-distribution with $p(k-1)$ and $N - k(p+1)$ degrees of freedom. $\hat{\sigma}_w^2$ is the estimator of the disturbance variance computed from the residuals of the individual regression functions in each group:

$$\hat{\sigma}_w^2 = \frac{1}{N - k(p+1)}\left[\sum_{j=1}^{k}\sum_{i=1}^{N_j}(x_{ij} - \bar{x}_{.j})^2 - \sum_{j=1}^{k}\mathbf{E}'_{uxj}\mathbf{E}_{uuj}^{-1}\mathbf{E}_{uxj}\right] \tag{29}$$

We should reject the hypothesis (23) if

$$F > F_{\alpha;\, p(k-1),\, N-k(p+1)} \tag{30}$$

for an appropriate choice of α.

Example 10-4

We shall test the assumption of a common regression coefficient vector for the instructor evaluation data of Example 10-3. The regression coefficients and sums of squares had these values in the five courses:

	Course				
	1	2	3	4	5
$\hat{\beta}_{1j}$	−0.0467	0.4839	0.4318	3.1333	−2.0833
$\hat{\beta}_{2j}$	1.5703	1.5383	1.4653	−2.1863	1.2115
$\mathbf{E}'_{uxj}\mathbf{E}_{uuj}^{-1}\mathbf{E}_{uxj}$	0.9690	2.9733	2.9344	0.5827	0.3878

Then

$$\text{SSP} = 7.8472 - 7.3572 = 0.49$$

$$\hat{\sigma}_w^2 = \frac{8.7005 - 7.8472}{11} = 0.07756$$

and $F = 0.79$. We cannot reject the hypothesis of a common function at any significance level. Curiously, our estimate of the disturbance variance is larger than that in Example 10-3 with parallel regression planes because we have lost so many degrees of freedom in fitting the separate planes for each course.

10-4 READINGS AND REFERENCES

The analysis of covariance was first proposed by R. A. Fisher (1932). Cochran (1957) has described the method in a paper introducing a special issue of *Biometrics* devoted to the analysis of covariance. Searle (1971) has developed the theory of estimation and testing with concomitant variables. Bliss (1970, Chapter 20) has given an extensive treatment of numerous applications in the life sciences.

The analysis of covariance can be extended easily to higher-way layouts. Ostle (1954) has given computational formulas for randomized blocks, latin squares, and factorial designs. Winer (1971) has illustrated the analysis of covariance for factorial layouts and repeated measurements.

Multiple comparison and simultaneous confidence interval methods for the analysis of covariance have been described by Miller (1981, pp. 60–62) and Scheffé (1959).

Bliss, C. I. (1970): *Statistics in Biology*, vol. 2, McGraw-Hill Book Company, New York.

Cochran, W. G. (1957): Analysis of covariance: its nature and uses, *Biometrics*, vol. 13, pp. 261–281.

Fisher, R. A. (1932): *Statistical Methods for Research Workers*, 4th ed., Oliver & Boyd Ltd., Edinburgh.

Miller, R. G. (1981): *Simultaneous Statistical Inference*, 2nd ed., Springer-Verlag, New York.

Ostle, B. (1954): *Statistics in Research*, Iowa State University Press, Ames, Iowa.

Scheffé, H. (1959): *The Analysis of Variance*, John Wiley & Sons, Inc., New York.

Searle, S. R. (1971): *Linear Models*, John Wiley & Sons, Inc., New York.

Winer, B. J. (1971): *Statistical Principles in Experimental Design*, 2nd ed., McGraw-Hill Book Company, New York.

10-5 EXERCISES

1. An experiment was conducted to investigate the effect of nutrition upon the sexual maturation of guppy fish (*Poecilia reticulatia*). Among other measures the weight of the fish at maturation was chosen as the response variable, and the initial weight of the fish was used as a concomitant variable. The treatments were the amounts of food given to the fish, with the first treatment consisting of unlimited, or *ad libitum*, food. The data from the first four treatments are shown in Table 10-7; one outlying observation pair in treatment 2 was deleted before our analysis.

TABLE 10-7 Maturation and Initial Weights (mg) of Guppy Fish

	Weight of feeding group:							
	1 (ad lib.)		2		3		4	
	Maturation	Initial	Maturation	Initial	Maturation	Initial	Maturation	Initial
	70	29	49	35	68	33	59	33
	74	32	61	26	70	35	53	36
	52	26	55	29	60	28	54	26
	63	30	69	32	53	29	48	30
			51	23	59	32	54	33
			38	26	48	23	53	25
			64	31	46	26	37	23
					54	29	46	28
Mean	64.75	29.25	55.29	28.86	57.25	29.375	50.5	29.25
Within-group sum of squares	278.75	18.75	653.43	102.86	529.5	105.875	318	143.5
Within-group sum of products	64.25		81.29		211.25		128	
$\hat{\beta}_j$	3.4267		0.7903		1.9953		0.8920	

SOURCE: Data reproduced with the kind permission of David Reznick.

(a) Verify that the unadjusted treatment and error sums of squares for maturation weight are

$$\text{SST}_{xx} = 563.73$$

$$\text{SSE}_{xx} = 1779.68$$

and their F-statistic equals 2.43. Should you conclude that the maturation mean weights differ for different amounts of food?

(b) The estimate of the common regression coefficient and its sum of squares are

$$\hat{\beta} = 1.3068 \quad \frac{(\text{SPE}_{xu})^2}{\text{SSE}_{uu}} = 633.50$$

The error sum of squares for the covariate model is

$$\text{SSE} = \text{SSE}_{xx} - \frac{(\text{SPE}_{xu})^2}{\text{SSE}_{uu}} = 1146.18$$

and the sum of squares for the concomitant variable regression without distinguishing the treatment groups is

$$\text{SSC} = 639.36$$

Show that the F-statistic for the adjusted treatment effects is equal to 3.57 and that for the common regression coefficient is 12.16. What should you infer about the effect of feeding on maturation weight and about the effectiveness of initial weight as a covariate?

(c) Verify that these adjusted treatment means follow from the covariate model:

Treatment	Adjusted mean weight
1	64.67
2	55.72
3	57.00
4	50.42

(d) Does the significant difference among the treatment effects found in (b) appear to be due to greater differences among the adjusted means or a reduction in the estimate of the error mean square?

(e) The sum of squares for parallelism of the four within-treatment regression coefficients is

$$\text{SSP} = 820.08 - \frac{(\text{SPE}_{xu})^2}{\text{SSE}_{uu}}$$

$$= 186.58$$

Show that the hypothesis of a common slope cannot be rejected at any conventional level.

2. Use the data and statistics of Example 10-3 for an analysis of covariance of the instructor means with the variable U_2 as the single covariate.

3. Two groups of animals received compounds A and B, respectively, in their food. At the end of three months of that diet all animals were weighed and sacrificed. Among the observations recorded at sacrifice were liver weights. The following body and liver weights were obtained:

	Compound A		Compound B	
	Body weight	Liver weight	Body weight	Liver weight
	255	10.0	259	12.6
	234	11.9	264	13.0
	232	12.1	278	13.5
	254	12.3	279	13.6
	264	12.4	300	13.9
	271	12.8	296	13.9
	288	12.8	309	14.4
	290	13.4	320	15.2
	291	13.6	311	15.3
	311	14.0	294	15.5
	300	14.6	398	15.6
	326	14.8	334	15.6
	299	14.8	337	15.8
	296	15.1	213	17.1
	303	16.5	329	17.8
Mean	280.933	13.407	301.400	14.853

SOURCE: Data kindly provided by S. Michael Free and reproduced with the permission of the SmithKline Beckman Corporation.

Test the hypothesis of equal effects of the compounds on liver weight when body weight has been used as a covariate.

4. From student evaluation forms collected in three graduate statistics courses taught by the same faculty member these variables were chosen for analysis:

X: overall evaluation of the instructor

U_1: contribution of the course to new skills

U_2: overall evaluation of the course

U_3: linear trend variable (year 0, 1, or 2)

These statistics were computed from the variables:

Course A ($N_1 = 20$)				
Variable	X	U_1	U_2	U_3
Sum	32	34	36	14

Sums of squares and products matrix:

$$\begin{bmatrix} 58 & 57 & 63 & 24 \\ & 68 & 69 & 23 \\ & & 74 & 26 \\ & & & 14 \end{bmatrix}$$

Course B ($N_2 = 15$)				
Variable	X	U_1	U_2	U_3
Sum	21	24	24	14

Sums of squares and products matrix:

$$\begin{bmatrix} 35 & 40 & 40 & 17 \\ & 50 & 50 & 19 \\ & & 50 & 19 \\ & & & 24 \end{bmatrix}$$

Course C ($N_3 = 30$)				
Variable	X	U_1	U_2	U_3
Sum	42	37	40	31

Sums of squares and products matrix:

$$\begin{bmatrix} 66 & 55 & 60 & 42 \\ & 51 & 52 & 36 \\ & & 60 & 41 \\ & & & 49 \end{bmatrix}$$

(a) Test the hypothesis of equal course means for the instructor overall evaluation after adjustment for the three concomitant variables.

(b) The four variables are very highly correlated. Repeat the analysis of covariance using only U_1 or U_2 as a single covariate.

appendix A

Tables and Charts

TABLE 1 NORMAL PROBABILITIES

The table contains the probabilities that a standard normal random variable is between zero and z:

$$P(0 \leq Z \leq z) = \frac{1}{\sqrt{2\pi}} \int_0^z e^{-1/2x^2}\, dx$$
$$= P(-z \leq Z \leq 0)$$

from the symmetry of the normal density function. The upper $100\alpha\%$ critical value z_α defined by

$$\alpha = P(Z > z_\alpha)$$
$$= 0.5 - P(0 \leq Z \leq z_\alpha)$$

is the value of z corresponding to the probability $0.5 - \alpha$ in the body of the table. From the symmetry of the density the lower $100\alpha\%$ critical value is $-z_\alpha$.

TABLE 2 PERCENTAGE POINTS OF THE CHI-SQUARED DISTRIBUTION

The upper percentage points $\chi^2_{\alpha;n}$ of the chi-squared distribution are defined by

$$\alpha = P(\chi^2 > \chi^2_{\alpha:n})$$
$$= \frac{1}{2^{n/2}\Gamma(n/2)} \int_{\chi^2_{\alpha;n}}^{\infty} x^{(n-2)/2} e^{-x/2}\, dx$$

The lower 100α% point is given by

$$\alpha = P(\chi^2 < \chi^2_{1-\alpha;\,n})$$

Table 2 contains common lower and upper percentage points for degrees of freedom n from 1 to 30.

TABLE 3 UPPER CRITICAL VALUES OF THE t-DISTRIBUTION

The upper 100α% critical value $t_{\alpha:\,n}$ of the Student–Fisher t-distribution with n degrees of freedom is defined by

$$P(t > t_{\alpha;\,n}) = \frac{\Gamma[(n+1)/2]}{\sqrt{\pi n}\,\Gamma(n/2)} \int_{t_{\alpha;\,n}}^{\infty} \frac{dx}{(1 + x^2/n)^{(n+1)/2}}$$

The lower percentage points follow from the symmetry of the t density as

$$t_{1-\alpha;\,n} = -t_{\alpha;\,n}$$

Some upper percentage points are given in Table 3.

TABLE 4 UPPER CRITICAL VALUES OF THE *F*-DISTRIBUTION

The upper 100α percentage point of the F-variate with m and n degrees of freedom is defined by

$$\begin{aligned} \alpha &= P(F > F_{\alpha;\,m,n}) \\ &= 1 - G(F) \end{aligned}$$

where

$$G(F) = \frac{\Gamma[(m+n)/2]}{\Gamma(m/2)\Gamma(n/2)} \left(\frac{m}{n}\right)^{m/2} \int_0^F x^{(m-2)/2} \left(1 + \frac{m}{n}x\right)^{-(m+n)/2} dx$$

is the cumulative distribution function. Upper $\alpha = 0.05$ and $\alpha = 0.01$ percentage points are given in the table. The lower percentage points can be found by the relationship

$$F_{1-\alpha;\,m,n} = \frac{1}{F_{\alpha;\,n,m}}$$

TABLE 5 UPPER CRITICAL VALUES OF *r*

The critical values $r_{\alpha;\,n}$ of the correlation coefficient r based on n degrees of freedom and the bivariate normal model with population correlation $\rho = 0$ are given in the table. The critical values are defined by

$$\begin{aligned} \alpha &= P(r > r_{\alpha;\,n}) \\ &= \frac{\Gamma[(n+1)/2]}{\sqrt{\pi}\,\Gamma(n/2)} \int_{r_{\alpha;\,n}}^{1} (1 - r^2)^{(n-2)/2}\, dr \end{aligned}$$

Because the density function is symmetric about zero the left-hand critical values are given by

$$r_{1-\alpha;\, n} = -r_{\alpha;\, n}$$

If the ordinary correlation coefficient were computed from N independent pairs of observations, $n = N - 2$.

TABLE 6 FISHER z-TRANSFORMATION AND ITS INVERSE

The Fisher z, or inverse hyperbolic tangent transformation

$$z = \tanh^{-1} r = \tfrac{1}{2} \ln \frac{1 + r}{1 - r}$$

of the correlation`coefficient, is given in the first part of the table. The inverse z-transformation

$$r = \frac{e^z - e^{-z}}{e^z + e^{-z}}$$

has been tabled in the second part.

CHARTS 1–10 POWER FUNCTIONS FOR THE F-TEST

These charts contain power probabilities given by the noncentral F-distribution. Each chart corresponds to a value $\nu_1 \equiv m$ of the first degrees of freedom parameter of the distribution and contains a family of curves for $\alpha = 0.05$ and another for $\alpha = 0.01$. The individual curves are for the second degrees of freedom parameter n. To find the power of an F-test based on m and n degrees of freedom we choose the proper chart and value of α, locate the value of the noncentrality parameter ϕ on the horizontal axis, and read up to the curve for the second degrees of freedom. The power probability of rejecting the null hypothesis when the alternative is true is found by reading across to the vertical scale.

TABLE 7 PERCENTAGE POINTS OF THE STUDENTIZED RANGE

The two parts of the table contain upper 0.05 and 0.01 critical values $q_{\alpha;\, k, \nu}$ for the Studentized range

$$q = \frac{x_{\max} - x_{\min}}{s}$$

The numerator is the range of k independent normal variates with a common mean and the same standard deviation σ, whereas s is an independent estimate of σ based on ν degrees of freedom.

TABLE 8 CRITICAL VALUES FOR THE DURBIN–WATSON TEST

The two parts of the table give critical values for the Durbin–Watson test of autocorrelation at the 0.05 or 0.01 level. N is the number of equally spaced points in the time series. p is the number of independent variables in the fitted model; the tables do not encompass the case of $p = 0$ or residuals from the mean. This decision rule is used with the computed Durbin–Watson statistic d:

Accept the hypothesis of no autocorrelation if $d > d_U$.
Reject the hypothesis of no autocorrelation if $d < d_L$.
The test is inconclusive if $d_L \leq d \leq d_U$.

TABLE 9 ORTHOGONAL POLYNOMIALS

Values of orthogonal polynomials up to the sixth degree for $N = 3$ to $N = 14$ equally spaced points are given in the table. In the notation of Chapter 4

$$\phi_j \equiv \phi_j(x) \qquad x = t - \frac{N+1}{2}$$

denotes the value of the jth-degree polynomial for a given N at the tth point. For $N = 13$ and $N = 14$ only the first seven values are given; to obtain the others we must use the relations

$$\phi_j(x) = \phi_j(-x) \qquad j \text{ even}$$
$$\phi_j(x) = -\phi_j(-x) \quad j \text{ odd}$$

The row below the ϕ_j values gives their sums of squares

$$\sum_x \phi_j^2(x)$$

The last row contains the multiplicative constants λ_j introduced to make the ϕ_j whole numbers in the table.

TABLE 1 Normal Probabilities

z	.00	.01	.02	.03	.04	.05	.06	.07	.08	.09
0.0	.0000	.0040	.0080	.0120	.0160	.0199	.0239	.0279	.0319	.0359
0.1	.0398	.0438	.0478	.0517	.0557	.0596	.0636	.0675	.0714	.0753
0.2	.0793	.0832	.0871	.0910	.0948	.0987	.1026	.1064	.1103	.1141
0.3	.1179	.1217	.1255	.1293	.1331	.1368	.1406	.1443	.1480	.1517
0.4	.1554	.1591	.1628	.1664	.1700	.1736	.1772	.1808	.1844	.1879
0.5	.1915	.1950	.1985	.2019	.2054	.2088	.2123	.2157	.2190	.2224
0.6	.2257	.2291	.2324	.2357	.2389	.2422	.2454	.2486	.2517	.2549
0.7	.2580	.2611	.2642	.2673	.2704	.2734	.2764	.2794	.2823	.2852
0.8	.2881	.2910	.2939	.2967	.2995	.3023	.3051	.3078	.3106	.3133
0.9	.3159	.3186	.3212	.3238	.3264	.3289	.3315	.3340	.3365	.3389
1.0	.3413	.3438	.3461	.3485	.3508	.3531	.3554	.3577	.3599	.3621
1.1	.3643	.3665	.3686	.3708	.3729	.3749	.3770	.3790	.3810	.3830
1.2	.3849	.3869	.3888	.3907	.3925	.3944	.3962	.3980	.3997	.4015
1.3	.4032	.4049	.4066	.4082	.4099	.4115	.4131	.4147	.4162	.4177
1.4	.4192	.4207	.4222	.4236	.4251	.4265	.4279	.4292	.4306	.4319
1.5	.4332	.4345	.4357	.4370	.4382	.4394	.4406	.4418	.4429	.4441
1.6	.4452	.4463	.4474	.4484	.4495	.4505	.4515	.4525	.4535	.4545
1.7	.4554	.4564	.4573	.4582	.4591	.4599	.4608	.4616	.4625	.4633
1.8	.4641	.4649	.4656	.4664	.4671	.4678	.4686	.4693	.4699	.4706
1.9	.4713	.4719	.4726	.4732	.4738	.4744	.4750	.4756	.4761	.4767
2.0	.4772	.4778	.4783	.4788	.4793	.4798	.4803	.4808	.4812	.4817
2.1	.4821	.4826	.4830	.4834	.4838	.4842	.4846	.4850	.4854	.4857
2.2	.4861	.4864	.4868	.4871	.4875	.4878	.4881	.4884	.4887	.4890
2.3	.4893	.4896	.4898	.4901	.4904	.4906	.4909	.4911	.4913	.4916
2.4	.4918	.4920	.4922	.4925	.4927	.4929	.4931	.4932	.4934	.4936
2.5	.4938	.4940	.4941	.4943	.4945	.4946	.4948	.4949	.4951	.4952
2.6	.4953	.4955	.4956	.4957	.4959	.4960	.4961	.4962	.4963	.4964
2.7	.4965	.4966	.4967	.4968	.4969	.4970	.4971	.4972	.4973	.4974
2.8	.4974	.4975	.4976	.4977	.4977	.4978	.4979	.4979	.4980	.4981
2.9	.4981	.4982	.4982	.4983	.4984	.4984	.4985	.4985	.4986	.4986
3.0	.4987	.4987	.4987	.4988	.4988	.4989	.4989	.4989	.4990	.4990

SOURCE: John E. Freund, *Mathematical Statistics*, 2nd ed., 1962, p. 366. Reprinted by permission of Prentice-Hall, Inc., Englewood Cliffs, N.J.

TABLE 2 Percentage Points of the Chi-Squared Distribution

n	$\alpha = .995$	$\alpha = .99$	$\alpha = .975$	$\alpha = .95$	$\alpha = .05$	$\alpha = .025$	$\alpha = .01$	$\alpha = .005$	ν
1	.0000393	.000157	.000982	.00393	3.841	5.024	6.635	7.879	1
2	.0100	.0201	.0506	.103	5.991	7.378	9.210	10.597	2
3	.0717	.115	.216	.352	7.815	9.348	11.345	12.838	3
4	.207	.297	.484	.711	9.488	11.143	13.277	14.860	4
5	.412	.554	.831	1.145	11.070	12.832	15.086	16.750	5
6	.676	.872	1.237	1.635	12.592	14.449	16.812	18.548	6
7	.989	1.239	1.690	2.167	14.067	16.013	18.475	20.278	7
8	1.344	1.646	2.180	2.733	15.507	17.535	20.090	21.955	8
9	1.735	2.088	2.700	3.325	16.919	19.023	21.666	23.589	9
10	2.156	2.558	3.247	3.940	18.307	20.483	23.209	25.188	10
11	2.603	3.053	3.816	4.575	19.675	21.920	24.725	26.757	11
12	3.074	3.571	4.404	5.226	21.026	23.337	26.217	28.300	12
13	3.565	4.107	5.009	5.892	22.362	24.736	27.688	29.819	13
14	4.075	4.660	5.629	6.571	23.685	26.119	29.141	31.319	14
15	4.601	5.229	6.262	7.261	24.996	27.488	30.578	32.801	15
16	5.142	5.812	6.908	7.962	26.296	28.845	32.000	34.267	16
17	5.697	6.408	7.564	8.672	27.587	30.191	33.409	35.718	17
18	6.265	7.015	8.231	9.390	28.869	31.526	34.805	37.156	18
19	6.844	7.633	8.907	10.117	30.144	32.852	36.191	38.582	19
20	7.434	8.260	9.591	10.851	31.410	34.170	37.566	39.997	20
21	8.034	8.897	10.283	11.591	32.671	35.479	38.932	41.401	21
22	8.643	9.542	10.982	12.338	33.924	36.781	40.289	42.796	22
23	9.260	10.196	11.689	13.091	35.172	38.076	41.638	44.181	23
24	9.886	10.856	12.401	13.848	36.415	39.364	42.980	45.558	24
25	10.520	11.524	13.120	14.611	37.652	40.646	44.314	46.928	25
26	11.160	12.198	13.844	15.379	38.885	41.923	45.642	48.290	26
27	11.808	12.879	14.573	16.151	40.113	43.194	46.963	49.645	27
28	12.461	13.565	15.308	16.928	41.337	44.461	48.278	50.993	28
29	13.121	14.256	16.047	17.708	42.557	45.722	49.588	52.336	29
30	13.787	14.953	16.791	18.493	43.773	46.979	50.892	53.672	30

SOURCE: Based on C. M. Thompson: Tables of percentage points of the incomplete beta function and of the chi-square distribution, *Biometrika*, vol. 32 (1941), pp. 187–191, with the kind permission of the *Biometrika* Trustees.

TABLE 3 Upper Critical Values of the t-Distribution

$t_{\alpha; n}$

n	$\alpha = .10$	$\alpha = .05$	$\alpha = .025$	$\alpha = .01$	$\alpha = .005$	ν
1	3.078	6.314	12.706	31.821	63.657	1
2	1.886	2.920	4.303	6.965	9.925	2
3	1.638	2.353	3.182	4.541	5.841	3
4	1.533	2.132	2.776	3.747	4.604	4
5	1.476	2.015	2.571	3.365	4.032	5
6	1.440	1.943	2.447	3.143	3.707	6
7	1.415	1.895	2.365	2.998	3.499	7
8	1.397	1.860	2.306	2.896	3.355	8
9	1.383	1.833	2.262	2.821	3.250	9
10	1.372	1.812	2.228	2.764	3.169	10
11	1.363	1.796	2.201	2.718	3.106	11
12	1.356	1.782	2.179	2.681	3.055	12
13	1.350	1.771	2.160	2.650	3.012	13
14	1.345	1.761	2.145	2.624	2.977	14
15	1.341	1.753	2.131	2.602	2.947	15
16	1.337	1.746	2.120	2.583	2.921	16
17	1.333	1.740	2.110	2.567	2.898	17
18	1.330	1.734	2.101	2.552	2.878	18
19	1.328	1.729	2.093	2.539	2.861	19
20	1.325	1.725	2.086	2.528	2.845	20
21	1.323	1.721	2.080	2.518	2.831	21
22	1.321	1.717	2.074	2.508	2.819	22
23	1.319	1.714	2.069	2.500	2.807	23
24	1.318	1.711	2.064	2.492	2.797	24
25	1.316	1.708	2.060	2.485	2.787	25
26	1.315	1.706	2.056	2.479	2.779	26
27	1.314	1.703	2.052	2.473	2.771	27
28	1.313	1.701	2.048	2.467	2.763	28
29	1.311	1.699	2.045	2.462	2.756	29
∞	1.282	1.645	1.960	2.326	2.576	inf.

SOURCE: Abridged with permission of Macmillan Publishing Co., Inc., from *Statistical Methods for Research Workers*, 14th ed., by R. A. Fisher. Copyright 1970 University of Adelaide.

TABLE 4 Upper Critical Values of the F-Distribution

a. Values of $F_{0.05;m,n}$

m = Degrees of freedom for numerator; n = Degrees of freedom for denominator

n	1	2	3	4	5	6	7	8	9	10	12	15	20	24	30	40	60	120	∞
1	161	200	216	225	230	234	237	239	241	242	244	246	248	249	250	251	252	253	254
2	18.5	19.0	19.2	19.2	19.3	19.3	19.4	19.4	19.4	19.4	19.4	19.4	19.4	19.5	19.5	19.5	19.5	19.5	19.5
3	10.1	9.55	9.28	9.12	9.01	8.94	8.89	8.85	8.81	8.79	8.74	8.70	8.66	8.64	8.62	8.59	8.57	8.55	8.53
4	7.71	6.94	6.59	6.39	6.26	6.16	6.09	6.04	6.00	5.96	5.91	5.86	5.80	5.77	5.75	5.72	5.69	5.66	5.63
5	6.61	5.79	5.41	5.19	5.05	4.95	4.88	4.82	4.77	4.74	4.68	4.62	4.56	4.53	4.50	4.46	4.43	4.40	4.37
6	5.99	5.14	4.76	4.53	4.39	4.28	4.21	4.15	4.10	4.06	4.00	3.94	3.87	3.84	3.81	3.77	3.74	3.70	3.67
7	5.59	4.74	4.35	4.12	3.97	3.87	3.79	3.73	3.68	3.64	3.57	3.51	3.44	3.41	3.38	3.34	3.30	3.27	3.23
8	5.32	4.46	4.07	3.84	3.69	3.58	3.50	3.44	3.39	3.35	3.28	3.22	3.15	3.12	3.08	3.04	3.01	2.97	2.93
9	5.12	4.26	3.86	3.63	3.48	3.37	3.29	3.23	3.18	3.14	3.07	3.01	2.94	2.90	2.86	2.83	2.79	2.75	2.71
10	4.96	4.10	3.71	3.48	3.33	3.22	3.14	3.07	3.02	2.98	2.91	2.85	2.77	2.74	2.70	2.66	2.62	2.58	2.54
11	4.84	3.98	3.59	3.36	3.20	3.09	3.01	2.95	2.90	2.85	2.79	2.72	2.65	2.61	2.57	2.53	2.49	2.45	2.40
12	4.75	3.89	3.49	3.26	3.11	3.00	2.91	2.85	2.80	2.75	2.69	2.62	2.54	2.51	2.47	2.43	2.38	2.34	2.30
13	4.67	3.81	3.41	3.18	3.03	2.92	2.83	2.77	2.71	2.67	2.60	2.53	2.46	2.42	2.38	2.34	2.30	2.25	2.21
14	4.60	3.74	3.34	3.11	2.96	2.85	2.76	2.70	2.65	2.60	2.53	2.46	2.39	2.35	2.31	2.27	2.22	2.18	2.13
15	4.54	3.68	3.29	3.06	2.90	2.79	2.71	2.64	2.59	2.54	2.48	2.40	2.33	2.29	2.25	2.20	2.16	2.11	2.07
16	4.49	3.63	3.24	3.01	2.85	2.74	2.66	2.59	2.54	2.49	2.42	2.35	2.28	2.24	2.19	2.15	2.11	2.06	2.01
17	4.45	3.59	3.20	2.96	2.81	2.70	2.61	2.55	2.49	2.45	2.38	2.31	2.23	2.19	2.15	2.10	2.06	2.01	1.96
18	4.41	3.55	3.16	2.93	2.77	2.66	2.58	2.51	2.46	2.41	2.34	2.27	2.19	2.15	2.11	2.06	2.02	1.97	1.92
19	4.38	3.52	3.13	2.90	2.74	2.63	2.54	2.48	2.42	2.38	2.31	2.23	2.16	2.11	2.07	2.03	1.98	1.93	1.88
20	4.35	3.49	3.10	2.87	2.71	2.60	2.51	2.45	2.39	2.35	2.28	2.20	2.12	2.08	2.04	1.99	1.95	1.90	1.84
21	4.32	3.47	3.07	2.84	2.68	2.57	2.49	2.42	2.37	2.32	2.25	2.18	2.10	2.05	2.01	1.96	1.92	1.87	1.81
22	4.30	3.44	3.05	2.82	2.66	2.55	2.46	2.40	2.34	2.30	2.23	2.15	2.07	2.03	1.98	1.94	1.89	1.84	1.78
23	4.28	3.42	3.03	2.80	2.64	2.53	2.44	2.37	2.32	2.27	2.20	2.13	2.05	2.01	1.96	1.91	1.86	1.81	1.76
24	4.26	3.40	3.01	2.78	2.62	2.51	2.42	2.36	2.30	2.25	2.18	2.11	2.03	1.98	1.94	1.89	1.84	1.79	1.73
25	4.24	3.39	2.99	2.76	2.60	2.49	2.40	2.34	2.28	2.24	2.16	2.09	2.01	1.96	1.92	1.87	1.82	1.77	1.71
30	4.17	3.32	2.92	2.69	2.53	2.42	2.33	2.27	2.21	2.16	2.09	2.01	1.93	1.89	1.84	1.79	1.74	1.68	1.62
40	4.08	3.23	2.84	2.61	2.45	2.34	2.25	2.18	2.12	2.08	2.00	1.92	1.84	1.79	1.74	1.69	1.64	1.58	1.51
60	4.00	3.15	2.76	2.53	2.37	2.25	2.17	2.10	2.04	1.99	1.92	1.84	1.75	1.70	1.65	1.59	1.53	1.47	1.39
120	3.92	3.07	2.68	2.45	2.29	2.18	2.09	2.02	1.96	1.91	1.83	1.75	1.66	1.61	1.55	1.50	1.43	1.35	1.25
∞	3.84	3.00	2.60	2.37	2.21	2.10	2.01	1.94	1.88	1.83	1.75	1.67	1.57	1.52	1.46	1.39	1.32	1.22	1.00

b. Values of $F_{0.01; m,n}$

m = Degrees of freedom for numerator

n = Degrees of freedom for denominator	1	2	3	4	5	6	7	8	9	10	12	15	20	24	30	40	60	120	∞
1	4,052	5,000	5,403	5,625	5,764	5,859	5,928	5,982	6,023	6,056	6,106	6,157	6,209	6,235	6,261	6,287	6,313	6,339	6,366
2	98.5	99.0	99.2	99.2	99.3	99.3	99.4	99.4	99.4	99.4	99.4	99.4	99.4	99.5	99.5	99.5	99.5	99.5	99.5
3	34.1	30.8	29.5	28.7	28.2	27.9	27.7	27.5	27.3	27.2	27.1	26.9	26.7	26.6	26.5	26.4	26.3	26.2	26.1
4	21.2	18.0	16.7	16.0	15.5	15.2	15.0	14.8	14.7	14.5	14.4	14.2	14.0	13.9	13.8	13.7	13.7	13.6	13.5
5	16.3	13.3	12.1	11.4	11.0	10.7	10.5	10.3	10.2	10.1	9.89	9.72	9.55	9.47	9.38	9.29	9.20	9.11	9.02
6	13.7	10.9	9.78	9.15	8.75	8.47	8.26	8.10	7.98	7.87	7.72	7.56	7.40	7.31	7.23	7.14	7.06	6.97	6.88
7	12.2	9.55	8.45	7.85	7.46	7.19	6.99	6.84	6.72	6.62	6.47	6.31	6.16	6.07	5.99	5.91	5.82	5.74	5.65
8	11.3	8.65	7.59	7.01	6.63	6.37	6.18	6.03	5.91	5.81	5.67	5.52	5.36	5.28	5.20	5.12	5.03	4.95	4.86
9	10.6	8.02	6.99	6.42	6.06	5.80	5.61	5.47	5.35	5.26	5.11	4.96	4.81	4.73	4.65	4.57	4.48	4.40	4.31
10	10.0	7.56	6.55	5.99	5.64	5.39	5.20	5.06	4.94	4.85	4.71	4.56	4.41	4.33	4.25	4.17	4.08	4.00	3.91
11	9.65	7.21	6.22	5.67	5.32	5.07	4.89	4.74	4.63	4.54	4.40	4.25	4.10	4.02	3.94	3.86	3.78	3.69	3.60
12	9.33	6.93	5.95	5.41	5.06	4.82	4.64	4.50	4.39	4.30	4.16	4.01	3.86	3.78	3.70	3.62	3.54	3.45	3.36
13	9.07	6.70	5.74	5.21	4.86	4.62	4.44	4.30	4.19	4.10	3.96	3.82	3.66	3.59	3.51	3.43	3.34	3.25	3.17
14	8.86	6.51	5.56	5.04	4.70	4.46	4.28	4.14	4.03	3.94	3.80	3.66	3.51	3.43	3.35	3.27	3.18	3.09	3.00
15	8.68	6.36	5.42	4.89	4.56	4.32	4.14	4.00	3.89	3.80	3.67	3.52	3.37	3.29	3.21	3.13	3.05	2.96	2.87
16	8.53	6.23	5.29	4.77	4.44	4.20	4.03	3.89	3.78	3.69	3.55	3.41	3.26	3.18	3.10	3.02	2.93	2.84	2.75
17	8.40	6.11	5.19	4.67	4.34	4.10	3.93	3.79	3.68	3.59	3.46	3.31	3.16	3.08	3.00	2.92	2.83	2.75	2.65
18	8.29	6.01	5.09	4.58	4.25	4.01	3.84	3.71	3.60	3.51	3.37	3.23	3.08	3.00	2.92	2.84	2.75	2.66	2.57
19	8.19	5.93	5.01	4.50	4.17	3.94	3.77	3.63	3.52	3.43	3.30	3.15	3.00	2.92	2.84	2.76	2.67	2.58	2.49
20	8.10	5.85	4.94	4.43	4.10	3.87	3.70	3.56	3.46	3.37	3.23	3.09	2.94	2.86	2.78	2.69	2.61	2.52	2.42
21	8.02	5.78	4.87	4.37	4.04	3.81	3.64	3.51	3.40	3.31	3.17	3.03	2.88	2.80	2.72	2.64	2.55	2.46	2.36
22	7.95	5.72	4.82	4.31	3.99	3.76	3.59	3.45	3.35	3.26	3.12	2.98	2.83	2.75	2.67	2.58	2.50	2.40	2.31
23	7.88	5.66	4.76	4.26	3.94	3.71	3.54	3.41	3.30	3.21	3.07	2.93	2.78	2.70	2.62	2.54	2.45	2.35	2.26
24	7.82	5.61	4.72	4.22	3.90	3.67	3.50	3.36	3.26	3.17	3.03	2.89	2.74	2.66	2.58	2.49	2.40	2.31	2.21
25	7.77	5.57	4.68	4.18	3.86	3.63	3.46	3.32	3.22	3.13	2.99	2.85	2.70	2.62	2.53	2.45	2.36	2.27	2.17
30	7.56	5.39	4.51	4.02	3.70	3.47	3.30	3.17	3.07	2.98	2.84	2.70	2.55	2.47	2.39	2.30	2.21	2.11	2.01
40	7.31	5.18	4.31	3.83	3.51	3.29	3.12	2.99	2.89	2.80	2.66	2.52	2.37	2.29	2.20	2.11	2.02	1.92	1.80
60	7.08	4.98	4.13	3.65	3.34	3.12	2.95	2.82	2.72	2.63	2.50	2.35	2.20	2.12	2.03	1.94	1.84	1.73	1.60
120	6.85	4.79	3.95	3.48	3.17	2.96	2.79	2.66	2.56	2.47	2.34	2.19	2.03	1.95	1.86	1.76	1.66	1.53	1.38
∞	6.63	4.61	3.78	3.32	3.02	2.80	2.64	2.51	2.41	2.32	2.18	2.04	1.88	1.79	1.70	1.59	1.47	1.32	1.00

SOURCE: M. Merington and C. M. Thompson: Tables of percentage points of the inverted beta (*F*) distribution, *Biometrika*, vol. 33 (1943), pp. 73–88, with the kind permission of the *Biometrika* Trustees.

TABLE 5 Upper Critical Values of r

n	$r_{\alpha;\,n}$				
	0.10	0.05	0.025	0.01	0.005
1	0.951	0.988	0.997	1	1
2	0.800	0.900	0.950	0.980	0.990
3	0.687	0.805	0.878	0.934	0.959
4	0.608	0.729	0.811	0.882	0.917
5	0.551	0.669	0.755	0.833	0.875
6	0.507	0.621	0.707	0.789	0.834
7	0.472	0.582	0.666	0.750	0.798
8	0.443	0.549	0.632	0.715	0.765
9	0.419	0.521	0.602	0.685	0.735
10	0.398	0.497	0.576	0.658	0.708
11	0.380	0.476	0.553	0.634	0.684
12	0.365	0.457	0.532	0.612	0.661
13	0.351	0.441	0.514	0.592	0.641
14	0.338	0.426	0.497	0.574	0.623
15	0.327	0.412	0.482	0.558	0.606
16	0.317	0.400	0.468	0.542	0.590
17	0.308	0.389	0.456	0.529	0.575
18	0.299	0.378	0.444	0.515	0.561
19	0.291	0.369	0.433	0.503	0.549
20	0.284	0.360	0.423	0.492	0.537
21	0.277	0.352	0.413	0.482	0.526
22	0.271	0.344	0.404	0.472	0.515
23	0.265	0.337	0.396	0.462	0.505
24	0.260	0.330	0.388	0.453	0.496
25	0.255	0.323	0.381	0.445	0.487
26	0.250	0.317	0.374	0.437	0.479
27	0.245	0.311	0.367	0.430	0.471
28	0.241	0.306	0.361	0.423	0.463
29	0.237	0.301	0.355	0.416	0.456
30	0.233	0.296	0.349	0.409	0.449
40	0.202	0.257	0.304	0.358	0.393
60	0.165	0.211	0.250	0.295	0.325
120	0.117	0.150	0.178	0.210	0.232

TABLE 6 Fisher z-Transformation and Its Inverse

a. $z = \frac{1}{2} \ln [(1 + r)/(1 - r)]$

r	0.00	0.01	0.02	0.03	0.04	0.05	0.06	0.07	0.08	0.09
0.0	0.0000	0.0100	0.0200	0.0300	0.0400	0.0500	0.0601	0.0701	0.0802	0.0902
0.1	0.1003	0.1104	0.1206	0.1307	0.1409	0.1511	0.1614	0.1717	0.1820	0.1923
0.2	0.2027	0.2132	0.2237	0.2342	0.2448	0.2554	0.2661	0.2769	0.2877	0.2986
0.3	0.3095	0.3205	0.3316	0.3428	0.3541	0.3654	0.3769	0.3884	0.4001	0.4118
0.4	0.4236	0.4356	0.4477	0.4599	0.4722	0.4847	0.4973	0.5101	0.5230	0.5361
0.5	0.5493	0.5627	0.5763	0.5901	0.6042	0.6184	0.6328	0.6475	0.6625	0.6777
0.6	0.6931	0.7089	0.7250	0.7414	0.7582	0.7753	0.7928	0.8107	0.8291	0.8480
0.7	0.8673	0.8872	0.9076	0.9287	0.9505	0.9730	0.9962	1.0203	1.0454	1.0714
0.8	1.0986	1.1270	1.1568	1.1881	1.2212	1.2562	1.2933	1.3331	1.3758	1.4219
0.9	1.4722	1.5275	1.5890	1.6584	1.7380	1.8318	1.9459	2.0923	2.2976	2.6467

b. $r = (e^{z} - e^{-z})/(e^{z} + e^{-z})$

z	0.00	0.01	0.02	0.03	0.04	0.05	0.06	0.07	0.08	0.09
0.0	0.000	0.010	0.020	0.030	0.040	0.050	0.060	0.070	0.080	0.090
0.1	0.100	0.110	0.119	0.129	0.139	0.149	0.159	0.168	0.178	0.188
0.2	0.197	0.207	0.217	0.226	0.235	0.245	0.254	0.264	0.273	0.282
0.3	0.291	0.300	0.310	0.319	0.327	0.336	0.345	0.354	0.363	0.371
0.4	0.380	0.388	0.397	0.405	0.414	0.422	0.430	0.438	0.446	0.454
0.5	0.462	0.470	0.478	0.485	0.493	0.501	0.508	0.515	0.523	0.530
0.6	0.537	0.544	0.551	0.558	0.565	0.572	0.578	0.585	0.592	0.598
0.7	0.604	0.611	0.617	0.623	0.629	0.635	0.641	0.647	0.653	0.658
0.8	0.664	0.670	0.675	0.680	0.686	0.691	0.696	0.701	0.706	0.711
0.9	0.716	0.721	0.726	0.731	0.735	0.740	0.744	0.749	0.753	0.757
1.0	0.762	0.766	0.770	0.774	0.778	0.782	0.786	0.789	0.793	0.797
1.1	0.800	0.804	0.808	0.811	0.814	0.818	0.821	0.824	0.827	0.831
1.2	0.834	0.837	0.840	0.843	0.845	0.848	0.851	0.854	0.856	0.859
1.3	0.862	0.864	0.867	0.869	0.872	0.874	0.876	0.879	0.881	0.883
1.4	0.885	0.887	0.890	0.892	0.894	0.896	0.898	0.900	0.901	0.903
1.5	0.905	0.907	0.909	0.910	0.912	0.914	0.915	0.917	0.919	0.920
1.6	0.922	0.923	0.925	0.926	0.927	0.929	0.930	0.932	0.933	0.934
1.7	0.935	0.937	0.938	0.939	0.940	0.941	0.943	0.944	0.945	0.946
1.8	0.947	0.948	0.949	0.950	0.951	0.952	0.953	0.954	0.954	0.955
1.9	0.956	0.957	0.958	0.959	0.960	0.960	0.961	0.962	0.963	0.963

CHARTS 1-10 Power Functions for the *F*-Test

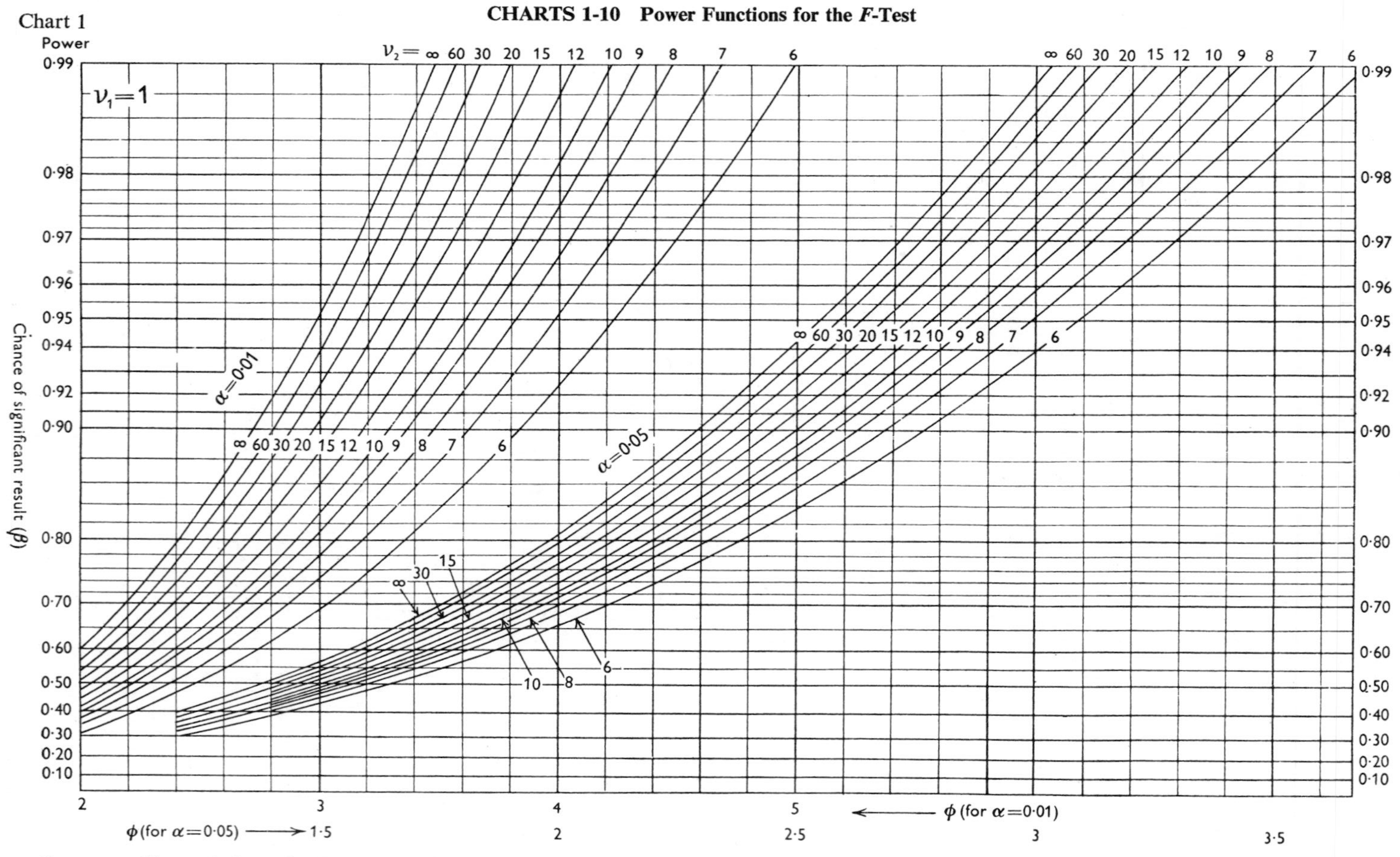

SOURCE: Charts 1 through 10 reprinted from E. S. Pearson and H. O. Hartley (eds.): *Biometrika Tables for Statisticians*, vol. 2, Cambridge University Press, 1972, with the kind permission of the *Biometrika* Trustees.

Chart 2

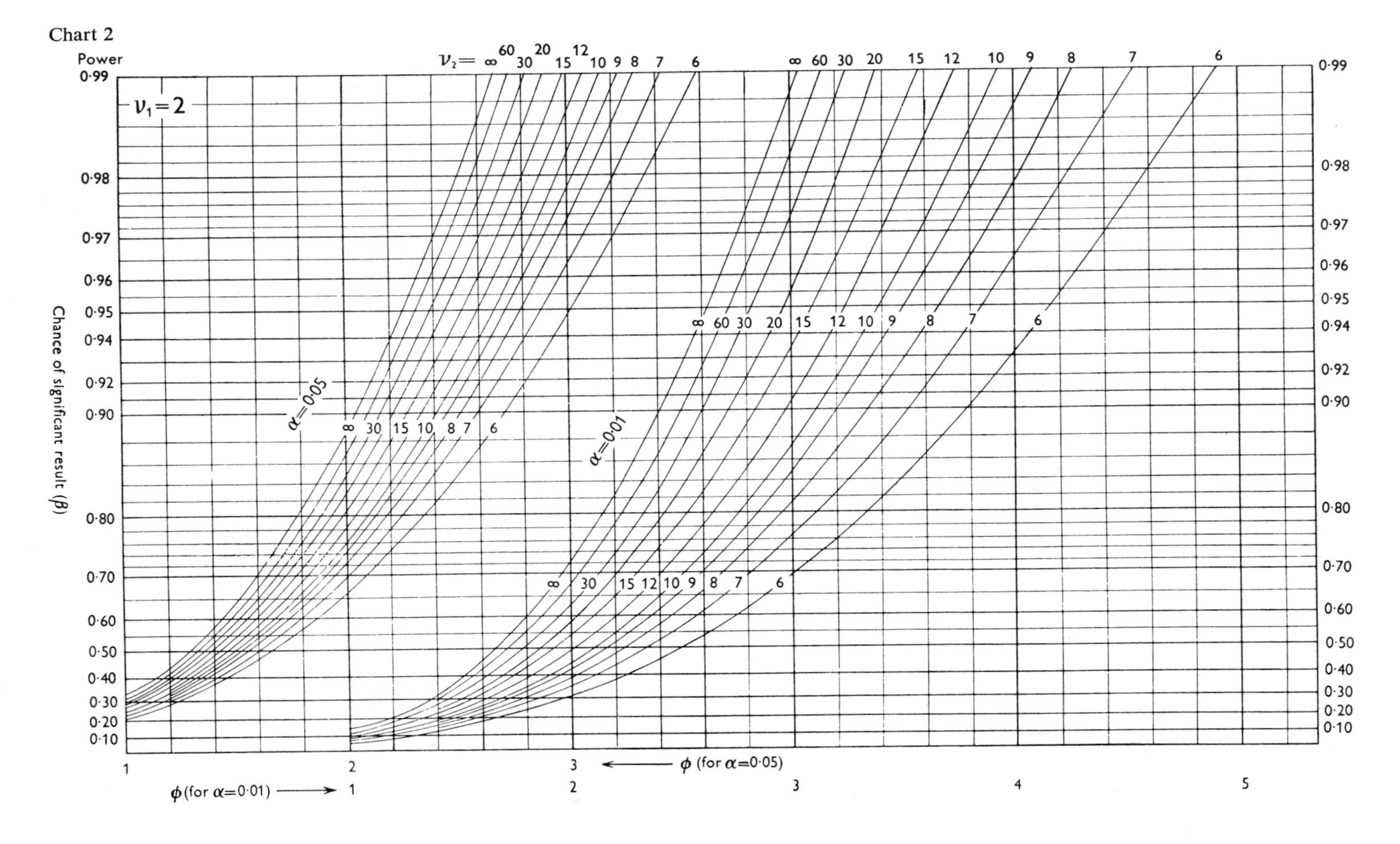

Power
Chance of significant result (β)
$\nu_1 = 2$
$\nu_2 =$ ∞ 60 30 20 15 12 10 9 8 7 6
∞ 60 30 20 15 12 10 9 8 7 6
$\alpha = 0{\cdot}05$
∞ 30 15 10 8 7 6
$\alpha = 0{\cdot}01$
∞ 60 30 20 15 12 10 9 8 7 6
∞ 30 15 12 10 9 8 7 6
0·99
0·98
0·97
0·96
0·95
0·94
0·92
0·90
0·80
0·70
0·60
0·50
0·40
0·30
0·20
0·10
ϕ (for $\alpha = 0{\cdot}01$) ⟶
1 2 3
⟵ ϕ (for $\alpha = 0{\cdot}05$)
1 2 3 4 5

Chart 3

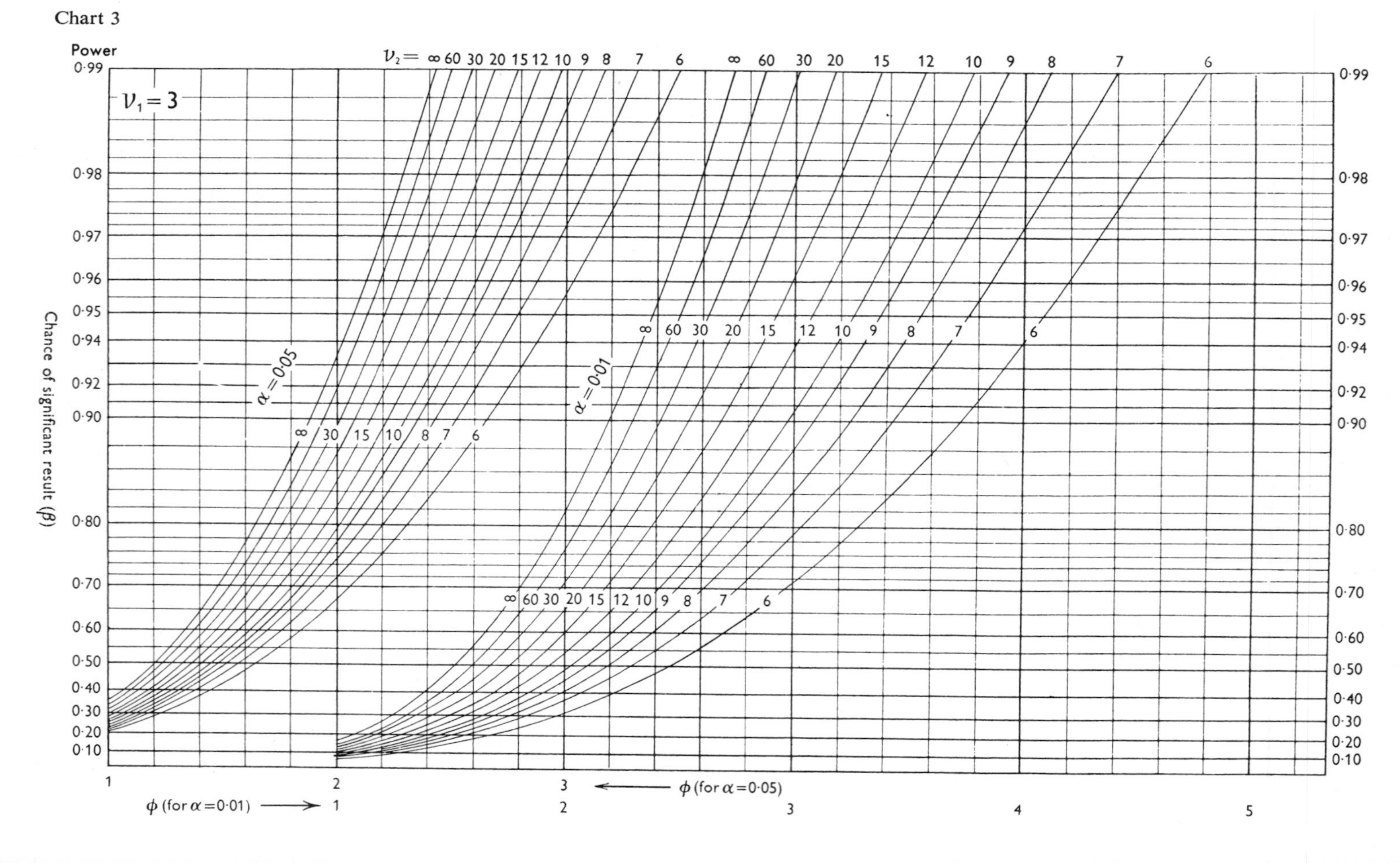
Power
$\nu_1 = 3$
$\nu_2 =$ ∞ 60 30 20 15 12 10 9 8 7 6 ∞ 60 30 20 15 12 10 9 8 7 6
$\alpha = 0{\cdot}05$
∞ 30 15 10 8 7 6
$\alpha = 0{\cdot}01$
∞ 60 30 20 15 12 10 9 8 7 6
∞ 60 30 20 15 12 10 9 8 7 6
0·99 0·98 0·97 0·96 0·95 0·94 0·92 0·90 0·80 0·70 0·60 0·50 0·40 0·30 0·20 0·10
Chance of significant result (β)
1 2 3 $\leftarrow$ ϕ (for $\alpha = 0{\cdot}05$)
ϕ (for $\alpha = 0{\cdot}01$) $\rightarrow$ 1 2 3 4 5

Chart 4

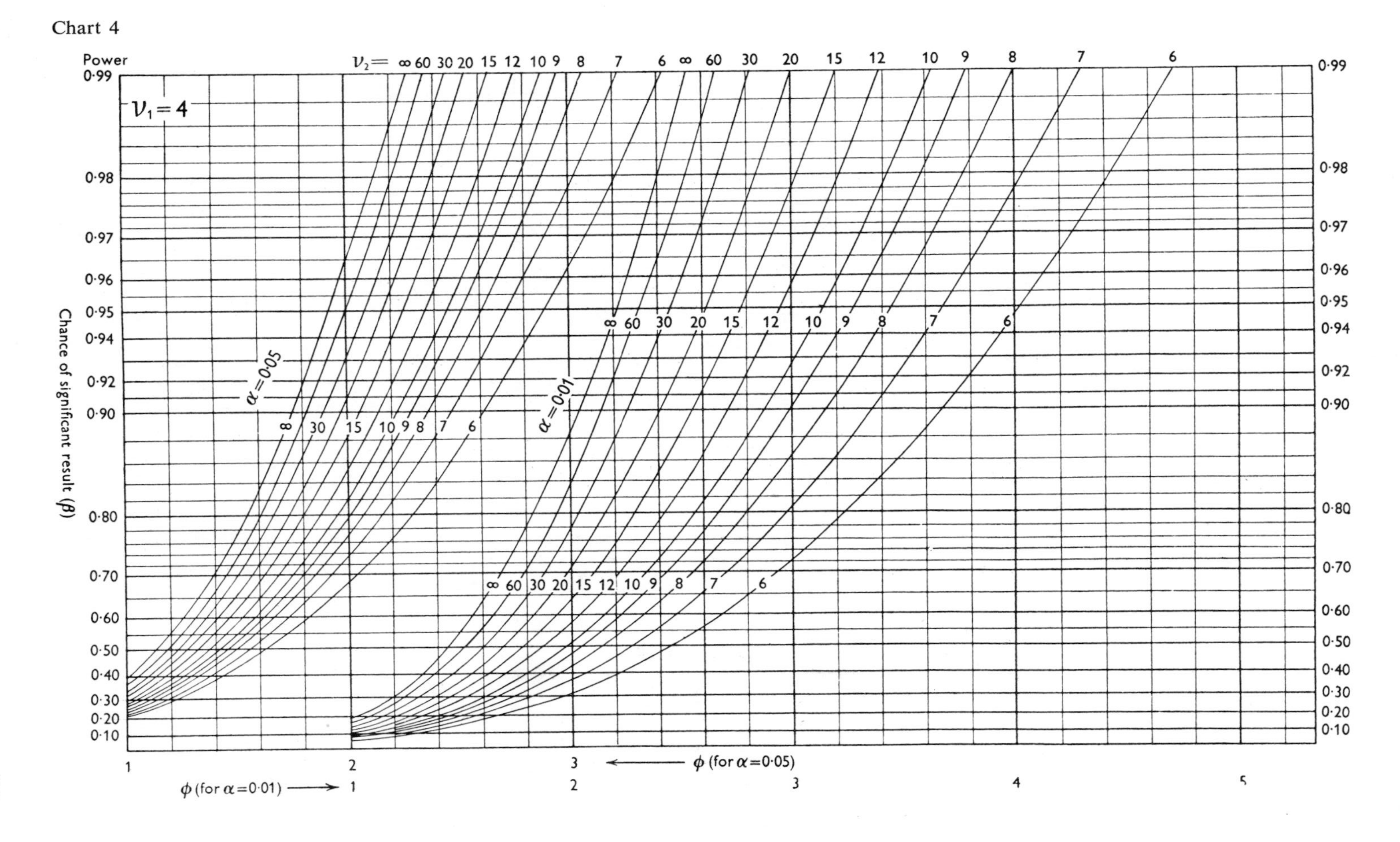

$\nu_1 = 4$
$\nu_2 =$ ∞ 60 30 20 15 12 10 9 8 7 6 ∞ 60 30 20 15 12 10 9 8 7 6
α = 0·05
α = 0·01
Power
0·99 0·98 0·97 0·96 0·95 0·94 0·92 0·90 0·80 0·70 0·60 0·50 0·40 0·30 0·20 0·10
Chance of significant result (β)
φ (for α=0·01) ⟶
1 2 3
φ (for α=0·05) ⟵
1 2 3 4 5

Chart 5

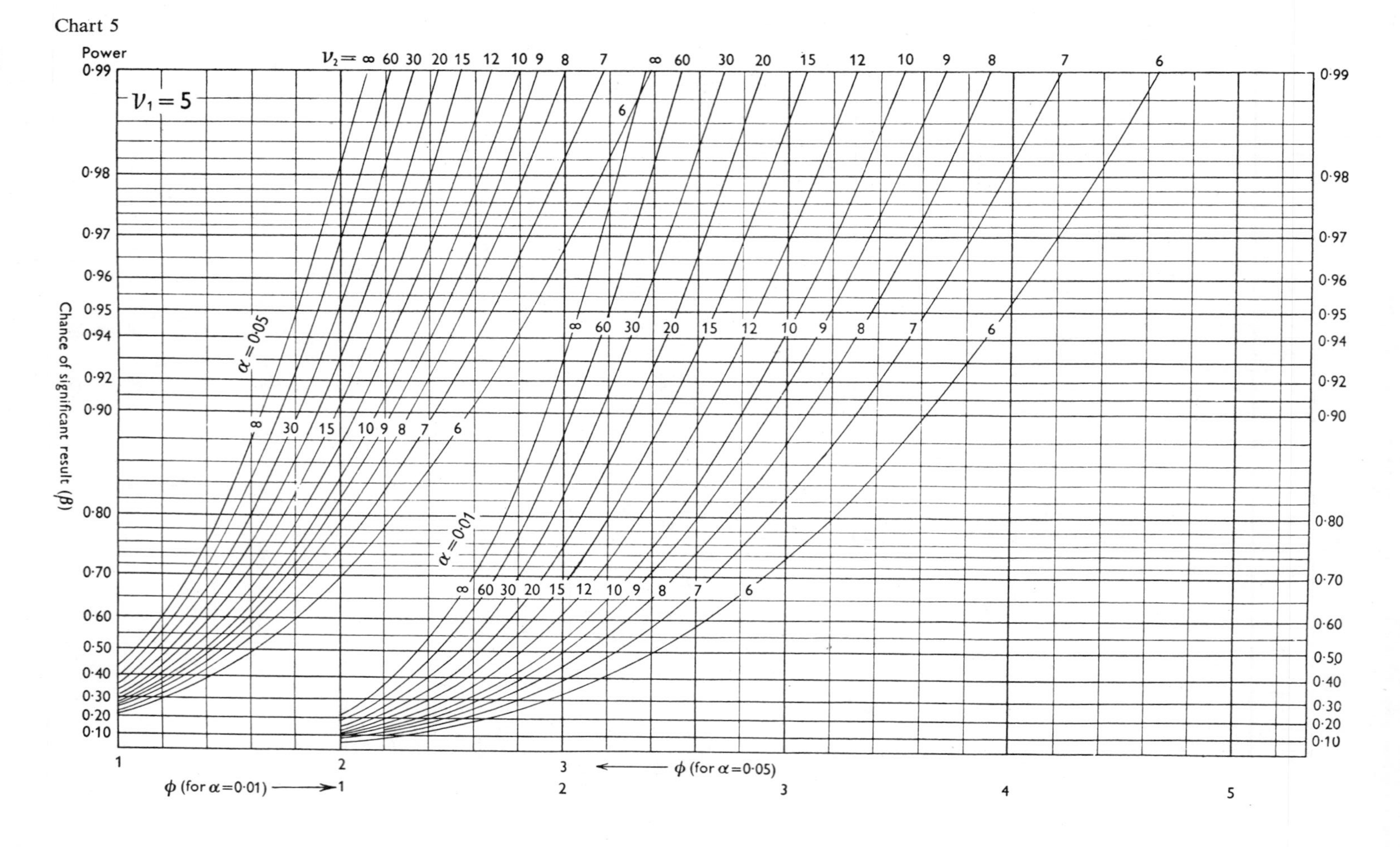
Power
$\nu_1 = 5$
$\nu_2 = \infty$ 60 30 20 15 12 10 9 8 7 ∞ 60 30 20 15 12 10 9 8 7 6
6
$\alpha = 0{\cdot}05$
∞ 30 15 10 9 8 7 6
∞ 60 30 20 15 12 10 9 8 7 6
$\alpha = 0{\cdot}01$
∞ 60 30 20 15 12 10 9 8 7 6
0·99
0·98
0·97
0·96
0·95
0·94
0·92
0·90
0·80
0·70
0·60
0·50
0·40
0·30
0·20
0·10
Chance of significant result (β)
1 2 3 4 5
ϕ (for $\alpha = 0{\cdot}01$) ⟶
← ϕ (for $\alpha = 0{\cdot}05$)

Chart 6

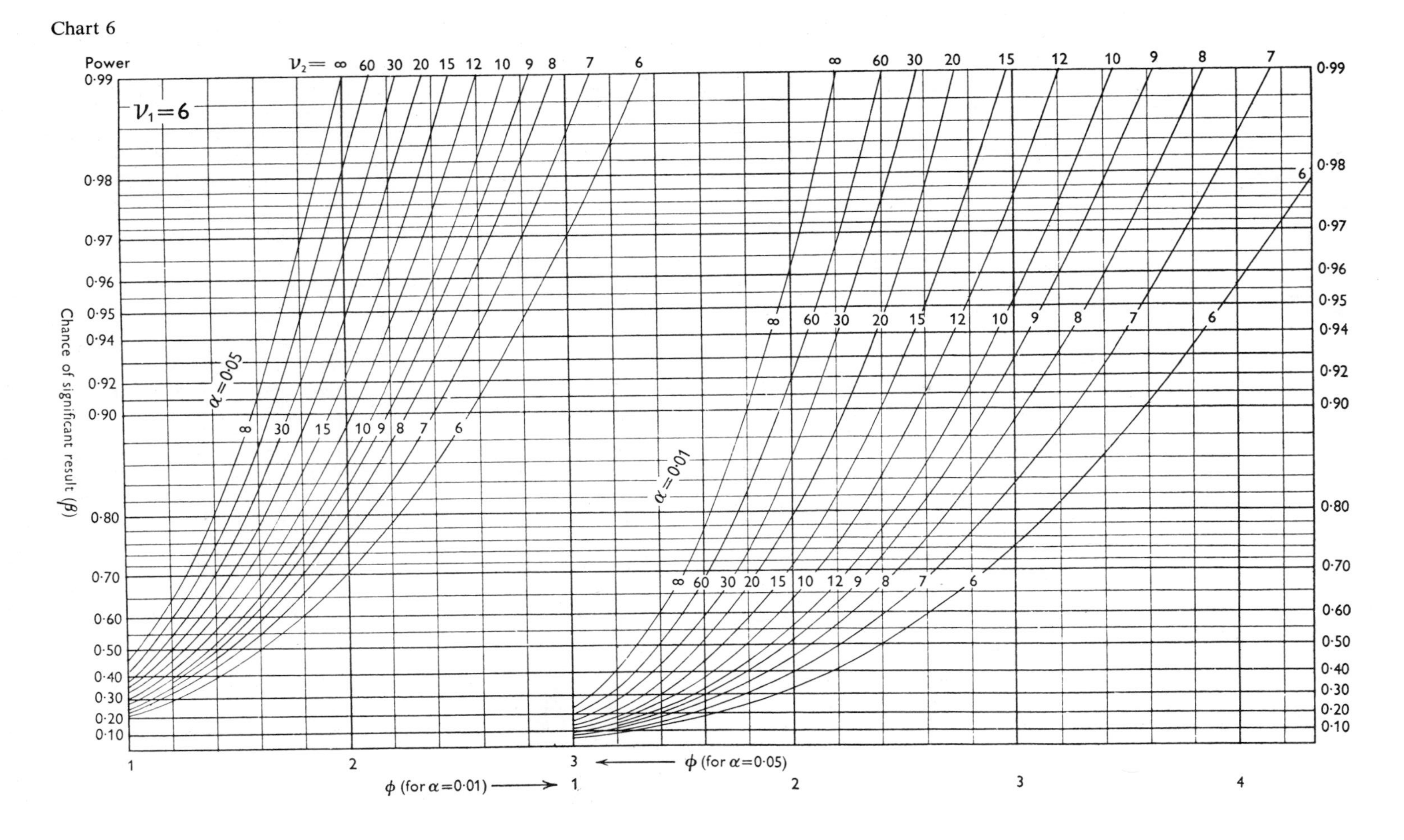

$\nu_1 = 6$
$\nu_2 = \infty$ 60 30 20 15 12 10 9 8 7 6
Power
0·99 0·98 0·97 0·96 0·95 0·94 0·92 0·90 0·80 0·70 0·60 0·50 0·40 0·30 0·20 0·10
Chance of significant result (β)
$\alpha = 0{\cdot}05$
$\alpha = 0{\cdot}01$
ϕ (for $\alpha = 0{\cdot}01$) ⟶
⟵ ϕ (for $\alpha = 0{\cdot}05$)
1 2 3
1 2 3 4

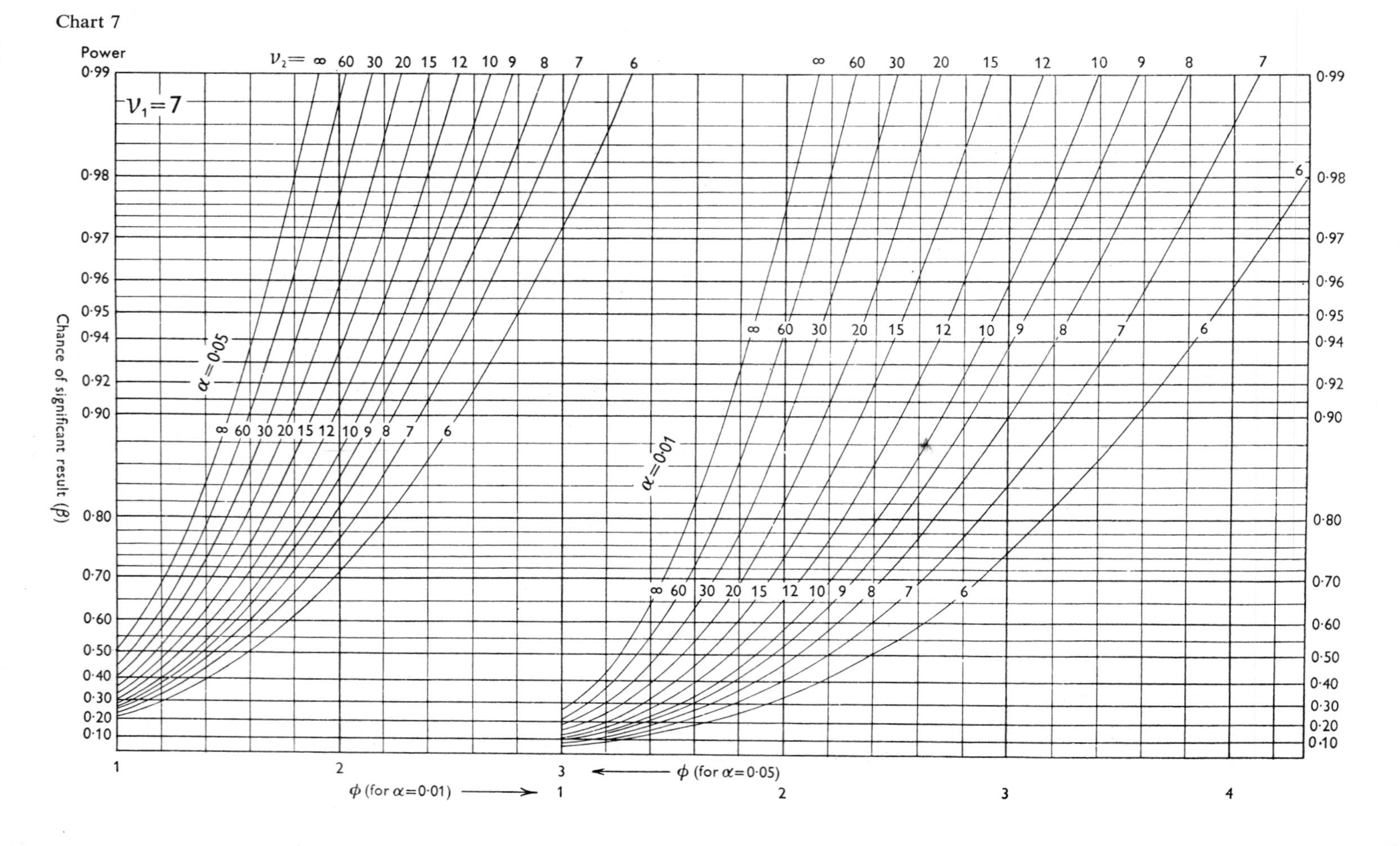
Chart 7
$\nu_1 = 7$
Power
Chance of significant result (β)
$\nu_2 = \infty$ 60 30 20 15 12 10 9 8 7 6
∞ 60 30 20 15 12 10 9 8 7 6
0·99 0·98 0·97 0·96 0·95 0·94 0·92 0·90 0·80 0·70 0·60 0·50 0·40 0·30 0·20 0·10
$\alpha = 0{\cdot}05$
$\alpha = 0{\cdot}01$
1 2 3 ← ϕ (for $\alpha = 0{\cdot}05$)
ϕ (for $\alpha = 0{\cdot}01$) → 1 2 3 4

Chart 8

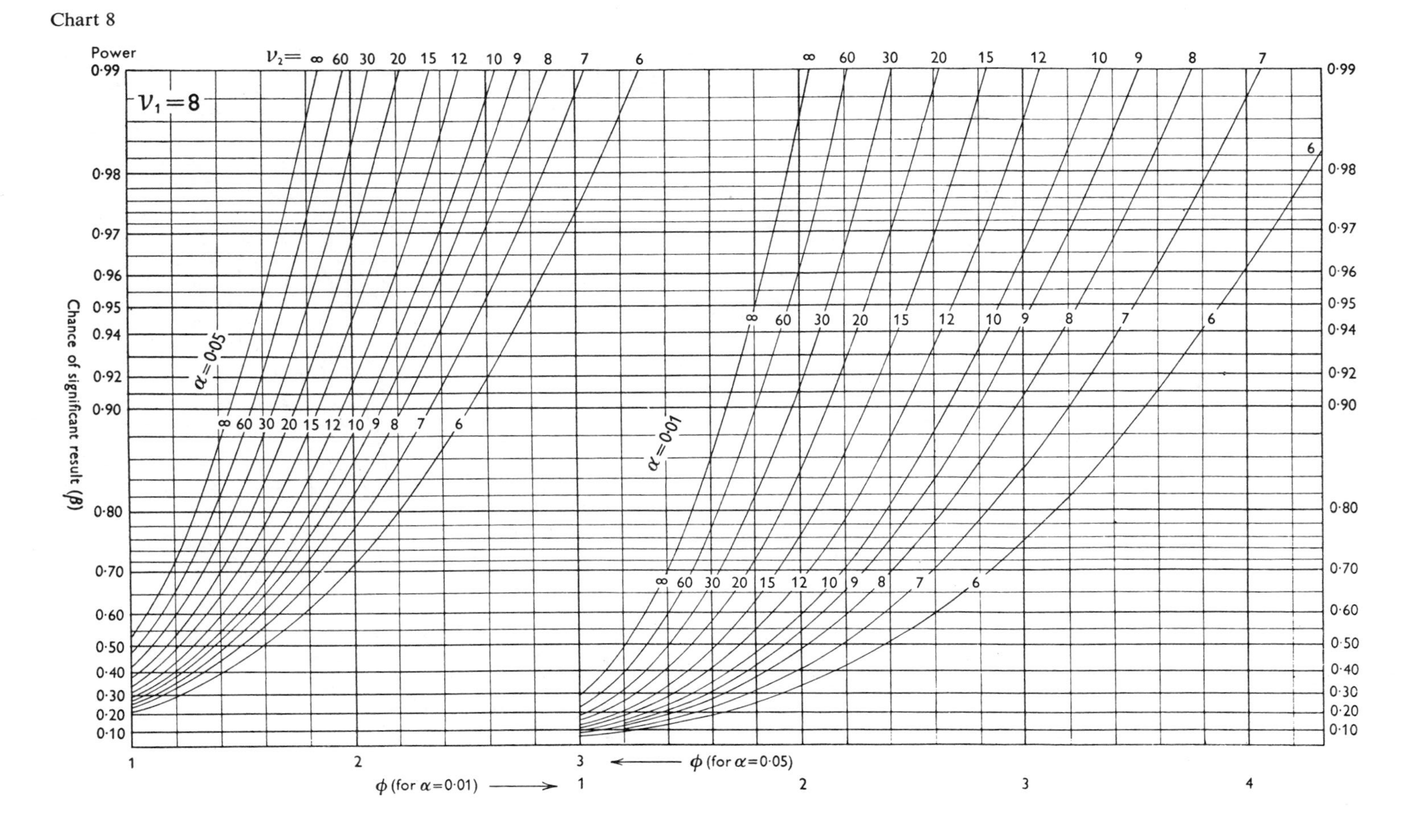
Power
Chance of significant result (β)
ν₁ = 8
ν₂ = ∞ 60 30 20 15 12 10 9 8 7 6
α = 0·05
α = 0·01
∞ 60 30 20 15 12 10 9 8 7 6
0·99
0·98
0·97
0·96
0·95
0·94
0·92
0·90
0·80
0·70
0·60
0·50
0·40
0·30
0·20
0·10
1
2
3
φ (for α = 0·01) →
← φ (for α = 0·05)
4

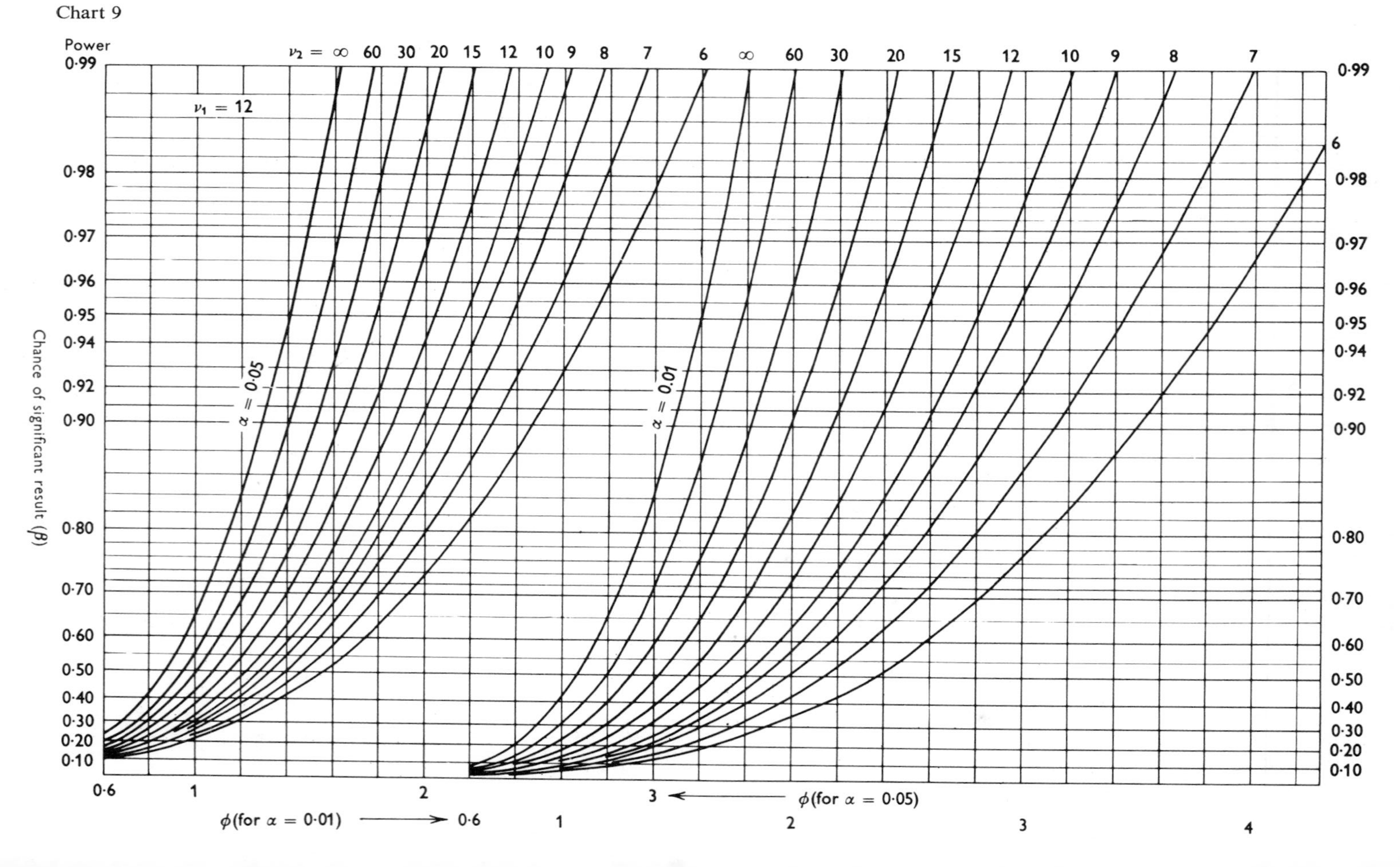
Chart 9
Power
0·99
0·98
0·97
0·96
0·95
0·94
0·92
0·90
0·80
0·70
0·60
0·50
0·40
0·30
0·20
0·10
Chance of significant result (β)
$\nu_2 = \infty$ 60 30 20 15 12 10 9 8 7 6 ∞ 60 30 20 15 12 10 9 8 7
6
$\nu_1 = 12$
$\alpha = 0·05$
$\alpha = 0·01$
0·6 1 2 3
ϕ(for $\alpha = 0·01$) ⟶ 0·6 1 2 3 4
3 ⟵ ϕ(for $\alpha = 0·05$)

Chart 10

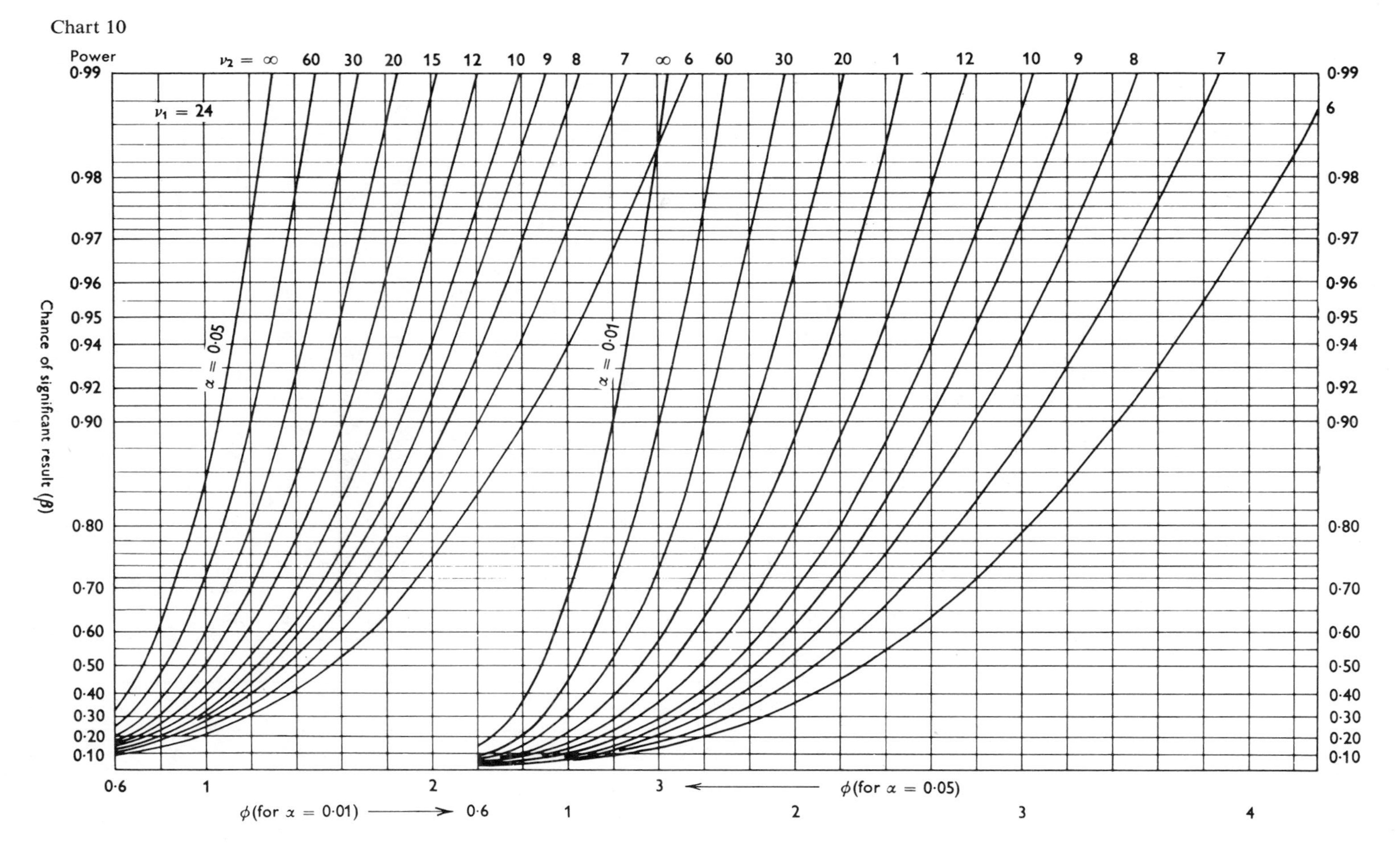

Power
ν₂ = ∞ 60 30 20 15 12 10 9 8 7 ∞ 6 60 30 20 1 12 10 9 8 7
ν₁ = 24
6
0·99 0·98 0·97 0·96 0·95 0·94 0·92 0·90 0·80 0·70 0·60 0·50 0·40 0·30 0·20 0·10
Chance of significant result (β)
α = 0·05
α = 0·01
0·6 1 2 3
φ (for α = 0·05)
φ (for α = 0·01)
0·6 1 2 3 4

TABLE 7 Percentage Points of the Studentized Range

a. Upper 0.05 critical values $q_{0.05;\,k,\nu}$

ν \ k	2	3	4	5	6	7	8	9	10	11	12	13	14	15	16	17	18	19	20
1	18·0	27·0	32·8	37·1	40·4	43·1	45·4	47·4	49·1	50·6	52·0	53·2	54·3	55·4	56·3	57·2	58·0	58·8	59·6
2	6·09	8·3	9·8	10·9	11·7	12·4	13·0	13·5	14·0	14·4	14·7	15·1	15·4	15·7	15·9	16·1	16·4	16·6	16·8
3	4·50	5·91	6·82	7·50	8·04	8·48	8·85	9·18	9·46	9·72	9·95	10·15	10·35	10·52	10·69	10·84	10·98	11·11	11·24
4	3·93	5·04	5·76	6·29	6·71	7·05	7·35	7·60	7·83	8·03	8·21	8·37	8·52	8·66	8·79	8·91	9·03	9·13	9·23
5	3·64	4·60	5·22	5·67	6·03	6·33	6·58	6·80	6·99	7·17	7·32	7·47	7·60	7·72	7·83	7·93	8·03	8·12	8·21
6	3·46	4·34	4·90	5·31	5·63	5·89	6·12	6·32	6·49	6·65	6·79	6·92	7·03	7·14	7·24	7·34	7·43	7·51	7·59
7	3·34	4·16	4·68	5·06	5·36	5·61	5·82	6·00	6·16	6·30	6·43	6·55	6·66	6·76	6·85	6·94	7·02	7·09	7·17
8	3·26	4·04	4·53	4·89	5·17	5·40	5·60	5·77	5·92	6·05	6·18	6·29	6·39	6·48	6·57	6·65	6·73	6·80	6·87
9	3·20	3·95	4·42	4·76	5·02	5·24	5·43	5·60	5·74	5·87	5·98	6·09	6·19	6·28	6·36	6·44	6·51	6·58	6·64
10	3·15	3·88	4·33	4·65	4·91	5·12	5·30	5·46	5·60	5·72	5·83	5·93	6·03	6·11	6·20	6·27	6·34	6·40	6·47
11	3·11	3·82	4·26	4·57	4·82	5·03	5·20	5·35	5·49	5·61	5·71	5·81	5·90	5·99	6·06	6·14	6·20	6·26	6·33
12	3·08	3·77	4·20	4·51	4·75	4·95	5·12	5·27	5·40	5·51	5·62	5·71	5·80	5·88	5·95	6·03	6·09	6·15	6·21
13	3·06	3·73	4·15	4·45	4·69	4·88	5·05	5·19	5·32	5·43	5·53	5·63	5·71	5·79	5·86	5·93	6·00	6·05	6·11
14	3·03	3·70	4·11	4·41	4·64	4·83	4·99	5·13	5·25	5·36	5·46	5·55	5·64	5·72	5·79	5·85	5·92	5·97	6·03
15	3·01	3·67	4·08	4·37	4·60	4·78	4·94	5·08	5·20	5·31	5·40	5·49	5·58	5·65	5·72	5·79	5·85	5·90	5·96
16	3·00	3·65	4·05	4·33	4·56	4·74	4·90	5·03	5·15	5·26	5·35	5·44	5·52	5·59	5·66	5·72	5·79	5·84	5·90
17	2·98	3·63	4·02	4·30	4·52	4·71	4·86	4·99	5·11	5·21	5·31	5·39	5·47	5·55	5·61	5·68	5·74	5·79	5·84
18	2·97	3·61	4·00	4·28	4·49	4·67	4·82	4·96	5·07	5·17	5·27	5·35	5·43	5·50	5·57	5·63	5·69	5·74	5·79
19	2·96	3·59	3·98	4·25	4·47	4·65	4·79	4·92	5·04	5·14	5·23	5·32	5·39	5·46	5·53	5·59	5·65	5·70	5·75
20	2·95	3·58	3·96	4·23	4·45	4·62	4·77	4·90	5·01	5·11	5·20	5·28	5·36	5·43	5·49	5·55	5·61	5·66	5·71
24	2·92	3·53	3·90	4·17	4·37	4·54	4·68	4·81	4·92	5·01	5·10	5·18	5·25	5·32	5·38	5·44	5·50	5·54	5·59
30	2·89	3·49	3·84	4·10	4·30	4·46	4·60	4·72	4·83	4·92	5·00	5·08	5·15	5·21	5·27	5·33	5·38	5·43	5·48
40	2·86	3·44	3·79	4·04	4·23	4·39	4·52	4·63	4·74	4·82	4·91	4·98	5·05	5·11	5·16	5·22	5·27	5·31	5·36
60	2·83	3·40	3·74	3·98	4·16	4·31	4·44	4·55	4·65	4·73	4·81	4·88	4·94	5·00	5·06	5·11	5·16	5·20	5·24
120	2·80	3·36	3·69	3·92	4·10	4·24	4·36	4·48	4·56	4·64	4·72	4·78	4·84	4·90	4·95	5·00	5·05	5·09	5·13
∞	2·77	3·31	3·63	3·86	4·03	4·17	4·29	4·39	4·47	4·55	4·62	4·68	4·74	4·80	4·85	4·89	4·93	4·97	5·01

b. Upper 0.01 critical values $q_{0.01; k,\nu}$

ν \ k	2	3	4	5	6	7	8	9	10	11	12	13	14	15	16	17	18	19	20
1	90·0	135	164	186	202	216	227	237	246	253	260	266	272	277	282	286	290	294	298
2	14·0	19·0	22·3	24·7	26·6	28·2	29·5	30·7	31·7	32·6	33·4	34·1	34·8	35·4	36·0	36·5	37·0	37·5	37·9
3	8·26	10·6	12·2	13·3	14·2	15·0	15·6	16·2	16·7	17·1	17·5	17·9	18·2	18·5	18·8	19·1	19·3	19·5	19·8
4	6·51	8·12	9·17	9·96	10·6	11·1	11·5	11·9	12·3	12·6	12·8	13·1	13·3	13·5	13·7	13·9	14·1	14·2	14·4
5	5·70	6·97	7·80	8·42	8·91	9·32	9·67	9·97	10·24	10·48	10·70	10·89	11·08	11·24	11·40	11·55	11·68	11·81	11·93
6	5·24	6·33	7·03	7·56	7·97	8·32	8·61	8·87	9·10	9·30	9·49	9·65	9·81	9·95	10·08	10·21	10·32	10·43	10·54
7	4·95	5·92	6·54	7·01	7·37	7·68	7·94	8·17	8·37	8·55	8·71	8·86	9·00	9·12	9·24	9·35	9·46	9·55	9·65
8	4·74	5·63	6·20	6·63	6·96	7·24	7·47	7·68	7·87	8·03	8·18	8·31	8·44	8·55	8·66	8·76	8·85	8·94	9·03
9	4·60	5·43	5·96	6·35	6·66	6·91	7·13	7·32	7·49	7·65	7·78	7·91	8·03	8·13	8·23	8·32	8·41	8·49	8·57
10	4·48	5·27	5·77	6·14	6·43	6·67	6·87	7·05	7·21	7·36	7·48	7·60	7·71	7·81	7·91	7·99	8·07	8·15	8·22
11	4·39	5·14	5·62	5·97	6·25	6·48	6·67	6·84	6·99	7·13	7·25	7·36	7·46	7·56	7·65	7·73	7·81	7·88	7·95
12	4·32	5·04	5·50	5·84	6·10	6·32	6·51	6·67	6·81	6·94	7·06	7·17	7·26	7·36	7·44	7·52	7·59	7·66	7·73
13	4·26	4·96	5·40	5·73	5·98	6·19	6·37	6·53	6·67	6·79	6·90	7·01	7·10	7·19	7·27	7·34	7·42	7·48	7·55
14	4·21	4·89	5·32	5·63	5·88	6·08	6·26	6·41	6·54	6·66	6·77	6·87	6·96	7·05	7·12	7·20	7·27	7·33	7·39
15	4·17	4·83	5·25	5·56	5·80	5·99	6·16	6·31	6·44	6·55	6·66	6·76	6·84	6·93	7·00	7·07	7·14	7·20	7·26
16	4·13	4·78	5·19	5·49	5·72	5·92	6·08	6·22	6·35	6·46	6·56	6·66	6·74	6·82	6·90	6·97	7·03	7·09	7·15
17	4·10	4·74	5·14	5·43	5·66	5·85	6·01	6·15	6·27	6·38	6·48	6·57	6·66	6·73	6·80	6·87	6·94	7·00	7·05
18	4·07	4·70	5·09	5·38	5·60	5·79	5·94	6·08	6·20	6·31	6·41	6·50	6·58	6·65	6·72	6·79	6·85	6·91	6·96
19	4·05	4·67	5·05	5·33	5·55	5·73	5·89	6·02	6·14	6·25	6·34	6·43	6·51	6·58	6·65	6·72	6·78	6·84	6·89
20	4·02	4·64	5·02	5·29	5·51	5·69	5·84	5·97	6·09	6·19	6·29	6·37	6·45	6·52	6·59	6·65	6·71	6·76	6·82
24	3·96	4·54	4·91	5·17	5·37	5·54	5·69	5·81	5·92	6·02	6·11	6·19	6·26	6·33	6·39	6·45	6·51	6·56	6·61
30	3·89	4·45	4·80	5·05	5·24	5·40	5·54	5·65	5·76	5·85	5·93	6·01	6·08	6·14	6·20	6·26	6·31	6·36	6·41
40	3·82	4·37	4·70	4·93	5·11	5·27	5·39	5·50	5·60	5·69	5·77	5·84	5·90	5·96	6·02	6·07	6·12	6·17	6·21
60	3·76	4·28	4·60	4·82	4·99	5·13	5·25	5·36	5·45	5·53	5·60	5·67	5·73	5·79	5·84	5·89	5·93	5·98	6·02
120	3·70	4·20	4·50	4·71	4·87	5·01	5·12	5·21	5·30	5·38	5·44	5·51	5·56	5·61	5·66	5·71	5·75	5·79	5·83
∞	3·64	4·12	4·40	4·60	4·76	4·88	4·99	5·08	5·16	5·23	5·29	5·35	5·40	5·45	5·49	5·54	5·57	5·61	5·65

SOURCE: E. S. Pearson and H. O. Hartley (eds.): *Biometrika Tables for Statisticians*, vol. 1, Cambridge University Press, 1956, with the kind permission of the *Biometrika* Trustees.

TABLE 8 Critical Values for the Durbin–Watson Test

a. $\alpha = 0.05$

N	$p=1$		$p=2$		$p=3$		$p=4$		$p=5$	
	d_L	d_U	d_L	d_U	d_L	d_U	d_L	d_U	d_L	d_U
15	1·08	1·36	0·95	1·54	0·82	1·75	0·69	1·97	0·56	2·21
16	1·10	1·37	0·98	1·54	0·86	1·73	0·74	1·93	0·62	2·15
17	1·13	1·38	1·02	1·54	0·90	1·71	0·78	1·90	0·67	2·10
18	1·16	1·39	1·05	1·53	0·93	1·69	0·82	1·87	0·71	2·06
19	1·18	1·40	1·08	1·53	0·97	1·68	0·86	1·85	0·75	2·02
20	1·20	1·41	1·10	1·54	1·00	1·68	0·90	1·83	0·79	1·99
21	1·22	1·42	1·13	1·54	1·03	1·67	0·93	1·81	0·83	1·96
22	1·24	1·43	1·15	1·54	1·05	1·66	0·96	1·80	0·86	1·94
23	1·26	1·44	1·17	1·54	1·08	1·66	0·99	1·79	0·90	1·92
24	1·27	1·45	1·19	1·55	1·10	1·66	1·01	1·78	0·93	1·90
25	1·29	1·45	1·21	1·55	1·12	1·66	1·04	1·77	0·95	1·89
26	1·30	1·46	1·22	1·55	1·14	1·65	1·06	1·76	0·98	1·88
27	1·32	1·47	1·24	1·56	1·16	1·65	1·08	1·76	1·01	1·86
28	1·33	1·48	1·26	1·56	1·18	1·65	1·10	1·75	1·03	1·85
29	1·34	1·48	1·27	1·56	1·20	1·65	1·12	1·74	1·05	1·84
30	1·35	1·49	1·28	1·57	1·21	1·65	1·14	1·74	1·07	1·83
31	1·36	1·50	1·30	1·57	1·23	1·65	1·16	1·74	1·09	1·83
32	1·37	1·50	1·31	1·57	1·24	1·65	1·18	1·73	1·11	1·82
33	1·38	1·51	1·32	1·58	1·26	1·65	1·19	1·73	1·13	1·81
34	1·39	1·51	1·33	1·58	1·27	1·65	1·21	1·73	1·15	1·81
35	1·40	1·52	1·34	1·58	1·28	1·65	1·22	1·73	1·16	1·80
36	1·41	1·52	1·35	1·59	1·29	1·65	1·24	1·73	1·18	1·80
37	1·42	1·53	1·36	1·59	1·31	1·66	1·25	1·72	1·19	1·80
38	1·43	1·54	1·37	1·59	1·32	1·66	1·26	1·72	1·21	1·79
39	1·43	1·54	1·38	1·60	1·33	1·66	1·27	1·72	1·22	1·79
40	1·44	1·54	1·39	1·60	1·34	1·66	1·29	1·72	1·23	1·79
45	1·48	1·57	1·43	1·62	1·38	1·67	1·34	1·72	1·29	1·78
50	1·50	1·59	1·46	1·63	1·42	1·67	1·38	1·72	1·34	1·77
55	1·53	1·60	1·49	1·64	1·45	1·68	1·41	1·72	1·38	1·77
60	1·55	1·62	1·51	1·65	1·48	1·69	1·44	1·73	1·41	1·77
65	1·57	1·63	1·54	1·66	1·50	1·70	1·47	1·73	1·44	1·77
70	1·58	1·64	1·55	1·67	1·52	1·70	1·49	1·74	1·46	1·77
75	1·60	1·65	1·57	1·68	1·54	1·71	1·51	1·74	1·49	1·77
80	1·61	1·66	1·59	1·69	1·56	1·72	1·53	1·74	1·51	1·77
85	1·62	1·67	1·60	1·70	1·57	1·72	1·55	1·75	1·52	1·77
90	1·63	1·68	1·61	1·70	1·59	1·73	1·57	1·75	1·54	1·78
95	1·64	1·69	1·62	1·71	1·60	1·73	1·58	1·75	1·56	1·78
100	1·65	1·69	1·63	1·72	1·61	1·74	1·59	1·76	1·57	1·78

SOURCE: J. Durbin and G. S. Watson: Testing for serial correlation in least-squares regression, *Biometrika*, vol. 38 (1951), pp. 159–178, with the kind permission of the authors and the *Biometrika* Trustees.

TABLE 8 (*cont.*)

b. $\alpha = 0.01$

N	p = 1		p = 2		p = 3		p = 4		p = 5	
	d_L	d_U	d_L	d_U	d_L	d_U	d_L	d_U	d_L	d_U
15	0·81	1·07	0·70	1·25	0·59	1·46	0·49	1·70	0·39	1·96
16	0·84	1·09	0·74	1·25	0·63	1·44	0·53	1·66	0·44	1·90
17	0·87	1·10	0·77	1·25	0·67	1·43	0·57	1·63	0·48	1·85
18	0·90	1·12	0·80	1·26	0·71	1·42	0·61	1·60	0·52	1·80
19	0·93	1·13	0·83	1·26	0·74	1·41	0·65	1·58	0·56	1·77
20	0·95	1·15	0·86	1·27	0·77	1·41	0·68	1·57	0·60	1·74
21	0·97	1·16	0·89	1·27	0·80	1·41	0·72	1·55	0·63	1·71
22	1·00	1·17	0·91	1·28	0·83	1·40	0·75	1·54	0·66	1·69
23	1·02	1·19	0·94	1·29	0·86	1·40	0·77	1·53	0·70	1·67
24	1·04	1·20	0·96	1·30	0·88	1·41	0·80	1·53	0·72	1·66
25	1·05	1·21	0·98	1·30	0·90	1·41	0·83	1·52	0·75	1·65
26	1·07	1·22	1·00	1·31	0·93	1·41	0·85	1·52	0·78	1·64
27	1·09	1·23	1·02	1·32	0·95	1·41	0·88	1·51	0·81	1·63
28	1·10	1·24	1·04	1·32	0·97	1·41	0·90	1·51	0·83	1·62
29	1·12	1·25	1·05	1·33	0·99	1·42	0·92	1·51	0·85	1·61
30	1·13	1·26	1·07	1·34	1·01	1·42	0·94	1·51	0·88	1·61
31	1·15	1·27	1·08	1·34	1·02	1·42	0·96	1·51	0·90	1·60
32	1·16	1·28	1·10	1·35	1·04	1·43	0·98	1·51	0·92	1·60
33	1·17	1·29	1·11	1·36	1·05	1·43	1·00	1·51	0·94	1·59
34	1·18	1·30	1·13	1·36	1·07	1·43	1·01	1·51	0·95	1·59
35	1·19	1·31	1·14	1·37	1·08	1·44	1·03	1·51	0·97	1·59
36	1·21	1·32	1·15	1·38	1·10	1·44	1·04	1·51	0·99	1·59
37	1·22	1·32	1·16	1·38	1·11	1·45	1·06	1·51	1·00	1·59
38	1·23	1·33	1·18	1·39	1·12	1·45	1·07	1·52	1·02	1·58
39	1·24	1·34	1·19	1·39	1·14	1·45	1·09	1·52	1·03	1·58
40	1·25	1·34	1·20	1·40	1·15	1·46	1·10	1·52	1·05	1·58
45	1·29	1·38	1·24	1·42	1·20	1·48	1·16	1·53	1·11	1·58
50	1·32	1·40	1·28	1·45	1·24	1·49	1·20	1·54	1·16	1·59
55	1·36	1·43	1·32	1·47	1·28	1·51	1·25	1·55	1·21	1·59
60	1·38	1·45	1·35	1·48	1·32	1·52	1·28	1·56	1·25	1·60
65	1·41	1·47	1·38	1·50	1·35	1·53	1·31	1·57	1·28	1·61
70	1·43	1·49	1·40	1·52	1·37	1·55	1·34	1·58	1·31	1·61
75	1·45	1·50	1·42	1·53	1·39	1·56	1·37	1·59	1·34	1·62
80	1·47	1·52	1·44	1·54	1·42	1·57	1·39	1·60	1·36	1·62
85	1·48	1·53	1·46	1·55	1·43	1·58	1·41	1·60	1·39	1·63
90	1·50	1·54	1·47	1·56	1·45	1·59	1·43	1·61	1·41	1·64
95	1·51	1·55	1·49	1·57	1·47	1·60	1·45	1·62	1·42	1·64
100	1·52	1·56	1·50	1·58	1·48	1·60	1·46	1·63	1·44	1·65

TABLE 9 Orthogonal Polynomials

N=3		N=4			N=5				N=6					N=7					
ϕ_1	ϕ_2	ϕ_1	ϕ_2	ϕ_3	ϕ_1	ϕ_2	ϕ_3	ϕ_4	ϕ_1	ϕ_2	ϕ_3	ϕ_4	ϕ_5	ϕ_1	ϕ_2	ϕ_3	ϕ_4	ϕ_5	ϕ_6
−1	1	−3	1	−1	−2	2	−1	1	−5	5	−5	1	−1	−3	5	−1	3	−1	1
0	−2	−1	−1	3	−1	−1	2	−4	−3	−1	7	−3	5	−2	0	1	−7	4	−6
1	1	1	−1	−3	0	−2	0	6	−1	−4	4	2	−10	−1	−3	1	1	−5	15
		3	1	1	1	−1	−2	−4	1	−4	−4	2	10	0	−4	0	6	0	−20
					2	2	1	1	3	−1	−7	−3	−5	1	−3	−1	1	5	15
									5	5	5	1	1	2	0	−1	−7	−4	−6
														3	5	1	3	1	1
2	6	20	4	20	10	14	10	70	70	84	180	28	252	28	84	6	154	84	924
1	3	2	1	$\frac{10}{3}$	1	1	$\frac{5}{6}$	$\frac{35}{12}$	2	$\frac{3}{2}$	$\frac{5}{3}$	$\frac{7}{12}$	$\frac{21}{10}$	1	1	$\frac{1}{6}$	$\frac{7}{12}$	$\frac{7}{20}$	$\frac{77}{60}$

N=8						N=9						N=10					
ϕ_1	ϕ_2	ϕ_3	ϕ_4	ϕ_5	ϕ_6	ϕ_1	ϕ_2	ϕ_3	ϕ_4	ϕ_5	ϕ_6	ϕ_1	ϕ_2	ϕ_3	ϕ_4	ϕ_5	ϕ_6
−7	7	−7	7	−7	1	−4	28	−14	14	−4	4	−9	6	−42	18	−6	3
−5	1	5	−13	23	−5	−3	7	7	−21	11	−17	−7	2	14	−22	14	−11
−3	−3	7	−3	−17	9	−2	−8	13	−11	−4	22	−5	−1	35	−17	−1	10
−1	−5	3	9	−15	−5	−1	−17	9	9	−9	1	−3	−3	31	3	−11	6
												−1	−4	12	18	−6	−8
1	−5	−3	9	15	−5	0	−20	0	18	0	−20						
3	−3	−7	−3	17	9							1	−4	−12	18	6	−8
5	1	−5	−13	−23	−5	1	−17	−9	9	9	1	3	−3	−31	3	11	6
7	7	7	7	7	1	2	−8	−13	−11	4	22	5	−1	−35	−17	1	10
						3	7	−7	−21	−11	−17	7	2	−14	−22	−14	−11
						4	28	14	14	4	4	9	6	42	18	6	3
168	168	264	616	2,184	264	60	2,772	990	2,002	468	1,980	330	132	8,580	2,860	780	660
2	1	$\frac{2}{3}$	$\frac{7}{12}$	$\frac{7}{10}$	$\frac{11}{60}$	1	3	$\frac{5}{6}$	$\frac{7}{12}$	$\frac{3}{20}$	$\frac{11}{60}$	2	$\frac{1}{2}$	$\frac{5}{3}$	$\frac{5}{12}$	$\frac{1}{10}$	$\frac{11}{240}$

SOURCE: E. S. Pearson and H. O. Hartley (eds.): *Biometrika Tables for Statisticians*. vol. 1, Cambridge University Press, 1956, with the kind permission of the *Biometrika* Trustees.

TABLE 9 (*cont.*)

$N=11$						$N=12$					
ϕ_1	ϕ_2	ϕ_3	ϕ_4	ϕ_5	ϕ_6	ϕ_1	ϕ_2	ϕ_3	ϕ_4	ϕ_5	ϕ_6
−5	15	−30	6	−3	15	−11	55	−33	33	−33	11
−4	6	6	−6	6	−48	−9	25	3	−27	57	−31
−3	−1	22	−6	1	29	−7	1	21	−33	21	11
−2	−6	23	−1	−4	36	−5	−17	25	−13	−29	25
−1	−9	14	4	−4	−12	−3	−29	19	12	−44	4
						−1	−35	7	28	−20	−20
0	−10	0	6	0	−40						
						1	−35	−7	28	20	−20
1	−9	−14	4	4	−12	3	−29	−19	12	44	4
2	−6	−23	−1	4	36	5	−17	−25	−13	29	25
3	−1	−22	−6	−1	29	7	1	−21	−33	−21	11
4	6	−6	−6	−6	−48	9	25	−3	−27	−57	−31
5	15	30	6	3	15	11	55	33	33	33	11
110	858	4,290	286	156	11,220	572	12,012	5,148	8,008	15,912	4,488
1	1	$\frac{5}{6}$	$\frac{1}{12}$	$\frac{1}{40}$	$\frac{11}{120}$	2	3	$\frac{2}{3}$	$\frac{7}{24}$	$\frac{3}{20}$	$\frac{11}{360}$

$N=13$						$N=14$					
ϕ_1	ϕ_2	ϕ_3	ϕ_4	ϕ_5	ϕ_6	ϕ_1	ϕ_2	ϕ_3	ϕ_4	ϕ_5	ϕ_6
−6	22	−11	99	−22	22	−13	13	−143	143	−143	143
−5	11	0	−66	33	−55	−11	7	−11	−77	187	−319
−4	2	6	−96	18	8	−9	2	66	−132	132	−11
−3	−5	8	−54	−11	43	−7	−2	98	−92	−28	227
−2	−10	7	11	−26	22	−5	−5	95	−13	−139	185
−1	−13	4	64	−20	−20	−3	−7	67	63	−145	−25
0	−14	0	84	0	−40	−1	−8	24	108	−60	−200
182	2,002	572	68,068	6,188	14,212	910	728	97,240	136,136	235,144	497,420
1	1	$\frac{1}{6}$	$\frac{7}{12}$	$\frac{7}{120}$	$\frac{11}{360}$	2	$\frac{1}{2}$	$\frac{5}{3}$	$\frac{7}{12}$	$\frac{7}{30}$	$\frac{77}{720}$

appendix B

Elements of Matrix Algebra

B-1 MATRICES AND VECTORS

A *matrix* is merely a table of numbers or algebraic symbols. Unlike most tables, matrices do not have row or column labels or headings, and are usually referred to as *rectangular arrays* of numbers or variables. The dimensions of the array are its number of rows r and its number of columns c given in that order. For example, the verbal and quantitative Graduate Record Examination scores for five applicants to a statistics graduate program might be represented by the 5×3 matrix

$$\mathbf{X} = \begin{bmatrix} 1 & 760 & 800 \\ 2 & 740 & 760 \\ 3 & 650 & 800 \\ 4 & 660 & 780 \\ 5 & 480 & 700 \end{bmatrix} \tag{1}$$

The applicants are the five rows of the matrix. The first column is the order in which the applications were received by the admissions office and is the row index of the matrix. The second column contains the verbal GRE scores, and the third contains the quantitative scores. With those conventions and labels in mind we could perform some elementary statistical analyses on the data stored in **X**. We shall see that the statistics can be represented conveniently by the notation and concepts of matrix algebra.

Now let us represent a general $r \times c$ matrix as

$$\mathbf{X} = \begin{bmatrix} x_{11} & \cdots & x_{1c} \\ \cdot & \cdots & \cdot \\ x_{r1} & \cdots & x_{rc} \end{bmatrix} \tag{2}$$

The subscripts of the element x_{ij} always denote its position in the ith row and jth column in that order. Every position of the matrix is filled with a number or symbol. Such symbols, or equivalently, the 1×1 matrix, are called *scalars*.

If the matrix has only one column, for example,

$$\mathbf{y} = \begin{bmatrix} 1 \\ 2 \\ 3 \\ 4 \end{bmatrix} \tag{3}$$

it is called a *column vector*, as opposed to a matrix with a single row, or *row vector*

$$\mathbf{x}' = [0, \quad 1, \quad 2] \tag{4}$$

All of our vectors will have the column form, or dimensions $r \times 1$, unless identified by a prime to indicate transposition to a $1 \times r$ row vector. In either case the physical or geometrical meaning of the vector is the same: It denotes the coordinates of a point in Euclidean space. The line from the origin of the Cartesian coordinate system of the space to the point has direction and length (magnitude) and so represents a vector in the physical sense. Similarly, the elements of the vector might be plotted for a graphical representation of data: The GRE scores of the matrix (1) could be summarized in a scatterplot of the vectors

$$\mathbf{x}_1' = [760, \quad 800] \quad \cdots \quad \mathbf{x}_5' = [480, \quad 700] \tag{5}$$

Certain matrices and vectors have useful special properties. A *square* matrix has as many rows as columns:

$$\mathbf{A} = \begin{bmatrix} 2 & -1 & 0 \\ -1 & 4 & 3 \\ 0 & 3 & 5 \end{bmatrix} \tag{6}$$

is not only square, but is also *symmetric:* Its element a_{ij} is equal to the element a_{ji} for all pairs of subscripts. The symmetry is about the *main diagonal* elements of the matrix, or those in the same row and column. *Diagonal* matrices contain nonzero elements on their main diagonal and zeros in every other position:

$$\mathbf{D}_1 = \begin{bmatrix} 3 & 0 \\ 0 & 4 \end{bmatrix} \qquad \mathbf{D}_2 = \begin{bmatrix} 1 & 0 & 0 & 0 \\ 0 & 2 & 0 & 0 \\ 0 & 0 & 3 & 0 \\ 0 & 0 & 0 & 4 \end{bmatrix} \tag{7}$$

The *identity* matrix is especially important. It has ones in its diagonal positions and zeros elsewhere. The 4×4 identity matrix is

$$\mathbf{I} = \begin{bmatrix} 1 & 0 & 0 & 0 \\ 0 & 1 & 0 & 0 \\ 0 & 0 & 1 & 0 \\ 0 & 0 & 0 & 1 \end{bmatrix} \tag{8}$$

The *null* matrix or vector consists of zeros in every position:

$$\mathbf{O} = \begin{bmatrix} 0 \\ 0 \\ 0 \end{bmatrix} \qquad \mathbf{O} = \begin{bmatrix} 0 & 0 \\ 0 & 0 \end{bmatrix} \tag{9}$$

We shall often need the vector of ones:

$$\mathbf{j}' = [1, \ldots, 1] \tag{10}$$

and the matrix of ones:

$$\mathbf{J} = \begin{bmatrix} 1 & \cdots & 1 \\ \cdots & \cdots & \cdots \\ 1 & \cdots & 1 \end{bmatrix} \tag{11}$$

Partitioned matrices have their elements grouped into submatrices or vectors so that their elements are matrices of smaller dimensions. For example, we might write

$$\mathbf{B} = \begin{bmatrix} 3 & 1 & 1 & 1 \\ 1 & 1 & 0 & 0 \\ 1 & 0 & 1 & 0 \\ 1 & 0 & 0 & 1 \end{bmatrix} \tag{12}$$

as

$$\mathbf{B} = \begin{bmatrix} 3 & \mathbf{j}' \\ \mathbf{j} & \mathbf{I} \end{bmatrix} \tag{13}$$

in which $\mathbf{j}' = [1, \quad 1, \quad 1]$, and $\mathbf{I}$ is the 3×3 identity matrix. Partitioned matrices arise when we wish to divide the independent variables of a regression model into two sets for hypothesis tests on subsets of the regression coefficients.

B-2 OPERATIONS WITH MATRICES

We have already seen that a column vector can be converted to a row vector (and vice versa) by the operation of *transposition*. Similarly, the *transpose* of an $r \times c$ matrix is a $c \times r$ matrix whose rows are the columns of the original. We denote a transposed matrix by a prime. For example, the transpose of the matrix (1) of the previous section is the 3×5 matrix

$$\mathbf{X}' = \begin{bmatrix} 1 & 2 & 3 & 4 & 5 \\ 760 & 740 & 650 & 660 & 480 \\ 800 & 760 & 800 & 780 & 700 \end{bmatrix} \tag{1}$$

Matrices of comparable dimensions are added and subtracted by adding and subtracting corresponding elements. If

$$A = \begin{bmatrix} 2 & 1 \\ 0 & 5 \end{bmatrix} \qquad B = \begin{bmatrix} 3 & -1 \\ -1 & 3 \end{bmatrix} \tag{2}$$

then

$$A + B = \begin{bmatrix} 2+3 & 1-1 \\ 0-1 & 5+3 \end{bmatrix} = \begin{bmatrix} 5 & 0 \\ -1 & 8 \end{bmatrix}$$
$$B - A = \begin{bmatrix} 3-2 & -1-1 \\ -1-0 & 3-5 \end{bmatrix} = \begin{bmatrix} 1 & -2 \\ -1 & -2 \end{bmatrix} \tag{3}$$

If the matrices differ in their numbers of rows or columns, their sums or differences are undefined.

Multiplication of any matrix by a scalar simply multiplies each element by the scalar. If $c = 10$ and

$$\mathbf{W} = \begin{bmatrix} 1 & 1 & 1 & 1 \\ -3 & -1 & 1 & 3 \end{bmatrix} \tag{4}$$

then

$$c\mathbf{W} = \begin{bmatrix} 10 & 10 & 10 & 10 \\ -30 & -10 & 10 & 30 \end{bmatrix} \tag{5}$$

Multiplication of two matrices is more involved. First, the dimensions must be conformable: The number of columns in the first matrix of the product must be equal to the number of rows in the second. Let the matrices **A** and **B** have respective dimensions $m \times n$ and $n \times p$. Then the ijth element of $\mathbf{C} = \mathbf{AB}$ is equal to the sum of the products of corresponding elements in the ith row of **A** and the jth column of B, or

$$c_{ij} = \sum_{h=1}^{n} a_{ih} b_{hj} \qquad i = 1, \ldots, m; \quad j = 1, \ldots, p \tag{6}$$

We must calculate such a sum for each position of **C**. As an example, let

$$\mathbf{A} = \begin{bmatrix} 1 & 1 & 1 \\ -1 & 0 & 1 \\ 1 & -2 & 1 \end{bmatrix} \qquad \mathbf{B} = \begin{bmatrix} 1 & 2 \\ 1 & -1 \\ 1 & -1 \end{bmatrix} \tag{7}$$

$$\mathbf{AB} = \begin{bmatrix} 1(1) + 1(1) + 1(1) & 1(2) + 1(-1) + 1(-1) \\ -1(1) + 0(1) + 1(1) & -1(2) + 0(-1) + 1(-1) \\ 1(1) - 2(1) + 1(1) & 1(2) - 2(-1) + 1(-1) \end{bmatrix}$$
$$= \begin{bmatrix} 3 & 0 \\ 0 & -3 \\ 0 & 3 \end{bmatrix} \tag{8}$$

Similarly, if $\mathbf{X}$ is the data matrix (1) of Section B-1 and $\mathbf{a}' = (1/5)[1, 1, 1, 1, 1]$, we can write the averages of the three columns of $\mathbf{X}$ in the vector

$$\bar{\mathbf{x}} = \mathbf{X}'\mathbf{a} = \begin{bmatrix} 3 \\ 658 \\ 768 \end{bmatrix} \tag{9}$$

The matrix

$$\mathbf{X}'\mathbf{X} = \begin{bmatrix} 55 & 9230 & 11{,}340 \\ 9{,}230 & 2{,}213{,}700 & 2{,}541{,}200 \\ 11{,}340 & 2{,}541{,}200 & 2{,}956{,}000 \end{bmatrix} \tag{10}$$

contains the sums of squares of the elements in the three columns of $\mathbf{X}$ in its diagonal positions and the three sums of products as the off-diagonal entries. Note that $\mathbf{X}'\mathbf{X}$ is symmetric, as are all such matrices of sums of squares and products.

Matrix multiplication is not necessarily commutative: The product $\mathbf{AB}$ may not equal $\mathbf{BA}$. If

$$\mathbf{A} = \begin{bmatrix} 2 & -1 \\ -1 & 4 \end{bmatrix} \qquad \mathbf{B} = \begin{bmatrix} 2 & 1 \\ 1 & 3 \end{bmatrix} \tag{11}$$

then

$$\mathbf{AB} = \begin{bmatrix} 3 & -1 \\ 2 & 11 \end{bmatrix} \qquad \mathbf{BA} = \begin{bmatrix} 3 & 2 \\ -1 & 11 \end{bmatrix} \tag{12}$$

Sums, averages, and other linear functions of data appear everywhere in statistics and can be written conveniently in matrix or vector notation. The mean

$$\bar{x} = \frac{1}{N}\sum_{i=1}^{N} x_i \tag{13}$$

of N observations can be represented by

$$\bar{x} = \frac{1}{N}\mathbf{x}'\mathbf{j} \tag{14}$$

where $\mathbf{x}' = [x_1, \ldots, x_N]$ is a vector consisting of the data and $\mathbf{j}$ is the $N \times 1$ vector of ones introduced in (10) of Section B-1. The vector product $\mathbf{x}'\mathbf{j}$ is the *inner product* of the two vectors. The sum of squares of the x_i can be represented by the inner product of $\mathbf{x}$ with itself, or

$$\mathbf{x}'\mathbf{x} = \sum_{i=1}^{N} x_i^2 \tag{15}$$

Statistical analysis also depends heavily on variances, sums of squares about the mean, and other measures of dispersion. We can write the variance of the x_i in vector form as

$$\begin{aligned} s^2 &= \frac{1}{N-1}(\mathbf{x} - \bar{x}\mathbf{j})'(\mathbf{x} - \bar{x}\mathbf{j}) \\ &= \frac{1}{N-1}(\mathbf{x}'\mathbf{x} - N\bar{x}^2) \end{aligned} \tag{16}$$

Two vectors $\mathbf{x}$ and $\mathbf{y}$ with the same number of elements are said to be *orthogonal* if

$$\mathbf{x}'\mathbf{y} = 0 \tag{17}$$

The vectors are called orthogonal because their lines to the origin are perpendicular. For example, these vectors are mutually orthogonal in a three-dimensional coordinate system:

$$\mathbf{x} = \begin{bmatrix} 1 \\ 1 \\ 1 \end{bmatrix} \qquad \mathbf{y} = \begin{bmatrix} -2 \\ 1 \\ 1 \end{bmatrix} \qquad \mathbf{z} = \begin{bmatrix} 0 \\ 1 \\ -1 \end{bmatrix} \tag{18}$$

Every square matrix has two unique scalar functions of its elements. The first is the *trace*, or the sum of its diagonal elements. The second is the *determinant* of the matrix, denoted by $|\mathbf{A}|$. If $\mathbf{A}$ is a 2×2 matrix its determinant is

$$\begin{vmatrix} a_{11} & a_{12} \\ a_{21} & a_{22} \end{vmatrix} = a_{11}a_{22} - a_{12}a_{21} \tag{19}$$

while the 3×3 matrix has determinant

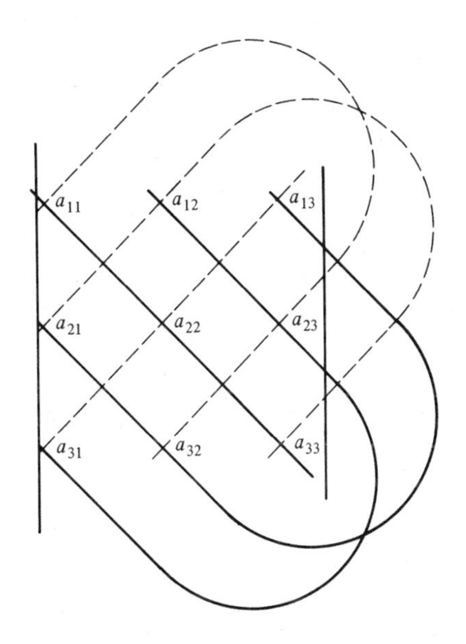

$$\begin{aligned} &= a_{11}a_{22}a_{33} + a_{12}a_{23}a_{31} + a_{13}a_{21}a_{32} \\ &\quad -a_{13}a_{22}a_{31} - a_{11}a_{23}a_{32} - a_{12}a_{21}a_{33} \end{aligned} \tag{20}$$

The positive products are denoted by the solid lines and the negative products by the dashed ones. Determinants of 4×4 and larger matrices can be computed by cofactors and other methods; those procedures are described in all books on linear algebra and matrices. Where determinants are required or computed in this text we shall merely take the method for granted.

Every matrix has a *rank*, or the number of linearly independent rows and columns of the matrix. A column (or row) is said to be linearly independent of another set of columns (rows) if it *cannot* be written as a linear function of those columns (rows). For example, the 5×5 matrix

$$\mathbf{X} = \begin{bmatrix} 1 & -2 & -1 & 0 & -1 \\ 1 & -1 & -1 & 3 & 0 \\ 1 & 0 & 0 & 6 & 0 \\ 1 & 1 & 1 & 9 & 0 \\ 1 & 2 & 1 & 12 & 1 \end{bmatrix} \tag{21}$$

has rank three because

$$\text{Column 4} = 6(\text{column 1}) + 3(\text{column 2})$$
$$\text{Column 5} = \text{column 2} - \text{column 3}$$

and it is possible to show that columns 1, 2, and 3 are mutually linearly independent. It must also follow that only three linearly independent rows exist in the matrix: The rank obtained by examining rows for linear independence is identical with that found for columns.

A square matrix with rank equal to its dimensions is said to be *nonsingular* or of *full rank*. The determinant of a nonsingular matrix has a nonzero value. The determinant of a *singular* matrix is equal to zero.

B-3 THE INVERSE MATRIX

Every nonsingular square matrix $\mathbf{A}$ has a unique inverse $\mathbf{A}^{-1}$ with the property that

$$\mathbf{A}^{-1}\mathbf{A} = \mathbf{A}\mathbf{A}^{-1} = \mathbf{I} \tag{1}$$

For example, the inverse of

$$\mathbf{A} = \begin{bmatrix} 4 & 2 \\ 6 & 5 \end{bmatrix}$$

is

$$\mathbf{A}^{-1} = \begin{bmatrix} \frac{5}{8} & -\frac{1}{4} \\ -\frac{3}{4} & \frac{1}{2} \end{bmatrix} \tag{2}$$

and in general the inverse of any nonsingular 2×2 matrix $\mathbf{A}$ is

$$\mathbf{A}^{-1} = \frac{1}{a_{11}a_{22} - a_{12}a_{21}} \begin{bmatrix} a_{22} & -a_{12} \\ -a_{21} & a_{11} \end{bmatrix} \tag{3}$$

The inverses of larger matrices can be defined in terms of cofactors [see, for example, Hohn (1964, Chapter 3), Morrison (1976, Chapter 2), or any text on linear algebra], although that definition is usually not an efficient numerical method for computing an inverse in an applied problem. Computing algorithms for inverses can be found in current texts on numerical analysis methods for computers. The inverse matrices in our examples were obtained from the APL/SF language subroutine on a DEC-System 1090 computer.

Certain matrices have simple inverses. The inverse of a diagonal matrix is also a diagonal matrix whose elements are the reciprocals of the original diagonal terms:

$$\begin{bmatrix} 2 & 0 & 0 \\ 0 & 5 & 0 \\ 0 & 0 & \frac{1}{2} \end{bmatrix}^{-1} = \begin{bmatrix} \frac{1}{2} & 0 & 0 \\ 0 & \frac{1}{5} & 0 \\ 0 & 0 & 2 \end{bmatrix} \tag{4}$$

An *orthogonal* matrix has mutually orthogonal rows *normalized* to have sums of squares equal to 1. The columns are also orthogonal and of unit length. Then the inverse of an orthogonal matrix is merely its transpose. For example,

$$\mathbf{P} = \begin{bmatrix} 1/\sqrt{3} & 1/\sqrt{3} & 1/\sqrt{3} \\ -2/\sqrt{6} & 1/\sqrt{6} & 1/\sqrt{6} \\ 0 & 1/\sqrt{2} & -1/\sqrt{2} \end{bmatrix} \tag{5}$$

is an orthogonal matrix, and one can easily verify that

$$\mathbf{PP}' = \mathbf{P}'\mathbf{P} = \mathbf{I} \tag{6}$$

Inverse matrices have these important properties:

1. The inverse of a transposed matrix is the transpose of the original matrix inverse:
$$(\mathbf{A}')^{-1} = (\mathbf{A}^{-1})' \tag{7}$$
2. The inverse of a product of nonsingular matrices is equal to the product of the inverses of the matrices in reversed order.
$$(\mathbf{ABCD})^{-1} = \mathbf{D}^{-1}\mathbf{C}^{-1}\mathbf{B}^{-1}\mathbf{A}^{-1} \tag{8}$$
3. The inverse of a symmetric matrix is also symmetric.
4. The inverse of a partitioned matrix can be expressed in terms of its submatrices [see, for example, Morrison (1976, Section 2.11)].
5. If the $p \times p$ matrix $\mathbf{A}$ is nonsingular, $\mathbf{b}$ is a $p \times 1$ vector, and c is a scalar,
$$(\mathbf{A} + c\mathbf{bb}')^{-1} = \mathbf{A}^{-1} - \frac{c}{1 + c\mathbf{b}'\mathbf{A}^{-1}\mathbf{b}}\mathbf{A}^{-1}\mathbf{bb}'\mathbf{A}^{-1} \tag{9}$$
This useful identity is due to Bartlett (1951); a current derivation can be found in Morrison (1976, Section 2.11).

B-4 QUADRATIC FORMS AND CHARACTERISTIC ROOTS

Statistical methods based on the normal distribution not only use linear functions of data as measures of population means or their differences but just as frequently need sums of squares and other quadratic functions as estimates of variation. Such general sums of squares are called *quadratic forms* in the observations or random variables of a model. The most general quadratic form in the variables $x_1, \ldots, x_N$ is the expression

$$\mathbf{x}'\mathbf{A}\mathbf{x} = \sum_{i=1}^{N} \sum_{j=1}^{N} a_{ij}x_ix_j \tag{1}$$

where

$$\mathbf{A} = \begin{bmatrix} a_{11} & \cdots & a_{1N} \\ \cdots & \cdots & \cdots \\ a_{1N} & \cdots & a_{NN} \end{bmatrix} \tag{2}$$

is the symmetric matrix giving the coefficients of the form, and

$$\mathbf{x}' = [x_1, \ldots, x_N] \tag{3}$$

is the row vector containing the variables. The algebraic study of quadratic forms is equivalent to finding the properties of the matrix **A**.

In our development of regression and the analysis of variance we shall only be concerned with *positive definite* and *positive semidefinite* quadratic forms and matrices. A form and its matrix are positive definite if

$$\mathbf{x}'\mathbf{A}\mathbf{x} > 0 \tag{4}$$

for all nonnull vectors **x**. The matrix of a positive definite quadratic form must be nonsingular. If

$$\mathbf{x}'\mathbf{A}\mathbf{x} \geq 0 \tag{5}$$

the form and matrix are positive semidefinite. For example, the uncorrected sum of squares

$$x_1^2 + x_2^2 + x_3^2 + x_4^2 \tag{6}$$

is a positive definite quadratic form with A equal to the 4×4 identity matrix. The sum of squared deviations of the x_i about their mean

$$\sum_{i=1}^{4} (x_i - \bar{x})^2 = \sum_{i=1}^{4} x_i^2 - \frac{1}{4}\left(\sum_{i=1}^{4} x_i\right)^2 \tag{7}$$

is a quadratic form with matrix

$$\mathbf{A} = \frac{1}{4}\begin{bmatrix} 3 & -1 & -1 & -1 \\ -1 & 3 & -1 & -1 \\ -1 & -1 & 3 & -1 \\ -1 & -1 & -1 & 3 \end{bmatrix} \tag{8}$$

We can show that the rank of **A** is only three, so that it cannot be positive definite. By other criteria [see, for example, Morrison (1976, Section 2.9)] **A** can be shown to be positive semidefinite. Then the sum of squares, and the variance of the x_i, cannot be negative.

The *characteristic roots* (or *latent roots* or *eigenvalues*) of the $p \times p$ matrix **A** are the p roots of the determinantal equation

$$|\mathbf{A} - \lambda \mathbf{I}| = 0 \tag{9}$$

The equation is a pth-degree polynomial in λ. The product of the characteristic roots $\lambda_1, \ldots, \lambda_p$ is the determinant of **A**, while the sum of the roots is the trace of **A**. Each root has a characteristic vector $\mathbf{a}_i$ defined by the system of equations

$$[\mathbf{A} - \lambda_i \mathbf{I}]\mathbf{a}_i = \mathbf{0} \tag{10}$$

For example, the characteristic roots of the matrix (8) can be computed to be

$$\lambda_1 = \lambda_2 = \lambda_3 = 1 \qquad \lambda_4 = 0 \tag{11}$$

The characteristic vector for the last root is proportional to

$$\mathbf{a}_4' = [1, \quad 1, \quad 1, \quad 1] \tag{12}$$

and the other three characteristic vectors are any four-element vectors orthogonal to it.

Many properties of square symmetric matrices are indicated by their characteristic roots. A positive definite matrix must have only positive roots. An $n \times n$ positive semidefinite matrix of rank r will have r positive roots and $n - r$ zero roots. If the characteristic roots λ_i and λ_j are different, their associated characteristic vectors a_i and a_j must be orthogonal.

A quadratic form can be transformed to a sum of squares of new variables by a transformation provided by the characteristic vectors of the matrix $\mathbf{A}$ of the form. We begin by dividing each vector by the square root of its inner product, so that each vector has been *normalized* to length one:

$$\mathbf{a}_i'\mathbf{a}_i = 1 \qquad i = 1, \ldots, n \tag{13}$$

We form the orthogonal matrix $\mathbf{P}$ with the normalized characteristic vectors as its columns:

$$\mathbf{P} = [\mathbf{a}_1 \quad \cdots \quad \mathbf{a}_n] \tag{14}$$

Then

$$\mathbf{P}'\mathbf{A}\mathbf{P} = \mathbf{D} \tag{15}$$

where

$$\mathbf{D} = \begin{bmatrix} \lambda_1 & \cdots & 0 \\ \cdot & \cdots & \cdot \\ 0 & \cdots & \lambda_n \end{bmatrix} \tag{16}$$

is the diagonal matrix containing the successive characteristic roots of $\mathbf{A}$ on its diagonal. In that way we can reduce the quadratic form $\mathbf{x}'\mathbf{A}\mathbf{x}$ to a sum of squares. Make the orthogonal transformation

$$\mathbf{x} = \mathbf{P}\mathbf{z} \tag{17}$$

so that the form becomes

$$\begin{aligned} \mathbf{x}'\mathbf{A}\mathbf{x} &= \mathbf{z}'\mathbf{P}'\mathbf{A}\mathbf{P}\mathbf{z} \\ &= \mathbf{z}'\mathbf{D}\mathbf{z} \\ &= \sum_{i=1}^{n} \lambda_i \mathbf{z}_i^2 \end{aligned} \tag{18}$$

or a weighted sum of squares of the new variables.

Let us reduce the quadratic form (7) to a sum of squares. We shall take

$$\begin{aligned} \mathbf{a}_1' &= [\tfrac{1}{2}, \tfrac{1}{2}, -\tfrac{1}{2}, -\tfrac{1}{2}] \\ \mathbf{a}_2' &= [-\tfrac{1}{2}, \tfrac{1}{2}, -\tfrac{1}{2}, \tfrac{1}{2}] \\ \mathbf{a}_3' &= [\tfrac{1}{2}, -\tfrac{1}{2}, -\tfrac{1}{2}, \tfrac{1}{2}] \end{aligned} \tag{19}$$

as the characteristic vectors corresponding to the first three characteristic roots $\lambda_1 = \lambda_2 = \lambda_3 = 1$ of (11). From them and the normalized form of the fourth characteristic vector (12) we have the columns of the transformation matrix $\mathbf{P}$:

$$\mathbf{P} = \begin{bmatrix} \frac{1}{2} & -\frac{1}{2} & \frac{1}{2} & \frac{1}{2} \\ \frac{1}{2} & \frac{1}{2} & -\frac{1}{2} & \frac{1}{2} \\ -\frac{1}{2} & -\frac{1}{2} & -\frac{1}{2} & \frac{1}{2} \\ -\frac{1}{2} & \frac{1}{2} & \frac{1}{2} & \frac{1}{2} \end{bmatrix} \tag{20}$$

We may verify that

$$\mathbf{P}'\mathbf{A}\mathbf{P} = \frac{1}{16}\begin{bmatrix} 1 & 1 & -1 & -1 \\ -1 & 1 & -1 & 1 \\ 1 & -1 & -1 & 1 \\ 1 & 1 & 1 & 1 \end{bmatrix}\begin{bmatrix} 3 & -1 & -1 & -1 \\ -1 & 3 & -1 & -1 \\ -1 & -1 & 3 & -1 \\ -1 & -1 & -1 & 3 \end{bmatrix}\begin{bmatrix} 1 & -1 & 1 & 1 \\ 1 & 1 & -1 & 1 \\ -1 & -1 & -1 & 1 \\ -1 & 1 & 1 & 1 \end{bmatrix}$$

$$= \begin{bmatrix} 1 & 0 & 0 & 0 \\ 0 & 1 & 0 & 0 \\ 0 & 0 & 1 & 0 \\ 0 & 0 & 0 & 0 \end{bmatrix} \tag{21}$$

or the diagonal matrix of characteristic roots. The transformation

$$\begin{bmatrix} x_1 \\ x_2 \\ x_3 \\ x_4 \end{bmatrix} = \frac{1}{2}\begin{bmatrix} 1 & -1 & 1 & 1 \\ 1 & 1 & -1 & 1 \\ -1 & -1 & -1 & 1 \\ -1 & 1 & 1 & 1 \end{bmatrix}\begin{bmatrix} z_1 \\ z_2 \\ z_3 \\ z_4 \end{bmatrix} \tag{22}$$

applied to the quadratic form leads to

$$\mathbf{x}'\mathbf{A}\mathbf{x} = z_1^2 + z_2^2 + z_3^2 \tag{23}$$

or the sum of squares of three new variables. If the x_i are independently distributed as normal random variables with a common mean μ and variance σ^2, then

$$\frac{\mathbf{x}'\mathbf{A}\mathbf{x}}{\sigma^2} = \frac{z_1^2 + z_2^2 + z_3^2}{\sigma^2} \tag{24}$$

is a chi-squared variate with three degrees of freedom, for the z_i/σ are independent standard normal variates. We shall say more about such statistical properties of quadratic forms in Appendix C.

B-5 REFERENCES

BARTLETT, M. S. (1951): An inverse adjustment arising in discriminant analysis, *Annals of Mathematical Statistics*, vol. 22, pp. 107–111.

HOHN, F. E. (1964): *Elementary Matrix Algebra*, 2nd ed., Macmillan Publishing Co., Inc., New York.

MORRISON, D. F. (1976): *Multivariate Statistical Methods*, 2nd ed., McGraw-Hill Book Company, New York.

appendix C

Some Results in Mathematical Statistics

C-1 RANDOM VARIABLES

As its name suggests, a *random variable* or *variate* is a variable that takes on values according to some probability rule. The rule is called a *probability* or *frequency function* if the variable has only a finite number of outcomes or at most a countable infinity of values. Such variates are called *discrete* and include the binomial model for the number of heads in a series of coin tosses, the card configurations in bridge or poker hands, and the number of accidents or other events in some time period. Continuous random variables have values defined for a continuous interval or other region in some geometric space. Because we have used only continuous random variables in this text we restrict our attention to their properties in this appendix.

Let X be a continuous random variable that can assume values on the real number line. We may describe X by its *density function* $f(x)$, a mathematical function that measures the relative frequency of X at the point x. The probability that X takes a value in the interval beginning at a and ending at b can be expressed in terms of the density function as the integral

$$P(a \leq X \leq b) = \int_a^b f(x)\, dx \tag{1}$$

The probability is equal to the area beneath the curve defined by $f(x)$ between the points a and b. We are assuming, of course, that $f(x)$ is a continuous function in that region and that its integral exists. The *distribution function* of X is

$$F(x) = P(X \leq x) = \int_{-\infty}^{x} f(u)\, du \tag{2}$$

or the probability that X is less than or equal to the number x. The distribution function is more useful than the density and is usually the function tabled to describe the particular random variable.

The average value of a random variable is its *mean*

$$E(X) = \int_{-\infty}^{\infty} xf(x)\,dx \tag{3}$$

$E(X)$ is the average of X continuously weighted by the density function. It is also called the *expectation* or *first moment* of X, for it is the horizontal center of gravity of an object with the shape of $f(x)$. If the density is symmetric, its point of symmetry is $E(X)$. Then $E(X)$ is also the *median*, or the value m such that

$$P(x \leq m) = \int_{-\infty}^{m} f(x)\,dx = \int_{m}^{\infty} f(x)\,dx = 0.5 \tag{4}$$

The mean and median are general measures of the location of the central portion of the density.

The *variance*

$$\begin{aligned}\text{Var}\,(X) &= \int_{-\infty}^{\infty} [x - E(X)]^2 f(x)\,dx \\ &= \int_{-\infty}^{\infty} x^2 f(x)\,dx - [E(X)]^2\end{aligned} \tag{5}$$

of the random variable X measures variation about the expected value. Variates with large variances have densities with greater dispersion. Because the variance is in squared units of X a more informative measure of variation is the standard deviation

$$\sigma = \sqrt{\text{Var}\,(X)} \tag{6}$$

It is often convenient to use the expectation and variance operators

$$E(X) \qquad \text{Var}\,(X)$$

to calculate or manipulate those parameters. The expectation operator applied to a linear function of the variate produces the mean of the function according to these rules:

$$\begin{aligned}E(cX) &= cE(X) \\ E(a + cX) &= a + cE(X) \\ E(c) &= c\end{aligned} \tag{7}$$

Similarly, the variance operator has these properties:

$$\begin{aligned}\text{Var}\,(c) &= 0 \\ \text{Var}\,(cX) &= c^2\,\text{Var}\,(X) \\ \text{Var}\,(a + cX) &= c^2\,\text{Var}\,(X)\end{aligned} \tag{8}$$

In each case a and c are nonrandom constants.

If the several random variables $X_1, \ldots, X_N$ have expectations $\mu_1 \ldots, \mu_N$, we may apply the expectation operator to their weighted sum

$$Y = a_1 X_1 + \cdots + a_N X_N \tag{9}$$

to obtain

$$\begin{aligned} E(Y) &= E(a_1 X_1 + \cdots + a_N X_N) \\ &= a_1 E(X_1) + \cdots + a_N E(X_N) \\ &= a_1 \mu_1 + \cdots + a_N \mu_N \end{aligned} \tag{10}$$

If in addition the X_i are independent variates, or ones whose joint distribution function

$$F(x_1, \ldots, x_N) = P(X_1 \leq x_1, \ldots, X_N \leq x_N) = F_1(x_1) \cdots F_N(x_N) \tag{11}$$

is the product of the individual distributions $F_i(x_i)$, then

$$\text{Var}\,(a_1 X_1 + \cdots + a_N X_N) = a_1^2 \,\text{Var}\,(X_1) + \cdots + a_N^2 \,\text{Var}\,(X_N) \tag{12}$$

In particular, the variance of a sum is the sum of the individual variances:

$$\text{Var}\,(X_1 + \cdots + X_N) = \text{Var}\,(X_1) + \cdots + \text{Var}\,(X_N) \tag{13}$$

These operators and their properties are useful for finding the means and variances of averages and other linear functions of random variables.

The dependent random variables $X_1, \ldots, X_N$ are described by their joint distribution function or their joint density function

$$f(x_1, \ldots, x_N) \tag{14}$$

The marginal density functions of the single X_i or any sets of them are found by integrating the joint density over the regions of the remaining variates. The degree of covariation (and to a large extent, the amount of dependence) of the random variables X_i and X_j is given by their *covariance*

$$\begin{aligned} \text{Cov}\,(X_i, X_j) &= \int_{-\infty}^{\infty} \int_{-\infty}^{\infty} [x_i - E(X_i)][x_j - E(X_j)] f_{ij}(x_i, x_j)\, dx_i\, dx_j \\ &= \int_{-\infty}^{\infty} \int_{-\infty}^{\infty} x_i x_j f_{ij}(x_i, x_j)\, dx_i\, dx_j - [E(X_i)][E(X_j)] \\ &= E(X_i X_j) - [E(X_i)][E(X_j)] \end{aligned} \tag{15}$$

If X_i and X_j are independent, their joint density $f_{ij}(x_i, x_j)$ factors into the product of the marginal densities, and

$$\text{Cov}\,(X_i, X_j) = [E(X_i)][E(X_j)] - [E(X_i)][E(X_j)] = 0 \tag{16}$$

Although independent variates must have a covariance of zero, zero covariance does not necessarily imply independence.

The covariance operator Cov (X, Y) is useful for finding the covariances of linear functions of random variables. The operator has these properties:

$$\begin{aligned} &\text{Cov}\,(X, a) = 0 \\ &\text{Cov}\,(aX + b, cY + d) = ac\,\text{Cov}\,(X, Y) \\ &\text{Cov}\,(X, X) = \text{Var}\,(X) \end{aligned} \tag{17}$$

where a, b, c, and d are constants. As with the variance, the covariance is unaffected by adding constants to its variates or by translating their scale origins.

The variances of sums or other linear functions of dependent random variables must contain covariance terms:

$$\text{Var}(X+Y) = \text{Cov}(X+Y, X+Y)$$
$$= \text{Var}(X) + \text{Var}(Y) + 2\,\text{Cov}(X, Y) \tag{18}$$
$$\text{Var}(X-Y) = \text{Var}(X) + \text{Var}(Y) - 2\,\text{Cov}(X, Y) \tag{19}$$

Also,

$$\text{Cov}(X+Y, X-Y) = \text{Var}(X) - \text{Var}(Y) \tag{20}$$

Means, variances, and covariances are examples of the *moments* of a random variable. The mean is the first moment of the variate about its scale origin, whereas the variance in the second moment of X about its mean. The covariance is the first *product moment* of two variates about their means. Variances and covariances are called central moments because they are computed from deviations about the mean. The kth moment of the variate X is

$$\mu_k' = E(X^k) = \int_{-\infty}^{\infty} x^k f(x)\,dx \tag{21}$$

The prime denotes a moment about the origin, as opposed to a central moment such as the variance calculated about the mean. Moments of a variate describe the location, dispersion, and shape of the density function, and in certain cases the complete sequence of moments defines the density uniquely.

The *moment generating function* of the variate X is the function

$$M(\theta) = E(e^{\theta x})$$
$$= \int_{-\infty}^{\infty} e^{\theta x} f(x)\,dx \tag{22}$$

of the positive variable θ. $M(\theta)$ can be used to generate moments: The kth moment is equal to

$$\mu_k' = \left.\frac{d^k M(\theta)}{d\theta^k}\right|_{\theta=0} \tag{23}$$

The moment generating function has another more important property: If it exists, it defines the density function uniquely. If we can recognize $M(\theta)$ for a particular function of a random variable, we know the distribution of the random function. Moment generating functions are especially useful for identifying the distributions of quadratic forms as chi-squared.

C-2 NORMAL RANDOM VARIABLES

Throughout this text we have used the normal distribution to describe random variation or disturbances in our models for least squares and the analysis of variance. The normal random variable X has density function

$$f(x) = \frac{1}{\sqrt{2\pi}\,\sigma} \exp\left\{-\frac{1}{2}\frac{(x-\mu)^2}{\sigma^2}\right\} \qquad -\infty < x < \infty \tag{1}$$

The density is symmetric about the mean $E(X) = \mu$. The variance of X is equal to the second parameter σ^2. The transformation

$$Z = \frac{X - \mu}{\sigma} \tag{2}$$

gives the standard normal random variable with mean zero and unit variance or standard deviation. The probabilities

$$P(0 \leq Z \leq z) = \frac{1}{\sqrt{2\pi}} \int_0^z e^{-1/2x^2}\, dx \tag{3}$$

for the distribution function of Z are given in Table 1, Appendix A.

Normal random variables have many useful properties that we have used repeatedly in our development of linear statistical methods. Probably the most important one is this:

> Linear functions of normal random variables are themselves normally distributed.

If $X_1, \ldots, X_N$ are independent normal random variables with respective means $\mu_1, \ldots, \mu_N$ and variances $\sigma_1^2, \ldots, \sigma_N^2$, the linear compound

$$Y = a_1 X_1 + \cdots + a_N X_N \tag{4}$$

is a normal variate with parameters

$$\begin{aligned} E(Y) &= a_1\mu_1 + \cdots + a_N\mu_N \\ \operatorname{Var}(Y) &= a_1^2\sigma_1^2 + \cdots + a_N^2\sigma_N^2 \end{aligned} \tag{5}$$

For example, if the X_i have a common mean μ and the same variance σ^2, their sum $X_1 + \cdots + X_N$ is normal with mean $N\mu$ and variance $N\sigma^2$. Their mean

$$\bar{X} = \frac{1}{N}(X_1 + \cdots + X_N) \tag{6}$$

is normal with parameters μ and σ^2/N. The difference $X_i - X_j$ of any two of the variates is normally distributed with a zero mean and variance $2\sigma^2$.

Multivariate normal variates. We defined a p-dimensional multinormal random vector in Section 2-3 as one with density function

$$f(\mathbf{x}) = (2\pi)^{-p/2} |\mathbf{\Sigma}|^{-1/2} \exp\{-\tfrac{1}{2}[(\mathbf{x} - \boldsymbol{\mu})'\mathbf{\Sigma}^{-1}(\mathbf{x} - \boldsymbol{\mu})]\} \tag{7}$$

The parameters of the density are the mean vector $E(\mathbf{X}) = \boldsymbol{\mu}$ and the covariance matrix $\mathbf{\Sigma}$. The covariance matrix is symmetric and positive definite, and its properties were developed in Chapter 2 for the treatment of multiple partial correlation. Multinormal random vectors with positive semidefinite covariance matrices also occur if their variates are linearly related. Such *singular* multinormal vectors must be described by their distribution functions, for their densities are undefined. As in the case of independent normal random variables this property holds:

> Linear functions of multinormal random variables are multinormally distributed.

If the new variate is

$$\mathbf{Y} = \mathbf{AX} \tag{8}$$

its joint distribution is multinormal with mean vector and covariance matrix

$$E(\mathbf{Y}) = \mathbf{A}\boldsymbol{\mu} \qquad \text{Cov}\,(\mathbf{Y}, \mathbf{Y}') = \mathbf{A}\boldsymbol{\Sigma}\mathbf{A}' \tag{9}$$

Let us list some special cases of the general linear transformation on multinormal vectors:

1. The ith variate of $\mathbf{X}$ is normal with mean μ_i and variance σ_{ii}. The parameters are the appropriate ith element of $\boldsymbol{\mu}$ and the ith diagonal term of $\boldsymbol{\Sigma}$.
2. The pair of elements X_i and X_j of $\mathbf{X}$ will have the bivariate normal distribution described by (1), Section 1-6, with parameters

$$\begin{bmatrix} \mu_1 \\ \mu_2 \end{bmatrix} \qquad \begin{bmatrix} \sigma_{ii} & \sigma_{ij} \\ \sigma_{ij} & \sigma_{jj} \end{bmatrix}$$

   Similarly, any set of variates from $\mathbf{X}$ has a multinormal distribution with parameters appropriately chosen from $\boldsymbol{\mu}$ and $\boldsymbol{\Sigma}$.
3. The single linear compound

$$Y = \mathbf{a}'\mathbf{X}$$

   is univariate normal with mean $E(Y) = \mathbf{a}'\boldsymbol{\mu}$ and covariance matrix

$$\mathbf{a}'\boldsymbol{\Sigma}\mathbf{a} = \sum_{i=1}^{p} \sum_{j=1}^{p} a_i a_j \sigma_{ij}$$

4. The pair of linear compounds

$$Y = \mathbf{a}'\mathbf{X} \qquad Z = \mathbf{b}'\mathbf{X}$$

   has the bivariate normal distribution with mean vector and covariance matrix

$$\begin{bmatrix} \mathbf{a}'\boldsymbol{\mu} \\ \mathbf{b}'\boldsymbol{\mu} \end{bmatrix} \qquad \begin{bmatrix} \mathbf{a}'\boldsymbol{\Sigma}\mathbf{a} & \mathbf{a}'\boldsymbol{\Sigma}\mathbf{b} \\ \mathbf{a}'\boldsymbol{\Sigma}\mathbf{b} & \mathbf{b}'\boldsymbol{\Sigma}\mathbf{b} \end{bmatrix}$$

5. If $\mathbf{X}$ has the covariance matrix $\sigma^2\mathbf{I}$, or independent elements, and $\mathbf{P}$ is an orthogonal matrix, $\mathbf{Y} = \mathbf{PX}$ has independent components with the same variance σ^2.

C-3 THE CHI-SQUARED, *t*-, AND *F*-DISTRIBUTIONS

Chi-squared random variables. The chi-squared variate with degrees of freedom parameter ν was defined by (2), Section 3-2, as the sum of the squares of ν independent standard normal random variables. The density function is

$$f(\chi^2) = \frac{1}{2^{\nu/2}\Gamma(\nu/2)}(\chi^2)^{(\nu-2)/2}e^{-\chi^2/2} \qquad 0 \leq \chi^2 < \infty \tag{1}$$

Percentage points of the distribution of χ^2 are given in Table 2, Appendix A.

The mean and variance of χ^2 are

$$E(\chi^2) = \nu \qquad \text{Var}(\chi^2) = 2\nu \tag{2}$$

and its moment generating function is

$$M(\theta) = (1 - 2\theta)^{-\nu/2} \tag{3}$$

We may see from $M(\theta)$ that the sum of k independent chi-squared random variable with respective degrees of freedom $\nu_1, \ldots, \nu_k$ is itself chi-squared with degrees of freedom parameter

$$\nu = \nu_1 + \cdots + \nu_k \tag{4}$$

A noncentral chi-squared variate was defined in (3), Section 3-2, as the sum of the squares of independently and normally distributed random variables with variance one but respective means $\mu_1, \ldots, \mu_\nu$. The noncentral chi-squared density function is

$$p(\chi'^2) = e^{-\lambda/2} \sum_{j=0}^{\infty} \frac{(\lambda/2)^j (\chi'^2)^{j+\nu/2-1} e^{-\chi'^2/2}}{j!\,\Gamma(\nu/2 + j) 2^{\nu/2+j}} \qquad 0 \le \chi'^2 < \infty \tag{5}$$

The density depends upon the degrees of freedom ν and the noncentrality parameter

$$\lambda = \sum_{i=1}^{\nu} \mu_i^2 \tag{6}$$

If $\lambda = 0$ the function reduces to the central, or ordinary, chi-squared density (1). The mean and variance of the non-central chi-squared variate are

$$\begin{aligned} E(\chi'^2) &= \nu + \lambda \\ \text{Var}(\chi'^2) &= 2\nu + 4\lambda \end{aligned} \tag{7}$$

The moment generating function is

$$M(\theta) = (1 - 2\theta)^{-\nu/2} \exp \frac{\lambda\theta}{1 - 2\theta} \tag{8}$$

We note that the product of the moment generating functions of k independent noncentral chi-squared variates with degrees of freedom $\nu_1, \ldots, \nu_k$ and noncentralities $\lambda_1, \ldots, \lambda_k$ is

$$M(\theta) = (1 - 2\theta)^{-(\Sigma \nu_i)/2} \exp \left[\frac{(\sum \lambda_i)\theta}{1 - 2\theta} \right] \tag{9}$$

or the moment generating function of a noncentral chi-squared with parameters

$$\nu = \sum_{i=1}^{k} \nu_i \qquad \lambda = \sum_{i=1}^{k} \lambda_i \tag{10}$$

Hence Theorem 3-2 holds.

The *t*-distribution. The Student–Fisher t-variate is defined as the quotient

$$t = \frac{z}{\sqrt{\chi^2/n}} \tag{11}$$

of a standard normal random variable z and the square root of an independent chi-squared variate divided by its degrees of freedom n. The density function of t is

$$f(t) = \frac{\Gamma[(n+1)/2]}{\sqrt{\pi n}\,\Gamma(n/2)}\left(1 + \frac{t^2}{n}\right)^{-(n+1)/2} \qquad -\infty < t < \infty \tag{12}$$

The density has the single degrees of freedom parameter n. As n increases without limit the function tends to a standard normal density. Upper percentage points of t are given in Table 3, Appendix A.

Central and noncentral F-variates. The ratio

$$\begin{aligned} F &= \frac{\chi_1^2/m}{\chi_2^2/n} \\ &= (n/m)(\chi_1^2/\chi_2^2) \end{aligned} \tag{13}$$

of two independent central chi-squared variates divided by their respective degrees of freedom has the F-, or variance ratio, distribution with density function

$$g(F) = \frac{\Gamma[(m+n)/2]}{\Gamma(m/2)\Gamma(n/2)}\left(\frac{m}{n}\right)^{m/2} F^{(m-2)/2}\left(1 + \frac{m}{n}F\right)^{-(m+n)/2} \qquad 0 \leq F < \infty \tag{14}$$

Table 4, Appendix A, contains the upper 100α percentage points $F_{\alpha;\, m,n}$ of the distribution function. The variates t and F are connected by the relation

$$F(1, n) = t^2(n) \tag{15}$$

or the square of a Student–Fisher t with n degrees of freedom is equal to an F with one and n degrees of freedom.

If the numerator chi-squared variate in the definition (13) is noncentral with noncentrality parameter λ, the ratio has the noncentral F-distribution with density function

$$\begin{aligned} g(F') = e^{-\lambda/2} \sum_{j=0}^{\infty} \frac{(\lambda/2)^j \Gamma\left(\dfrac{m+n}{2} + j\right)}{j!\,\Gamma(m/2 + j)\Gamma(n/2)}\left(\frac{m}{n}\right)^{m/2+j} (F')^{(m+2j-2)/2} \\ \cdot \left(1 + \frac{m}{n}F'\right)^{-(m+n+2j)/2} \qquad 0 \leq F' < \infty \end{aligned} \tag{16}$$

When $\lambda = 0$ the series reduces to the central density (14). The noncentral F-distribution probabilities are shown in Charts 1 through 10, Appendix A.

C-4 DISTRIBUTIONS OF QUADRATIC FORMS

In this section we give some further results on the quadratic form distributions introduced in Section 3-2.

In Section B-4 we showed that the quadratic form $\mathbf{x}'\mathbf{A}\mathbf{x}$ could be reduced to the weighted sum of squares

$$\mathbf{x}'\mathbf{A}\mathbf{x} = \sum_{i=1}^{n} \lambda_i z_i^2 \tag{1}$$

by the orthogonal transformation $\mathbf{x} = \mathbf{Pz}$. The λ_i are the characteristic roots of $\mathbf{A}$. If $\mathbf{A}$ is idempotent of rank r, it will have r characteristic roots of one and $n - r$ roots equal to zero. If the elements of $\mathbf{x}$ are independent standard normal variates, the z_i will also be independent and standard normal by the results in Section C-2, and

$$\mathbf{x}'\mathbf{A}\mathbf{x} = \sum_{i=1}^{r} z_i^2 \tag{2}$$

will have the central chi-squared distribution with r degrees of freedom.

The quadratic form

$$(\mathbf{x} - \boldsymbol{\mu})'\boldsymbol{\Sigma}^{-1}(\mathbf{x} - \boldsymbol{\mu}) \tag{3}$$

in the exponent of the multinormal density (7), Section C-2, has the chi-squared distribution with p degrees of freedom. One direct proof of that property follows from the moment generating function of the form. It is

$$\begin{aligned} M(\theta) &= E[\exp\{\theta(\mathbf{x} - \boldsymbol{\mu})'\boldsymbol{\Sigma}^{-1}(\mathbf{x} - \boldsymbol{\mu})\}] \\ &= \frac{1}{(2\pi)^{p/2}\,|\boldsymbol{\Sigma}|^{1/2}} \int_{-\infty}^{\infty} \cdots \int_{-\infty}^{\infty} \exp\left\{-\frac{1}{2}(1 - 2\theta)(\mathbf{x} - \boldsymbol{\mu})'\boldsymbol{\Sigma}^{-1}(\mathbf{x} - \boldsymbol{\mu})\right\} dx_1 \\ &\qquad \cdots dx_p \\ &= (1 - 2\theta)^{-p/2} \end{aligned} \tag{4}$$

or that of the chi-squared variate with p degrees of freedom. Alternatively, we note that the product of the quadratic form and covariance matrices is $\boldsymbol{\Sigma}^{-1}\boldsymbol{\Sigma} = \mathbf{I}$, or an idempotent matrix of full rank p. The distribution then follows from Theorem 3-1.

If the random vector $\mathbf{x}$ has expectation $E(\mathbf{x}) = \boldsymbol{\mu}$ and covariance matrix $\boldsymbol{\Sigma}$, the quadratic form $\mathbf{x}'\mathbf{A}\mathbf{x}$ has expected value

$$E(\mathbf{x}'\mathbf{A}\mathbf{x}) = \operatorname{tr} \boldsymbol{\Sigma}\mathbf{A} + \boldsymbol{\mu}'\mathbf{A}\boldsymbol{\mu} \tag{5}$$

The expectation does not require that $\mathbf{x}$ be multinormal. We may use that moment to find the expected value of

$$\text{SSE} = \mathbf{y}'(\mathbf{T} - \mathbf{X}(\mathbf{X}'\mathbf{X})^{-1}\mathbf{X}')\mathbf{y} \tag{6}$$

introduced in (28), Section 2-2, for the ordinary least squares model. $\mathbf{X}$ is the $N \times (p + 1)$ matrix of values of the fixed independent variables and may or may not be expressed in centered, or deviation, form. In the present case

$$\boldsymbol{\mu} = E(\mathbf{y}) = \mathbf{x}\boldsymbol{\beta} \qquad \boldsymbol{\Sigma} = \operatorname{Cov}(\mathbf{y}, \mathbf{y}') = \sigma^2\mathbf{I} \tag{7}$$

Then

$$\begin{aligned} E(\text{SSE}) &= \operatorname{tr}[\sigma^2\mathbf{I}(\mathbf{I} - \mathbf{X}(\mathbf{X}'\mathbf{X})^{-1}\mathbf{X}')] + \boldsymbol{\beta}'\mathbf{X}'(\mathbf{I} - \mathbf{X}(\mathbf{X}'\mathbf{X})^{-1}\mathbf{X}')\mathbf{X}\boldsymbol{\beta} \\ &= \sigma^2(N - \operatorname{tr}[\mathbf{X}'\mathbf{X}(\mathbf{X}'\mathbf{X})^{-1}]) + \boldsymbol{\beta}'\mathbf{X}'\mathbf{X}\boldsymbol{\beta} - \boldsymbol{\beta}'\mathbf{X}'\mathbf{X}(\mathbf{X}'\mathbf{X})^{-1}\mathbf{X}'\mathbf{X}\boldsymbol{\beta} \\ &= \sigma^2(N - p - 1) \end{aligned} \tag{8}$$

and the divisor for the unbiased estimator (29) in Section 2-2 follows.

If the random vector $\mathbf{x}$ is multinormal with parameters $\boldsymbol{\mu}$ and $\boldsymbol{\Sigma}$, the variance of its general quadratic form is

$$\operatorname{Var}(\mathbf{x}'\mathbf{A}\mathbf{x}) = 2\operatorname{tr}(\boldsymbol{\Sigma}\mathbf{A})^2 + 4\boldsymbol{\mu}'\mathbf{A}\boldsymbol{\Sigma}\mathbf{A}\boldsymbol{\mu} \tag{9}$$

C-5 FURTHER REFERENCES

ANDERSON, T. W. (1958): *An Introduction to Multivariate Statistical Analysis*, John Wiley & Sons, Inc., New York. (Multivariate normal distribution.)

FREUND, J. E., and R. E. WALPOLE (1980): *Mathematical Statistics*, 3rd ed., Prentice-Hall, Inc., Englewood Cliffs, N. J. (Introduction to statistical theory.)

GRAYBILL, F. A. (1961): *An Introduction to Linear Statistical Models*, vol. 1, McGraw-Hill Book Company, New York. (Multivariate normal distribution and quadratic forms.)

GRAYBILL, F. A. (1976): *Theory and Application of the Linear Model*, Duxbury Press, North Scituate, Mass. (Multivariate normal distribution, quadratic forms.)

HOGG, R. V., and A. T. CRAIG (1978): *Introduction to Mathematical Statistics*, 4th ed., Macmillan Publishing Co., Inc., New York. (Random variables and moments, multivariate normal variates, quadratic forms.)

LANCASTER, H. O. (1969): *The Chi-Squared Distribution*, John Wiley & Sons, Inc., New York. (Noncentral chi-squared, quadratic forms.)

MOOD, A. M., F. A. GRAYBILL, and D. C. BOES (1974): *Introduction to the Theory of Statistics*, 3rd ed., McGraw-Hill Book Company, New York. (Random variables and mathematical statistics.)

MORRISON, D. F. (1976): *Multivariate Statistical Methods*, 2nd ed., McGraw-Hill Book Company, New York. (Multivariate normal variates.)

SEARLE, S. R. (1971): *Linear Models*, John Wiley & Sons, Inc., New York. (Quadratic forms.)

Solutions to Odd-Numbered Exercises

CHAPTER 1

1. a. $\hat{y} = -19.58 + 1.15x$, $r^2 = 0.60$, $t_\beta = 4.23$
   b. Apparently not. The averages of the initial and second assays are different, and the line seems largely determined by the extreme levels of the second child.
   c.

Children omitted	Equation	r^2
2	$\hat{y} = 13.75 + 0.40x$	0.07
2, 10	$\hat{y} = -48.21 + 1.83x$	0.45

   The regression lines are greatly affected by the extreme levels of subjects 2 and 10.
3. Virtually none. Only 0.64 of 1% of the variation in test scores is explained by a linear regression relationship with the chemical level.
5. Set 1 gives a conventional scatterplot; the least squares line is appropriate. Set 2 plots as a smooth curve; the linear model is incorrect. Set 3 consists of a linear relationship and a single outlier. The least squares line in set 4 is due to the values in the eighth observation pair.
7. Verbal comprehension: $u = 1.21$; hypothesis tenable
   Visual concentration: $u = -3.40$; reject the equal-correlation hypothesis.
   Closure: $u = 1.67$; the null hypothesis cannot be rejected in favor of the two-sided alternative at the 0.05 level.
9.
$$\frac{dh(\rho)}{d\rho} = \frac{c}{1-\rho^2} = \frac{c}{2}\left(\frac{1}{1+\rho} + \frac{1}{1-\rho}\right)$$
$$h(\rho) = \frac{c}{2}\ln\frac{1+\rho}{1-\rho}$$
$$= c\tanh^{-1}\rho$$
11. Omission of the four extreme pairs has almost no effect on the correlation coefficient. Dropping pairs 26 and 29 changes the linear shape of the plot and decreases r slightly. If

the general linear configuration is retained r may increase when the outlier pairs 26, 27, or 28 are omitted.

13. The trend in average starting salaries appears to be taking a nonlinear turn.

15. a. 2.84%
b. No. $t = 0.57$.
c. $\hat{y} = 18.89 + 0.1442x$; $\hat{x} = 20.35 + 0.1967y$. The product of the slopes is equal to $r^2 = 0.0284$.
d. Moderately: $r = 0.5722$.

17. a. $\hat{y} = 0.493 + 0.0552x$
b. $r^2 = 0.8559$. With the exception of observations 3, 23, and 24, the residuals were generally less than one standard deviation in magnitude.
c. $0.0426 \leq \beta \leq 0.0678$

19. $r_{uv} = -0.1561$. Since $|r_{uv}|$ does not exceed any conventional critical value, we may conclude that the hypothesis of equal part I and part III variances is tenable.

21. a. A linear function will probably suffice.
b. $\hat{y} = -2.13 + 0.9131x$
$t_{\text{intercept}} = -1.00$; the hypothesis of a zero intercept is tenable.
c. $0.0767 \leq \beta \leq 1.7495$
d. $\hat{\beta} = 0.8486$
e. $0.7976 \leq \beta \leq 0.8996$
f. Since one is not contained in the interval of (e), the hypothesis is not tenable.

23. The slope coefficient of the standard scores is

$$\frac{\sum u_i v_i}{\sum u_i^2} = \frac{s_x^2 \sum (x_i - \bar{x})(y_i - \bar{y})}{s_x s_y \sum (x_i - \bar{x})^2}$$

$$= \frac{\sum (x_i - \bar{x})(y_i - \bar{y})}{(N-1) s_x s_y}$$

$$= r_{xy}$$

25. a. The straight line fits the data poorly for the first and last two years.
b. $\hat{y}_t = 25.62 + 1.6044(t - 7)$, $t = 1, \ldots, 13$
$r^2 = 0.8991$

c.

t	1	2	3	4	5	6	7
Residual	4.01	1.41	0.80	−1.80	−2.41	−3.01	−1.62

t	8	9	10	11	12	13
Residual	−0.22	−1.82	−0.43	0.97	2.36	1.76

d. $\hat{y}_t = 29.375 + 2.2976(t - 9.5)$, $t = 6, \ldots, 13$
e. $1.9834 \leq \beta \leq 2.6118$
f. Segmental model: SSE $= 1.2 + 4.1548 = 5.3548$, $r^2 = 0.9897$
The segmental model explains a greater proportion of variation than the linear model in (b).

CHAPTER 2

1. a. $\hat{y}_i = 0.5931 + 0.9689x_{i1} - 0.0017x_{i2}$
b. $R^2 = 0.8428$

c.

Variable	$\hat{\beta}_j$	t_j
X_1	0.9689	5.13
X_2	−0.0017	−1.08

Variable X_2 probably can be omitted.

d. $r_{Y,X_2} = -0.1168$; $\hat{\beta}_2$ not significant.
H_1 does not appear to be supported by the data.

3. a. $\hat{y}_i = 19{,}408 + 1770x_{i1} + 260.9x_{i2} - 305.6x_{i3}$

b. $t_{0.025;29} = 2.045$; reject the hypotheses of zero coefficients for gender and years of service. Educational level does not have a significant coefficient.

c. $R^2 = 0.2743$

d.

Variable	Coefficient	t
X_1: sex	3432	2.03
X_2: service	401.4	3.09

$R^2 = 0.2573$
Yes.

e. Measures of managerial responsibility; dummy variables for job classes.

5. a. Partial correlation of performance test scores with verbal tests fixed is 0.1732. Partial correlation of verbal test scores with performance tests fixed is −0.6658.

b. The nearest conservative one-sided 0.05 critical value is $r_{0.05;40} = 0.257$. The hypothesis is tenable for the first partial correlation but not for the second. The respective 95% confidence intervals are

$$-0.130 \leq \rho_{P.V} \leq 0.447$$
$$-0.804 \leq \rho_{V.P} \leq -0.460$$

7. a. $r_{12.3} = 0.65$

b. $N = 16$ observation triplets

9. Partial correlations:

$$\begin{bmatrix} 1 & -0.1728 & 0.3118 \\ & 1 & -0.3397 \\ & & 1 \end{bmatrix}$$

11. Write

$$R^2 = \frac{\text{SSR}}{\text{SSR} + \text{SSE}} = \frac{pF}{N - p - 1 + pF}$$

and find $E(R^2)$ from the F-distribution with p and $N - p - 1$ degrees of freedom. Then $E(R^2) = p/(N - 1)$, or the same expectation as in the multinormal model.

13. a.

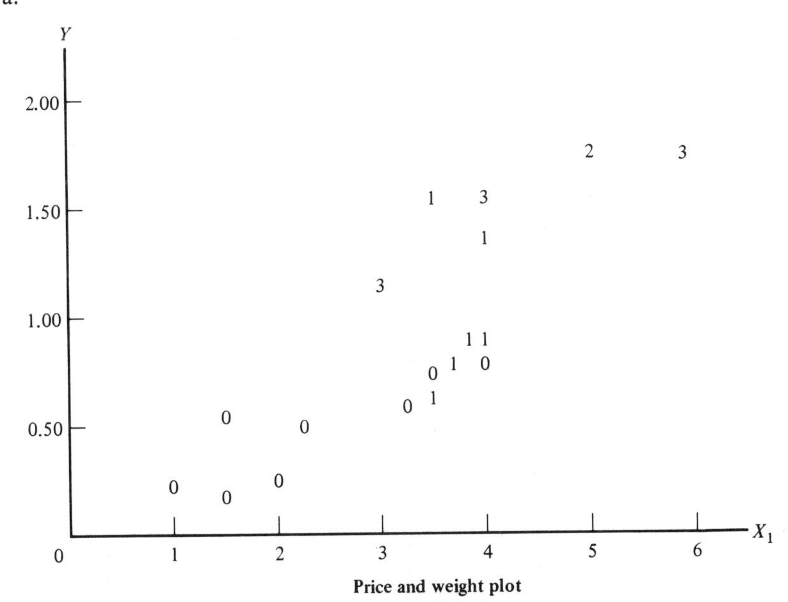

Price and weight plot

b.

Variable	Mean	$\hat{\beta}_j$	t_j
Y	0.89	—	—
X_1	3.32	0.2192	3.79
X_2	0.94	0.2109	3.32

SSE $= 0.8070$ SSR $= 3.6220$ $R^2 = 0.8178$

$$\hat{y}_i = -0.03 + 0.22x_{i1} + 0.21x_{i2}$$

c. Both t_j-statistics exceed the 0.05 critical value. The statistic for the simultaneous test is $F = 33.664$.

15. a. $\hat{y}_t = 19.45 - 0.424t + 0.1449t^2$

b. SSE $= 10.5855$, $\hat{\sigma}^2 = 1.0586$

c. The common variance assumption is probably valid for such a short time series. With the exception of the eighth year the quadratic model seems to fit the data.

d. 97.97%

e. The rounded scatterplot does not suggest autocorrelation.

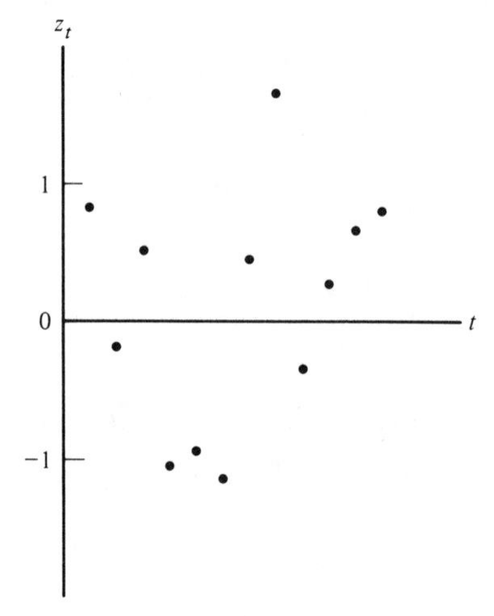

Residual plot

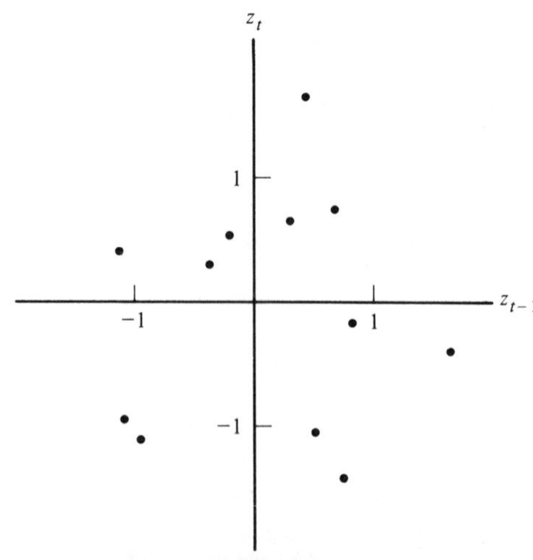

Plot of successive residuals

17. a. $\hat{\beta}_1$: standard deviation and t are 0.08597, 3.47.
$\hat{\beta}_2$: standard deviation and t are 0.08606, 7.34.

b. $0.049 \leq \beta_1 \leq 0.547$
$0.383 \leq \beta_2 \leq 0.881$

c. $F = 114.9$; 2, 17 d.f. Reject the hypothesis.

d. Correlation matrix of Y, X_1, and X_2:

$$\begin{bmatrix} 1 & 0.8439 & 0.9390 \\ & 1 & 0.7411 \\ & & 1 \end{bmatrix} \qquad \begin{aligned} r_{YX_1 \cdot X_2} &= 0.6410 \\ r_{YX_2 \cdot X_1} &= 0.8706 \end{aligned}$$

19.

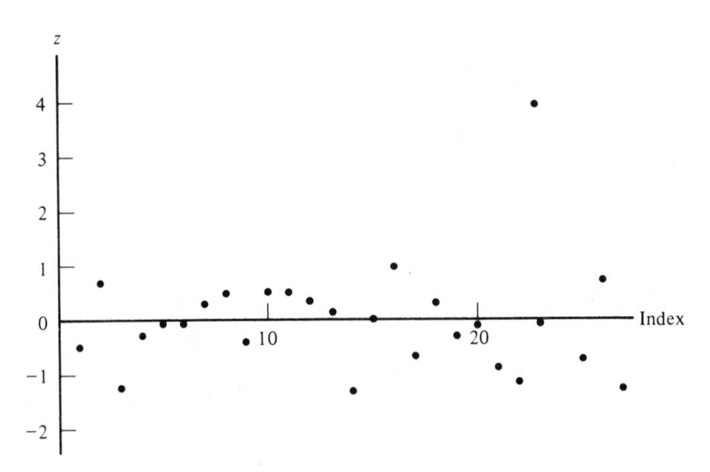

Residuals for the three-predictor model

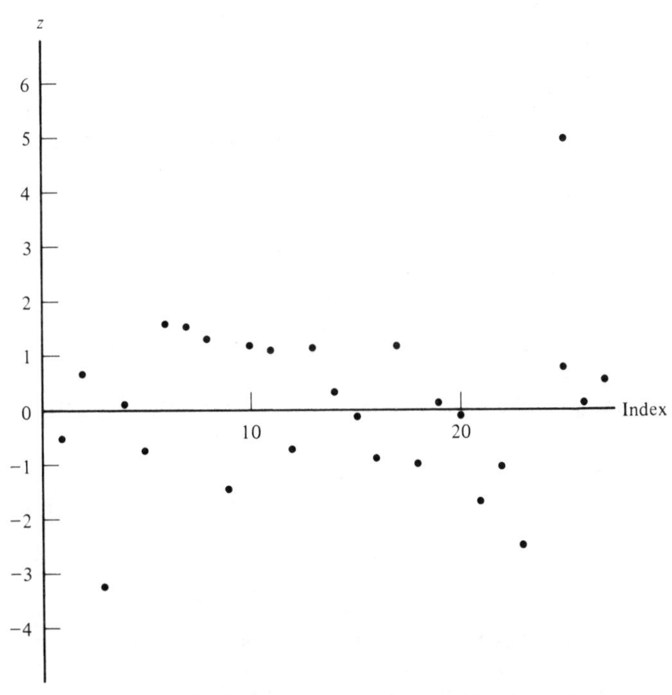

Residuals for the simple linear model

The three-predictor model residuals still have some runs of positive values, but they have smaller variation and fewer outliers.

21. $r_{X_1X_2 \cdot Y} = -0.28$

23. a. $R^2 = 0.6955$
b. X_1 and X_4 might be eliminated.
c. 652
d. $F = 12.56$; d.f. $= 4, 22$. The hypothesis should be rejected.
e. $\hat{y} = -1338.5 + 6.82x_2 + 0.151x_3$
$= 267$ for the player described in (c).
f. $R^2 = 0.6515$
g. $r^2_{YX_2} = 0.3003$, $r^2_{YX_3} = 0.3132$. On the basis of explained variation X_2 and X_3 provide more information together.

25. $$\begin{aligned}\text{Cov}\,(\mathbf{X}\hat{\boldsymbol{\beta}}, \mathbf{y}' - \hat{\boldsymbol{\beta}}'\mathbf{X}') &= \text{Cov}\,(\mathbf{X}(\mathbf{X}'\mathbf{X})^{-1}\mathbf{X}'\mathbf{y}, \mathbf{y}' - \mathbf{y}'\mathbf{X}(\mathbf{X}'\mathbf{X})^{-1}\mathbf{X}') \\ &= \mathbf{X}(\mathbf{X}'\mathbf{X})^{-1}\mathbf{X}'(\sigma^2\mathbf{I} - \sigma^2\mathbf{I}) \\ &= \mathbf{0}\end{aligned}$$

27. Some of the extreme residuals have been reduced in magnitude. The variance of the residuals appears smaller than in Figure 2-2(c).

CHAPTER 3

1. a. $F = 10.38$; reject the hypothesis at the 0.01 level.
b. Variables 2 – 7 have regression coefficients significant at the 0.05 level or smaller. Variables 1 and 8 do not have coefficients significantly different from zero.

c.

Method	0.05 Critical value	Variables with significant coefficients
Bonferroni	2.74	2, 5, 6
Scheffé	3.94	2

In both cases we have used infinite second degrees of freedom.
d. Clearly, variables 1 and 8 should be dropped. The simultaneous tests are probably too conservative for variable selection.
e. Variables 1 and 3 do not have significant simple correlations with blood lead level. The signs of the other correlations and regression coefficients are in agreement. The individual correlations explain very small proportions of blood lead variation.
f. $\hat{y} = 23.19$

3. a.

Variable	Regression coefficient	SD (coefficient)	t
U_1: race	2.7517	0.93588	2.94
U_2: race × sex	2.2555	1.22208	1.85
U_3: surface lead	0.07669	0.014052	5.46

$SS_{total} = 20{,}142.7027$ $SSR = 2637.1139$ $SSE = 17{,}505.5888$
$R^2 = 0.1309$

b.

Variable values	95% confidence limit Lower	95% confidence limit Upper	95% prediction limit Lower	95% prediction limit Upper
[0, 0, 10]	18.99	20.70	6.90	32.79
[0, 0, 24.02]	20.16	21.68	7.98	33.86
[1, 1, 40]	25.69	28.61	14.11	40.19

The confidence intervals are sufficiently narrow for useful predictions of means, but the prediction limits seem too wide for reliable estimation of individual blood lead levels.

5. Analysis of variance for course A:

Source	SS	d.f.	MS	F
Linear trend	0.1393	1	0.1393	0.34
Fit	0.1315	2	0.0658	0.14
Within years	5.6667	12	0.4722	—
Error	5.7982	14	0.4142	—
Total	5.9375	15	—	—

Test for equal regression coefficients in courses A and B:
$\hat{\beta}_A = 0.0826$ $\quad \hat{\beta}_B = -0.2443$ $\quad \hat{\beta}_W = -0.1386$ $\quad$ SSP $= 1.4777$
$\hat{\sigma}^2_{\text{SSE}} = 0.5187$ $\quad \hat{\sigma}^2_{\text{pure}} = 0.509$
(a) $F_{\text{SSE}} = 2.85$; 1, 48 d.f. (b) $F_{\text{pure}} = 2.90$; 1, 44 d.f.
Both F-statistics are only significant at the 0.10 level.

7. Merely verify that $\mathbf{b}'\mathbf{A} = [0, \ldots, 0]$.

9.

Sample	N_j	$\Sigma(x_i - \bar{x})^2$	$\Sigma(x_i - \bar{x})(y_i - \bar{y})$	$\Sigma(y_i - \bar{y})^2$	SSR	SSE
1	17	22,945.65	22,808.29	22,797.58	22,669.78	127.15
2	57	38,594.49	31,598.73	27,470.88	25,871.04	1,599.84
3	48	1,291.71	1,744.02	3,449.92	2,354.70	1,095.21
Sum	122	62,831.85	56,151.04	53,718.38	50,895.52	2,822.20

$\hat{\beta}_{1W} = 0.8937$ $\quad$ SSP $= 714.94$ $\quad \hat{\sigma}^2 = 24.3293$ $\quad F = 14.69$
Reject the hypothesis of parallel lines.

11.

Source	SS	d.f.	MS	F
Line	803.92	1	803.92	38.47
Lack of fit	24.61	3	8.20	0.38
Within	999.47	46	21.73	—
Error	1024.08	49	20.90	—
Total	1828.00	50	—	—

Fitted line: $\hat{y} = 4.47 + 0.80x$

13.

$$\left(\begin{bmatrix} (1/v_1)(1 - d/v_1) & \cdots & -d/v_1 v_k \\ \vdots & & \vdots \\ -d/v_1 v_k & \cdots & (1/v_k)(1 - d/v_k) \end{bmatrix} \begin{bmatrix} v_1 & \cdots & 0 \\ \vdots & & \vdots \\ 0 & \cdots & v_k \end{bmatrix}\right)^2$$

$$= \begin{bmatrix} 1 - d/v_1 & \cdots & -d/v_1 \\ \vdots & & \vdots \\ -d/v_k & \cdots & 1 - d/v_k \end{bmatrix}$$

so that the product matrix is idempotent.

15. a. $\sum (y_i - \bar{y})^2 = 861{,}157$

$$\mathbf{y}'\mathbf{X}_1 = [107.15, \quad 40{,}336.95, \quad 1{,}892{,}200, \quad 6093.46]$$

$$\mathbf{X}_1'\mathbf{X}_1 = \begin{bmatrix} 0.0377 & 1.46 & 508 & 1.73 \\ & 6291.86 & -16{,}879 & 42.41 \\ & & 13{,}275{,}942 & 48{,}468.26 \\ & & & 405.63 \end{bmatrix}$$

b. 95% Scheffé confidence intervals:

Variable	$\hat{\beta}_j$	S.D. (coefficient)	t	Interval limits
X_1	1432.38	940.29	1.52	−1725.65, 4590.41
X_2	6.43	1.42	4.52	1.66, 11.20
X_3	0.117	0.056	2.09	−0.071, 0.305
X_4	−5.71	7.28	−0.78	−30.16, 18.74

c. $\text{SSR}(X_1, X_2, X_3, X_4) = 598.895$ $\quad \text{SSR}(X_2, X_3)_{\text{unadjusted}} = 561{,}121$
$\text{SSR}(X_1, X_4)_{\text{adjusted}} = 37{,}774$ $\quad F = 1.58$ with d.f. 2, 22
The null hypothesis is tenable.

d. $\text{SSR}(X_1, X_3, X_4)_{\text{adjusted}} = 340{,}296$ $\quad F = 9.52$ with d.f. 3, 22
The null hypothesis should be rejected at the 0.01 level.

17. a. Analysis of variance for Y and X_1:

Source	SS	d.f.	MS	F
Line	6.3588	1	6.3588	44.53
Error	2.5707	18	0.1428	—
Within	1.5367	5	0.3073	—
Lack of fit	1.0340	13	0.0795	0.26
Total	8.9295	19	—	—

b. Analysis of variance for Y and X_2:

Source	SS	d.f.	MS	F
Line	7.8737	1	7.8737	134.24
Error	1.0558	18	0.0586	—
Within	0.4300	4	0.1075	—
Lack of fit	0.6258	14	0.0447	0.42
Total	8.9295	19	—	—

The linear models fit the data adequately.

19. a. Weather Code 0 ($N_1 = 16$)
Matrix of sums of squares and products about means:

$$\begin{array}{c} \\ Y \\ X_1 \\ X_2 \end{array} \begin{array}{c} \begin{array}{ccc} Y & X_1 & X_2 \end{array} \\ \begin{bmatrix} 248.31 & 3812.86 & 39.76 \\ & 66{,}575 & 466.5 \\ & & 20.9375 \end{bmatrix} \end{array}$$

$\hat{y} = 0.1287 + 0.0521x_1 + 0.7384x_2$
$SSR_0 = 228.0038$ $SSE_0 = 20.3082$ $R_0^2 = 0.9182$ $\hat{\sigma}^2 = 1.5622$

b. Weather Code 1 ($N_2 = 10$)

$$\begin{array}{c} \\ Y \\ X_1 \\ X_2 \end{array} \begin{array}{ccc} Y & X_1 & X_2 \\ \end{array} \begin{bmatrix} 167.14 & 2889.44 & 19.34 \\ & 52{,}560 & 252 \\ & & 8.4 \end{bmatrix}$$

$\hat{y} = -2.3213 + 0.05132x_1 + 0.7626x_2$
$SSR_1 = 163.0268$ $SSE_1 = 4.1166$ $R_1^2 = 0.9754$ $\hat{\sigma}^2 = 0.5881$
Inspection of the coefficients of X_1 and X_2 reveals virtually no difference due to weather conditions. However, the poor weather regression plane is below that for good weather.

21. 99% Scheffé simultaneous confidence intervals:

Variable	$\hat{\beta}_j$	Interval limits
1	0.6527	0.1946, 1.1108
2	−0.9485	−1.0153, −0.8817
3	0.2623	−0.1459, 0.6705

The third variable might be omitted.

23. a. Regression coefficients:

	Group		
Variable	A	B	C
X_1	2.8	3.6	7.6
X_2	1	1	0.5714

$SSR_A = 92.4$ $SSR_B = 143.6$ $SSR_C = 582.17$
$SSE_A = 4.4$ $SSE_B = 9.2$ $SSE_C = 2.63$
$SSP = 133.9810$ $\hat{\sigma}^2 = 2.7048$ $F = 12.38$
The hypothesis of parallel regression planes should be rejected at the 0.01 level.

b. Scheffé 0.01 critical value: $\sqrt{4F_{0.01;4,6}} = 6.0498$
Comparison of the regression coefficients of X_1:

A, B: $t = -1.09$ (not significant)
B, C: $t = -5.44$ (not significant)
A, C: $t = -6.53$ (significant in the Scheffé sense at the 0.01 level)

The three coefficients of X_2 can be seen by inspection to be not significantly different.

c. Because the independent variables have a common mean of zero in each group the test reduces to a one-way analysis of variance based on $\bar{y}_1, \bar{y}_2, \bar{y}_3$, and the within-groups error variance $\hat{\sigma}^2 = 2.7048$;

$$SSC = 109.2 \qquad F = 20.19 \qquad F_{0.05;2,6} = 5.14$$

The hypothesis of equal intercepts should be rejected. The Scheffé critical value is $\sqrt{2(5.14)} = 3.21$. The intercepts for A and B are not significant in the Scheffé sense at the 0.05 level; those for B and C, as well as A and C, are statistically significant.

25.

$$F = \frac{N-p-1}{p}\,\frac{R^2}{1-R^2}$$

$$R^2_{\alpha;\,p,N-p-1} = \frac{p}{N-p-1}F_{\alpha;\,p,N-p-1}\Big/\left(1+\frac{p}{N-p-1}F_{\alpha;\,p,N-p-1}\right)$$

CHAPTER 4

1. a. The significant coefficient of the squared number of fishermen may indicate a diminishing-returns effect: Intensive fishing leads to lower total catch.

b. These residuals were obtained from the fitted model (successively from 1932 to 1975):

93	−205	163	593	−457	−582	183	−44	−85	−39	−451	248
133	850	189	362	−385	−149	−265	544	−25	−97	−378	−148
−723	681	−451	−114	200	−871	552	545	211	−452	445	−523
1077	455	−345	−328	−453	−154	−319	519				

Some residuals have greater magnitudes than their counterparts in Table 4-5.

3. a. Analysis of variance for a fourth-degree orthogonal polynomial model:

Polynomial	SS	d.f.	MS	F
Linear	176.09	1	176.09	4.76
Quadratic	207.79	1	207.79	5.62
Cubic	99.60	1	99.60	2.69
Quartic	129.38	1	129.38	3.50
Error	702.47	19	36.97	—

Successive polynomials fitted to the data:

(1) Linear model analysis of variance:

Source	SS	d.f.	MS	F
Linear	176.09	1	176.09	3.40
Error	1139.25	22	51.78	—

(2) Quadratic model:

Source	SS	d.f.	MS	F
Linear	176.09	1	176.09	3.97
Quadratic	207.79	1	207.79	4.68
Error	931.46	21	44.36	—

(3) Cubic model:

Source	SS	d.f.	MS	F
Linear	176.09	1	176.09	4.23
Quadratic	207.79	1	207.79	5.00
Cubic	99.60	1	99.60	2.39
Error	831.86	20	41.59	—

b. First 10 observations: all t- or F-statistics are less than one. Any polynomial model seems inappropriate.
   Last 14 observations: The linear coefficient has test statistic $t = -4.07$. The higher-degree polynomials do not have significant coefficients.
c. The F-statistics for the first four orthogonal polynomial coefficients were generally smaller than those found in (a). Condensing the data seems to have deemphasized its trends.
d. Residual mean squares for the polynomials fitted in (c):

Model	$\hat{\sigma}^2$
Linear	39.94
Quadratic	33.98
Cubic	32.77
Quartic	27.78

These are smaller than the mean squares found from the bimonthly data.

5. *Model 1.* The residual variance seems to increase with larger predicted values. Residuals of the larger y_t are predominantly positive: The model underpredicts catch.

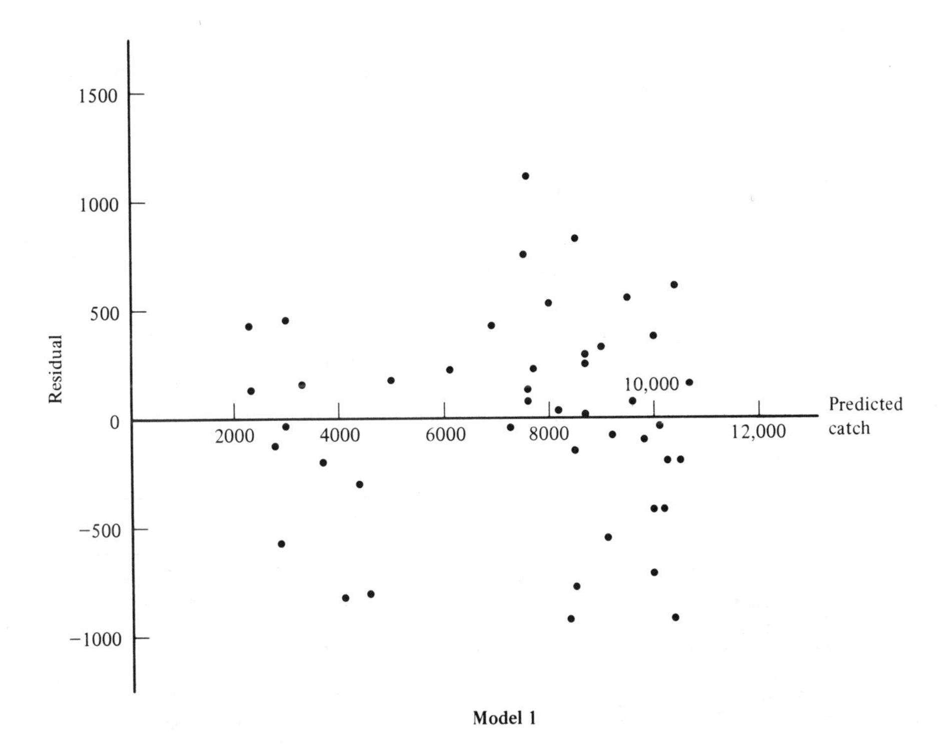

Model 1

Model 2. The variation in the residuals is slightly greater than that of Model 1. The shapes of the plots are similar, but the numbers of positive and negative residuals for the larger y_t are nearly equal.

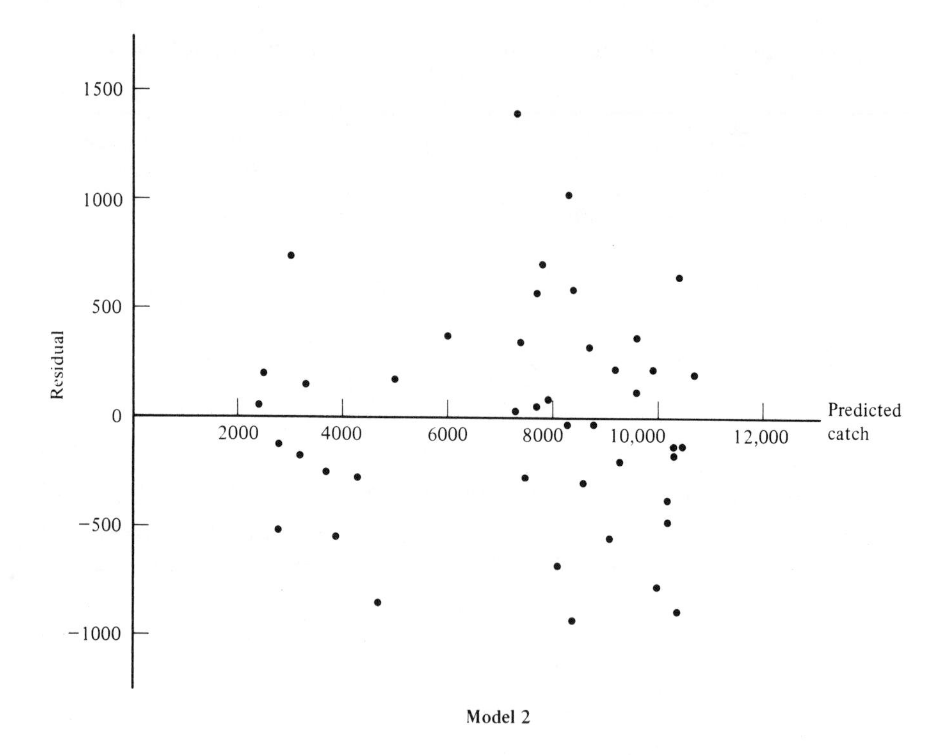

Model 2

7. Because the sample sizes in each year are so great we have kept a large number of decimal places in the estimates and in certain intermediate computations.

$$\mathbf{X'X} = \begin{bmatrix} 2{,}012{,}664 & -124{,}544 \\ -124{,}544 & 3{,}976{,}914 \end{bmatrix} \qquad \mathbf{X'y} = \begin{bmatrix} 908{,}570{,}598 \\ -83{,}175{,}892 \end{bmatrix}$$

$$\begin{bmatrix} \hat{\beta}_0 \\ \hat{\beta}_1 \end{bmatrix} = \begin{bmatrix} 451.0066597 \\ -6.7906218 \end{bmatrix}$$

$\hat{y}_t = 471.37852 - 6.79062t \qquad t = 1, \ldots, 5$

Analysis of variance:

Source	SS	d.f.	MS	F
Grand mean	$4.101529721 \times 10^{11}$	1	—	—
Slope (adjusted for intercept)	183,234,435	1	183,234,435	15,292
Intercept and slope	$4.103362065 \times 10^{11}$	2	—	—
Lack of fit	644,710	3	214,903	17.93
Within years	$2.411684839 \times 10^{10}$	2,012,659	11,982.58	—
Error	$2.41174931 \times 10^{10}$	2,012,662	11,982.88	—
Total	$4.344536996 \times 10^{11}$	2,012,664	—	—

The mean squares not needed for the F-ratios have been omitted. Because of the enormous sample sizes the linear model does not adequately explain the trend in the yearly means.

CHAPTER 5

1. t- and R^2-statistics for selected subset regression analyses:

Variable	1	2	3	4	5	6	7	8	9
1	5.5	4.8	5.3	5.0	4.0	3.3	3.7	5.7	7.2
2	3.5	3.5	3.5	3.2	3.5	3.1	—	3.3	3.6
3	3.6	3.3	3.7	3.6	2.8	—	—	3.5	3.1
4	—	−0.6	—	—	—	—	—	—	—
5	5.6	5.2	5.5	5.5	5.3	5.2	5.5	3.0	—
6	−3.8	−3.7	−3.7	−3.4	−3.4	−3.9	−4.2	—	—
7	−3.8	−3.5	−3.6	−3.8	−3.7	−4.2	−3.7	—	—
8	−2.1	−2.4	−2.4	—	—	—	—	—	—
9	−1.2	—	—	—	—	—	—	—	—
10	3.4	3.2	3.2	2.7	—	—	—	—	—
R^2	0.965	0.963	0.963	0.953	0.940	0.922	0.893	0.894	0.860

Variable	10	11	12	13	14	15	16	17	18	19
1	4.8	6.8	—	6.4	10.1	—	9.3	—	—	—
2	3.0	—	4.4	3.3	—	6.1	—	6.0	—	—
3	—	3.2	2.5	—	2.8	1.9	—	—	1.6	—
5	2.5	3.2	4.5	—	—	—	—	—	—	6.0
R^2	0.846	0.851	0.767	0.812	0.797	0.597	0.743	0.547	0.079	0.544

CHAPTER 6

1. Successive orthogonal polynomial models:

a. Linear model:

Polynomial	Coefficient	t
Intercept	23.987	24.85
Linear	0.3051	5.11

$\hat{\sigma}^2 = 26.090$ $\quad R^2 = 0.5007$ $\quad$ Durbin–Watson $d = 0.39$
The hypothesis of no autocorrelation should be rejected at the 0.01 level.

b. Quadratic model:

Polynomial	Coefficient	t
Intercept	23.987	48.61
Linear	0.3051	9.99
Quadratic	−0.0731	−8.63

$\hat{\sigma}^2 = 6.818$ $\quad R^2 = 0.8745$ $\quad$ Durbin–Watson $d = 1.49$
The hypothesis of no autocorrelation is tenable at the 0.01 level, but the test is inconclusive with an 0.05 error rate.

c. Cubic model:

Polynomial	Coefficient	t
Intercept	23.987	48.41
Linear	0.3051	9.95
Quadratic	−0.0731	−8.60
Cubic	−0.001615	−0.89

$\hat{\sigma}^2 = 6.874$ $R^2 = 0.8786$ Durbin–Watson $d = 1.55$
The cubic polynomial should be omitted. The Durbin–Watson test is still inconclusive at the 0.05 level. Fitting a second-degree polynomial by ordinary least squares is probably justified.

3. a.

$$\begin{bmatrix} \hat{\alpha} \\ \hat{\beta} \end{bmatrix} = \begin{bmatrix} (11y_1 + 2y_2 + 8y_3 + 2y_4 + 11y_5)/34 \\ (y_5 - y_1)/4 \end{bmatrix}$$

b. Yes—the estimator is merely the average successive difference of the y_i.
c. The inverse is

$$\frac{1}{3c}\begin{bmatrix} 4 & 2 & 0 & 0 & 0 \\ 2 & 5 & 2 & 0 & 0 \\ 0 & 2 & 5 & 2 & 0 \\ 0 & 0 & 2 & 5 & 2 \\ 0 & 0 & 0 & 2 & 4 \end{bmatrix}$$

CHAPTER 7

1.

Variable	Coefficient	t
X_1: sex	0.9536	1.73
X_2: years service	0.4476	4.42
X_3: degree	0.3544	0.45
X_4: graduate degree	−0.7449	−1.31

SSE = 150.7828 SSR = 126.5879 $R^2 = 0.4564$
Male gender has a positive effect on salary. College degree does not have a significant effect, although a graduate degree has a negative net effect.

5. a. Matrices of residuals of the observed and reproduced correlations:
(1) Single-factor model:

$$\begin{bmatrix} — & 0.21 & 0.03 & -0.05 & -0.14 \\ & — & 0.04 & -0.08 & -0.07 \\ & & — & -0.02 & -0.02 \\ & & & — & 0.09 \\ & & & & — \end{bmatrix}$$

(2) Two-factor model:

$$\begin{bmatrix} — & 0.31 & -0.01 & 0.03 & -0.03 \\ & — & 0.01 & -0.02 & 0.02 \\ & & — & 0.01 & -0.01 \\ & & & — & 0.00 \\ & & & & — \end{bmatrix}$$

b. The inner product of the vectors is not zero but 0.1225.

7. The first, third, and fourth independent variables are nearly collinear: The determinant of their correlation matrix is equal to 0.0288.

9. a.

	Admitted		Not admitted	
	Q	V	Q	V
Mean	72.83	59.83	65.75	42.00
Sum of squares and products matrix	$\begin{bmatrix} 186.83 & -52.17 \\ -52.17 & 382.83 \end{bmatrix}$		$\begin{bmatrix} 755.5 & -746 \\ -746 & 1420 \end{bmatrix}$	

Within-groups covariance matrix:

$$\begin{bmatrix} 78.5278 & -66.5139 \\ -66.5139 & 150.2361 \end{bmatrix}$$

Linear discriminant function coefficients: $\mathbf{a}' = [0.305, \quad 0.25]$
Application of the function to the test scores of the groups:

	Actual Groups	
Classification by LDF	Admitted	Not admitted
Admitted	5	1
Not admitted	1	7
Total	6	8

b. Group covariance matrices:

Admitted:

$$\begin{bmatrix} 37.3667 & -10.6667 \\ -10.4333 & 76.5667 \end{bmatrix}$$

Not admitted:

$$\begin{bmatrix} 107.9286 & -106.5714 \\ -106.5714 & 202.8571 \end{bmatrix}$$

The assumption may not be valid, although the small sample sizes would make determination of a significant difference difficult.

CHAPTER 8

1. a. Analysis of variance:

Source	SS	d.f.	MS	F
TA	9.8652	4	2.4663	3.46
Error	155.9741	219	0.7122	—
Total	165.8393	223	—	—

b. Conservative Scheffé critical value (120 error d.f.): 3.130; Bonferroni critical value (10 paired comparisons): 2.81

Comparison	t-statistic
A, B	1.78 (not significant)
A, C	2.32 (not significant)
A, D	3.14 (significant by both criteria)
A, E	3.42 (significant by both criteria)
B, E	1.43 (not significant)

Inspection shows the other comparisons to be insignificant.

c. We might compare A with the weighted mean of C, D, and E: $t = -3.51$ (significant in the Scheffé sense at the 0.05 level). However, the substantive meaning of the contrast is unclear.

d. The coefficient vector is proportional to

$$[-15.09, \quad -3.04, \quad 0.15, \quad 7.23, \quad 10.74]$$

$t_{max} = 3.722$, $t^2_{max} = 4F_{computed}$

3. a. Analysis of variance:

Source	SS	d.f.	MS	F
Education	599	2	299.5	8.66
Error	6229	180	34.6	—

The hypothesis of equal levels should be rejected at the 0.01 Type I error rate.

b.

Comparison	t
Elementary vs. high school	−3.38 (significant)
Elementary vs. college	4.11 (significant)
High school vs. college	1.38 (not significant)

0.05 Scheffé critical value: 2.48

5. a. Dry consistency analysis of variance:

Source	SS	d.f.	MS	F
Laboratories	1,051,927	5	210,385	2.33
Error	2,712,156	30	90,405	—

$F_{0.05;5,30} = 2.53$; the equal-laboratory effects hypothesis is barely tenable at the 0.05 level.

Wet consistency analysis of variance:

Source	SS	d.f.	MS	F
Laboratories	2,776,388	5	555,278	44.85
Error	371,436	30	12,381	—

The equal-effects hypothesis should be rejected at any reasonable level.

b. Dry consistency: $\hat{\sigma}^2_L = 19{,}997$
Wet consistency: $\hat{\sigma}^2_L = 90{,}483$

c. Dry consistency: $\hat{\rho} = 0.1426$; $0 \leq \rho \leq 0.612$
Wet consistency: $\hat{\rho} = 0.8457$; $0.736 \leq \rho \leq 0.971$

7. Drug and placebo group sample sizes:

	Power			
	$\alpha = 0.05$		$\alpha = 0.01$	
$(\mu_1 - \mu_2)/\sigma$	0.95	0.99	0.95	0.99
0.5	110	148	146	195
0.6	73	103	100	135
0.7	54	76	74	100
0.8	41	58	57	76
0.9	33	47	46	60
1.0	27	38	37	49

9. The random-effects model power can be obtained from the charts given by Bowker and Lieberman (1972, pp. 398–402). In the following table, $\alpha = 0.05$.

			Effects	
k	N	σ_T^2/σ^2	Fixed	Random
2	10	0	0.05	0.05
		0.5	0.85	0.40
		1	0.98	0.54
		2	1	0.66
2	20	0	0.05	0.05
		0.5	0.99	0.54
4	10	0	0.05	0.05
		0.5	0.96	0.70
		1	>0.99	0.84
		2	>0.99	0.93
4	20	0	0.05	0.05
		0.5	>0.99	0.86

11.

a.

Source	SS	d.f.	MS	F
Groups	2,140.53	3	713.51	2.52
Error	13,606	48	283.46	

The group means are only significantly different at the 0.10 level.

b. (1) Left–Right Comparisons: (a) Weighted contrast coefficients: $[\frac{11}{19}, \frac{8}{19}, -\frac{13}{33}, -\frac{20}{33}]$, $t = 2.55$; the a priori hypothesis of equal hemisphere effects is untenable at the 0.05 level. (b) Unweighted contrast: $[\frac{1}{2}, \frac{1}{2}, -\frac{1}{2}, -\frac{1}{2}]$, $t = 2.66$; the hypothesis is untenable. (2) Male–Female Comparisons: The weighted-mean contrast with coefficients $[\frac{11}{24}, -\frac{8}{28}, \frac{13}{24}, -\frac{20}{28}]$ has $t = 1.37$; the sex difference does not appear to be significant.

c. The contrast for comparing male–female differences on the left and right sides has coefficient vector $[1, -1, -1, 1]$; $t = 0.69$. The contrast appears to have a zero population mean.

d. The three unweighted contrast vectors are mutually orthogonal, but the unequal group sizes lead to correlated contrast estimates. The weighted contrasts are not mutually orthogonal.

13.

Exercise	Contrast coefficients	Contrast value	S.D.	t_{max}
3	113.93, −26.40, −87.53	604.7	144	4.20
11	−49.69, −99.34, 67.38, 81.66	2141.3	779	2.75
12	−28.20, 23.47, 19.00, −17.07, 2.80	376.3	108	3.49

15. P(negative estimate) $= P[W < \sigma_T^2/(n\sigma_T^2 + \sigma^2)]$, where W is a central F random variable with $k - 1$ and $k(n - 1)$ degrees of freedom.

CHAPTER 9

1. a. Analysis of variance:

Source	SS	d.f.	MS	F
Levels	90.2141	5	18.0428	63.53
Blocks	15.0081	1	15.0081	52.85
Error	1.4200	5	0.284	—
Total	106.6422	11	—	—

b. Critical values for comparison of two means:

Tukey: 2.272
Scheffé: 2.678
Bonferroni: 2.798

The following pairs of levels were significantly different by each method:

1, 4   2, 5   3, 5   4, 5
1, 5   2, 6   3, 6   4, 6

c. SSN $= 0.01921$, $F = 0.055$

3. Cross-classified treatments and their totals:

	Male		
Female	IL	NC	Sum
IL	6.76	6.88	13.64
NC	10.33	9.32	19.65
Sum	17.09	16.20	33.29

Analysis of variance:

Source	SS	d.f.	MS	F
Females	1.5050	1	1.5050	374
Males	0.0330	1	0.0330	8.20
Interaction	0.0532	1	0.0532	13.22
Within	0.0805	20	0.004025	—
Total	2.3964	23	—	—

The significant interaction suggests that the male contrasts should be examined separately for each female state. The male effect seems to be due to the difference for North Carolina females.

5. We must verify that the expectations of the estimators are their respective parameters, or $E(\hat{\alpha}_i) = \alpha_i$ and $E(\hat{\beta}_j) = \beta_j$.

7. a. $E(\hat{\mu}_{ij.}) = \mu + \alpha_i + \beta_j$
   b. $\text{Var}\,(\hat{\mu}_{ij.}) = (\sigma^2/n)(1/c + 1/r - 1/rc)$

9. Analysis of variance:

Source	SS	d.f.	MS	F
Rows	0.885	3	0.295	1.02
Columns	38.350	3	12.7833	44.25
Error	2.555	9	0.2889	—
Total	41.790	15	—	—

The Tukey multiple comparisons critical value is $q_{0.05;4,9} = 4.42$. Since $\hat{\sigma} = 0.5375$, a mean difference must exceed 1.188 to be significant in the Tukey sense. Only the column comparison 3 ,4 was not significant.

11. The F-ratio for the Tukey test is equal to 2.33. The hypothesis should not be rejected.

13.

Interaction	Treatment contrast	Block differences
AB	$(1) + c + ab + abc - a - b - bc - ac$	1 and 2 − (3 and 4)
BC	$(1) + a + abc + bc - b - c - ab - ac$	1 and 3 − (2 and 4)
AC	$(1) + abc + b + ac - c - ab - a - bc$	1 and 4 − (2 and 3)

15. Analysis of variance:

Source	SS	d.f.	MS	F
Rows	641.59	2	320.8	8.65
Columns	0.25	1	0.25	0.01
Interaction	141.00	2	70.50	1.90
Error	6603.72	178	37.10	—

17. a. Mean blood pH:

Arterial	Jugular
7.40	7.35

Analysis of variance:

Source	SS	d.f.	MS	F
Subjects	0.056242	12	0.004687	8.85
Columns	0.027692	1	0.027692	52.27
Interaction	0.006358	12	0.0005298	0.93
Error	0.0148	26	0.0005692	—
Total	0.105092	51	—	—

b. Analysis of variance:

Source	SS	d.f.	MS	F
Subjects	0.056242	12	0.0046868	9.21
Columns	0.030538	3	0.0101793	20.01
Interaction	0.018312	36	0.0005086	—
Total	0.105092	51	—	—

Column means:

Arterial		Jugular	
1	2	1	2
7.405	7.392	7.361	7.344

0.05 Scheffé critical value = 2.93

Comparisons significant in the Scheffé sense	t
A1, J1	4.97
A1, J2	6.90
A2, J1	3.50
A2, J2	5.43

19. $\mathbf{a}'\mathbf{C}\bar{\mathbf{x}} = \bar{\mathbf{x}}'\mathbf{C}'(\mathbf{CSC}')^{-1}\mathbf{C}\bar{\mathbf{x}}$
The greatest statistic is

$$t = \frac{\bar{\mathbf{x}}'\mathbf{C}'(\mathbf{CSC}')^{-1}\mathbf{C}\bar{\mathbf{x}}}{\sqrt{(1/N)\bar{\mathbf{x}}'\mathbf{C}'(\mathbf{CSC}')^{-1}\mathbf{C}\bar{\mathbf{x}}}} = \pm\sqrt{T^2}$$

21. a. Analysis of variance:

Source	SS	d.f.	MS	F
Treatments	5768.67	5	1153.73	20.01
Dogs	1809.00	4	452.25	7.84
Interaction	1153.00	20	57.65	—
Total	8730.67	29	—	—

b. Scheffé 0.01 critical value = 4.53

Contrast	t
1. Standard and test	3.99
2. (S3 − S1) − (T3 − T1)	−1.12
3. T1 − S1	2.50
4. T3 − S3	4.08

The standard–test contrast is really an a priori comparison and should be tested with an ordinary t critical value. It is significant in that sense. The other contrasts are not significant in the Scheffé sense.

23. a. Analysis of variance:

Source	SS	d.f.	MS	F
All nested treatments (litters)	4.7969	13	0.3690	27.79
Primary treatments (locations)	2.3921	1	2.3921	180.16
Nested treatments (litters)	2.4049	12	0.2004	15.09
Within litters	1.6730	126	0.0132	—

We shall restrict multiple comparisons to pairs of means within each primary (grandfather locality) condition. The 0.05 Tukey conservative critical value is $q_{0.05;7,120} = 4.24$. Litters with nonsignificantly different means are indicated by the lines joining them:

Illinois litters

2	1	6	5	7	4	3
1.0	1.07	1.08	1.12	1.2	1.39	1.40

North Carolina litters

6	4	7	3	1	2	5
0.73	0.79	0.88	0.96	1.0	1.02	1.05

b. The females may have been selected at random from a larger set. A random-effects model permitting generalizations beyond the seven litters in each primary treatment might be more appropriate and useful.

CHAPTER 10

1. a. Since the 0.05 critical value is approximately equal to 3.03, the hypothesis should not be rejected.

b. Reject the hypothesis of equal treatment effects after the covariate adjustment; reject the hypothesis of a zero regression coefficient.

d. The adjusted treatment mean square is 185.96, or virtually the same as the unadjusted mean square. The unadjusted and adjusted error mean squares are 77.38 and 52.1, respectively, so that the significant F-ratio appears to be due to the reduced error mean square.

e. $F = 1.19$.

3.

Source	SS	d.f.	MS	F
Covariate (after mean)	23.75	1	23.75	12.4
Treatments (after mean and covariates)	7.24	1	7.24	3.77
Error	51.82	27	1.92	—
Total	82.81	29	—	—

Index

A

B

C

D

E

F

I

J

K

L

M

N

O

P

Q

R

S

T

U

V